Wood and Wood-Based Materials for Building

Wood and wood-based materials are taking centre stage in contemporary architecture — from striking multi-storey structures to innovative urban designs. Understanding their properties is essential to avoid costly mistakes and to apply standards such as Eurocode 5 with confidence.

This book offers a clear, practical overview of the key physical and mechanical characteristics of major wood species and wood-based products — including moisture behaviour, elasticity, creep, and strength. It also addresses crucial topics such as protection, fire performance, bonding, and non-destructive monitoring of timber structures.

Whether you are an architect, civil engineer, or a professional in the wood industry, this is your go-to reference for designing and building with timber. It is equally valuable as an introduction for mechanical engineers, chemists, automation specialists, and vocational students entering the field.

Peter Niemz is an Emeritus Professor at the Institute for Building Materials, ETH Zürich, Switzerland, where he previously served as Head of the Wood Physics group. He is an Elected Fellow of the International Academy of Wood Science.

Dick Sandberg is a Mechanical Engineer and Professor of Design and Fabrication of Timber Constructions at the Norwegian University of Science and Technology in Gjøvik, Norway, and an Elected Fellow of the International Academy of Wood Science.

Wood and Wood-Based Materials for Building

A Compendium for Civil Engineers and Architects

Edited by

Peter Niemz and Dick Sandberg

CRC Press
Taylor & Francis Group
Boca Raton London New York

CRC Press is an imprint of the
Taylor & Francis Group, an **informa** business

Cover image: Peter Niemz: Mjøstårnet (2019), Brumunddal, Norway 2023. The building was erected on the shore of Lake Mjøsa, Norway's biggest lake. The company Moelven AS played a key role in the construction of the tower. The company produced and installed glulam columns, beams and diagonals for the primary load-bearing system. CLT was used for elevator shafts, balconies, and "Trä8" deck slabs for the floors up to level 11. Mjøstårnet is ratified as the world's tallest timber building by the Council on Tall Buildings and Urban Habitat (CTBUH), as well as by Guinness World Records

First edition published 2026
by CRC Press
4 Park Square, Milton Park, Abingdon, Oxon, OX14 4RN

and by CRC Press
2385 NW Executive Center Drive, Suite 320, Boca Raton FL 33431

CRC Press is an imprint of Informa UK Limited

British Library Cataloguing-in-Publication Data
A catalogue record for this book is available from the British Library

ISBN: 978-1-032-53431-2 (hbk)
ISBN: 978-1-032-53360-5 (pbk)
ISBN: 978-1-003-41199-4 (ebk)

DOI: 10.1201/9781003411994

Typeset in Times
by SPi Technologies India Pvt Ltd (Straive)

Contents

Chapter 7   Other Important Physical Properties of Wood and Wood-Based
Materials...177

P. Niemz and W. Sonderegger

Chapter 8   Elastic and Inelastic Properties of Wood and Wood-based Materials....220

P. Niemz and W. Sonderegger

Preface

Timber construction has gained significant importance in recent decades, especially in Europe, after the regulations for construction-material use changed in the mid-1990s. During this period, 90% of all single-family homes in regions like Scandinavia had (and still have) a timber frame, and most of these houses had a wooden façade and joinery-construction elements – but the authorities allowed a maximum of two-storey buildings to be built in timber. New-built large-scale timber buildings in free-form design were at this time unusual worldwide. Today, multi-storey administrative and residential buildings are being built in timber; in Scandinavia, approximately 20% of all new multi-storey buildings are in timber, and in Winterthur, Switzerland, a 30-storey building with a height of 100 m is under construction. Still, even considerably taller timber buildings are planned in several countries.

New types of wooden materials such as cross-laminated timber and various hybrid composites are entering the market, and advancements in computer-aided design and CNC-controlled, high-precision production make construction principles in timber possible that would be nearly impossible or not cost-effective to realise with carpentry (manual means). In addition, timber construction can significantly increase the degree of prefabrication and automatisation of the production processes.

In the past, timber construction was often local and concentrated in certain regions, for example, Scandinavia, the mountainous regions of Europe, southern Chile and large parts of the USA. Today, however, due to the increase in high-rise timber building, this approach is becoming more widespread.

The sharp rise in high-rise buildings has also led more architects and engineers to turn their attention to timber construction, having previously often used traditional building materials such as bricks, concrete and steel. Modern construction principles such as cubic architecture with no or an exceedingly small roof overhang are also used for timber buildings in the same way as for other building materials, which in many cases has resulted in material degradation, moisture related issues and short maintenance intervals. In addition, there are new materials, such as thermally modified timber (TMT), which present challenges due to limited user experience. The earlier applied non-timber construction methods are often transferred directly to timber construction. Due to its bio-based nature, wood differs from traditional non-bio-based materials, resulting in improper building design that sometimes leads to damage, which could have been avoided with a deeper knowledge of wood as a construction material. For example, prolonged exposure to moisture can cause significant issues such as fungal attack and swelling of the wood material. This must be prevented in design and construction by implementing specific knowledge of the material intended to be used. In addition, use of sensor technology, including Internet of Things (IoT) solutions is becoming more prevalent to detect such issues.

The book is intended to introduce wood as a construction material for architects and engineers. Its focus is on wood-material knowledge, which is necessary for practical application. Therefore, it is intended as an introduction for professionals working with the practical use of wood and wood-based materials. It is also designed to provide architects, construction engineers and wood and wood-based material manu-facturers with a common discussion platform.

For structural engineers, the legal normative regulations are groundbreaking (*e.g.* Eurocode 5). This field of knowledge is not included in the present book, as many specialist books on timber construction exist.

Some parts of the fundamentals of this book are based on the books *Physik des Holzes* und *der Holzwerkstoffe* by Niemz and Sonderegger (Hanser Verlag 2022), the chapters on wood physics in *Springer Handbook of Wood Science and Technology* by Niemz, Teischinger and Sandberg (Springer Nature 2023), and the book *Wood Reimagined: Sustainable Architecture with Engineered Wood Products* by Miloshevska Janakieska, Kitek Kuzman and Sandberg (International Balkan University, North Macedonia 2024). The authors would like to thank these publishers for giving their permission to partially reproduce the content, in some cases in a sig-nificantly reduced form. Specialists in wood protection, fire protection and surface treatments have contributed in new chapters to present the edge of knowledge in these fields.

The authors are grateful for any comments and additions.

January, 2025

Prof. Dr. Peter Niemz
Zürich, Switzerland

Prof. Dr. Dick Sandberg
Gjøvik, Norway

About the Editors

Prof. Dr.-Ing.habil. Dr.h.c. Peter Niemz studied at the Faculty of Mechanical Engineering at the Technical University of Dresden (TU Dresden). From 1972 onwards, he worked at the Institute of Wood Technology Dresden (IHD) and TU Dresden. From 1993 to 1996, he served as a Full Professor at the Universidad Austral de Chile in Valdivia before joining the Swiss Federal Institute of Technology (ETH Zurich) in 1997. He was Professor of Wood Physics from 2002 until his retirement in 2015. Since 2002, he has been a Fellow of the International Academy of Wood Science. He is a member of the editorial board of numerous wood-related journals, including *Holzforschung*.

Dick Sandberg is Chaired Professor of Design and Fabrication of Timber Construction at the Norwegian University of Science and Technology (NTNU). He received his Doctorate in Mechanical Engineering from the Royal Institute of Technology (KTH), Stockholm, Sweden. He has also been a wood science, wood processing and timber engineering professor at KTH, Linnaeus University, Luleå University of Technology in Sweden and the Czech University of Life Sciences Prague. His research covers material properties, wood scanning technology, wood machining, timber engineering and wood production systems, and studies the connections between different links in the forestry-wood chain. He has a background as a specialist and manager of several companies in the wood industry sphere and the former Swedish Institute of Wood Technology Research, and he has been a national delegate of several European COST (Cooperation in Science and Technology) Actions. He is Editor-in-Chief of the journal *Wood Material Science & Engineering*. He is an Elected Fellow of the International Academy of Wood Science.

Contributors

Christian Brischke
Johann Heinrich von Thünen Institute,
  Federal Research Institute for Rural
  Areas, Forestry and Fisheries,
  Institute of Wood Research
Leuschnerstraße 91, Hamburg-
  Bergedorf, Germany

Boris Forsthuber
Holzforschung Austria
Wien, Austria

Gerhard Grüll
Holzforschung Austria
Wien, Austria

Miha Humar
University of Ljubljana, Biotechnical
  Faculty, Department for Wood
  Science and Technology
Ljubljana, Slovenia

Manja Kitek Kuzman
University of Ljubljana, Biotechnical
  Faculty, Department for Wood
  Science and Technology
Ljubljana, Slovenia

Michael Klippel
Institute of Structural Engineering,
  Timber Structures (IBK)
ETH Zürich, Zurich, Switzerland

Peter Niemz
Wood Physics (today Wood Material
  Science), Institute for Building
  Materials
ETH Zürich, Zurich, Switzerland

Yannick Plüss
Institute of Structural Engineering,
  Timber Structures (IBK)
ETH Zürich, Zurich, Switzerland

Dick Sandberg
Department of Manufacturing and Civil
  Engineering, Norwegian University
  of Science and Technology (NTNU)
Gjøvik, Norway

Walter Sonderegger
Swiss Wood Solutions AG
Altdorf, Switzerland

Michael Truskaller
Holzforschung Austria
Wien, Austria

1 General Aspects of Using Wood in Construction

P. Niemz and D. Sandberg

1.1 A BRIEF HISTORY OF THE USE OF WOOD IN CONSTRUCTION

Wood is a biological material with a complex structure. Because of its unique advantages, it has been used in its natural state for thousands of years. However, wood can behave quite differently as circumstances change. From the growing seed to its use as sawn timber, it undergoes an "ageing" process that changes its properties considerably but never loses its cellular structure and excellent properties. This means that wood can play a wide range of roles in construction.

Because wood is found locally in many regions of the world and is easy to process with pretty simple tools, it is one of the earliest construction materials used by humans. It has a high strength-to-weight ratio, is an excellent thermal isolator to heat transport, and can also be used to generate heat for cocking, warming up during cold nights, or just giving light to scare off wild animals that may be difficult to defend against. The different properties of the various types of wood were recognised and exploited by humans at an early stage. Material-specific construction principles such as roof overhangs and selecting suitable types of wood according to strength and durability were applied and further developed based on practical experience and local conditions. Scientific studies and the development of calculation methods to support selection of suitable wood for a specific construction took place later. The achievements of early timber construction are evidenced by numerous and often centuries-old timber buildings worldwide that still exist, such as the stave churches in Norway, the timber temples in Japan, and timber churches in Chiloe in southern Chile (Figure 1.1). However, wood was also used for other constructions and purposes (Figure 1.2). Until today, remaining timber buildings are often related to religious traditions.

At a local level, there has historically always been a great deal of competition for forest-based raw materials, even if there was a surplus globally. The reason was that the enormous distances in the forest-rich areas made it difficult for manpower and logistics to reach and bring out the material. The local availability, however, meant that timber was used for construction and other daily life purposes. On a higher level, the regional or national leadership needed timber, for example warships and mining. In wet regions such as Holland, Northern Italy, and many other areas worldwide, the foundations for entire cities and villages were laid using wooden piles driven into the ground before houses could be built. The regulation of who was allowed to fell a specific type of tree could be tightly regulated by, for example, the king.

DOI: 10.1201/9781003411994-1

FIGURE 1.1 Examples of traditional timber buildings are evidence of more than 1000 years of worldwide timber-construction tradition: (a) A stave church in Lom, Norway (completed approximately 1170). (b) Kerimäki Church in Finland (1847) is the largest timber church in the world, with a seating capacity of more than 3000, and altogether, there can be 5000 people at a time in the church. (c) Todaiji Temple in Narra, Japan (752), and (d) Church of San Francisco (1912), Castro, Chiloé Islands, Chile.

(Courtesy of P. Niemz and D. Sandberg.)

As is well known, the economic uses of forest resources changed radically as industrialisation spread throughout the Western world. The need for steel and other metals increased globally, but all types of material manufacturing that needed wood or charcoal for production increased, for example, the glass industry. There was also a great demand for timber in the construction industry when cities multiplied. In the beginning, wood was mainly used to build houses. To ensure transportation, large quantities of round timber were transported from the forest to the sawmills by rafting, for example, in the regions of the Bavarian and Black Forests in Germany and Scandinavia. The rivers were adapted for this task and often ruled by specific companies for this purpose, and specially constructed canals (*e.g.*, the Schwarzenberg Canal in the Bohemian and Bavarian Forests) were also used.

FIGURE 1.2   A covered timber bridge over the Rhine between Bad Säckingen in Germany and Stein in Switzerland. The timber bridge is Europe's longest covered (203.7 m) and was first mentioned in documents in 1272. Wars and floods destroyed it several times.

(Courtesy of P. Niemz.)

The heavy use of forests led to high deforestation, at least locally. The problem was recognised, and sustainable forestry methods were gradually established. Only as many trees were allowed to be removed from the forest as could be regrown or reforested artificially. The concept of sustainability in forestry goes back to the Saxony (Germany) mining engineer Johann (Hannß) Carl von Carlowitz (1645–1714), who wrote the first comprehensive treatise on forestry and coined the term sustainability in his book *Sylvicultura oeconomica* (the economic news and instructions for the natural growing of wild trees), published in 1713.

Systematic studies of timber construction have presumably existed in different forms since ancient times. However, in early Modern Europe, the military mainly studied it, and one early specialised forestry schools (so-called Forstmeister master schools) were founded by the Naval Cadet Corps Academy in St. Petersburg, Russia (1803). This and many other schools of this kind were inspired by the first formal education programs for foresters starting in Wernigerode, Germany, as early as 1763 by Hans Dietrich von Zanthier (1717–1768), later transferred to the close-situated city Ilsenburg, ending with his death (Fernow 1911 in Niemz and Sandberg 2025). The German schools were initially mainly located in mining faculties because the mining business required large quantities of wood for smelting. A Forestry College was founded by Heinrich Cotta (1763–1844) in Tharandt 1811 and still exists as a part of Dresden University of Technology (TU Dresden) in Germany (Radkau 2007, Niemz *et al.* 2023).

Especially in times of shortage, wood was often promoted and used as a construction material and a raw material for chemicals (Köstler *et al.* 1960). Figure 1.3 shows the development of wood utilisation over time. There was an apparent decline in the use of wood until around 1900, after which it increased again due to the shortage of raw materials during the First and Second World Wars. Before the advent of plastics, paints and adhesives for paper were often made from tree resins such as rosin, and biobased raw materials, especially wood, frequently served as the basis for the chemical industry (Stephan 2012). Today, the trend is returning to using renewable raw materials.

[see figure]
Wood and Wood-Based Materials for Building

Wood utilisation and developement in wood construction since 1750

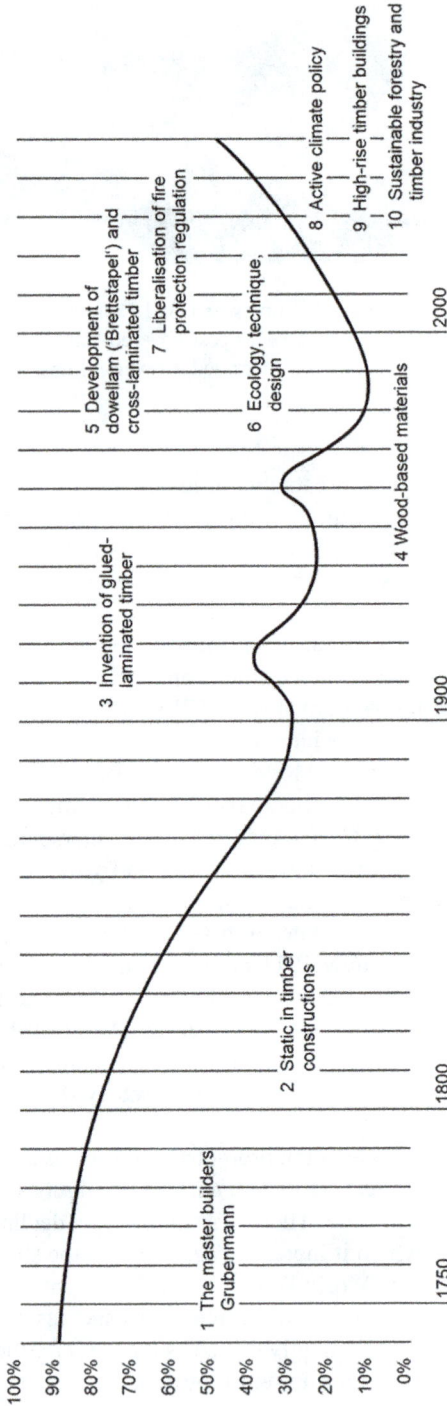

y-axis: 100%, 90%, 80%, 70%, 60%, 50%, 40%, 30%, 20%, 10%, 0%

x-axis: 1750, 1800, 1900, 2000

1 The master builders Grubenmann
2 Static in timber constructions
3 Invention of glued-laminated timber
4 Wood-based materials
5 Development of dowellam ('Brettstapel') and cross-laminated timber
6 Ecology, technique, design
7 Liberalisation of fire protection regulation
8 Active climate policy
9 High-rise timber buildings
10 Sustainable forestry and timber industry

FIGURE 1.3 Use of wood in general and developments in timber construction (based on Kolb *et al.* 2024). (1) Widely extended timber bridges and constructions built by the master builder Hans Ulrich Grubenmann (1709–1783). (2) Load-bearing constructions will be determined by static analysis. (3) The development and patenting of glued-laminated timber by Karl Friedrich Otto Hetzer (1846–1911). (4) Application of various wood-based materials, for example particleboard. (5) In the 1980s, various sawn-timber EWPs, such as cross-laminated timber, were developed and implemented. (6) Until 2000, common influences from ecology, technique, and design led to increasing wood applications in the second half of the 20th century. Since 2000, joining technology and wood-based panel materials have allowed multi-storey constructions. (7) Due to the deregulation of the fire protection requirements in Europe in the early 1990s, the restriction on height for timber buildings that were common in many European countries is no longer valid. Wood has become a standard building material. The application fields of steel, concrete, and wood were identical. (8) 2020–2025: An environmentally friendly and sustainable climate policy relies on forest and wood resources. Wood became famous as a climate-neutral building material. (9) On all continents, large-volume and high-rise timber buildings are constructed, and (10) in future, the tightening of the available construction materials forces reflections. Renewable raw materials like wood will be applied more and more.

Petroleum-based raw materials dominated from the 1950s and 1960s. Adhesives, varnishes, and paints were traditionally based on casein and gluten, which then were displaced by petroleum-based alternatives. Today, traditional adhesives and coatings are niche products for special applications such as musical instruments. It was only when environmental awareness increased that there was a rise in the use of natural raw materials, which continues to this day. However, the use of so-called biobased materials still needs to be improved in many fields, for example, the cost needs to be lowered for price competition, and the homogeneity refinement of the biobased raw materials used, for example for adhesives, needs to be improved (Dunky and Mital 2023). Despite many approaches, cheap biobased raw materials for adhesives, such as lignin and tannins, account for comparatively small amounts of the total consumption of adhesives in the wood industry. Synthetic adhesives are predominantly used. The trend towards biobased adhesives is positive, but the use is still low as natural raw-material properties often vary greatly and are seasonal. The quantities required are too high compared to availability.

Over the last 20 years, timber construction has developed from traditional carpentry to highly industrialised work. Although flat-pack building components or smaller full-size buildings have been manufactured off-site in factories for at least 150 years, this type of production was mainly "moving the carpenters indoors," *i.e.*, the manufacturing processes were industrialised to a minimum, limited to transport and lifting devices. Today, highly industrialised and automatised factories for manufacturing timber buildings exist. Figure 1.4 shows an overview of the development of the production processes in the timber construction industry in the last decades.

1.2 A NEW ERA FOR TIMBER CONSTRUCTION – CHANGES IN EUROPE WERE THE KICK-OFF

Although we have witnessed an ever-increasing dependence on steel and concrete in structural applications in the last 150 years, wood remains an essential feature in society and one of the few truly renewable resources available to us in volumes needed. With the development of technology, the use of timber has significantly improved, but only recently has timber been developed into a range of increasingly functional products based on a combination of performance and sustainability requirements (Miloshevska Janakieska *et al.* 2024). This has been possible because of new industrial processes that extend the size and modify the properties of natural wood, as well as the need to refine residues and lower-grade trees to produce more versatile and consistent products.

One of the critical features of wood processing is that the dimensions of the tree are changed into those of an engineered material with well-defined dimensions and properties adapted to the purpose for which it is to be used. This development has led to the elimination of strength-reducing defects in the wood structure and a reduction in the adverse effects of anisotropy and inhomogeneity. A range of well-defined

Traditional construction and woodworking

- Conventional timber-frame construction using dimension sawn timber
- Structural systems are constructed entirely or largely on-site
- Linear construction, which requires each production stage to be completed before the next can begin

Prefabricated timber components

- Sawn timber, wood-based panels and other types of EWPs are used to form pre-assembled components for wall, floors, and roof systems
- Constructed at off-site facilities
- Recognises via code evaluation reports; production facilities utilise factory inspection for quality assurance

Prefabrication of panelised systems

- Prefabricated components are assembled into large panels or complete assemblies before being shipped to a building site
- Can be delivered as complete exterior and interior wall panels or structural roof and floor systems
- Can be panelised at an off-site facility, or assembled adjacent to the jobsite when other work is done on-site

Prefabricated construction

- Components are assembled off-site and built into modular structures, which are transported to the site to be assembled into a complete building
- Temporary or permanent buildings
- Modules include interior and exterior finishes as well as mechanical, electrical and plumbing (MEP)
- 80% to 95% of the building completed before shipping
- Can be up to 50% faster than traditional construction

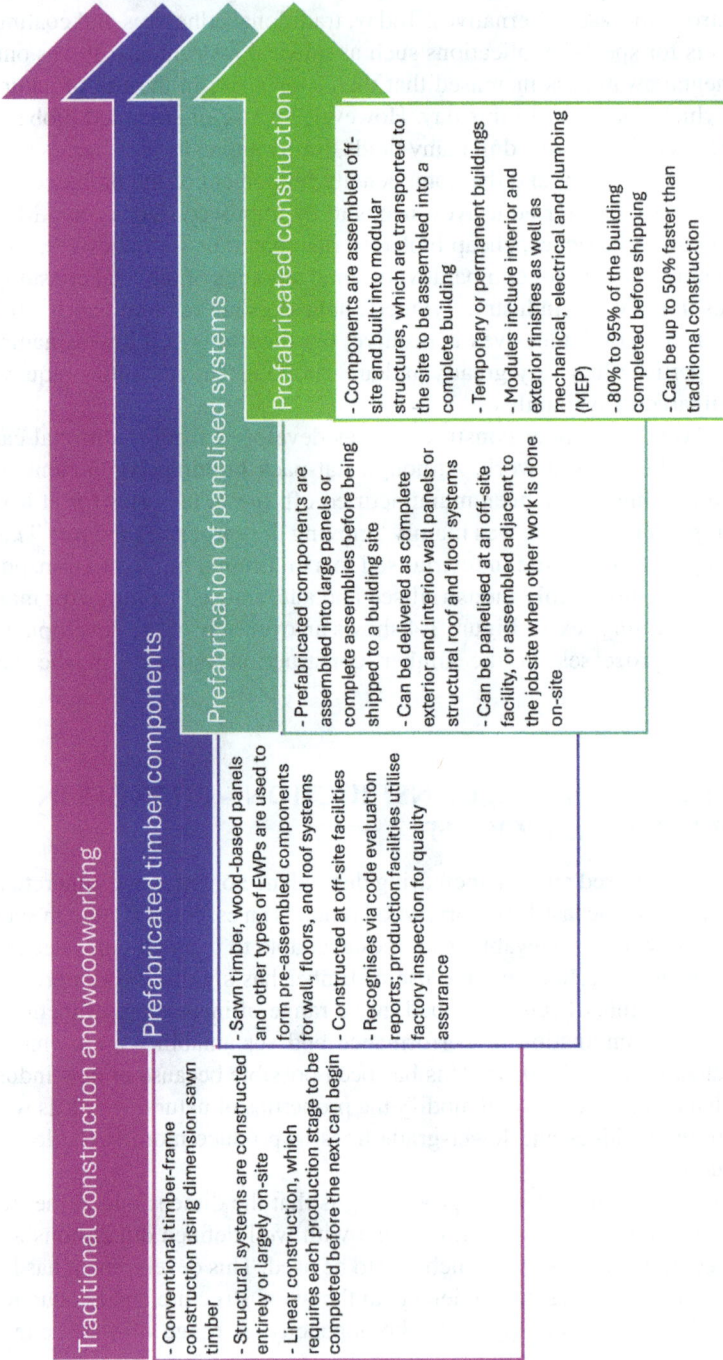

FIGURE 1.4 The production processes in the timber-house industry have developed over the last decades from craftsmanship to industry following advanced production principles used in other industry branches.

(Courtesy of D. Sandberg.)

wood products have been widely developed during the last century, and nowadays, they are commonly called engineered wood products (EWPs) (Kitek Kuzman *et al.* 2018).

EWPs are now considered high-performance structural wood-based materials, defined as construction components manufactured by bonding sawn timber, veneers, particles, strands and wood fibres. This definition is firmly based on the decisions of the North American Engineered Wood Association (APA) and its industrial members (APA 2010). In a broader sense, any wooden product of an engineering process with properties engineered for a specific use can be seen as an EWP. Currently, a dynamic innovation process continuously develops new engineered wood materials, components, and elements, ranging in dimensions from microstructures to large units. The growing timber-manufacturing industry also faces challenges related to the increasing geometrical complexity of architectural design. Complex and structurally efficient curved geometries are nowadays efficiently designed, but their realisation involves intensive manufacturing and excessive machining. Thus, engineered wood describes wood products that are engineered for structural applications such as cross-laminated timber (CLT), glued-laminated timber (GLT or glulam), laminated strand lumber (LSL), laminated veneer lumber (LVL), oriented strand board (OSB), parallel strand lumber (PSL), structural plywood, and various types of box beams, I-joists, and other engineered speciality products (Glavonjić *et al.* 2022, Sandberg *et al.* 2023).

Progressive construction practices increasingly focus on environmentally friendly materials, mainly wood and various EWPs. However, the debut of novel products in the construction industry may encounter limited awareness and heightened consumer uncertainty. Therefore, it is of utmost importance that the short- and long-term properties of both traditional and modern EWPs are documented based on scientific principles and communicated to the market to give foundational confidence in the use of timber in construction.

However, due to large city fires in Europe, fire protection measures, including legislation, were introduced in several European countries during the late 19th century to discourage or prohibit using timber to a large extent in multi-storey buildings. In several countries, only two-storey timber buildings were allowed in cities. In the late 1980s, a construction product directive from the European Commission stipulated function-based requirements for using products in building construction to remove technical barriers to trade in construction materials between member states in the European Union (European Commission 1988). The current renaissance in timber construction began in the mid-1990s, and there is no end in sight – building with timber is booming. In addition, the systematic development of specialised production methods and digital design and fabrication techniques have raised timber construction to a new building level without depleting resources. Wood products in myriad forms provide architects with an inspiring overview of many EWPs and a glimpse into the future of creating the next generation of wood-based construction solutions for use in tall timber buildings. Today, we are in a unique moment in architectural and building engineering history when changing world needs have asked us to question some of the fundamentals of how we built in the last century and how we shall build in the future.

In addition to this significant change in construction regulations, support for an increase in timber building comes from increased environmental awareness, the development of new materials and numerous newly started research and development projects on acoustics and fire behaviour, hybrid-material composites, timber façade systems, and, in particular, high-rise timber construction methods where new reliable steel connections for timber buildings and using hybrid timber-concrete composites also brought many technical and often visual advances. At the beginning of the new millennium, several examples of multi-storey residential and administrative buildings, industrial and commercial buildings (halls, shopping centres), and agricultural buildings (stables, storage rooms, *etc.*) could be studied in several countries. Different construction systems were tested and developed to gain knowledge of material consumption, recycling, reuse, and cost. Scandinavian and Central European countries, where the tradition of professional carpenters (in the German language called Zimmerer) in the construction and woodworking sectors still is common, were foreground countries in the new timber construction era.

Different types of construction principles can be identified today, for example,

- log and block-timber construction,
- timber-frame construction,
- mass-timber construction,
- modular timber construction, and
- free-form construction.

1.2.1 LOG AND BLOCK-TIMBER CONSTRUCTION

Log construction is one of the oldest known timber-building principles. In forest-rich regions such as the Alps, Russia, and Scandinavia, log construction was commonly employed for agricultural purposes, including hay barns, livestock housing, and residential buildings. Today, log houses are primarily used as holiday homes and cabins. Round timber or squared logs are stacked horizontally or vertically (as staves) to form walls. When stacked horizontally, the corners are securely joined using staggered connections. These corner joints are created using techniques such as notch or interlocking joints. The meeting logs typically extend beyond the corner joint for structural wood protection. In historical log houses, the ends of additional beams often interlock using dovetail or trapezoidal cuts. The length-wise connection of the logs prevents shifting between logs and is achieved through tongue and groove joints or wooden pegs and dowels. The traditional log construction was developed into prefabricated plank houses, and the wall construction was improved to achieve increased stability and thermal insulation. Brick-like timber (block) is also used today (Figure 1.5).

Typical features from log construction are (Kolb 2024):

- a high level of craftsmanship is required,
- unique wood selection,
- unique mechanical corner joints,

- rigid floor plan arrangement,
- a high consumption of wood per unit area, and
- setting dimensions must be observed (shrinkage/swelling perpendicular to the fibre direction).

Block-timber construction is a relatively new and evolving construction method that uses solid, rectangular timber blocks (wooden bricks) as the primary structural components (Figure 1.5). Unlike traditional timber framing or mass-timber construction techniques, where large panels or beams are the main construction component, block-timber construction assembles buildings using individual wooden blocks, similar to how bricks or masonry blocks are used in construction. Each block is typically machined or engineered for precision and ease of assembly. The blocks used in block-timber construction are usually solid or laminated wood assemblies designed to fit together without the need for nails, screws, or other fasteners. They are engineered for structural integrity and durability. Block-timber construction can be used for various building types, including residential, small commercial, temporary, modular, and low-rise buildings.

Many companies today offer residential and holiday homes constructed from logs or glued-laminated timber beams. Machine-produced squared or partly round logs are used for the walls. Such building components are dried to the correct moisture content and are dimensionally stable.

The exterior wall can be constructed in multiple layers to achieve low thermal transmittance (low U-value). Insulation with wooden cladding can be applied to the inside or outside of the beam layers, or an insulation layer can be placed between two log walls. The necessary wind-tightness can be achieved using sealing strips made of compressed foam between the logs or beams (Figure 1.6).

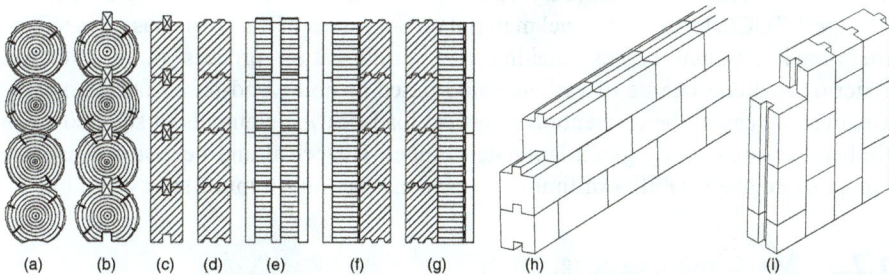

FIGURE 1.5 Log and block-timber construction variations: (a) traditional round-timber wall, (b) round timbers with prepared surfaces and tongue-and-groove joints between the logs (length-wise), (c, d) squared timber with length-wise tongue-and-groove joints. (e–g) Prefabricated or on-site prepared sandwich elements: (f) on the exterior side thermally insulated wall where the squared timber remains visible inside (to the right). The façade cladding is solid wood and can be ventilated depending on the situation. (g) The same as (f), but the squared timber remains visible outside. Walls of timber blocks with (h) horizontal or (i) vertical fibre orientation of the blocks.

(Courtesy of P. Niemz and D. Sandberg.)

(a)   (b)

FIGURE 1.6   A modern (a) log house made of (b) three-layer glued-together plank walls.

(Courtesy of P. Niemz.)

1.2.2   Timber-frame Construction

Timber-frame construction is a lightweight construction principle, as the log houses have a long history dating back thousands of years. Timber framing was a common building technique during the medieval period, for example, in Central Europe and Great Britain. Homes were typically constructed using a framework of heavy timber posts and beams, with infill panels made of "wattle and daub," a mixture of woven twigs and mud. This method allowed for flexible, open interiors and was well-suited for constructing smaller timber-framed houses, manor houses, and medieval timber-framed halls.

In recent years, timber-frame construction has gained attention due to its sustainability, energy efficiency, and speed of construction. Modern timber-frame homes often utilise EWPs and advanced construction techniques, making them structurally sound and meeting contemporary building standards for acoustics, fire safety, etc. (Figure 1.7). Cladding with panel material on both sides of the frame has replaced the traditional diagonal braces, making the prefabrication process more effective. Exterior cladding can be of various wood-based or non-wood-based façade materials. In some cases, the conventional vertical load-bearing elements of squared-sawn timber are replaced by glued-laminated timber. Historical and new structures also demonstrate the reliability of timber-frame construction for multi-storey buildings.

1.2.3   Mass-timber Construction

The term "mass timber" became widely used in the early 2000s to describe heavy EWP structures of, for example, CLT and glued-laminated timber and their use in mid-rise and high-rise buildings to replace steel and concrete to create a structure that provides durable, cost-effective performance with less embodied carbon, *i.e.*, the amount of carbon emitted during the construction of a building. The extraction of raw materials, the manufacturing and refinement of materials, transportation, installation, and disposal of old supplies can all produce embodied carbon emissions. The word "mass" refers to the large-scale and solid nature of the wood elements used in construction, emphasising that timber can be used in significant structural applications

FIGURE 1.7 Timber-frame building under construction with a sawn-timber frame covered by oriented-strand board (OSB). The residential building is a construction site in Moses Lake (WA), USA.

(Courtesy of D. Sandberg.)

rather than just as a framing or cladding material. Thus, mass-timber construction describes engineered timber products for building large, multi-storey, or complex structures, representing a modern evolution of timber construction in response to technological innovations and environmental considerations. Mass timber EWPs can be used to design everything from tall timber towers and long-span sports arenas to educational facilities, multi-family structures, office buildings, and more.

Because these products are lighter than other construction materials (concrete, steel), they lend themselves well to prefabricated wood building systems, allowing a speedier and more efficient process. Today, CLT is maybe the most common mass-timber product because of its versatility, and it is used to build elevator shafts, floors, roofs, and walls. Glued-laminated timber (glulam) is often used for headers, beams, columns, trusses, or load-bearing arches. Several other EWPs are used in prefabrication, including nailed-laminated timber, dowel-laminated timber, laminated strand lumber, laminated veneer lumber, and parallel strand lumber. Figure 1.8 shows examples of two high-rise mass-timber buildings.

1.2.4 MODULAR TIMBER CONSTRUCTION

Any building with significant sections built off-site, in a factory, and assembled on-site takes advantage of prefabricated construction. Usually called modular construction, off-site prefabrication can range from some distinct portions of a structure to nearly all the components of a complete building. Prefabricated timber systems are made from conventional light-frame or mass-timber systems, and finished construction often involves a mix of both. Modular construction may be categorised into permanent or relocatable buildings. Structurally, modular buildings are generally stronger than site-built construction because each module is engineered to withstand the rigours of transportation independently and craning onto foundations. Once together and sealed, the modules become one integrated wall, floor, and roof assembly.

(a)   (b)

FIGURE 1.8 Mass-timber construction: (a) Mjøstårnet in Brumunddal, Norway, completed in 2019. The building was erected on the shore of Lake Mjøsa, Norway's largest lake. Glued-laminated timber columns, beams, and diagonals for the primary load-bearing system, CLT for elevator shafts and balconies, and the so-called Tre8 lightweight wooden deck system (Moelven company). Mjøstårnet was ratified in 2019 as the world's tallest timber building (85.4 m). (b) Sara Kulturhus/Sara Cultural Centre, Skellefteå, Sweden, is a 74-m-high combined official building and hotel completed in 2021. Sara Kulturhus is named after Sara Lidman, one of the Skellefteå region's many famous authors. She has inspired the architects to create a cultural centre with courage and a desire to embrace innovation.

(Courtesy of D. Sandberg.)

Modular construction is promoted as an opportunity to combat rising interest rates and construction prices through greater efficiency, address skilled labour shortages, and reduce job-site waste. Studies on improving productivity and quality in the building industry typically recommend increasing the prefabricated portion (Anon. 2016). The benefits are clear: faster construction process, higher quality products, less waste, and lower unit cost. Other benefits include less noise, dust and site disruption, improved health and safety, continuity of employment, workforce upskilling, predictable product performance, and lower product operational costs (Anon. 2013). Removing approximately 80 to 95% of the building construction activity from the site location significantly reduces site disruption vehicular traffic and improves overall safety and security. Prefabrication in a factory results in high quality because the work is carried out under suitable conditions, and construction moisture is eliminated. Highly active businesses such as those in the education and healthcare markets require reduced on-site activity, and off-site construction elimi-nates a large part of ongoing construction hazards, ensuring an advantage when building new projects.

The units leaving the factory can have either the form of (1) flat-pack building components, usually floor, roof cassettes, or wall panels, and consist only of typically framing and sheathing structural components (panelised construction), (2) flat-pack building components including structural as well as other components such as fin-ishes, mechanical, electrical and plumbing engineering (MEP), etc. (prefabricated construction), and (3) volumetric units each consisting of a floor, walls, and ceiling, which typically include structural elements as well as the different degree of finishes, mechanical, MEP, etc. (prefabricated construction). The volumes produced can be used for small or substantial portions of a building, *i.e.*, a single room or an entire flat unit.

Architects and engineers have expressed the opinion that the designs of the building are to be compromised, and the modular systems do not have the flexibility or functionality to execute specific project typologies. Working with modular sys-tems, however, can be a great help since it is challenging to design traditionally and then translate the design of the building to an industrial context. With growing con-cern for sustainable forms, it is easier to adapt the construction and organisation of the building to the system's limits from the beginning. For example, the modules must be of a size that can be transported on roads. The modules also have a more rigid construction, which can be challenging if the building height is restricted. In addition, the system requires an early commitment to the project, with minimal scope for making changes later. As mentioned, the advantages of contemporary industrial timber constructions are a shorter time on the building site, less transport, less disruption for neighbours, reasonable cost control, and no drying time com-pared to in situ concrete, as shown in some recently awarded architectural projects (Figure 1.9).

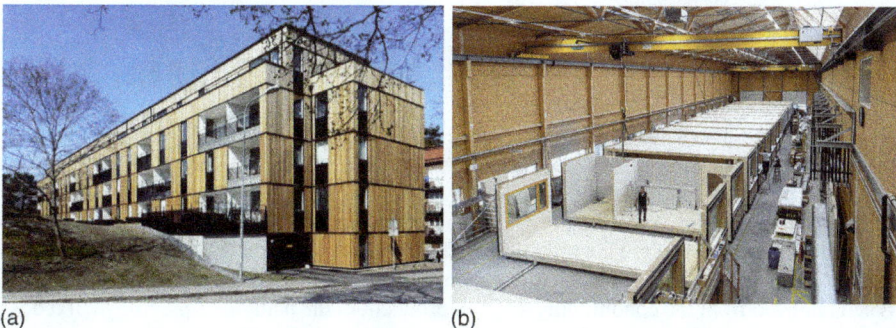

(a)   (b)

FIGURE 1.9   Examples of modular systems: (a) Skagershuset, 112 prefabricated box mod-ules built around a timber frame, completed in 2013, and (b) prefabrication of modules at Blumer-Lehmann in Switzerland.

(Courtesy of P. Niemz and D. Sandberg.)

1.2.5 FREE-FORM CONSTRUCTION: PRODUCTION WITH MINIMUM TOLERANCES AND MAXIMUM FLEXIBILITY

The framework for producing components is full of mathematically exact, parameterised models of the structure and its components, ensuring minimal tolerances in the construction, processing, and installation phases. 3D modelling, high-quality code, and error-free information for computer numerically controlled (CNC) machines are also critical for prototyping parts and managing 3D printing. These models are part of the entire process, from project development, feasibility studies, and design through the CAD/CAM processes to the construction and service life.

Depending on the type and complexity of the structure, specialists in computer-aided design (CAD) and computer-aided manufacturing (CAM) software are needed to convert graphical data into machine codes, generally for steering 5-axis CNC joinery machines. Programming expertise and handling of this equipment are required to ensure the flexible and precise production of double-curvature timber structures (Figure 1.10).

FIGURE 1.10 The Urbach tower in Urbach, Germany, is made of 14-m-long sections shaped by a self-shaping process to achieve curved timber elements. The self-shaping effect is only driven by the shrinking of the wood at a lowering moisture content. The curved CLT components for the tower's structure were designed and produced as flat panels that deform autonomously into predicted curved shapes when dried. The 5.0 m × 1.2 m Norway spruce wood "bilayer parts" were manufactured with a high moisture content and specific layups. When removed from the conditioning stage in production, the parts were precisely curved.

(ICD I IKTE Universität Stuttgart, Germany; courtesy of Blumer-Lehmann, Gossau, Switzerland.)

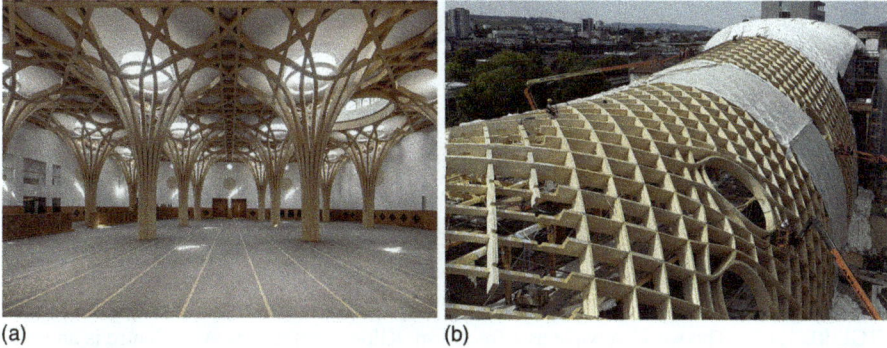

(a)   (b)

FIGURE 1.11 Examples of free-form timber construction: (a) the Cambridge Mosque, Cambridge, UK; a modular timber-construction system from design to production with 2746 components in 145 variants, and (b) the Omega administration building in Biel, Switzerland.

(Morley von Sternberg, courtesy of Blumer-Lehmann, Gossau, Switzerland.)

Modern design and production methods allow complex structures and buildings to become real. Free-form structures are distinguished by their cellular supporting structures and the unique nature of each component (Figure 1.11). They are exceptional – from the initial idea to the design, production, and installation, with the required quality in the specified time frame and in cost-effectiveness from the perspective of the investor and builders. Advanced timber structures save money at the construction site because they allow exact planning and quick assembly due to prefabrication. This is also an economic benefit for builders when the time between new construction and rental is short. For investors, it is essential that advanced financing and property marketing take less time and are accompanied by assured on-time completion. One of the main advantages of a modern production concept, in-factory rather than on-site, for timber structures is that the construction method is primarily dry, and there is a reduced risk of damage to the structure due to moisture.

Public buildings can be a model for others: sustainability, energy efficiency, and value conservation are all essential requirements. Timber structures are often at the forefront of architectural design on bold scales and with commanding aesthetics. Kilden Performing Arts *Centre* is located on the waterfront in Kristiansand, Norway. The first thing visitors note about the building at the harbour entrance is its unique, exciting façade. A wooden wave of 3500 m², 100 m wide, and overhanging by 35 m towers above the visitors' heads in all its massive woodenness. Multiply curved, this wave presents itself in the best possible free-form manner (Figure 1.12).

FIGURE 1.12   The wooden wave as a free form. Kilden Performing Arts Centre is an opera house and theatre in Kristiansand, Norway. Initially planned as a steel construction, the façade was completed in timber. The timber façade extends from the floor in the centre's interior via the glazed façade to the exterior. A particular challenge was the different interfaces, climatic zones, and heavy loads. Thanks to prefabricated façade elements and curved glued-laminated timber, the façade was assembled quickly and efficiently. The key to the seamless progress of the project was a detailed 3D model. The façade elements were prepared in Switzerland and transported to the construction site in southern Norway. The façade is a timber construction kit with 14,309 individual components.

(Courtesy of D. Sandberg.)

1.3   TRENDS IN THE DEVELOPMENT OF TIMBER CONSTRUCTION

Timber multi-storey construction has gathered momentum recently, especially in European countries. Construction of the first experimental high-rise timber buildings was completed in the mid-1990s, and today, the trust in new timber buildings is growing. The number of projects, quality and importance, and rising interest from different groups and customers show this trend. There are two reasons for this: economic and ecological. Timber multi-storey buildings are progressing in performance, construction methods, and decreasing building costs. It is becoming increasingly widespread, and the number, quality, and types of timber buildings indicate progress towards becoming a common construction practice in the middle-rise building environment of many countries. However, the use of wood for multi-storey building construction varies widely among countries. Positive aspects of timber as a structural material include its strength, environmental friendliness, simple handling, and appropriateness for industrial use. However, knowledge gaps have reduced structural engineers' and architects' use of timber in construction, a lack of knowledge that is now being regained.

Today, there is certainly a trend towards globalisation in the timber construction industry, but cultures vary significantly from country to country. The particularity of the Central European region is its culture of craftsmanship. Timber construction companies have primarily evolved from family carpentry businesses and not from industrial companies as in many other countries. This is quite different from Northern Europe, for example. Craft companies have many more opportunities to realise larger, smaller, and more complex projects and are ideal partners for architects. They need partners who can realise their buildings and have little use for prefabricated building systems.

Renewable raw materials are increasingly in demand today, and there is plenty of green capital to invest in sustainability. Wood is an excellent building material that can compensate for conventional non-biobased building materials. This means that if the primary material of a building is made of renewable raw materials, there are significant reductions in carbon dioxide (CO_2) emissions. Whether this material remains visible and architecturally effective is a secondary question. Of course, it would be ideal to experience this type of construction and feel the qualities of this material in the interior. If not, *i.e.*, if fire regulations, etc., prevent this, it is still not a disadvantage in sustainability.

From not being present, nearly, in new large-scale construction since the Second World War, the number of newly built high-rise buildings in Europe has increased steadily. The reasons for the increase in timber construction include:

- Increased ecological awareness. Trees remove CO_2 from the atmosphere during growth through photosynthesis and produce oxygen. One cubic metre of wood weighs an average of 435 kg when absolutely dry, half of which consists of carbon. This corresponds to almost 800 kg of CO_2, which is retained in the atmosphere over the life cycle of a wood-based product when the raw material is used as a construction material (Rüter and Hafner 2022). Increased use of wood in buildings can bind CO_2 in the long term.
- A comparatively lower energy consumption is required for producing timber compared to other materials (Figure 1.13).
- The development of cross-laminated timber, glued-laminated timber, and laminated veneer lumber with a high load-bearing capacity and reduced variability of wood properties.

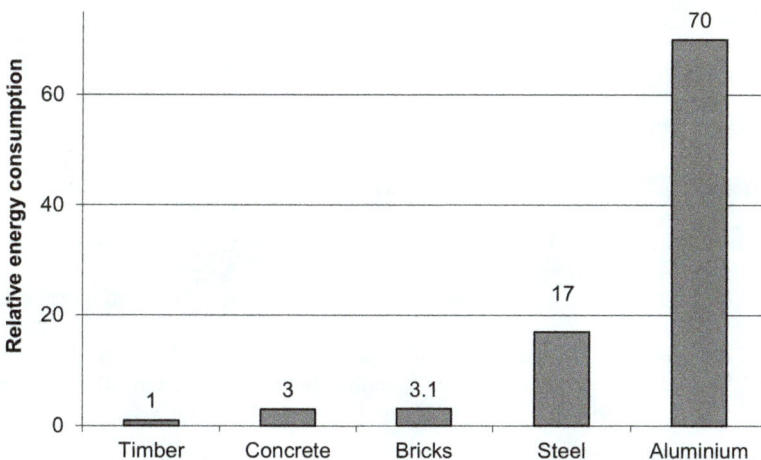

FIGURE 1.13 Relative energy consumption for the manufacture of different materials. (Courtesy of D. Sandberg.)

- The favourable fire behaviour of wood compared to steel. A charcoal layer at the surfaces of the burning timber is developed and reduces the speed of the further burning of the underlying by fire unaffected wood. The strength of burning timber elements is predictable as the charring rate is approximately 0.5–1.0 mm/minute in a fully developed fire. Beneath the carbon layer, however, the wood is affected by the elevated temperature, which may weaken it (*cf*. Chapter 13). Developing temperature-resistant adhesives and cross-laminated timber with optimised combustion behaviour gives opportunities to increase the load-bearing capacity under high temperatures further (Klippel 2013, Buchanan and Östman 2022).
- The development of low-emission adhesives (Dunky and Mital 2023).
- The natural material wood also offers numerous advantages from a structural engineering perspective, including being lighter than steel with the same load-bearing capacity.
- Wood has approximately the same compressive strength as concrete.
- Wood has a high thermal insulation capacity, significantly better than concrete and bricks (Figure 1.14).
- Wood creates a pleasant indoor climate, *i.e.*, changes in the relative humidity in a room with much wood slow down; wood gives away moisture to the air in the room when the relative humidity is low and takes up moisture when it is high.

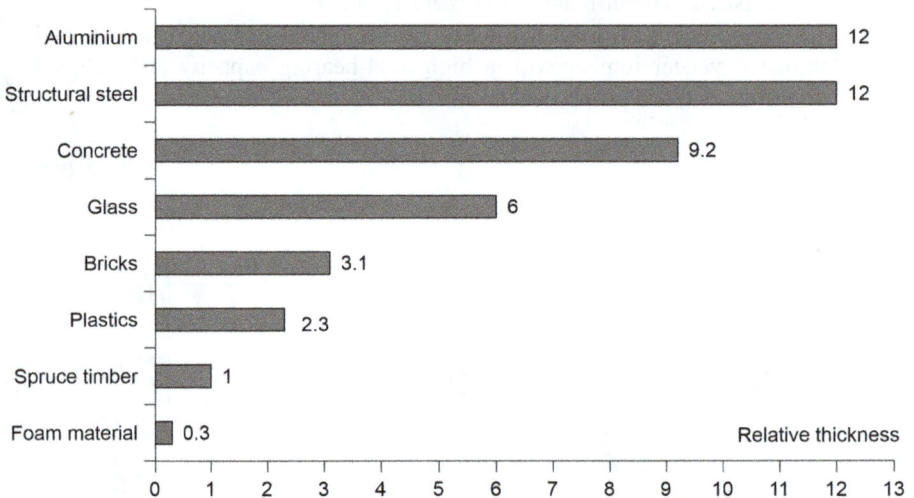

FIGURE 1.14   The relative thermal insulation performance of wood compared to other materials, *i.e.*, the thickness needed of various materials to reach the same insulation capacity as Norway spruce timber with a dry density of approximately 450 kg/m^3.

(Courtesy of D. Sandberg.)

- Timber construction allows high-quality and exact prefabrication, for example, with CNC machines and is easily combined with CAD software.
- Another advantage of timber construction is the comparatively short execution time required. For example, in the case of a medium-rise building for residential use, the total execution time may be two or three months shorter than when steel or concrete are used as the primary construction material. A timber building is also lighter, making the foundations simple and may be less costly.

There is an almost explosive demand for buildings made from renewable raw materials. The climate debate has reached the centre of the construction industry. New legislation and regulations on sustainability are a massive driver for timber construction. For this reason, many projects are currently being implemented worldwide. The question is not whether timber has a future; that has already been answered. The critical stage for increasing the volume of timber construction is how quickly the production capacity can respond to growing demand. This is linked to whether the expertise of designers and builders can develop at the same rate to serve this market. It is, however, an overhanging risk that new, less serious businesses without any experience get into the timber construction field, which can lead to significant mistakes and, as a result, immense structural damage, which could once again bring timber into disrepute. In conclusion, timber construction is the most modern form of building today. Buildings manufactured mainly in the workshop and assembled to a high standard on-site are the future of construction.

1.3.1 RECYCLING AND REUSE OF TIMBER IN SUSTAINABLE CONSTRUCTION

In recent years, the growing need for sustainable construction practices has become increasingly clear, driven by the urgent need to address climate challenges. The construction industry is a major contributor to global carbon emissions, responsible for approximately 38% of CO_2 emissions worldwide (World Economic Forum 2022). More specifically, the built environment accounts for around 70% of global CO_2 emissions, and construction and demolition activities contribute significantly to this figure (Benachio et al. 2020). We must rethink how we manage resources, and recycling and reusing materials, such as timber, are fundamental strategies for reducing the environmental footprint of the construction industry.

One of the most significant environmental impacts in construction comes from waste materials such as concrete, treated wood, and asbestos-based products, all of which have negative environmental consequences. Consequently, the construction sector must develop more sustainable methods for managing waste, reducing the consumption of new materials, and cutting emissions caused by the production of raw resources. The building and demolition sectors alone account for over 30% of global raw material consumption and 25% of all waste (Benachio et al. 2020).

1.3.2 RECYCLING AND REUSE: A KEY TO CIRCULAR CONSTRUCTION

Reusing and recycling building materials, including timber, is a key component of the circular economy, which seeks to limit the use of virgin resources while maintaining the quality of building processes. The main aim is to ensure that construction resources are not simply consumed and discarded but are continuously reused within closed loops, providing a sustainable alternative to the linear "take-make-dispose" model that dominates today's industry. The emphasis is on reducing the environmental impact and extending the lifecycle of materials, thereby supporting the development of resource-efficient and environmentally friendly construction practices.

The timber sector, in particular, offers excellent potential for both recycling and reuse. Timber, as a renewable material, can be recycled or reused multiple times when properly handled. During demolition or deconstruction, timber products should either be reused directly, recycled into new materials, or, in some cases, used for energy recovery. The energy produced through this process is often greater than the energy required to manufacture a new timber structure, making it a carbon-neutral energy source that can replace fossil fuels.

1.3.3 ENVIRONMENTAL BENEFITS OF REUSING TIMBER

Reusing timber in construction can have significant environmental advantages. According to current studies, the carbon-neutral energy generated from timber waste can offset a large amount of CO_2 emissions. The energy content of dry timber is about 4.5 kWh/kg or roughly 2000 kWh per cubic metre of timber products, making it a viable alternative to non-renewable energy sources. Furthermore, the recycling of timber helps reduce the need for deforestation, conserving natural habitats and preserving biodiversity.

The shift towards circular practices in the building industry aligns with the broader goal of mitigating climate change. By integrating timber reuse and recycling into construction processes, the sector can significantly reduce global carbon emissions and create a more sustainable future. Reusing timber can serve as a climate-positive strategy, as timber stored in buildings can act as a carbon sink, locking away carbon for the duration of the building's lifespan.

1.3.4 CHALLENGES AND OPPORTUNITIES IN TIMBER REUSE

While the potential for timber reuse is clear, several challenges remain. For instance, the reuse of structural elements such as beams, supports, and flooring in timber buildings is still underdeveloped in many regions. Timber reuse in structural components is not widespread, partly due to a lack of standardisation, knowledge, and testing protocols. To fully realise the benefits of timber reuse, there needs to be more comprehensive systems for testing, certifying, and guaranteeing the safety and performance of reused timber materials.

Moreover, reusing timber from older buildings requires careful evaluation to ensure that the materials meet the necessary safety standards for modern construction. Timber components that are well-preserved and protected from degradation can have an extended lifespan, contributing to the sustainability of future buildings. However, more research is needed into the most effective methods for restoring and reusing timber from older buildings in an environmentally and economically advantageous way compared to constructing new buildings.

1.3.5 CIRCULAR ECONOMY AND DESIGN FOR REUSE

Implementing circular principles in the design and construction phases is crucial to enabling greater reuse of materials. Circular economy strategies are often summed up in the three R: reduce, reuse, and recycle. In construction, this translates to designing buildings that allow for the easy disassembly and reuse of materials, choosing durable and easy-to-repair or recycle materials, and creating buildings that can be adapted or repurposed for different uses over time.

The key to successful circular construction is designing for disassembly. This means buildings should be designed to allow individual components, for example, timber beams, floorboards, indoor panelling, or exterior façades, to be easily removed and reused when the building is no longer used or rebuilt for other purposes. Additionally, flexibility and adaptability in design can ensure that buildings remain functional for more extended periods, reducing the need for demolition and new construction.

Recycling and reusing timber in construction represent vital steps towards creating a more sustainable built environment. The global construction industry must continue exploring circular economy principles, focusing on reducing resource consumption, and minimising waste through more innovative design and efficient reuse and recycling processes. Adopting these practices can significantly contribute to meeting global climate targets and reducing the environmental footprint of construction. As knowledge and technologies in this area continue to evolve, the widespread adoption of timber reuse and recycling can help create a more sustainable and climate-resilient future for the construction industry.

REFERENCES

Anon. (2013). UK Construction Industry Council. *Offsite Housing Review*. Available online: https://www.buildoffsite.com/content/uploads/2015/04/CIC-Offsite-Housing-Review. pdf, [Retrieved September 2024], (27 pp.).

Anon. (2016). Shaping the Future of Construction: A Breakthrough in Mindset and Technology. [Industry report]. Available online: https://www.weforum.org/publications/shaping-the-future-of-construction-a-breakthrough-in-mindset-and-technology/, [Retrieved September 15, 2024].

APA (2010). A Guide to Engineered Wood Products. *The Engineered Wood Association (APA)*. Tacoma (WA), USA.

Benachio G.L.F., Freitas M. Do C.D. & Tavares S.F. (2020). Circular economy in the construction industry: A systematic literature review. *Journal of Cleaner Production*, 260, Article ID: 121046.

Buchanan A. & Östman B. (2022). *Fire Safe Use of Wood in Buildings – Global Design Guide*. CRC Press, Boka Raton (FL), USA, (486 pp.).

Dunky M. & Mital K. (2023). *Biobased Adhesives. Sources, Characteristics and Applications*. Wiley, Hoboken (NJ), USA, (768 pp.).

European Commission (1988). Council directive of 21 December 1988 on the approximation of laws, regulations and administrative provisions of the Member States relating to construction products. *Official Journal of the European Communities*, L40, 12–26.

Fernow B.E. (1911). *A Brief History of Forestry in Europe, the United States and Other Countries*. 2nd Ed., University Press, Toronto, and Forestry Quarterly, Cambridge (Mass), USA, (506 pp.).

Glavonjić B., Kitek Kuzman M. & Sandberg D. (2022). Kompozitni Proizvodi od Drveta u Savremenoj Arhitekturi: Novi na čini korišćenja Drveta u Dudućnosti. [*Engineered Wood Products in Contemporary Architecture: New Ways to use Timber in Future.*] Luleå University of Technology, Skellefteå, Sweden, (272 pp.).

Kitek Kuzman M., Sandberg D. & Moutou Pitti R. (2018). Produits d'Ingénierie en Bois pour l'Architecture Contemporaine – Cas d´étude. [*Engineered Wood Products in Contemporary Architectural Use: Case Studies*]. University of Ljubljana, Biotechnical Faculty, Department of Wood Sciences and Technology, Ljubljana, Slovenia, (p. 173.).

Klippel M. (2013). *Fire Safety of Bonded Structural Elements*. Doctoral Thesis, ETH Zürich, Zurich, Switzerland.

Kolb J., Kolb H. & Müller A. (2024). *Holzbau mit System*. (4th Ed.), Birkhäuser Verlag, Basel, Switzerland, (464 pp.).

Köstler J.N., Kollmann F. & Massov V.V. (1960). *Denkschrift zur Lage der Forstwirtschaft und Holzforschung*. Steiner, Wiesbaden, Germany.

Miloshevska Janakieska M., Kitek Kuzman M. & Sandberg D. (2024). *Wood Reimagined: Sustainable Architecture with Engineered Wood Products*. Balkan University Press, International Balkan University, Skopje, North Macedonia, Architecture and Design Series, No. 4. (240 pp.).

Niemz P., Teischinger A. & Sandberg D. (Eds.) (2023). *Springer Handbook of Wood Science and Technology. Springer Nature*, Cham, Switzerland, (XXV+ 2154 pp.).

Sandberg D. & Niemz P. (2025). From forestry schools to wood physics as a scientific discipline: A review of historical milestones and future directions of wood science. *Holzforschung*.

Radkau J. (2007). Holz - Wie ein Naturstoff Geschichte schreibt. Oecom Verlag, München, Germany, (352 pp.).

Rüter S. & Hafner A. (2022). Verwendung von Holz in Gebäuden als Beitrag zum Klimaschutz. In: Sahling U. (Ed.), *Klimaschutz und Energiewende in Deutschland*. Springer Spektrum, Berlin, Germany, (pp. 795–807).

Sandberg D., Gorbachewa G., Lichtenegger H., Niemz P. & Teischinger A. (2023). Advanced Engineered Wood-material Concepts. In: Niemz P., Teischinger A. & Sandberg D. (Eds.), *Springer Handbook of Wood Science and Technology*. Springer Nature, Cham, Switzerland, (pp. 1835–1888).

Stephan G. (2012). *Die Gewinnung des Harzes der Kiefer* (*Pinus silvestris*). (3rd Ed.), Verlag Kessel, Remagen-Oberwinter, Germany, (p. 162.).

World Economic Forum (2022). *Accelerating the Decarbonization of Buildings: The Net-Zero Carbon Cities Building Value Framework*. Briefing Paper, World Economic Forum. Geneva Switzerland.

2 The Structure of Wood and Wood-based Materials and Its Modification

P. Niemz, W. Sonderegger, M. K. Kuzman, and D. Sandberg

2.1 TIMBER IN CONSTRUCTION

When used in construction, wood is often referred to as *timber* when it is solid, for example, in the form of a log or sawn timber; in more refined forms, it is referred to as wood-based materials, engineered wood products (EWPs), reconstituted wood, wood-based composites and wood-based products. Wood-based materials for constructive purposes may be divided into (Figure 2.1):

- sawn timber,
- components/EWPs made of logs or sawn timber, and
- wood-based materials/EWPs made of veneer, wood strands, chips, particles or fibres.

Round timber (roundwood, logs) is used to a limited extent in timber construction. Because cut fibres occur in sawn timber but not in logs, round timber has about 20% higher modulus of elasticity and strength, as described in the EN 14251 standard (CEN 2003a).

Sawn timber is made from round timber through the sawing of a log in the length direction. The log itself can be shredded for construction, as in traditional log houses, but it is more common to cut, join length-wise, plane and assemble sawn timber for various construction components. Further refinement results in various high-performance structural wood-based materials, *i.e.* construction components manufactured by bonding veneers, strands, particles or wood fibres. Veneer is a thin sheet of wood with a maximum thickness of 6 mm, but typical thicknesses are between 0.5 mm and 3.0 mm (ISO 2020). Veneer is produced in specially designed veneer mills through rotary cutting and to a lesser degree through slicing or sawing a log or flitch. The wood strands (flakes), particles and fibres either come from waste streams (recycled wood), for example, from the sawmill or veneer production or are produced by crushing low-quality virgin round timber from forestry. The wood particles may also come from recycled wood.

DOI: 10.1201/9781003411994-2

```
                          ┌─────────────────────────┐
                          │     Sawn timber &        │
                          │  Wood-based materials    │
                          └─────────────────────────┘
```

Sawn timber	EWPs from sawn timber, (logs)	EWPs based on veneer, particles, fibres
· Apperance graded · Strength graded · Specific dimensions · Sawn surfaces · Planed surfaces · Rounded edges · hardwood, softwood	· Components of sawn timber · Edge-glued panel · Glued-laminated timber · Cross-laminated timber · Fibre-reinforced timber · Lightweight material · Fibre-reinforced timber	· Plywood · Laminated veneer lumber · Parallel strand lumber · Particleboard · Insulation material · Medium-density fibreboard · High-density fibreboard

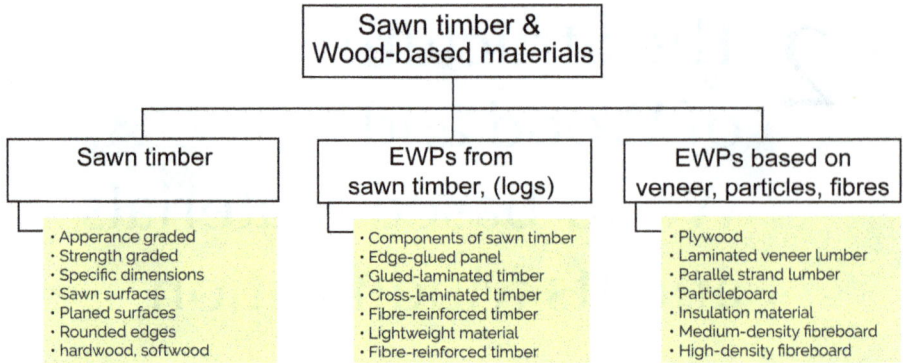

FIGURE 2.1   Classification of wood-based materials and some examples of products.

(Courtesy of D. Sandberg.)

EWPs are becoming increasingly important. Adhesives are the dominant way of bonding structural wood-based materials, and additives such as coatings, fire retardants, hydrophobic agents, *etc.*, are also used to enhance their properties. Wood-based materials made of fibres can sometimes bond through wood's inherent binding forces, for example, fibreboards manufactured in the so-called wet process. Metal fasteners such as nails are standard for mechanically joining different elements into a structure. Still, wooden nails and dowels of high-density hardwoods (sometimes also densified) are also used instead of adhesives in cross-laminated timber (CLT), glued-laminated timber (GLT, or glulam) and flooring elements made of stacked sawn timber. Rare but existing are mechanical connections such as wooden "dovetails" and tongue-and-groove connections to fix the layers.

Sawn-timber components are commonly used for construction purposes in Central Europe (in German-speaking regions named Konstruktionsvollholz®). These components are supplied with a defined wood moisture content of 15±3%, processed with smooth surfaces and to specific dimensions, and finger-jointed to any length. Specific visual criteria are also specified. In addition, grooves (well-defined cuts in the timber) reduce or eliminate cracking.

EWPs are increasingly treated with various environmentally friendly wood-modification methods to improve the above-ground durability to biological attack, shape stability of the material, *etc.* (see Section 2.3). The main reasons for the increased interest during the last decades in wood modification through ongoing research, the industry and society in general can be summarised as:

- a change (degradation) in wood properties because of changes in silvicultural practices and the way of processing the round timber and sawn timber,
- awareness of the use of rare species with outstanding properties, such as durability and appearance,
- awareness and restrictions by law of using environmentally unfriendly chemicals for increased durability and reduced maintenance of wood products,

- an increased interest from the industry to add value to sawn timber and by-products from the sawmill and refining processes further up in the value chain,
- EU and international policies supporting the development of a sustainable society, and
- the international dimension of climate change and related activities is mainly organised within the United Nations (UN) framework, such as the Paris Agreement under the United Nations Framework Convention on Climate Change.

Sawn timber is appearance or strength graded. Appearance grading is based on purely optical criteria of the wood surfaces and the sawn timber as a whole. Such graded timber is used in specific products such as furniture, flooring, parquet, solid-wood panels, *etc.* Strength grading is used for construction and is based on various physical properties of timber, such as density, dynamic modulus of elasticity, fibre-to-load orientation, and number, location, type and size of knots. It follows principles described in the EN 14081 standard (CEN 2019) or other standards. Mechanical strength grading (bending tests) is mainly used in larger companies. Sometimes, handheld or stationary devices based on the natural frequency of the sawn timber measurement or ultrasonic devices are used for strength grading in large and small companies.

Grading into so-called Strength Classes can significantly reduce the variability in strength and guarantee the strength properties of sawn timber and wood-based materials. A Strength Class is named C for softwoods sawn timber, and for hardwood sawn timber, it is named D. Glued-laminated timber strength classes are named GL (Table 2.1). Characteristic value is generally a value corresponding to a fractile of the statistical distribution of a timber property. The fractile is the 5th percentile for the

TABLE 2.1

Characteristic Strength Values for Sawn Timber According to the EN 338 Standard (CEN 2016a) and Glued-laminated Timber According to the EN 14080 Standard (CEN 2013a)

	Softwoods			Hardwoods			Glued-laminated Timber[a]	
Strength Class:	**C18** (MPa)	**C24** (MPa)	**C30** (MPa)	**D30** (MPa)	**D40** (MPa)	**D50** (MPa)	**GL 24h** (MPa)	**GL 30h** (MPa)
Bending ‖	18	24	30	30	40	50	24	30
Tension ‖	11	14	18	18	24	30	19.2	24
Tension ⊥	0.4	0.4	0.4	0.6	0.6	0.6	0.5	0.5
Compression ‖	18	21	23	23	27	30	24	30
Compression ⊥	2.2	2.5	2.7	5.3	5.5	6.2	2.5	2.5
Shear strength (torsion)	3.4	4.0	4.0	3.9	4.2	4.5	3.5	3.5
Rolling shear							1.2	1.2

‖ – in the fibre direction and ⊥ – perpendicular to the fibre direction of the sawn timber.
[a]Homogeneous lay-up of the beam.

modulus of elasticity, and the mean value is also a characteristic value. The EN 338 (CEN 2016a) standard lists characteristic values for stiffness and strength for hardwood and softwood. They are divided into classes C14 to C50 for softwood and D18 to D80 for hardwood. Among other things, the characteristic values for density and bending strength parallel and perpendicular to the fibre direction of the timber, *i.e.* to the grain, are fixed. Glulam has characteristic strength and stiffness classes according to the EN 14080 standard (CEN 2013a).

The strength properties of wood-based panel materials are also determined according to standards. Table 2.2 shows characteristic stiffness and strength values according to the EN 12369 standard (CEN 2001).

TABLE 2.2

Characteristic Strength Values for Wood-based Materials According to the EN 12369 Standard (CEN 2001)

Wood-based Materials	Density (kg/m³)	Bending f_m (f_p) (MPa)	Tension f_t (MPa)	Compression f_c (MPa)	Shear f_v (MPa)	f_r (MPa)
Solid wood panel ‖	410	12–35 (10–25)	6–16	10–16	2.5–4.0	1.2–1.6
Solid wood panel ⊥	410	5–9 (12)	6	10–16	2–5	1.4
Plywood ‖, ⊥	350…750	3–80	1.2–40	1.2–40	1.8–7.5	0.4–1.2
OSB/2 ‖, OSB/3 ‖	550	14.8–18.0	9.0–9.9	14.8–15.9	6.8	1.0
OSB/2 ⊥, OSB/3	550	7.4–9.0	6.8–7.2	12.4–12.9	6.8	1.0
OSB/4 ‖	550	21.0–24.5	10.9–11.9	17.0–18.1	6.9	1.1
OSB/4 ⊥	550	11.4–13.0	8.0–8.5	13.7–14.3	6.9	1.1
Particleboard, type P4	500–650	5.8–14.2	4.4–8.9	6.1–12.0	4.2–6.6	1.0–1.8
Particleboard, type P5	500–650	7.5–15.0	5.6–9.4	7.8–12.7	4.4–7.0	1.0–1.9
Particleboard, type P6	500–650	10.0–16.5	7.5–10.5	10.4–14.1	5.5–7.8	1.7–1.9
Particleboard, type P7	500–650	12.5–18.3	8.0–11.5	13.0–15.5	7.0–8.6	1.8–2.4
Fibreboard, HB.HLA2	800–900	32–37	23–27	24–28	16–19	2.5–3.0
Fibreboard, MBH.LA2	600–650	15–17	8–9	8–9	4.5–5.5	0.25–0.30
MDF.LA	500–650	19–21	10–13	10–13	5.0–6.5	–

‖ – in the direction of the central axis or main fibre direction of the surface layer; ⊥ – in the direction of the minor axis or perpendicular to the main fibre direction of the surface layer; f_m – strength when bending transversely to the board plane; f_p – strength when bending in the board plane; f_t, f_c – tensile and compressive strengths in parallel to the board plane; f_v – shearing transversely to the board plane; f_r – shearing in board plane.

2.2   WOOD IS A MATERIAL FROM NATURE

There are 3.5 billion hectares of forests worldwide with 250,000 wood species, but only about 300 are used commercially. The most significant regions for forestry are Scandinavia, Eastern Europe and South America.

About 150 million hectares of usable forests exist throughout Europe (excluding Russia), corresponding to a total usable stock of 21.4 billion m^3. Every year, 661 million m^3 of wood grows in this area, resulting in a round-timber production of about 327 million m^3, *i.e.* about 50% of the annual growth. Most of the European output occurs in Finland, France, Germany and Sweden. In several European countries, more forest grows than is harvested.

About half of the South American land is forested or about 832 million hectares. According to the FAO (2020), about 50% of forests are classified as primary forests, the most biodiverse forest. South America is among the world's leading regions in plantation forestry, with the second largest area of tree plantations globally (behind Asia), concentrated in Argentina, Brazil, Chile and Uruguay. However, even though 15% of the world's approximately 131 million hectares of tree plantations are in South America, they have neither helped to solve the region's deforestation problems nor significantly proven effective in reducing greenhouse gas emissions.

Besides its availability, carbon storage is an essential aspect of wood as a renewable resource. An average of 255 kg of carbon is stored in one cubic metre of wood, corresponding to a CO_2 equivalent of 0.935 tonnes. CO_2 storage in coniferous forests averages 157 tonnes/hectares annually (Burgert 2016). Wood stores carbon, both in the forest and when wood is used. Therefore, using wood or wood-based materials in long-life products contributes significantly to long-term carbon storage and helps achieve the United Nations Framework Convention on Climate Change goals.

There are 3.5 billion hectares of forests worldwide, with approximately 250,000 wood species, of which only approximately 300 are used commercially. The most significant forestry regions in the world, vital for global timber production, carbon storage and biodiversity conservation, are:

- Eastern Europe – countries such as Russia, Ukraine, Poland and Romania have extensive forested areas, contributing significantly to global timber production,
- Scandinavia – known for its vast forests in countries such as Sweden, Finland and Norway, which are major producers of timber and wood-based products,
- South America – particularly Argentina, Brazil, Chile and Uruguay, which have large areas of both natural forests and tree plantations, although deforestation remains a challenge in some regions,
- North America – Canada and the USA have significant forested areas, especially in regions like the Pacific Northwest and the boreal forests of Canada, and
- Asia – China and Indonesia are notable for their vast forest areas and for being leading regions for plantation forestry.

Europe (excluding Russia) has about 150 million hectares of usable forests, yielding a total usable stock of 21.4 billion m^3. Each year, 661 million m^3 of wood grows in this area, leading to round-timber production of approximately 327 million m^3, representing about 50% of the annual growth. Most European production occurs in Finland, France, Germany and Sweden. In several European countries, forest growth exceeds the rate of harvesting.

South America has approximately 832 million hectares of forested land, with around 50% of its forests classified as primary forests, the most biodiverse type (FAO, 2020). South America is a leading region in plantation forestry, with the second-largest area of tree plantations globally, after Asia, concentrated in Argentina, Brazil, Chile and Uruguay. However, despite accounting for 15% of the world's 131 million hectares of tree plantations, these have not significantly addressed the region's deforestation challenges nor proven effective in reducing greenhouse gas emissions.

North America's forested areas are vast and diverse, particularly in Canada and the USA. Canada's boreal forests, covering 60% of the country, are one of the largest forest biomes in the world, dominated by coniferous trees such as spruce and pine species. The Pacific Northwest, spanning parts of Canada and the USA, is renowned for its temperate rainforests, home to towering trees such as Douglas fir and Sitka spruce. These forests are rich in biodiversity and provide essential timber resources, though they face logging and climate change pressures.

In addition to its availability, carbon storage is a key aspect of wood as a renewable resource. On average, one cubic metre of wood stores 255 kg of carbon, equivalent to 0.935 tonnes of CO_2. CO_2 storage in coniferous forests averages 157 tonnes per hectare annually (Burgert 2016). Wood stores carbon in the forest, and its use makes it a significant contributor to long-term carbon storage. Therefore, using wood or wood-based materials in long-lasting products plays a crucial role in achieving the United Nations Framework Convention on Climate Change goals.

2.2.1   ORIGIN AND STRUCTURE OF WOOD

Evolution has resulted in two categories of trees: conifers (softwoods) and broad-leaved (hardwoods) trees, originating from the so-called seed plants. Many lineages of seed plants have appeared during evolution and most have long been extinct. Timber merchants label all conifers "softwoods" and all broad-leaved trees "hardwoods", even though some conifers are a lot harder than many hardwoods, and the softest woods of all are, in fact, hardwoods.

Wood is formed via photosynthesis (Figure 2.2):

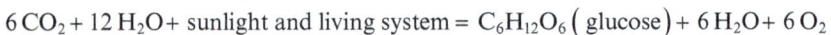

$$6\,CO_2 + 12\,H_2O + \text{sunlight and living system} = C_6H_{12}O_6\left(\text{glucose}\right) + 6\,H_2O + 6\,O_2$$

The elemental mass fractions of dry wood are approximately 50% carbon, 43% oxygen, 6% hydrogen, and 1% nitrogen and minerals.

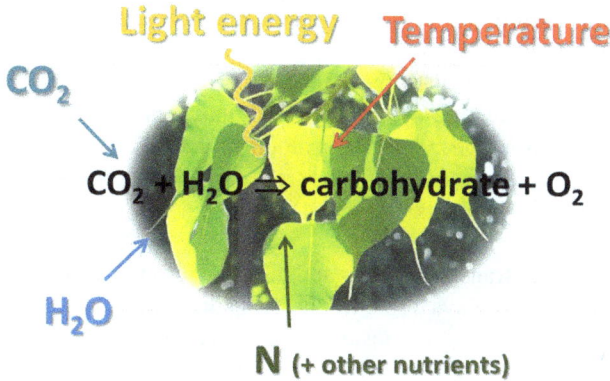

FIGURE 2.2　Photosynthesis.

(Schmitt et al. 2023, courtesy of Springer Nature.)

TABLE 2.3
Composition of Softwoods and Hardwoods

Constituents	Softwoods (%)	Hardwoods (%)
Cellulose	42–49	42–51
Hemicelluloses	24–30	27–40
Lignin	25–30	18–24
Extractives	2–9	1–10
Minerals	0.2–0.8	

Wood is a natural polymeric material located inside the tree's bark. Wood is sometimes called xylem. The main chemical components of wood are:

- cellulose,
- hemicelluloses,
- lignin and
- extractives.

Softwoods and hardwoods have specific differences in their chemical composition, as shown in Table 2.3. The handbooks of Fengel and Wegener (1984) and Mai and Zhang (2023) are recommended for detailed studies of wood's chemical structure. Wood extractives are, in contrast to cellulose, hemicelluloses and lignin, non-structural wood components. They are typically concentrated in the heartwood and are often produced by the standing tree as defensive compounds against environmental stresses. Extractives stored in the wood are essential for its sorption behaviour, swelling and bonding properties.

Wood is primarily adapted to the tree's stability and liquid (water) conduction requirements. The conduction system of a tree and other plants, *i.e.* the so-called vascular system, consists of xylem and phloem tissues. Xylem and phloem form a complex network running throughout the tree, carrying resources to different parts and disposing of waste products (Table 2.4):

- Xylem primarily transports water and mineral nutrients from the roots to the rest of the tree and plays a role in physical support.
- Phloem transports organic substances, such as sugars produced during photosynthesis, from the leaves to other parts of the plant.

The fluid transported in the tree's xylem cells or phloem sieve-tube elements is called *sap*. Filled with nutrients and minerals, sap is the blood of a tree. Xylem transports a mixture of water, minerals and hormones from the bottom to the top of the tree in a long string formation (tracheids or vessel elements). In the phloem, photosynthesis's sugars are fed back into the tree and leaves as much-needed food during growth.

TABLE 2.4
Comparison between the Two Transport Systems in a Tree: Xylem and Phloem

	Xylem	Phloem
Description	Xylem tissue consists of various specialised, water-conducting cells known as tracheary elements.	The phloem tissue is a tubular, elongated structure with transverse end walls containing sieve-like pore groups (sieve plates).
Features	The mature xylem is dead tissue. Lignin fibres make the xylem waterproof and keep it from collapsing under pressure.	Phloem is living tissue, although the sieve tube cells lack nuclei and contain little cytoplasm.
Functions	Transports water and minerals in support of photosynthesis and transpiration. The xylem also functions as structural support for the plant.	Transports organic molecules, such as sugars, amino acids, some plant hormones and mRNA (mRNA stands for messenger RiboNucleic Acid and is a single-stranded molecule that carries the instructions to make proteins)
Location	It occurs in roots, stems and leaves in the centre of the vascular bundle.	It occurs outside the xylem in the so-called inner bark of the tree, leaves, roots and fruits.
Mode of transport	Negative pressure powers the upward flow of fluid.	Turgor pressure from osmosis powers the flow of sap.
Sap movement	Fluid only moves upwards from the roots towards the stems and leaves.	Fluid movement is bidirectional, moving up or down depending on the plant's needs.
Tissues	Consists of tracheids, vessel elements, xylem fibres, xylem parenchyma and xylem sclerenchyma cells.	Consists of sieve tubes, companion cells, bast fibres, phloem fibres, intermediary cells and phloem parenchyma.

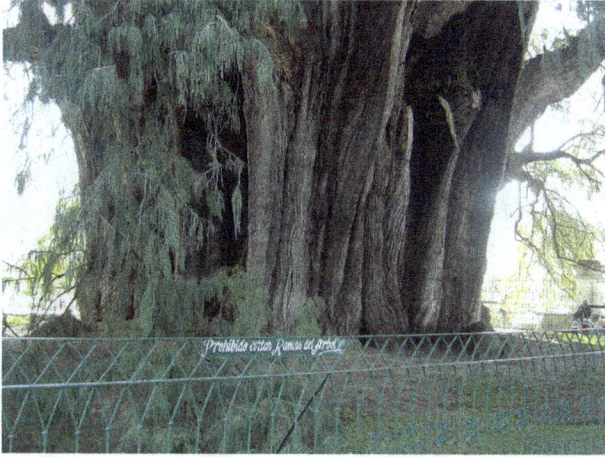

FIGURE 2.3   Tree of Thule in Mexico, a Montezuma bald cypress, is approximately 1600 years old, 14.5 m in diameter, has a volume of 817 m³ and a weight of 636 metric tonnes.

(Courtesy of P. Niemz.)

Trees adapt their annual growth in height and thickness to changing environmental conditions (soil, climate, altitude above sea level). Some tree species can grow over 100 m high and live over 6000 years. The tree of Thule in Mexico is a Montezuma bald cypress and is considered the thickest tree in the world (Figure 2.3). Well-known trees of considerable height are the giant sequoia in North America, commonly up to 80 m high, and the Patagonian cypress in southern South America, usually growing to 40–60 m but occasionally more than 70 m. The Patagonian cypress may be up to 3000 years old and reach a trunk diameter of 3.5 m.

Wood's structure results from the living tree's requirements to perform the roles of support, conduction and storage, which makes its characteristics different from those of man-made, non-bio-based materials. Wood is built up of elongated cells oriented almost in the direction of the trunk. The cells are a few millimetres long, spool-shaped and hollow. Wood has a hierarchical structure following five structural levels (Speck and Burgert 2011):

- integral level (stem structure),
- macroscopic level (tissue structure),
- microscopic level (cell structure),
- ultrastructural level (cell-wall structure) and
- biochemical level (biochemical composition of the cell wall).

Figure 2.4 shows the structure of conifer wood within a growth ring and at the transition to the bark, as well as the typical structure and composition of the cell walls.

Within the living tree, most of the cells are dead, *i.e.* the protoplasm is absent, leaving hollow cells with rigid walls. The support role enables a tree stem to remain erect despite the height to which the tree grows. Because of this height, wood must

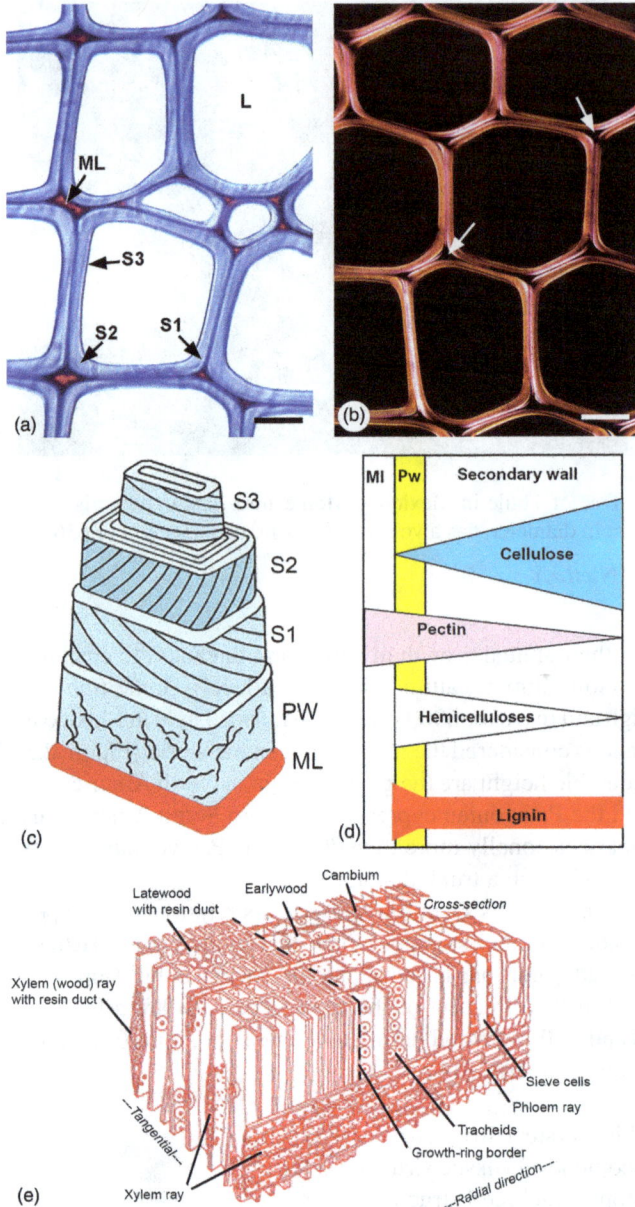

FIGURE 2.4 Structure of conifer wood and cell-wall composition (Schwarze 2008). (a) Transverse section of earlywood tracheids: ML is the middle lamella; S1, S2 and S3 are the outer, middle and inner secondary wall layers, respectively, and L is the lumen, *i.e.* the hollow void inside the cell. (b) Cross-section view of earlywood. The secondary walls appear bright, whereas the amorphous compounds in the middle lamella are dark (arrows). (c) The conventional cell-wall model distinguishes five layers (PW is the primary wall). (d) Diagram of the relative distribution of constituents within the different layers of the cell wall. (e) Schematic of the softwood structure.

(Courtesy of F.W.M.R. Schwarze.)

also perform the role of conduction, transporting water and minerals from the ground to the upper parts of the tree. Finally, nutrients are stored in certain parts of the wood until required by the living tree. The only living cells in the woody parts of the tree are the nutrient-storing cells. The inside bark and the cellular layer beneath it, called the cambium, also consist of living cells.

On the macroscopic level, some essential features of wood can be identified. However, it is necessary first to define the three principal directions and the three sections or planes of wood in general (Figure 2.5), *i.e.* the longitudinal (L), radial (R) and tangential (T) directions and the transversal or cross-section (RT), the radial section (LR) and the tangential section (LT).

The main macroscopic parts of the wood are:

- growth rings, *i.e.* layers of earlywood and latewood together,
- pith,
- rays,
- resin canals of conifer wood,
- vessels of broad-leaved wood,
- sapwood and heartwood,
- knots,
- reaction wood,
- juvenile wood, and
- texture and colour.

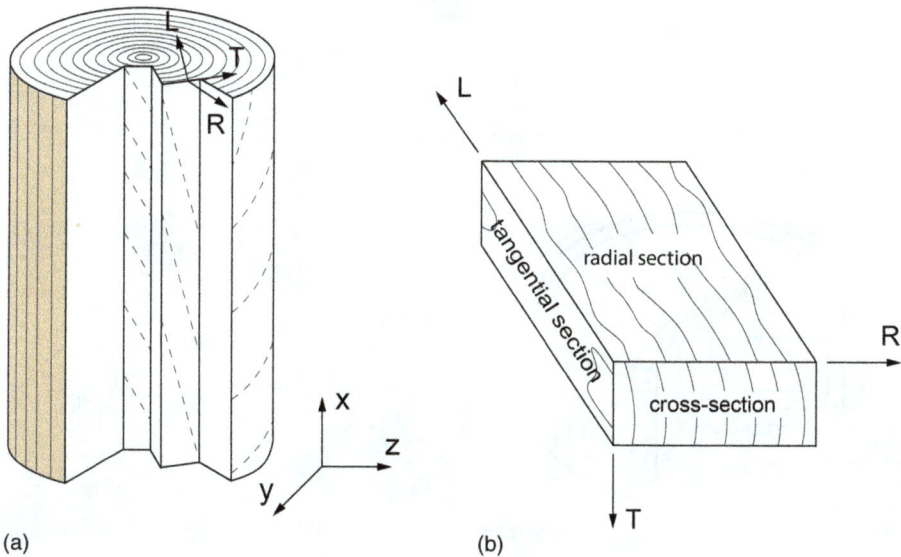

FIGURE 2.5 Definition of principal directions and sections of wood: (a) stem section shows the principal axes, viz. longitudinal (L), radial (R) and tangential (T), and (b) a piece of wood showing the definition of the principal sections in sawn timber: cross-section (RT), radial section (LR) and tangential section (LT).

(Courtesy of D. Sandberg and Taylor and Francis group.)

2.2.1.1 Growth Rings

The cross-section of a tree trunk is the cross-cut of the stem and has a close-to-cir-cular shape (Figure 2.6). Typical for the appearance of the cross-section is that the naked eye can see the individual growth rings of the tree as dark and light concentric rings. A single growth ring is divided into two distinct parts: earlywood and late-wood. The light-coloured earlywood, formed in the spring, has large cavities in the cells to transport large amounts of water. The cells formed later in the summer do not need the same transportation capacity and have smaller cavities and a higher density. These dark-coloured cells are called latewood and are mainly responsible for the strength of the wood.

2.2.1.2 Pith

In the centre of the stem, branches and roots is the pith, formed during the first year of growth and becomes a storage area for impurities deposited from the active xylem during the tree's growth. Pith consists of soft, spongy *parenchyma* cells and in most cases, it is soft. The shape of the pith varies between species and in diameter from about 0.5 mm to 8 mm. Freshly grown pith in young shoots is typically white or pale brown, but it usually darkens with age. It may be inconspicuous, but it is always at the centre of the stem, branches and roots. The roots have, however, only a small or no pith, and the anatomical structure is more variable.

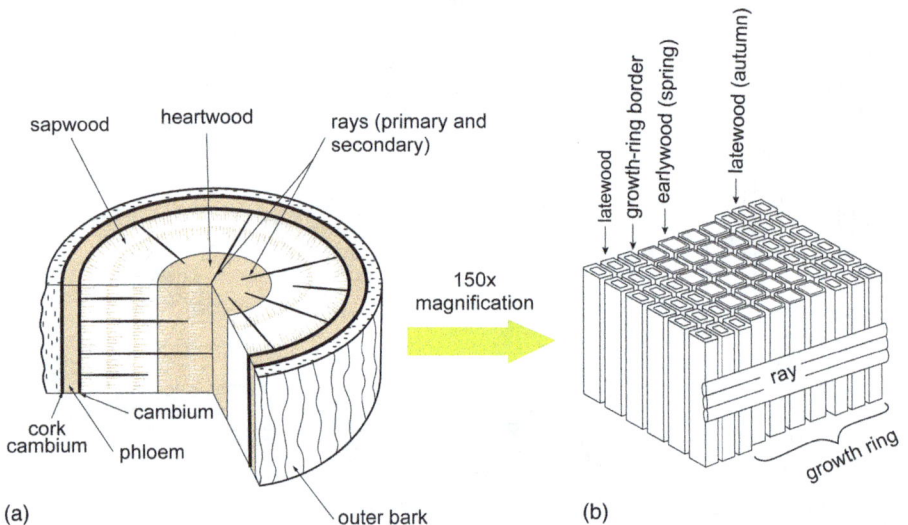

FIGURE 2.6 Major tissue types in a typical conifer tree: (a) macroscopic view and (b) enlargement of a growth ring, showing the relative difference in size between earlywood and latewood tracheids.

(Courtesy of D. Sandberg and Taylor and Francis group.)

2.2.1.3   Rays

Wood rays extend transversely from the inner bark towards the tree's centre at a right angle to the growth rings. The cambium forms the rays, extending in the xylem's radial direction (cf. Figure 2.6). Rays consist of nutrient-storing cells and provide a route for sap to be transported horizontally to or from the inner bark (phloem).

2.2.1.4   Resin Canals in Coniferous Wood

A characteristic feature of some conifer woods is their resin content, often sufficient to give them an unmistakable fragrance and make newly sawn timber sticky. Resin canals or resin ducts are tube-like intercellular spaces which transport resin in both the longitudinal and horizontal directions (Figure 2.7).

2.2.1.5   Vessels of Broad-leaved Wood

The fundamental anatomical difference between wood from coniferous and wood from broad-leaved trees is that broad-leaved woods contain specialised conducting cells called vessel elements. These vessel elements are generally much more prominent in diameter than other types of longitudinal cells, and the vessels are generally shorter than both broadleaved and conifer fibres. Several vessel elements are linked end-to-end along the stem to form long tube-like structures. The vessels' size and arrangement in the wood cross-section are used to classify wood from different broad-leaved trees, cf. Figure 2.10.

2.2.1.6   Sapwood and Heartwood

Water is transported from the roots into the crown of the living tree. The sapwood is the outer, water-conducting part of the trunk containing living ray cells, *i.e.* cells to store reserve nutrients. Young trees have only sapwood, but as they mature and no longer need the whole cross-section of the xylem part of the trunk for fluid transport, they develop heartwood, *i.e.* the water-conducting function ceases and the remaining living wood cells die. The cell walls are preserved and will help support the tree for

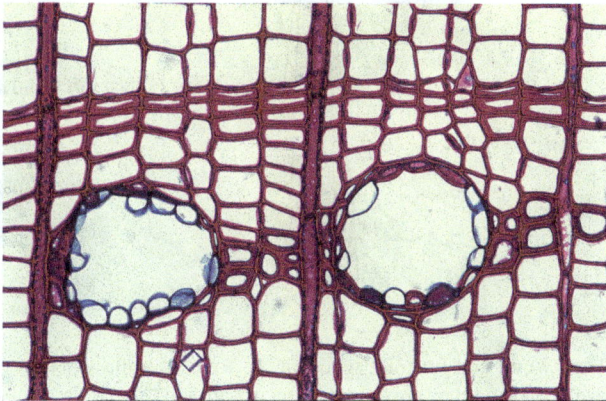

FIGURE 2.7   Cross-section view of two resin canals in Norway spruce.

(Courtesy of D. Sandberg.)

many years. Heartwood is the inner and central part of the trunk, which, in the living tree, contains only dead and non-water-transporting cells and in which the reserve materials have been removed or converted into extractives (Figure 2.8). The new wood cells thus created are added to the outer part of the sapwood, while the older sapwood cells adjacent to the heartwood gradually change to form new heartwood. The proportions of sapwood and heartwood vary according to species, the tree's age, position in the tree, rate of growth and environment.

Typical tree species with obligatory heartwood formation, which leads to a coloured core in the stem, are black locus, Douglas fir, European chestnut, oak and pine spp. The trees produce chemical substances in complex biochemical processes stored in the cell walls. These processes provide the wood with better protection against the degradation of microorganisms and achieve significantly higher durability. However, there are also tree species with light-coloured heartwood, such as spruce, birch, fir and willow, in which sapwood and heartwood appear (but are not) similar. There are also tree species that are known as facultative heartwood formers. These tree species, which include European ash and beech, usually have a light-coloured heartwood (Burgert 2016). In the case of some hardwoods, so-called tylosis develops and leads to the closure of water transport through the vessels. In softwoods, the pits partially or fully close after felling, which reduces or stops the liquid transport through the cells.

2.2.1.7 Juvenile Wood

Unlike heartwood that evolves in the lower parts of the trunk upwards, juvenile wood is formed nearest to the pith at all heights in young trees and only in the top regions of mature trees (Figure 2.8). Although it is not visible in the trunk cross-section, there

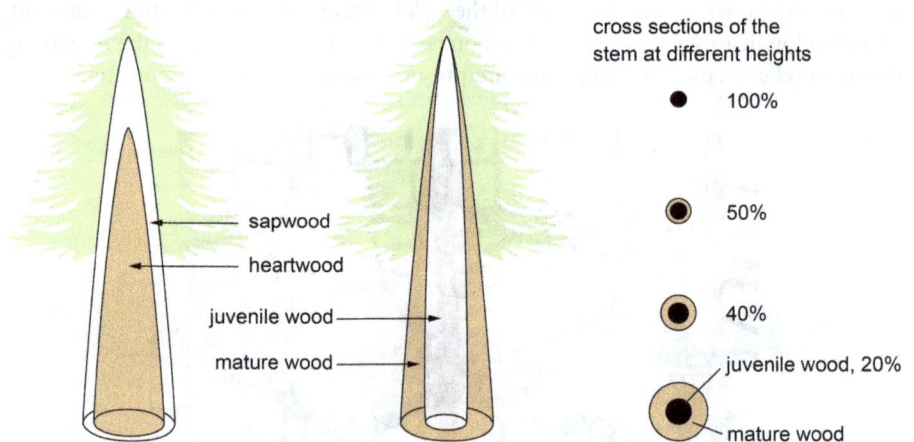

FIGURE 2.8 Heartwood and juvenile wood in the tree. Juvenile wood is part of both the heartwood (if developed) and the sapwood and occurs around the pith, roughly forming a cylinder up the tree. In contrast to heartwood, the cross-section-wise proportion of juvenile wood increases towards the top of the tree.

(Courtesy of D. Sandberg and Taylor and Francis group.)

is an essential pith-to-bark gradient in density that is unique for each species. The fact that a relatively pronounced change in density often occurs in conifers during approximately the first 15 to 30 years of growth gave rise to the term juvenile wood (Zobel and Sprauge 1998). This term can lead to confusion because this wood is found not just in young (juvenile) trees but near the pith in every tree, regardless of age. Sawn timber containing a high proportion of juvenile wood may have distortion issues because fibre and fibril orientation in juvenile wood differ from that in mature wood.

2.2.1.8   Reaction Wood

When a tree grows on a sloping land surface or is exposed to a dominant wind direction, the load on the stem is unbalanced. The tree then starts to produce abnormal wood, known as reaction wood, to compensate for the unbalanced load (Figure 2.9). The reaction wood formation is related to the straightening of leaning stems; the same happens in the branches and the area where the branches join the stem. Sawn timber containing a high proportion of reaction wood may have issues with distortion.

2.2.1.9   Knots

As the trunk grows, old and new branches form junctions called knots. Where the cambium is alive at these points, there is continuity of growth combined with a

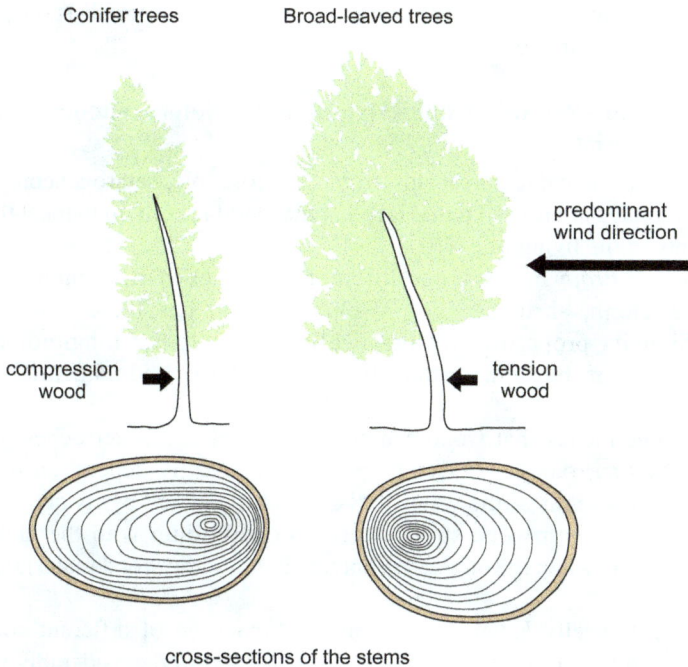

FIGURE 2.9   Formation of reaction wood in wood in conifer and broad-leaved trees due to a predominant wind direction over a long time.

(Courtesy of D. Sandberg and Taylor and Francis group.)

change in orientation and the knot is termed green or alive. The cambium is frequently dead on lower branches, and the trunk grows around the branch, enclosing its bark. These "black" or "dead" knots are liable to drop from sawn timber during sawing or further processing. An essential feature of knots is their deviant fibre orientation around and in the knot itself, affecting the appearance and properties of sawn timber.

2.2.1.10 Texture, Colour and Scent of Wood

The texture of wood is material-dependent, *i.e.* it depends on the type of wood and how it is built up. A piece of wood can show a significant variation in hue depending on the type of wood, the content of extractive substances, heartwood or sapwood, and age. For most types of wood, the growth-ring orientation in the cross-section of the wood is essential for the textural appearance.

Colour is one of the most conspicuous characteristics of wood, and although quite variable, it is one of the essential features used in identification and adding aesthetic value. Basic wood substances, *i.e.* cellulose and lignin, have a slight colour, so any distinctive colour is associated with extractives, mainly in the heartwood. A dark colour indicates heartwood, whereas a light colour can be either heartwood or sapwood. Some wood also changes colour with age or exposure to ultraviolet (UV) radiation.

Certain woods have distinctive odours. Many conifer woods, as well as numerous tropical woods, are known for their aromatic character. The smell is due to volatile extractives or resins in the wood.

2.2.1.11 Wood – An Anisotropic, Hygroscopic, Heterogeneous and Variable Material

Wood has four principal features: anisotropic, hygroscopic, heterogeneous and variable. These features are related to the fact that the wood is built up to meet the specific requirements of the living tree.

Wood's *anisotropy* means it has different properties in different directions, for example, strength, shrinkage and swelling. Wood has three main directions within which its properties are practically constant: the longitudinal (in the length direction of the stem), radial (from pith to bark) and tangential directions (Figure 2.5).

Hygroscopic means that wood can absorb and release water depending on the humidity of its surroundings. Wood's properties strongly depend on its moisture content, *i.e.* the weight of the water in the wood is divided by its dry weight. The amount of moisture in wood strives to reach equilibrium with the ambient air's relative humidity; when the moisture content changes, the wood material shrinks or swells.

Wood is a typically *heterogeneous* material made up of different components. Knots are a natural part of wood, but their properties differ considerably from those of knotless wood and can be both favourable and unfavourable.

Wood is *variable*. Many properties vary within the tree regardless of the main directions. For example, the density in many species decreases from the bark towards the centre and from the root towards the top.

2.2.2   GROWTH DYNAMICS

A tree's thickness growth is influenced by various parameters, including the differences between tree species, within a tree species, within a trunk, and even between individual tissue and cell types (Burgert 2016). Important factors are, for example, the genetic differences between the trees and environmental factors such as the climate or mechanical stress such as wind and slope. Trees are strongly influenced in their growth by temperature, water supply, nutrient supply and day length.

This is reflected in the growth rings and the properties of the wood. Therefore, the trees in temperate zones form well-known growth rings resulting from annual growth, with a pause in growth in the winter months (Figures 2.5 and 2.6). No yearly growth rings can be seen in the wood in zones without a seasonal rhythm, such as in the tropics. However, growth zones can be assigned in the woods of tropical areas, for example, if regular dry periods influence thickness growth. From the trends and constitution of the growth rings, the age of the trees, climate and chemical influences from the environment can be traced back. Today, growth-ring analysis is often used in climate research for climate analyses (Schweingruber 2007). Fritz Schweingruber (WSL Birmensdorf, Switzerland) and Dieter Eckstein (University of Hamburg) were pioneers of growth-ring research (Cufar *et al.* 2024).

Growth rings' chemical isotope analysis can also provide conclusions on environmental influences (Timofeeva 2017). The growth rings can measure light and heavy atoms (stable isotopes) of the chemical elements carbon (C) and oxygen (O). This information can be used to conclude past environmental conditions, the absorption of CO_2 and water evaporation. The unstable isotope carbon-14 enables dating historical, archaeological and subfossil wood.

There is a clear density differentiation within the growth rings, especially in coniferous wood. Earlywood has a lower density than latewood because it primarily provides nutrients and water during fast growth in the spring. Latewood is formed to provide mechanical stability and has a considerably higher density than earlywood. The growth dynamics apply to both softwoods and hardwoods. Figure 2.10 shows the density distribution across earlywood and latewood in Norway spruce.

In terms of evolution, conifers are significantly older than deciduous trees. This is why conifers have a relatively simple wood structure, while hardwoods have a much more differentiated and complex structure (Figure 2.11a–b). Hardwoods are distinguished between ring-porous, semi-ring porous and diffuse-porous species depending on the organisation and size of vessels (Figure 2.11c).

Figure 2.11d shows the density distribution over the growth rings for softwood, ring-porous hardwood and diffuse-porous hardwood. For softwoods, the average bulk density decreases with the width of the growth rings; for ring-porous hardwoods, it increases; and for diffuse-porous hardwoods, no clear tendency is recognisable. In addition to the growth-ring width, many other factors, such as fibre length and microfibril angle, influence the wood properties.

2.2.3   SOFTWOOD

Figure 2.12 shows the structure of a softwood. In addition to the axially aligned earlywood and latewood cells, the radial single-row rays and the axial resin canals can be seen. Coniferous woods consist of two cell types, approximately 90% of tracheids

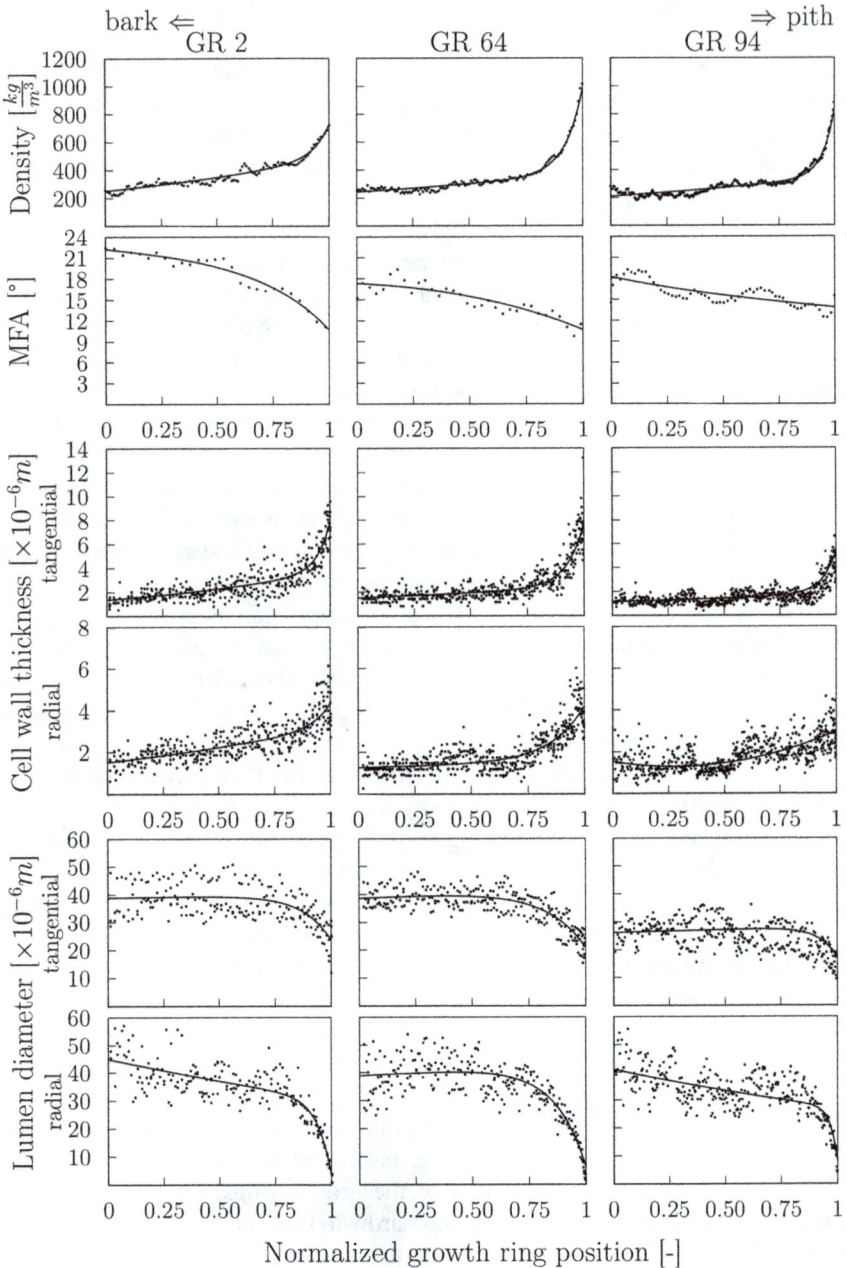

FIGURE 2.10 Density distribution and other parameters in growth rings (GR) from Norway spruce wood (Lanvermann 2014). MFA – microfibril angle.

(Courtesy of P. Niemz and C. Lanvermann.)

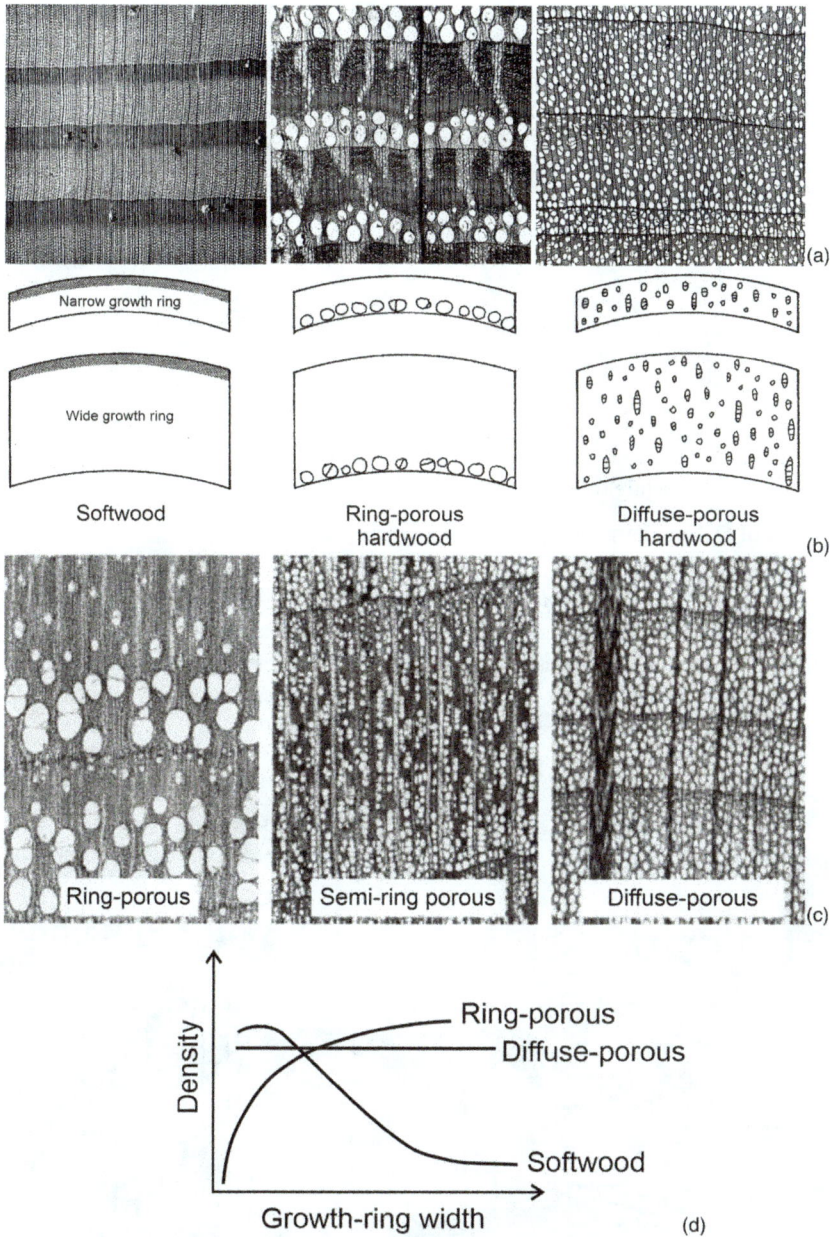

FIGURE 2.11 Density distributions over growth rings of softwood and hardwood: (a) Cross-section view of a conifers species, Scots pine (left), a ring-porous broad-leaved species, European oak (middle), and a diffuse-porous broad-leaved species, goat willow (right), (b) relationship between growth rate (growth-ring width) and microscopic structure in the cross-section of softwoods and hardwoods, (c) vessel structure in the cross-sections of three hardwoods (European ash, wild cherry, common beech) (Grosser 1977) and (d) schematic representation of the density distribution of softwoods and ring-porous and diffuse-porous hardwoods.

(Courtesy of P. Niemz and D. Sandberg.)

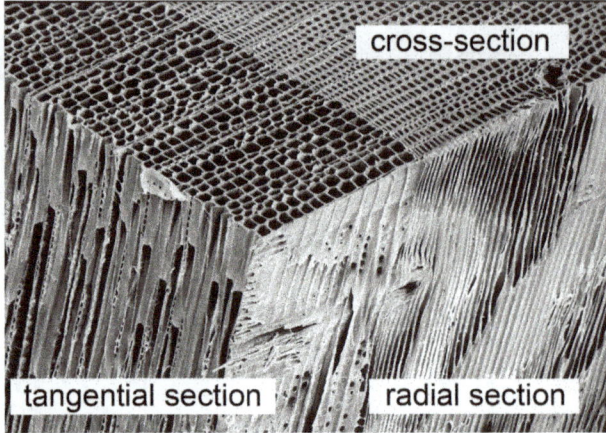

FIGURE 2.12   Structure of softwood.

(Courtesy of D. Sandberg and Taylor and Francis group.)

and approximately 10% of parenchyma cells, mainly found in the rays. Tracheids are approximately 3 mm long, with a diameter of approximately 30 μm. In earlywood, the tracheids are wide-lumened and have thin cell walls; in latewood, they are narrow-lumened and thick-walled. In this way, this cell type can take on both the function of water conduction and the function of strength.

Figure 2.13 shows the transitions between earlywood and latewood for various softwoods.

To ensure the conduction of sap and water within the tree, it is thus necessary for each tracheid to be functionally connected to other tracheids. The conduction between

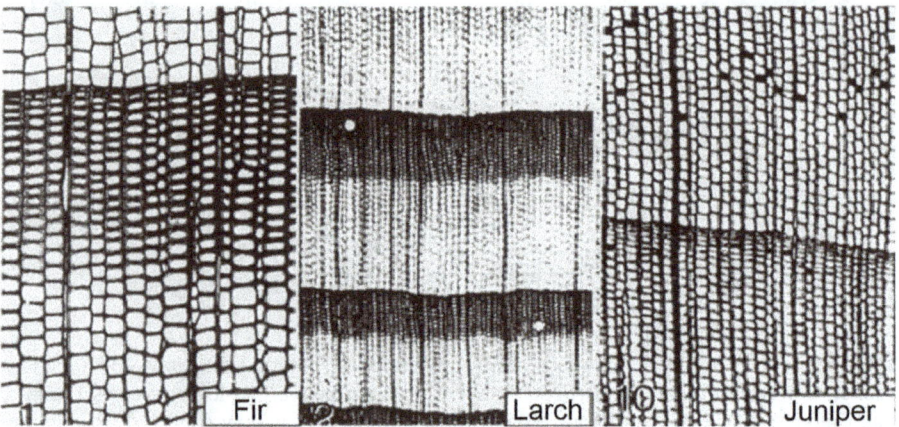

FIGURE 2.13   Characteristic transitions from earlywood to latewood in fir, European larch and juniper.

(Grosser 1977, courtesy of Springer.)

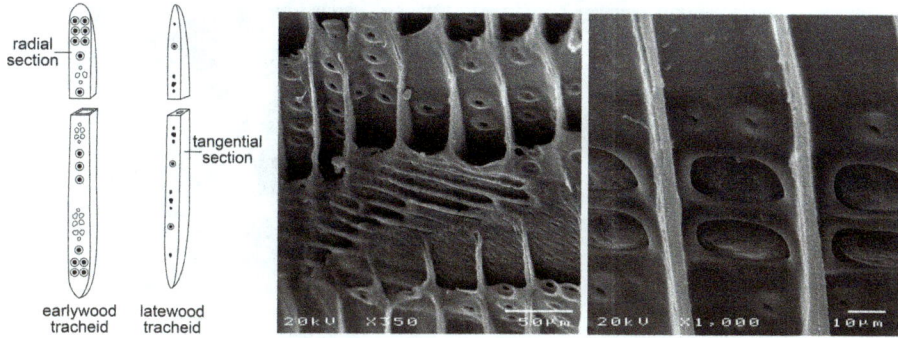

FIGURE 2.14 Diagrammatic representation of an earlywood and a latewood tracheid (left), and micrographs of the radial walls of Scots pine tracheids presenting bordered pits (middle) and (to the right) simple pits between tracheids and ray parenchyma cells (in the centre), and small bordered pits between tracheids and ray tracheids.

(Courtesy of D. Sandberg and Taylor and Francis group.)

tracheids occurs through pits in both the lateral and vertical directions (Figure 2.14). The pits of earlywood tracheids are large and circular, averaging about 200 pits per tracheid, whereas latewood tracheids have relatively small, slit-like pits and only 10 to 50 per tracheid (Trendelenburg and Mayer-Wegelin 1955).

Pits have two essential parts, the pit cavity and the pit membrane, the cavity being open internally towards the lumen of the cell and closed by the pit membrane. In the bordered pit pairs of most coniferous woods, the membrane has a thickening in the central zone called the *torus*, which is somewhat larger in diameter than the aperture and is impermeable to water. The membrane around the torus, the *margo*, is porous (Figure 2.15). When the torus is pressed against one of the apertures, the passage of water is prevented. This phenomenon is called an aspirated pit, which occurs when sapwood is transformed into heartwood or when the wood is dried. In heartwood, the pits are definitely blocked in this position (Figure 2.15).

2.2.4   HARDWOOD

The structure of broad-leaved wood is more complex than that of conifer wood. During its evolution, broad-leaved wood has developed particular types of cells from the tracheid: vessel elements for conduction and fibres for support (Figure 2.16). The axial reinforcing tissue consists of fibres. The main part of the water conduction takes place in the vessels, which can be several metres long.

2.2.5   THE MAIN SPECIES OF WOOD USED IN CONSTRUCTION

The properties of wood are strongly influenced by the dimensions due to the hierarchical structure, *i.e.* the size of the specimen tested influences the result. Certain features and defects in timber also influence properties, especially mechanical properties, compared to the small defect-free specimens commonly used in scientific

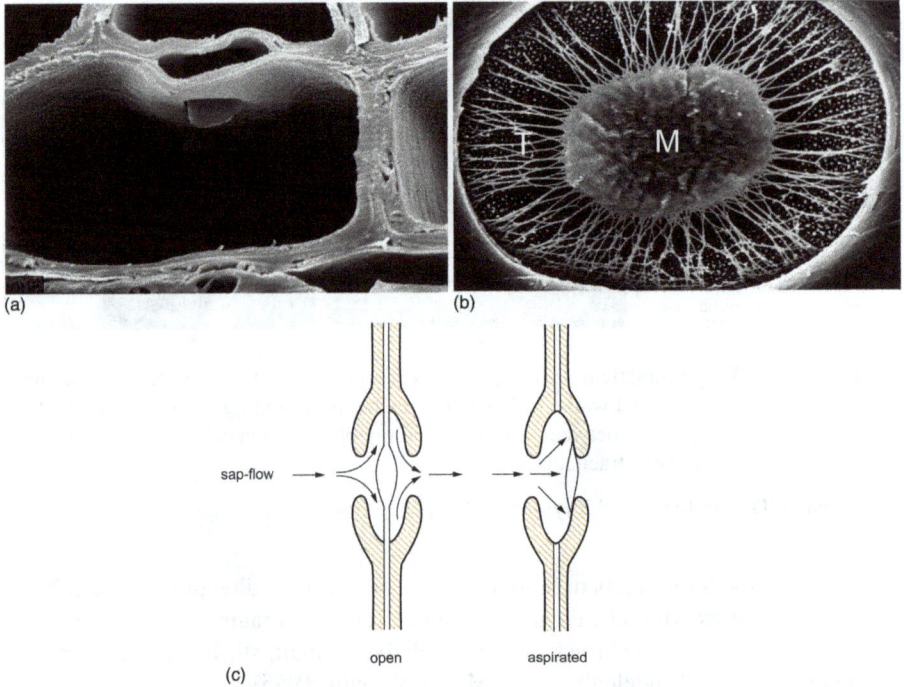

(a)

(b)

(c)

open   aspirated

sap-flow

FIGURE 2.15 Cross-section view through a cell wall of European silver fir containing (a) a pit and (b) the membrane of a bordered pit showing the torus (T) and the margo (M) through which water passes from one cell to the next. (c) Aspiration of pits is involved in heartwood formation in conifers and may also occur when drying the sapwood.

(Courtesy of D. Sandberg and Taylor and Francis group.)

work to determine properties. Fibres, fibril aggregates or crystalline regions have significantly higher strength than full-size sawn timber (Table 2.5).

Important wood properties are compiled in numerous reference books:

- Eigenschaften und Kenngrössen von Holzarten (Sell 1997)
- Wood Handbook (Ross 2021)
- Holzatlas (Wagenführ and Wagenführ 2022)
- Handbook of Wood Science and Technology (Niemz *et al.* 2023)
- Holzarten: Ansichten, Kennwerte und Kennwerte (Teischinger *et al.* 2023)
- Wood Material and Processing Data (Niemz *et al.* 2025)

These books have primary information about characteristics related to small, clear (defect-free) specimens. Calculation parameters for the use in construction are specified in Eurocode 5 (CEN 2004) and EN 338 (CEN 2016a). In particular, the *Springer Handbook of Wood Science and Technology* and the related *Data Handbook*

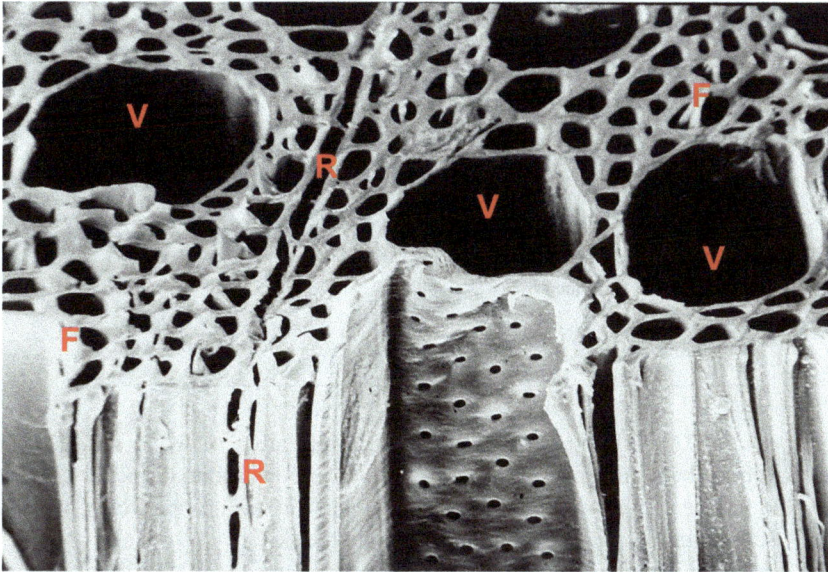

(a)

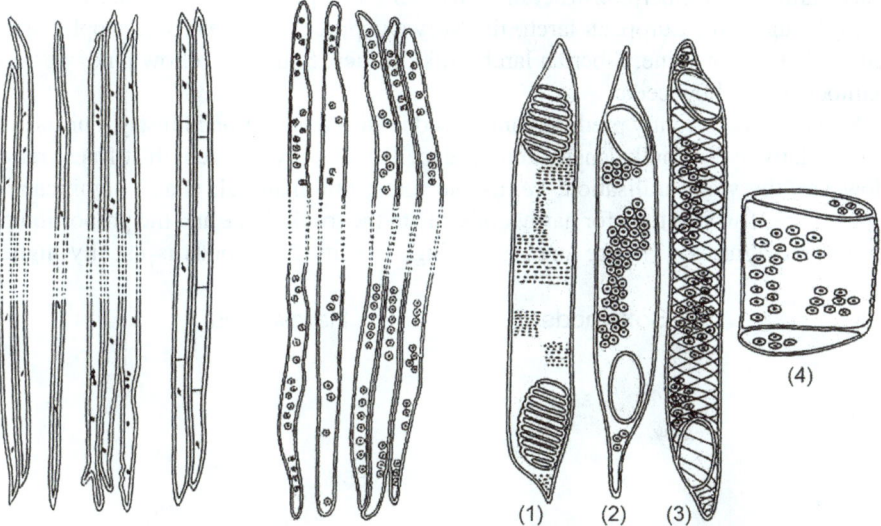

| Libriform wood fibre | Septate wood fibre | Fibre-tracheids | Vessels: (1) with scalariform perforation, (2) with simple perforation, (3) with spiral wall thickening, (4) short early-wood vessel with wide lumen |

(b)

FIGURE 2.16   Hardwood species: (a) SEM micrograph of a broad-leaved wood structure showing V – vessels, F – fibres and R – rays (courtesy of D. Sandberg and Taylor and Francis group), and (b) cell types.

(Grosser 1977, courtesy of Springer.)

TABLE 2.5

Modulus of Elasticity (MOE) and Strength of Wood (Niemz and Sonderegger 2021)

Specimen Type	MOE (N/mm²)	Tensile Strength (N/mm²)
Components (boards, beams)	11,000	25
Small clear specimens	11,000	90
Mechanically separated fibres	40,000	400
Fibril assemblies	70,000	700
Crystalline regions	130,000–250,000	8000–10,000

(Niemz *et al.* 2023, 2025) show an extensive compilation of wood species worldwide. Bamboo (becoming increasingly important in construction) and its characteristic value for the orthotropy of the mechanical and physical properties of wood and wood-based materials are also listed in these handbooks.

The EN 14080 standard (CEN 2013a) suggests some specific timber species for glued-laminated timber construction: Austrian black pine, Corsican pine, Dahurian larch, Douglas fir, European larch, fir, Norway spruce, maritime pine, poplar spp., radiata pine, Scots pine, Siberian larch, Sitka spruce, Southern yellow pine, western hemlock and yellow cedar.

Softwood is currently predominantly used in the construction industry. This is due to the relatively low bulk density and the excellent density-to-strength/stiffness ratio. However, the wood utilisation, *i.e.* usable forest raw material, is also significantly higher for softwoods than for hardwoods due to the trunk shape and the proportion of branches (Figure 2.17). The strength of high-density hardwoods is usually higher

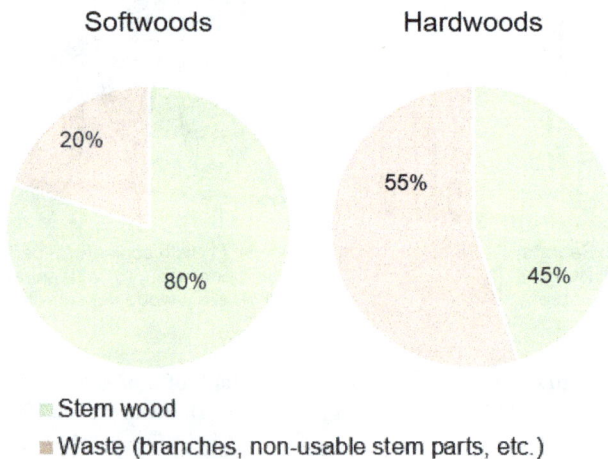

FIGURE 2.17   Usable part of a hardwood and softwood stem.

(Data from Krackler et al. 2010, courtesy of D. Sandberg.)

than that of commonly used construction softwoods. High-density hardwoods are used in construction to absorb compressive stresses for supports for transferring vertical loads and reinforcement of supports, for example, in the case of transverse compressive loads.

However, due to the lower wood utilisation and the changed bonding behaviour, the prices of glued-laminated hardwood timber are several times higher than those of softwoods.

Softwoods are predominantly used for cross-laminated timber and glued-laminated in the construction industry. The most important wood species are Douglas fir, European larch, fir, Norway spruce, radiata pine, Scots pine and Southern yellow pine. Hardwoods such as birch spp., eucalyptus spp., common ash, beech and oak are used due to their higher strength but to a lesser extent.

Due to the cross-wise bonding of the lamellae/layers in cross-laminated timber, it is challenging to meet the requirements for the delamination test according to the EN302-2 standard (CEN 2003b) when hardwoods are used due to their high swelling and shrinkage. However, to save space and cost in building construction, hardwoods may be used instead of softwoods in large cross-sectional glued-laminated timber.

For aesthetic reasons, glued-laminated Norwegian spruce timber may be coated with a thin layer of hardwood. This also applies to laminated veneer lumber (LVL) of Norway spruce. However, the difference in swelling and shrinkage of the hardwood and softwood/LVL materials may cause lamination of the bonding between them.

Due to their high hardness, high-density hardwoods are also used, particularly for parquet, stairs and partly LVL, so-called construction beech, so-called Baubuche.

Important selection criteria are strength (for softwood and hardwood EN 338 CEN (2016a), for glulam EN 14080 (CEN 2013a), durability according to EN 350 (2016b), and hardness and optical properties, including colour consistency and stability. Colour changes also occur in the room's interior, with ageing-related changes in the appearance of the wood surfaces, usually caused by oxidation processes in the wood. For example, the yellowing of light woods (*e.g.* birch, maple) or darkening (*e.g.* spruce, larch) is well known. Thermally modified timber becomes lighter. A good overview of parquet wood made of various wood species can be found in the study by Pitt (2010).

2.3 WOOD MODIFICATION

As a natural renewable resource, wood is generally a non-toxic, easily accessible and inexpensive biomass-derived material. Nevertheless, as wood is a natural product that originates from different individual trees, limits are imposed on its use, and the material needs to be transformed to acquire the desired functionality. Wood modification is implemented to improve the intrinsic properties of wood, produce new materials, and acquire the form and functionality desired by engineers without changing the eco-friendly characteristics of the wood material (Sandberg *et al.* 2013).

2.3.1 THE IDEA OF WOOD MODIFICATION

Wood modification is an all-encompassing term describing the application of chemical, mechanical, physical or biological methods to alter the properties of the material, *i.e.* processes adopted to improve the physical, mechanical or aesthetic properties of sawn timber, veneer or wood particles used in the production of EWPs. A wood-modification process produces a material that can be disposed of at the end of a product's life cycle without presenting any environmental hazards more significant than those associated with the disposal of unmodified wood. Modification is thus applied to overcome weak points of the wood material that are mainly related to moisture sensitivity, low dimensional stability, low hardness and wear resistance, low resistance to biodeterioration by fungi, termites, marine borers, and low resistance to UV radiation.

The modification of wood has been used to improve the properties of the service life of wood for numerous centuries, whether it is the charring or burning of surfaces or through the application of waxes or oils. An early example of what would today be construed as wood modification was undertaken by Alfred Nobel's father, Immanuel. His work, patented with Colonel Nikolai Aleksandrovich Ogarev in 1844, considered the impregnation of wood used in carriage wheels with a mixture of ferric sulphate and an acid, which was dried slowly in special boxes. After drying, linseed oil and varnish were applied to reduce moisture absorption further (Tolf 1976). The wood modification industry is undergoing significant developments, partly driven by environmental concerns regarding using wood treated with certain preservatives. Several reasonably new technologies, such as acetylation, furfurylation, thermal modification and different impregnation processes, have been successfully introduced on the market and demonstrate the potential of these modern technologies. To modify wood, four main types of process can be implemented according to Sandberg *et al.* (2021): (1) chemical treatment, (2) thermally based treatment, including thermo-hydro (TH) and thermo-hydro-mechanical (THM) treatments, (3) physical treatment with the use of electromagnetic irradiation, plasma or laser, and (4) other types of treatment for example, treatments based on biological processes (Figure 2.18).

Wood modification can involve active modification, which changes the chemical nature of the material or passive modification, in which the properties are altered without any alteration in the material's chemistry. Most active modification methods investigated to date have involved a chemical reaction with the cell-wall polymer hydroxyl groups. These hydroxyl groups play a key role in the wood–water interaction while simultaneously being the most reactive sites. In moist wood, the water molecules settle between the wood polymers, forming hydrogen bonds between the hydroxyl groups and individual water molecules. A change in the number of these water molecules results in shrinkage or swelling of the wood. All possible types of wood treatment affect the wood–water interaction mechanism. The main wood-treatment interaction mechanisms that may be responsible for new wood properties are summarised in Figure 2.19. Several wood-treatment interaction mechanisms tend to occur simultaneously. For example, in thermal modification, parts of the cell-wall polymers are altered, which may lead to

Wood modification

Chemical processes

Active modification
- Acetylation (Accoya™)
- Resin impregnation
  - Compreg™
  - PF-resin
  - Other type of chemicals
- Furfurylation
  - Kebony™
  - Other types, e.g Keywood™ Nobelwood™
- DMDHEU
  - Belmadur™, HartHolz™
- Silicate/silane-based

Passive modification
- Resin impregnation (Impreg)
  - Melamine resin
  - Other chemicals e.g. methacrylate
- Chitosan
- Natural oils/waxes/paraffins
- Polyethylene glycol
- Indurite™, Lignia™

Thermally-based processes

Thermo treatment
- Charring of wood surfaces

Thermo-hydro treatment
- Releasing internal stresses
- Softening
- Drying
- Ageing
- Thermal modification

Thermo-mechanical treatment
- Self-bonding of veneer
- Frictional wood welding

Thermo-hydro-mechanical treatment
- Bending
  - solid wood
  - laminated wood
- Moulding
  - surface
  - cross-sectional
- Bulk densification
- Surface densification

EPL processes*

Electromagnetic treatment
- Ultraviolet light (UV)
- Infrared light (IR)
- Microwave (MW)
- High-frequency (HF)

Plasma treatment
- Cold plasma
- Thermal plasma

Laser treatment
- CO_2 laser
- Excimer laser

* The use of electromagnetic radiation, plasma and laser for wood modification

Other processes

Biological treatment
- Degradation metabolites
- Fungal incising
- Enzyme treatment
- Bacterial incising
- Natural extracts
- Fungi antagonists

Biomimetics
- Water repellency
- Self cleaning

Mineralisation
- Silica-based
- Carbonate based

Novel media
- Supercritcal fluids
- Ionic liquids

New methods
- Graft polymerisation
- Click chemistry
- Rhodium catalysis of hydroxyls
- Modification of chitosan

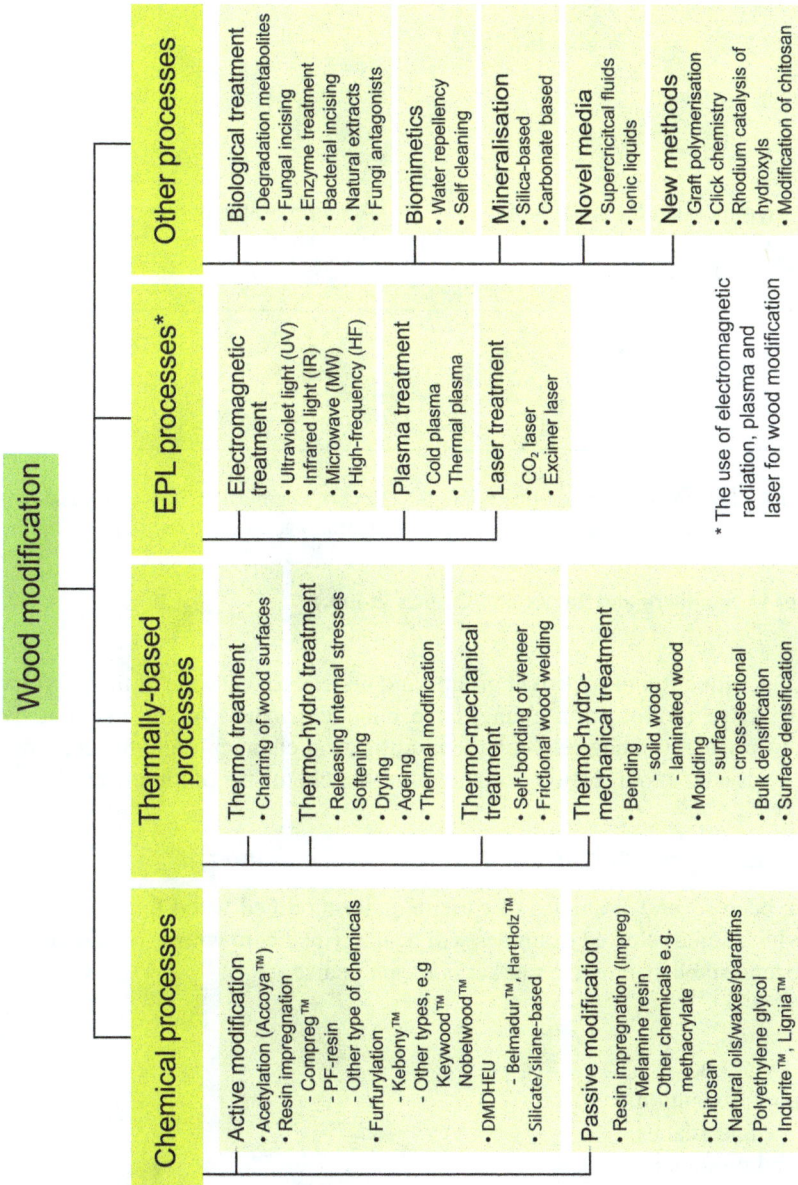

FIGURE 2.18 Overview of wood modification processes.

(Courtesy of D. Sandberg and Taylor and Francis group.)

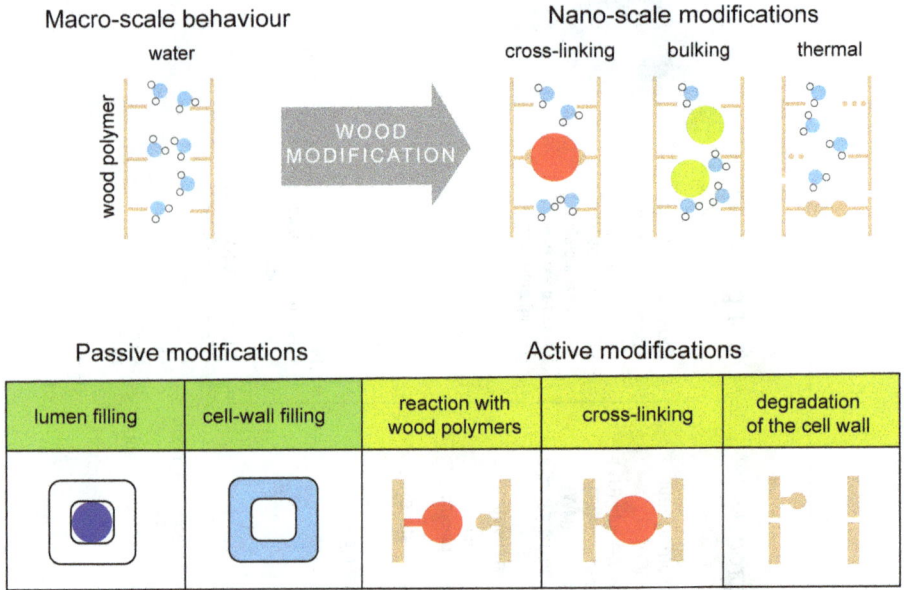

FIGURE 2.19   Schematic diagram illustrating the effect of chemical modification. The red, green and dark blue circles represent different modification agents acting within the cell or the cell wall.

(Courtesy of D. Sandberg and Taylor and Francis group.)

cross-linking, reduction of hydroxyl groups and undesired cleavage of the polymer chains. An active modification changes the chemical nature of the material by forming new covalent bonds between the modification chemicals and the cell-wall polymers. A passive modification does not alter the chemistry of the material.

2.3.2   SOME NEW MODIFICATION PRODUCTS ON THE MARKET

There have been several examples of commercially modified wood for at least 100 years, of which some selected examples will be described here with a focus on those that have been established on the market in recent years:

- acetylation,
- furfurylation,
- resin impregnation,
- silicate and silanes,
- thermal modification,
- aged wood,
- charred wood and
- densified wood.

As modified wood is mainly used for outdoor applications, the Use Classes concept is essential for classifying modified timber due to suitable use (Table 2.6).

TABLE 2.6

Applications of Different Types of Modified Wood Divided Into Use Classes, *cf.* Table 2.17

Product Type	Acetylation					DMDHEU					Furfurylation				
	Interior		Exterior			Interior		Exterior			Interior		Exterior		
Use Class	1	2	3	4	5	1	2	3	4	5	1	2	3	4	5
Indoor furniture	x					(x)	(x)				x	x			
Floor and non-structural interior uses	x	x				(x)	(x)				x	x			
Exterior joinery			x					(x)					x		
Cladding			x					x					x		
Decking			x					x							
Fencing			x					(x)					x		
Outdoor furniture			x	x				(x)					x		
Construction elements			x	x									x	(x)	
In-ground timber				x										(x)	
Products exposed to water				x	(x)									x	(x)

Product Type	Compreg™, Impreg™					Silicate, silane					Thermally-modified timber				
	Interior		Exterior			Interior		Exterior			Interior		Exterior		
	1	2	3	4	5	1	2	3	4	5	1	2	3	4	5
Indoor furniture	x	x			x						x	x			
Floor and non-structural interior uses	x	x			x	x	x				x	x			
Exterior joinery			x					x					x		
Cladding			x					x					x		
Decking			x					x					x		
Fencing								(x)					x		
Outdoor furniture			x					x					x		
Construction elements			x					x							
In-ground timber													(x)		
Products exposed to water			x	x									(x)		

Product Type	Aged wood					Charred wood					Densified wood				
	Interior		Exterior			Interior		Exterior			Interior		Exterior		
	1	2	3	4	5	1	2	3	4	5	1	2	3	4	5
Indoor furniture	x	x				x					x	x			
Floor and non-structural interior uses	x	x				x	x				x	x			
Exterior joinery								x					x		

(Continued)

TABLE 2.6 (Continued)

	Aged wood					Charred wood					Densified wood				
	Interior		Exterior			Interior		Exterior			Interior		Exterior		
	1	2	3	4	5	1	2	3	4	5	1	2	3	4	5
Cladding								x					(x)		
Decking													x		
Fencing								x					(x)		
Outdoor furniture								x					x		
Construction elements			(x)					x					x		
In-ground timber															
Products exposed to water								(x)							

Note: x = products have been produced by companies using the modified wood

    (x) = products maybe produced, based on pre-commercial trials, research etc.

(Courtesy of D. Sandberg.)

2.3.2.1 Acetylation

The main form of chemical modification has been acetylation, a process that uses acetic anhydride for modification reactions. The reaction of acetic anhydride with wood polymers results in the esterification of accessible hydroxyl groups in the cell wall with the formation of a by-product, acetic acid. This by-product is mainly removed from the modified wood, as the human nose is quite sensitive to the odour of acetic acid. Like untreated wood, acetylated wood comprises only carbon, hydrogen and oxygen and contains no toxic constituents. Commercially available acetylated wood exhibits high resistance against wood-decaying fungi. Wood properties such as dimensional stability (swelling/shrinkage) are improved, and equilibrium moisture content is lowered.

The acetylation process yields chemically modified timber and improves many physical, chemical, mechanical and biological material properties. Acetylated timber is mainly intended for above-ground applications such as bridges and structural, decking, marinas, shutters and louvres, cladding, sidings, façades, windows and doors, where preservative-treated timber is used today (Figure 2.20). The Accoya® products (Accsys Technologies company in the Netherlands) aim for service lives of 50 years in above-ground uses (Use Classes 1, 2 and 3). Even though moisture uptake is low and the dimensional stability is good, capillary water is however absorbed by the acetylated wood (*e.g.* radiata pine), and mould growth may be a problem with outdoor claddings and window frames. Acetylated wood has the same appearance as untreated wood but gets grey during weathering. Various products have been manufactured and tested using acetylated wood, summarised in Table 2.6.

The density of acetylated wood is changed compared to untreated wood, but the equilibrium moisture content at 20°C and 65% relative humidity is lower, *i.e.* 4.6%

FIGURE 2.20 Acetylated radiata pine: (a) lookout tower Pompejus, Architects: RO&AD Architecten, and (b) Moses bridge (2010), both structures in Fort De Roovere, Halsteren, The Netherlands.

(Courtesy of D. Sandberg.)

to 5.2% compared to approximately 12% for untreated wood. Acetylation has only a low impact on the strength properties of wood. The modulus of rupture (MOR) has been reported to increase in softwoods but decrease in hardwoods. The modulus of elasticity (MOE) was similar to or slightly less than that of regular radiata pine wood. However, the MOR was considerably higher under wet conditions than the control (Table 2.7).

2.3.2.2 Furfurylation

The furfurylation process is based on impregnating non-toxic furfuryl alcohol ($C_5H_6O_2$) into the wood structure. The furfuryl alcohol is deposited in the wood cavities and cell walls, leading to a permanent "bulking" of the cell wall, meaning the cells are swollen permanently. A key development aspect of the furfurylation process has been sourcing furfuryl alcohol from lesser-used biomass such as corn cobs. This has allowed the furfurylation process to be recognised as an environmentally friendly modification, as demonstrated by the "Swan" ecological label for the furfurylation process adopted by Kebony company in Norway.

TABLE 2.7

Dry and Wet Strength and Stiffness of Untreated and Acetylated Radiata Pine, from Sandberg *et al.* (2021)

Treatment	Dry strength MOR (MPa)	Dry stiffness MOE (MPa)	Wet strength MOR (MPa)	Difference (%)	Wet stiffness MOE (MPa)	Difference (%)
Untreated	63.6	10,540	39.4	−62	6760	−36
Acetylated	64.4	10,602	58.0	−10	9690	−8.6

Note: MOR – modulus of rupture, MOE – modulus of elasticity.

FIGURE 2.21   Kebony Clear™ and Kebony Character™ before and after weathering. (Courtesy of D. Sandberg and Taylor and Francis group.)

Two furfurylated wood products are produced (2025): Kebony Character™ and Kebony Clear™. The latter is highly furfurylated wood, dark and hard, mainly based on radiata pine and Southern yellow pine and maple used for flooring. Kebony Character™, produced from Scots pine, is more lightly loaded and presently used as decking, siding, roofing (where a waterproof membrane is required) and outdoor furniture (Figure 2.21). Furfurylated wood can also be used in cladding, street furniture and marine applications such as piers, walkways and boat decks (Table 2.6).

The density of Kebony Clear™ is 670 kg/m³ compared with a density of 480 kg/m³ for untreated radiata pine; for Kebony Character™, a density of 570 kg/m³ has been reported compared with 490 kg/m³ for untreated Scots pine. Other furfuryl alcohol mixtures can result in completely different properties (Table 2.8). Furfurylated wood has greater hardness, elasticity and rupture modulus than untreated wood, but it is also more brittle (has lower impact resistance). Wood's resistance to fungal decay can be significantly improved by furfurylation, and the biological durability has been upgraded to "Class 1".

TABLE 2.8

Properties of furfurylated timber (Esteves *et al.* 2009)

Treatment	Wood Species	MC (%)	Owen-dry Density (kg/m³)	MOE (GPa)	MOR (MPa)	ASE (%)
Furfurylation (furfuryl alcohol mixture (FA 70 mix) from Kebony™)	maritime pine	7.3	850	10.8	176	1.2–3.4

Note: MC – moisture content at 20 °C and 65% RH, MOE – modulus of elasticity, MOR – modulus of rupture, ASE – anti-swelling efficiency

2.3.2.3   Resin-impregnated Timber

Phenoplastic polymers were among the first commercially successful synthetic polymeric products. Although many phenoplastic materials have been replaced with less brittle thermoplastic polymers, they are still used to create water-resistant bonding in plywood, hardboards and other panel products, and they are an active component of modified wood products. Hard and dense products can also be created by curing wood impregnated with a thermoset resin like the Compreg™ and Impreg™ processes. Studies have shown that impregnating wood with phenol-formaldehyde resins improves dimensional stability and resistance to biodeterioration against fungi, termites and marine borers. If such a product is densified under process, a dimensionally stable high-density product with a high wear resistance is achieved (Compreg™). Impregnation modification of wood with melamine-formaldehyde (MF) resins has increased in recent decades, especially in Europe, with positive results for dimensional stability and biological resistance to brown-rot fungi. Table 2.6 summarises where products have been made using Compreg™ and Impreg™.

Resin-impregnated paper protects the surfaces of various particleboard, fibreboard, and laminated products. Impregnation with resins and additives, such as coloured pigments, enhances aesthetics and constructional possibilities for products such as tables and knife handles (Figure 2.22).

FIGURE 2.22   Applications of resin-impregnated materials and products: Impregnated wood for knife handle (left), rotten wood impregnated with clear resin and phosphorescent powder (middle), and plywood foliated with phenolic-resin paper (right).

(Courtesy of D. Sandberg and Taylor and Francis group.)

Tables 2.9 and 2.10 show some physical and mechanical properties of a compreg-type resin-impregnated and compressed-laminated veneer.

Of course, the properties of resin-impregnated wood depend on the resin and the amount of resin used. Table 2.11 compares various treatments.

TABLE 2.9

Physical Properties of Resin-impregnated and Compressed-laminated Wood and Insulating Wood[a] (Compreg type); Lignostone™ From the Röchling Company (Sandberg et al. 2021)

Property	Unit	DIN 7707[a]	Parallel Veneer KP 20212	Parallel Veneer KP 20214	Cross-wise Veneer KP 20222	Cross-wise Veneer KP 20224	Tangential Packed Veneer KP 20242	Tangential Packed Veneer KP 20244
Density	(kg/m³)		850	1250	950	1250	950	1250
Oil absorption	(%)		30	7	25	7	25	7
Moisture content	(%)		5	5	5	5	5	5
Operating temperature limit	(°C)		105	100	105	100	105	100
Volume resistivity	(Ω cm)		10^{12}	10^{12}	10^{12}	10^{12}	10^{12}	10^{12}
Electric strength	(kV/3 mm)	90 °C[b]	45–55	45–55	45–55	45–55	50–55	50–55
	(kV/25 mm)	90 °C[c]	70–90	70–90	70–90	70–90	80–90	80–90
Dissipation factor	(tan δ)	50 Hz, 25 °C	0.01	0.01	0.01	0.01	0.01	0.01
Relative permittivity	(ε_r)		3.7	4.1	3.7	4.1	3.7	4.1

[a] Name according to the DIN 7707-1 standard (DIN 1979)
[b] Perpendicular to the lamination
[c] Parallel to the lamination

TABLE 2.10

Mechanical Properties of Resin-impregnated and Compressed-laminated Wood and Insulating Wood[a] (Compreg Type); Lignostone™ from the Röchling Company (Sandberg et al. 2021)

Property	Unit	DIN 7707[a]	Parallel Veneer KP 20212	Parallel Veneer KP 20214	Cross-wise Veneer KP 20222	Cross-wise Veneer KP 20224	Tangential Packed Veneer KP 20242	Tangential Packed Veneer KP 20244
Density	(kg/m³)		850	1250	950	1250	950	1250
Flexural strength[b,c]	(MPa)	$\perp$[d]	140	200	110	130	130	180

(Continued)

TABLE 2.10 (Continued)

| Property | Unit | DIN 7707[a] | Parallel Veneer | | Cross-wise Veneer | | Tangential Packed Veneer | |
			KP 20212	KP 20214	KP 20222	KP 20224	KP 20242	KP 20244
Modulus of elasticity[b,c]	(GPa)	$\perp$[d]	11	16	9	11	11	13
Compressive strength	(MPa)	$\perp$[d]	100	120	200	230	120	140
	(MPa)	//[e]	55	90	70	90	80	100

[a] Name according to the DIN 7707-1 standard (DIN 1979)
[b] The fibres of the outside veneers must run in the longitudinal direction of the test specimen
[c] Parallel laminated types must be present in the tension zones of at least four longitudinal layers
[d] Perpendicular to the lamination
[e] Parallel to the lamination

As the name suggests, the main difference between Impreg and Compreg is the application of compressive forces before and during the curing process. Ibach (2010) gave an excellent overview of the properties of these two products, which is summarised in Table 2.12.

2.3.2.4 Silicate and Silanes

In recent years, there has been activity in the treatment of wood with inorganic silicon compounds, mainly based on condensation products of silicic acid (colloidal silicic acids, silicates, water glass) or tetraalkoxysilanes which undergo hydrolysis and condensation steps to form sols and finally gels (sol-gel technology). Water glass, an alkali silicate, has been shown to enhance the durability of wood, though some essential drawbacks were noted during subsequent analysis. Because of its high hygroscopicity and high pH values, increased moisture absorption and strength loss of wood were frequently observed. Wood treated with tetra alkoxysilanes showed enhanced dimensional stability, especially when the hydrolysis and the condensation of the silanes were controlled to react within the cell wall. Similarly, treatment with siloxanes increased the water repellence of wood but did not considerably influence the sorption behaviour of wood. The modification has shown protective effectiveness against wood-destroying basidiomycetes such as brown-rot fungi.

Much of the commercial development has been based on treatment methods with aqueous solutions, which are sold commercially for do-it-yourself application or by vacuum impregnation using conventional treatment facilities. Treatments, including soluble silicates and sealing surface treatments, give the products desirable properties, such as light grey wooden façades (Table 2.6).

Silicates, silanes and chitosan are mainly used as a treatment for decking, water bridges, poles, cladding and similar construction details for surface protection and reducing mould growth. Mechanical properties are not influenced by the treatment (Sandberg *et al.* 2021).

TABLE 2.11

Properties of Resin-impregnated Timber from Some Selected Processes

Treatment	Wood Species	MC (%)	Owen-dry Density (kg/m³)	MOE (GPa)	MOR (MPa)	Brinell Hardness (N/mm²)	ASE (%)
Treatment with thermosetting resins: melamine formaldehyde (MF), phenol-formaldehyde (PF) or urea-formaldehyde (UF).	Burflower tree			Increase over untreated: MF: 10% PF: 12% UF: 6%	MF: 32.8 PF: 33.4 UF: 30.9		PF: 70.6 MF: 68.2 UF: 48.5
Styrene impregnating with glycidyl methacrylate (GMA) as a crosslinker	Rubber wood			7.3	120		53
Treatment with phenol-formaldehyde (PF) resin (solid content of 20%) at T = 80–180°C and under compression at 2 MPa	Japanese cedar	oven-dry	950–1070	17.9–22.6	148–255		
Compreg™ (PF-resin impregnated and densified veneer)	Different species	low	1200–1400	17	up to 200	10–20 times increase compared to that of untreated wood	~70
Impreg™ (uncompressed phenol-, melamine- or urea-based resin-impregnated wood, cured under mild acidic or alkaline conditions)	Different species	low	15–20% greater than untreated wood	small changes compared to untreated wood		Increase considerably more than proportional to density increase	~70

Note: ASE – anti-swelling efficiency, MC – moisture content at 20°C and 65% RH, MOE – modulus of elasticity, MOR – modulus of rupture.

TABLE 2.12

Comparison of Impreg and Compreg Processed Timbers (Ibach 2010)

Property	Impreg	Compreg
Specific gravity	15–20% greater than untreated wood	Usually, 1.0 to 1.4
Equilibrium swelling and shrinkage	1/4 to 1/3 that of normal wood	¼ to 1/3 that of untreated wood at a right angle to the direction of compression, greater in the direction of compression but very slow to attain
Spring-back	None	Very small when properly made
Face checking	Practically eliminated	Practically eliminated for specific gravities less than 1.3
Grain raising	Greatly reduced	Significantly reduced for uniform-texture woods, considerable for contrasting grain woods
Surface finish	Similar to normal wood	Varnished-like appearance for specific gravities greater than about 1.0; cut surfaces can be given this surface by sanding and buffing
Permeability to water vapour	About 1/10 that of untreated wood	No data, but presumably much less than Impreg
Decay and termite resistance	Considerably better than untreated wood	Substantially better than untreated wood
Acid resistance	Considerably better than untreated wood	Better than Impreg because of impermeability
Alkali resistance	Same as normal wood	Somewhat better than untreated wood because of its impermeability
Fire resistance	Same as normal wood	Same as untreated wood for long exposures, somewhat better for short exposures
Heat resistance	Greatly increased	Greatly increased
Electrical conductivity	conductivity 1/10 that of normal wood at 30% RH; 1/1000 that of normal wood at 90% RH	Slightly more than Impreg at low relative humidity values due to entrapped water
Heat conductivity	Slightly increased	Increased about in proportion to a specific gravity increase
Compressive strength	Increased more than proportional to the density increase	Increased considerably more than proportional to a specific gravity increase
Tensile strength	Decreased significantly	Increased less than proportional to a specific gravity increase
Flexural strength	Increased less than proportional to the density increase	Increased less than proportional to specific gravity increase parallel to the grain, increased more perpendicular to the grain
Hardness	Increased considerably more than proportional to the density increase	10 to 20 times that of untreated wood

(Continued)

TABLE 2.12 (Continued)

Property	Impreg	Compreg
Impact strength: Toughness Izod impact	About 1/2 of the value for untreated wood, but very susceptible to the variables of manufacture About 1/5 of the value for untreated wood	1/2 to 3/4 of value for untreated wood, but very susceptible to the variables of manufacture 1/3 to 3/4 of the value for untreated wood
Abrasion resistance (tangential)	About 1/2 of the value for untreated wood	Increased about in proportion to a specific gravity increase
Machinability	Cuts cleaner than untreated wood but dulls tools more	Requires metalworking tools and metalworking tool speeds
Mouldability	Cannot be moulded but can be formed to single curvatures at the time of assembly	Can be moulded by compression and expansion moulding methods
Glue-ability	Same as normal wood	Same as untreated wood after light sanding or in the case of thick stock, machining surfaces plane

2.3.2.5 Thermal Modification

The modification of wood by heat without chemical additives and with a limited sup-ply of oxygen to prevent oxidative combustion, *i.e.* thermal modification, is a gener-ally accepted and commercialised procedure for improving some characteristics of wood (Jones *et al.* 2019). Thermally modified timber (TMT) is distinguished by an increased resistance against wood-destructive fungi, enhanced dimensional stability, lower equilibrium moistures and darker colour shade. As a rule, an increased inten-sity in the modification treatment (raised temperature and duration of treatment) will result in improved dimensional stability. The main characteristics of the thermally modified timber are its improved durability and dimensional stability. Target mar-kets are indoors and outdoors, requiring materials such as claddings, doors, flooring, garden products, windows and speciality products such as saunas and bathrooms (Figure 2.23).

The durability of TMT depends on the process conditions and ranges in Durability Classes II–IV. However, it is not comparable with that achieved with copper-chromi-um-containing preservatives. Therefore, it is not usually advised to use TMT in con-tact with soil and water.

However, the main drawback of TMT treated at high temperatures is its poorer mechanical properties (bending strength, impact strength). Therefore, TMT should not be used in load-bearing constructions or under circumstances where a high, sud-den impact can occur (Table 2.6). As with all uncoated wood in exterior use, TMT will get a grey hue.

TMT distinguishes itself from unmodified timber by its increased resistance to wood-destructive fungi. However, the properties of TMT are challenging to

FIGURE 2.23 Cladding of fire retardant-impregnated thermally modified timber (Scots pine) at construction (2019) and after five years of weathering (2024). Valle Wood, Helsfyr, Oslo, Norway.

(Courtesy of D. Sandberg.)

generalise as they are highly dependent on the species, the type of modification process and the process conditions. An excellent method to study the effectiveness of the thermal modification process is to measure the mass loss of the processed timber (Table 2.13). A high mass loss indicates a high improvement in the resistance to biological degradation.

TMT is generally not recommended for use in structures that come into direct contact with soil or water. Table 2.14 shows the uses of different classes of ThermoWood®, the most common thermal-modification process.

2.3.2.6   Aged Wood

Wood ageing further develops the classic thermal-modification processes currently used industrially. It is an accelerated process similar to thermal modification but uses a mild temperature range of 100 °C to 150 °C, sometimes under controlled relative humidity and pressure. Therefore, the negative effects that a classic thermal modification typically has on the strength and brittleness of wood are decreased.

The main application for aged wood is indoors, as its durability and hygroscopic properties are only slightly improved. Aged wood Norway spruce and common oak can typically be used as an alternative to old wood for the construction and renovation of chalets, typical building types, for example, in Switzerland. Other uses are for musical instruments, renovating old and manufacturing new instruments (Table 2.6).

TABLE 2.13

Average Values of the Summative Chemical Composition, Mass Loss and Actual Mass Loss of Unmodified and Thermally Modified Shining Gum in an Open and Closed System (Mai and Militz 2023). atm. – Atmospheric Pressure (1 atm. is Defined as the Pressure Exerted by the Earth's Atmosphere At Sea Level At 15 °C, Which Is Approximately 101.3 kPa)

	T (°C)	RH (%)	Max Pressure (bar)	Extractives (%)	Lignin (%)	Cellulose (%)	Hemicelluloses (%)	Mass Loss (%)	Real Mass Loss (%)
Reference	–	–		4.7	22.5	48.3	27.4	0	0
Closed system	150	30	1.4	6.0	21.0	50.1	23.5	1.3	2.6
	160	30	1.8	7.6	23.1	51.3	17.6	2.0	5.0
	170	30	2.3	8.4	22.8	53.9	14.2	2.5	6.3
	150	100	4.5	12.7	25.4	55.7	5.9	4.0	11.4
	160	100	6.1	13.2	26.5	53.7	6.5	10.7	18.6
	170	100	7.7	10.1	31.8	49.7	10.3	15.8	20.5
Open system	160		atm.	6.8	20.8	52.2	18.8	3.3	5.4
	180		atm.	9.0	23.5	48.6	18.7	3.5	7.9
	200		atm.	12.3	23.7	52.5	10.4	7.2	14.6
	210		atm.	12.7	27.4	49.3	10.8	11.2	18.7
	220		atm.	9.1	28.6	50.2	11.8	12.8	16.8
	230		atm.	7.4	36.9	40.1	18.0	16.9	19.3

Note: RH – relative humidity of the air during the processing and T – process temperature.

TABLE 2.14

Use Classes (UC) According to the EN 335 (CEN 2013b) Standard of TMT According to the ThermoWood® Process (International Thermowood Association 2021)

ThermoWood product	Species	Use class
Thermo-D	Norway spruce, Scots pine	UC 3
Thermo-D	European ash, ayous, frake	UC 3
Thermo-D	Hardwood spp.	UC 2
Thermo-S	Norway spruce, Scots pine	UC 2
Thermo-S	Hardwood spp.	UC 2

Note: UC 3 = exterior use, exposed to weather and UC 2 = outdoor use under the roof.

2.3.2.7 Charred Wood

The surface charring of wood is a process that burns and thereby modifies the wood surface in a controlled manner. The charring process mimics the combustion of wood. However, the char layer is an efficient insulator and a poor conductor of heat, and it effectively inhibits heat transfer to the active pyrolysing zone, thus retarding further combustion. The charring process changes the appearance and improves properties such as fire resistance and water uptake through carbonisation, therefore making the wood more durable. The burning also reduces sugars on the wood surface, reducing fungal growth. The charred surface protects against attack by insects and prevents the greying of wood because of weathering in exterior use.

Various engineered wood products (EWPs) with charred surfaces have become popular with architects for their appearance. They are mainly used for interior walls, exterior façades, decking and fences (Table 2.6). The degree of charring can be adapted according to the use and desired appearance. There are several producers of charred wood worldwide who either follow the Japanese yakisugi tradition or use their own charring processes and their own trademarks (Figure 2.24).

2.3.2.8 Densified Wood

Wood densification is a process whereby the density of wood is increased by compression of the wood cells in the transverse direction to reduce the void volume of the lumens by the impregnation of the cell wall or cell lumen with a synthetic resin (chemical bulking), or by a combination of THM densification and bulking (Figures 2.19 and 2.25). The purpose of densification is to improve the hardness, the abrasion resistance and the strength (Table 2.15). However, compression of the cells in the transverse direction may also be used to increase the bendability in moulding to manufacture, for example, shells and tubes (Sandberg *et al.* 2021).

It can be concluded that the use of wood densification of low-density woods as an alternative to expensive high-density woods has so far been very limited, although the

(a)   (b)

FIGURE 2.24 Charred wood: (a) detail of a facade "Shou Sugi Ban or Yakisugi", which means charred wood, Higashiyama district, Kyoto, Japan, and (b) Café Birgitta completed 2014, Helsinki, Finland.

(Courtesy of D. Sandberg.)

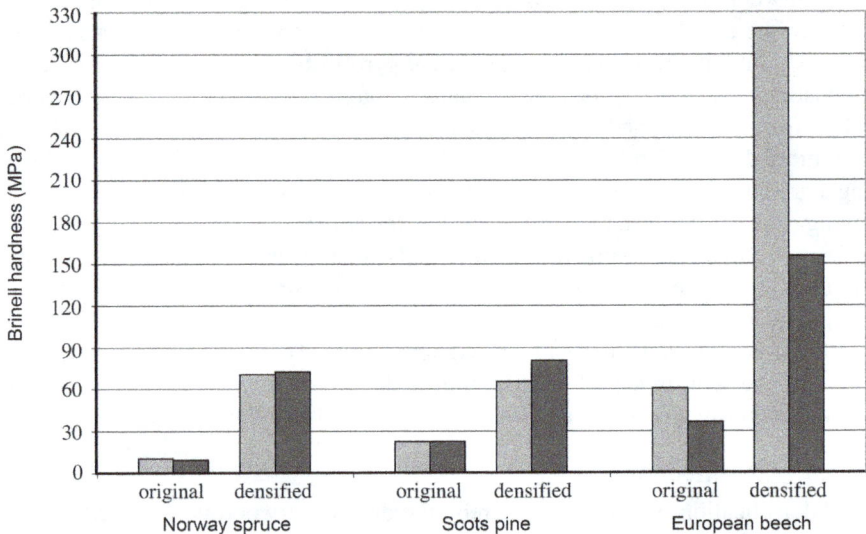

FIGURE 2.25 Average Brinell hardness of un-densified (original) and in an open system densified wood of Norway spruce, maritime pine, and common beech (Navi and Sandberg 2012). Light grey is the radial section and dark grey is the tangential section.

(Courtesy of D. Sandberg.)

possible applications may be many (Table 2.6). Several products are on the market, including but not limited to Insulam™, Lignostone™, Dehonit™, Ranprex™, and impregnated and densified wood for advanced use in, for example, musical instruments and credit cards (Sonowood™) instead of tropical species such as ebony.

TABLE 2.15

**Mechanical Properties of Common Beech Densified in
An Open System (Navi and Sandberg 2012)**

| Type of wood | Density (kg/m³) | Impact Resistance[a] | | Bending Properties | | MOE |
		Tangential (J/cm²)	Radial (J/cm²)	σ_{Rup}[b] (MPa)	σ_{max}[c] (MPa)	E_L[d] (MPa)
Beech for joinery purposes	630	3.8	8.5	68	29	6.9
Beech to be densified	820	8.2	10	117	53	11.0
Densified beech	1130	14	15	200	95	22.9

[a] Energy per unit area necessary to cause rupture under radial or tangential loading
[b] Crack opening under stress in mode I (radial loading)
[c] Linear elastic stress limit (radial loading)
[d] Modulus of elasticity (MOE) in longitudinal tension

Swiss Wood Solutions markets impregnated and densified wood (Sonowood) instead of tropical species such as ebony for advanced use in musical instruments. Table 2.16 gives some examples of properties for Sonowood.

2.3.3 Use Classes – The Use of Modified Wood

The development of wood modification has focused on the potential to alter the performance and reduce the risks of wood in service, particularly regarding dimensional stability and decay, both of which are strongly influenced by the presence of moisture. Wood in service in interior conditions is usually restricted to moisture contents below 10%, but design or exposure to high moisture conditions can significantly affect its performance.

TABLE 2.16

**Properties of Sonowood™ (Densified Solid Wood) from the
Swiss Wood Solutions Company (Sandberg et al. 2021)**

Property	Unit	Norway Maple	Norway Spruce
Density	(kg/m³)	1200–1400	1300–1400
Brinell hardness	(N/mm²)	> 80	> 100
Colour		Mocha	Caramel
Dimensional stability (difference in swelling)	(% per % unit MC change)	Height ~ 0.70 Width ~ 0.30	Height ~ 0.75 Width ~ 0.33
Sound velocity	(m/s)	> 4400	> 5500
Damping (logarithmic decrement)		~ 0.053	~ 0.040
Elastic modulus	(GPa)	> 23	> 39

Note: MC – moisture content.

The European standard EN 335 (CEN 2013b) defines five Use Classes representing different service situations to which wood and wood-based products can be exposed. This standard also indicates the biological agents relevant to each situation. A Use Class is not a "performance class" and does not give guidance for how long wood and wood-based products will last in service. The differences between the Use Classes are based on differences in environmental exposures that can make the wood or wood-based products susceptible to biological deterioration. The risk of decay increases, mainly if the exposure is prolonged. A more detailed description of the Use Classes is shown in Table 2.17, along with definitions of risks and typical product ranges as defined within a product guidance manual produced by the Wood Protection Association in the United Kingdom (Manuel 2012).

In terms of the application of modified wood, surveys and overviews undertaken by Jones *et al.* (2019) indicate that various wood modification treatments can be applied to the Use Classes as described in Table 2.18. The Use Classes where the modified wood products have been commercially demonstrated are marked in green, whilst those in yellow refer to those where studies indicate products may be used.

Decking for terraces, balconies, platforms and similar is the most critical application for modified wood such as thermally modified, acetylated and furfurylated timber, followed by façade cladding, including shading elements. These applications are

TABLE 2.17

Use Classes, Defined According to the European standard EN 335 (CEN 2013b) and Typical Product Ranges and Associated Risks According to the UK Wood Protection Association

Use Class	Biological Risk	Typical Service Exposure	Examples of Product Ranges
UC 1	Insect	Internal, with no risk of wetting.	All timbers in normal pitched roofs except tiling battens and valley gutter members.
UC 2	Fungi/ Insect	Internal, with risk of occasional wetting.	Floorboards, architraves, internal joinery, skirtings. All timbers in upper floors not built into solid external walls. Tiling battens, frame timbers in timber frame houses, timber in pitched roofs with high condensation risk, timbers in flat roofs, ground floor joists, sole plates (above the damp proof course), timber joists in upper floors built into external walls.
UC 3.1	Fungi/ Insect	External, above damp-proof course, coated.	External joinery including roof soffits and fascias, barge boards, etc., cladding, valley gutter timbers, external structural load-bearing timbers.
UC 3.2	Fungi/ Insect	External, above damp-proof course, uncoated.	Cladding, fence rails, gates, fence boards, agricultural timbers not in soil/ manure contact and garden decking timbers that are not in contact with the ground.

(Continued)

TABLE 2.17 (Continued)

Use Class	Biological Risk	Typical Service Exposure	Examples of Product Ranges
UC 4	Fungi/ Insect (incl. termites)	Timbers in permanent contact with the ground or below damp-proof course.	Fence posts, gravel boards, agricultural timbers in soil/ manure contact, poles, sleepers, playground equipment, motorway and highway fencing/ sound barriers and garden decking timbers that are in contact with the ground. Lock gates and canal linings. Cooling tower infrastructure (fresh water).
		Timbers in permanent contact with fresh water. Cooling tower packing. Timbers exposed to the particularly hazardous environment of cooling towers.	
UC 5	Marine borer/ Fungi	All components in permanent contact with sea water.	Marine piling, piers and jetties, dock gates, sea defences, ships hulls, and cooling tower packing (sea water).

(Courtesy of D. Sandberg.)

TABLE 2.18

Different Wood Modification Methods Are Applied to Products in Different Use Classes

Modification method/Use Class	UC 1	UC 2	UC 3.1	UC 3.2	UC 4	UC 5
Acetylation	yellow	green	green	green	green	yellow
DMDHEU	yellow	green	green	green		
Furfurylation	yellow	green	green	green	yellow	yellow
Resin: Compreg-type	green	green	green	green		
Resin: Impreg-type	green	green	green	green		
Silicates and silanes	green	green	green	green		
Thermal modification	yellow	green	green	green		
Aged wood	green	green	yellow			
Charred wood	green	green	yellow			
Densified wood	green	green	yellow	yellow		

Note: Green – commercially demonstrated and yellow – products may be used.
(Courtesy of D. Sandberg.)

well established and are the primary sales of modified wood products. Some suppliers have introduced modified timber for window scantlings, but current data on actual sales are unavailable. Considerable amounts of TMT are used for sash cores of polyurethane-coated windows. Modified wood is also used for garden furniture, playground devices, or other gardening and landscaping purposes, such as fences, poles or screens. Because of its aesthetic appearance, TMT is also used for interior applications, mainly for flooring and panelling. This is not the case with the same content for acetylated and furfurylated timber. For interiors, existing requirements on emission should be considered for TMT. Modified wood has gained market entry across a range of Use Classes, depending on the levels of treatment and how they alter the durability and moisture exclusion levels of the treated material.

2.4 ENGINEERED WOOD PRODUCTS (EWPS)

2.4.1 OVERVIEW

Wood is the basis for all types of EWPs, commonly called wood-based materials, reconstituted wood, wood-based composites and wood-based products. Using EWPs made of sawn timber, veneer or strands is possible today to produce load-bearing construction elements with spans much longer than the tree's height. The strength of the tree is preserved or even improved, especially in EWPs based on sawn timber. Panel-shaped EWPs have dimensions widely exceeding those of the tree and are excellent for covering large building surfaces.

EWPs are usually classified due to the size of the wood components of which the materials are made (Figure 2.26), *i.e.*:

- sawn timber (solid wood),
- veneer,
- strands, strips, chips or particles (particle-based materials),
- fibres and
- hybrid materials where different wood-based materials are combined, sometimes also with non-wood materials.

The size of the wood components in building up an EWP influences its properties, material and energy consumption (Figure 2.27). The degree of wood utilisation from

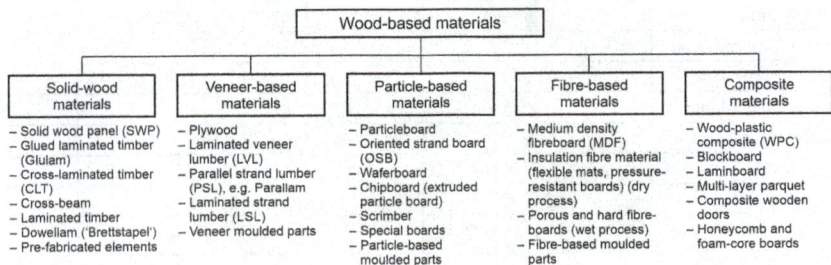

Wood-based materials				
Solid-wood materials	Veneer-based materials	Particle-based materials	Fibre-based materials	Composite materials
– Solid wood panel (SWP) – Glued laminated timber (Glulam) – Cross-laminated timber (CLT) – Cross-beam – Laminated timber – Dowellam ('Brettstapel') – Pre-fabricated elements	– Plywood – Laminated veneer lumber (LVL) – Parallel strand lumber (PSL), e.g. Parallam – Laminated strand lumber (LSL) – Veneer moulded parts	– Particleboard – Oriented strand board (OSB) – Waferboard – Chipboard (extruded particle board) – Scrimber – Special boards – Particle-based moulded parts	– Medium density fibreboard (MDF) – Insulation fibre material (flexible mats, pressure-resistant boards) (dry process) – Porous and hard fibre-boards (wet process) – Fibre-based moulded parts	– Wood-plastic composite (WPC) – Blockboard – Laminboard – Multi-layer parquet – Composite wooden doors – Honeycomb and foam-core boards

FIGURE 2.26 Categorisation of wood-based materials due to the size of the wood components of which the materials are made.

(Courtesy of Hanser Verlag.)

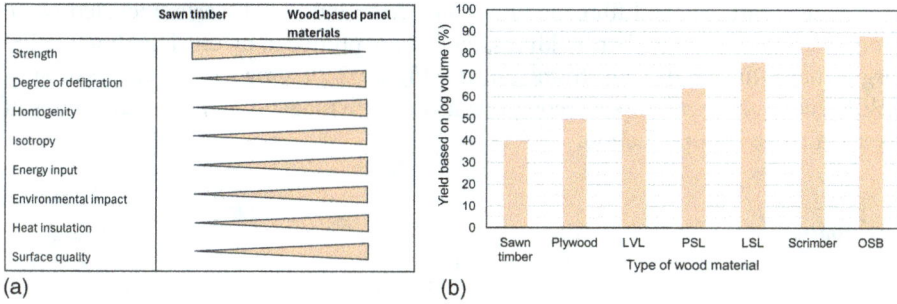

FIGURE 2.27   Size effect on wood properties: (a) influence of structure on various properties (Paulitsch 1989) and (b) material utilisation for various wood-based materials (data from Gong 2021). LVL – laminate veneer lumber, PSL – parallel strand lumber, LSL – laminated strand lumber, OSB – oriented strand board. Scrimber - a EWP made by crushing and reforming small-diameter logs or forest thinnings into panels or beams.

(Courtesy of D. Sandberg.)

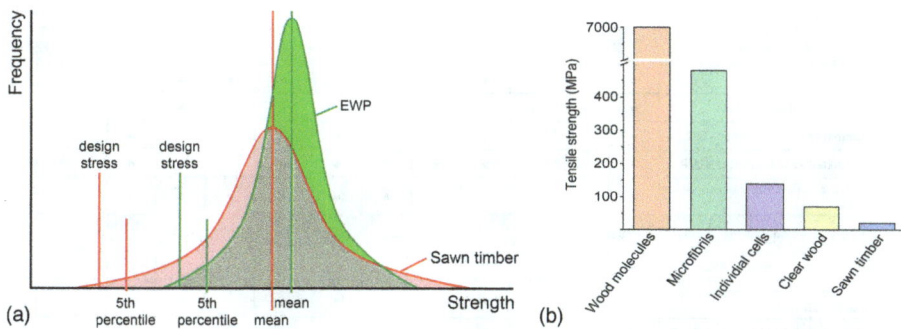

FIGURE 2.28   The homogenisation effect: (a) schematic distribution of strength of sawn timber and EWPs, and (b) tensile strength of wood at different structural levels from the molecular level to sawn timber.

(Data from Gong 2021, courtesy of P. Niemz and D. Sandberg.)

a stem depends on the type of material. Particle-based panel materials such as oriented strand board (OSB), particleboard and fibreboard have a significantly higher material utilisation than EWPs based on sawn timber and veneer (Figure 2.27b).

The production of EWPs generally results in a significant homogenisation compared to solid wood (Figure 2.28a). Bonding different components and adding hydrophobic agents, fire retardants, paint, *etc.*, creates various high-performance wood-based construction materials.

The chemical elementary structures (molecular level) have a higher strength than the wood itself, which varies significantly in its properties due to growth conditions, species and the structure of the wood and its different features, such as knots (Figure 2.28b).

Sawn timber and veneer are the main products from sawmills and veneer mills. Sawn timber is graded into Strength Classes with specified properties by visual or mechanical testing, which can be seen as a homogenisation of its properties.

In contrast, particles and fibres mainly come from industrial waste, recycled wood products, demolished timber buildings, *etc.*, or are produced by crushing low-quality virgin round timber directly from forestry. Figure 2.29 shows a systematisation of EWPs due to the type of wood components they are made of and some examples of their main use in construction.

EWPs	Structural			Joinery				Interior				Exterior
	Infrastructure	Frame parts	Envelope	Doors	Windows	Shadings	Stairs, rails	Mouldings	Flooring	Ceiling	Furniture	Walk paths, Platforms, Stairs, etc.
EWPs based on sawn timber												
Components of sawn timber	●	●	●	●	●	●	●	●	●	●	●	●
Edge-glued panel – EGP	●		●	●			●	●	●	●	●	(●)
Glued-laminated timber – GLT	●	●	(●)				(●)				●	●
Cross-laminated timber – CLT	●	●	●				●		(●)	(●)	(●)	●
Lightweight material – LWM		(●)		●	●	●	●	●	●	●	●	(●)
Fibre-reinforced timber – FRT	●	●										(●)
EWPs based on veneers												
Plywood – PW	●	●	●	●	●	●	●	●	●	●	●	
Mass plywood panel – MPP	●	●	(●)						(●)			
Laminated veneer lumber – LVL	●	●					(●)				●	
Laminated veneer product – LVP				●	●	●	●	(●)		(●)	●	
High-pressure laminated veneer – HPLV	●			●	(●)	●	(●)	●	●	●	●	●
EWPs based on strands, strips, chips or particles												
Parallel strand lumber – PSL	●	●										
Waferboard, oriented strand board – WB, OSB	(●)			(●)			(●)			●	(●)	
Laminated strand lumber – LSL	●	●		(●)								
Particleboard – PB	(●)			●			(●)	●	●	●	●	
Inorganic-bonded composite – IBC	(●)	●		(●)			(●)		(●)	●		
Wood-plastic composite – WPC	(●)		●	●	●	●	●	●	●	●	●	●
EWPs based on fibres												
Low-density fibreboard – LDF				●						●		
Medium-density fibreboard – MDF				●	●	●	●	●	(●)	●	●	
High-density fibreboard – HDF				●	(●)	(●)	(●)	(●)	(●)	(●)	●	
Hybrid and decorative EWPs												
I-joist and box-beam		●										
Wood sandwich panel			●	●			(●)		(●)	(●)	●	
Flexible EWPs				(●)			(●)	(●)		(●)	●	
Aged wood	(●)			●		●	●	●		●	●	●
Charred wood	(●)		●	●		●				●		●
Bark EWPs				(●)						●		
Bamboo		●	(●)	●	●	●	●	●	●	●	●	●
Cork				(●)					●	(●)	●	

Legend: ● = preferable use, (●) = possible use.

FIGURE 2.29   Overview of wood-engineered wood products (EWPs).

(After M. K. Kuzman and Sandberg 2023, courtesy of D. Sandberg.)

2.4.2 BONDING PRINCIPLES FOR EWPS

Bonding involves connecting two solid bodies using, for example, an adhesive that fills the gap and transmits forces between them. For wood, this means that the adhesive can penetrate beyond the surface and into cell lumens. Adhesives are the most common bonding agents when EWPs are manufactured. The chemical properties of the adhesive and wood are essential in bonding for cohesive strength. Bond-line performance is measured as the mechanical strength holding the wood surfaces together under various exposure conditions.

The type of adhesive has a strong influence on the weather resistance of wood-based materials. As an example, urea-resin-based particleboard is only suitable for use in facilities with low relative humidity, *i.e.* Use Class 1 (Section 2.3.3). Particleboard bonded by melamine-based, polymeric methyl diphenyl diisocyanate (PMDI) and phenolic-bade adhesives can be used in Use Class 2. In outdoor use (Use Class 3), particle- and fibre-based panels are not permitted at all, but cement-bonded particleboard allows unrestricted exterior use. Cement-bonded particleboard is also used for fire protection.

Sawn timber for constructive purposes, cross-laminated timber and glued-laminated timber are bonded with high-strength and moisture-resistant adhesives such as melamine urea formaldehyde (MUF), melamine formaldehyde (MF), phenol resorcin formaldehyde (PRF), one-component polyurethane (1C-PUR) and more rarely emulsion polymer isocyanate (EPI).

Various adhesive systems are used for edge-glued panels (EGP) made of solid wood for non-load-bearing construction. In furniture construction and interior design, urea-formaldehyde (UF) adhesive with a very low formaldehyde content dominates. In some cases, 1C-PUR is used. Formaldehyde is a chemical compound commonly included in adhesives to make them highly effective in bonding wood. However, formaldehyde is a volatile organic compound (VOC), and the wood products may off-gas over time. Prolonged exposure to formaldehyde vapours, especially in poorly ventilated indoor environments, can be harmful, with potential effects on respiratory health, causing irritation to the eyes, nose and throat, and increasing the risk of certain cancers with long-term exposure. Because of these concerns, formaldehyde emissions are regulated in many countries to ensure that wood furniture and building materials meet safety standards. Polymeric methyl diphenyl diisocyanate (PMDI) is a formaldehyde-free adhesive. Purely bio-based adhesives (*e.g.* lignin, soy- and tannin-based adhesives) have no or very low formaldehyde emissions but still have small market shares. For further readings about the use of adhesives, see Chapter 12.

Fibreboards manufactured by the wet process do not need adhesives, as the fibres bond together.

Mechanical fasteners instead of adhesives are sometimes used for cross-laminated timber and elements bonded to concrete. Aluminium nails, hardwood dowels or mechanical joints such as dovetail and tongue-and-groove join the layers; they are rare but used (Figure 2.30).

Stacking sawn timber flat-side by flat-side and nailing it together to form a thick panel (nail-laminated timber) is a type of EWP that has been used for a long time (see Section 2.4.3).

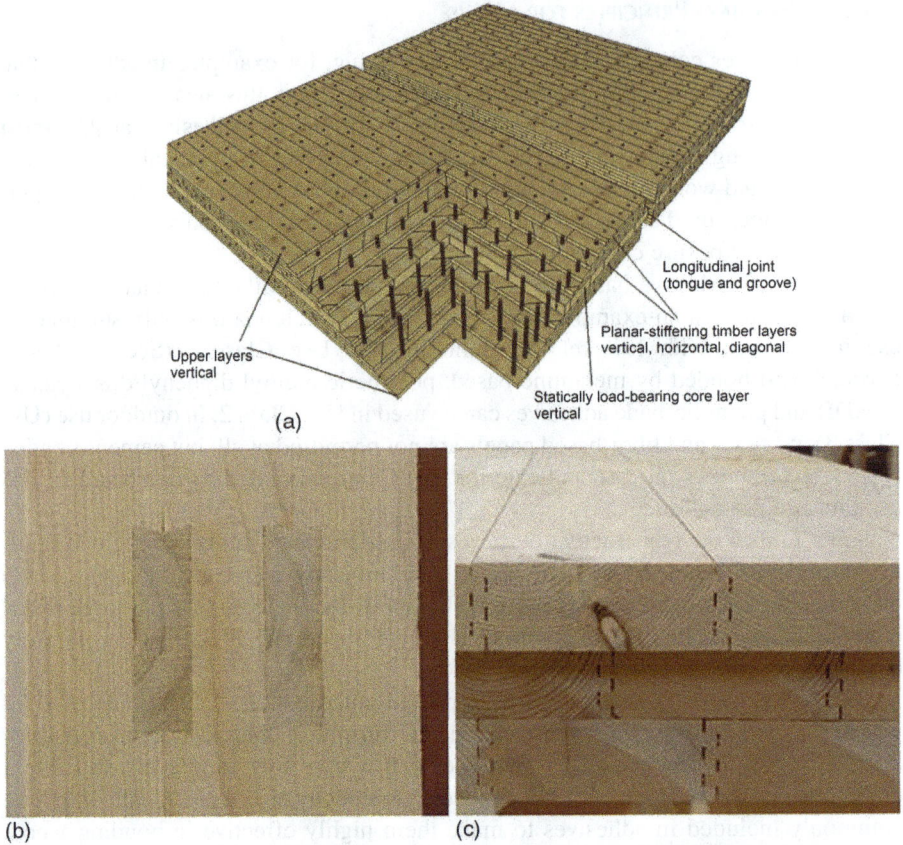

(a)

FIGURE 2.30 Non-adhesive bonding of timber: (a) mechanically connected cross-lami-nated timber (CLT) layers (Holzius, South Tyrol, Italy), (b) CLT layers connected by dovetail without an adhesive (SOLIGNO) and (c) tongue-and-groove joined lamellae for use in CLT.

(Appenzeller Holz, Nägeli Holz, Gais, Switzerland, courtesy of P. Niemz.)

Nail-laminated timber is often combined with a concrete layer on top to create a stiff element. Such a timber-flooring system improves the acoustic damping of the construction. The force transmission between concrete and wood is via screws, grooves in the wood surface or sometimes also by adhesives. The advantages of com-bining these two materials are that the concrete bears load under pressure while the wood reinforces tensile load-bearing purposes. Concrete is a mineral that does not burn, and its mass positively affects the construction's dynamic, acoustic and thermal behaviour. At the same time, the wood transmits the tensile forces and insulates. The visible wood surface is sometimes preferable from an aesthetic point. The con-crete-wood composite can be created either on-site or as an on-factory prefabricated unit. This hybrid EWP is particularly interesting in refurbishment, where new build-ing requirements occasionally change load conditions, sound and vibrational damp-ing requirements, *etc.*

2.4.3   EWPs Based on Sawn Timber

Wood forms the foundation for all EWPs. By utilising EWPs made from sawn timber, it is possible to create load-bearing construction elements with spans significantly longer than the length of the original sawlog, and even longer than the height of the tree itself. The strength of the wood is either preserved or even enhanced. Figure 2.31 presents a schematic classification and examples of components based on sawn timber.

2.4.3.1   Panel Components

Typical panel-shaped materials based on sawn timber are (Figure 2.32):

- edge-glued panels (single-layer) and
- cross-laminated timber (multi-layer).

An edge-glued panel (EGP) is a board of width-wise glued lamellae of sawn timber. The lamellae are sometimes joined lengthwise, especially in panels of lower-grade timber. EGPs are mainly used for furniture, exterior and interior joinery purposes, or

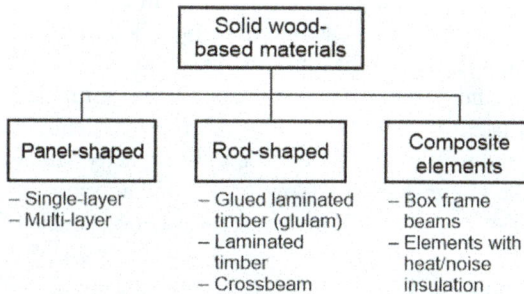

FIGURE 2.31   Systemisation of EWPs based on sawn timber.

(Courtesy of Hanser Verlag.)

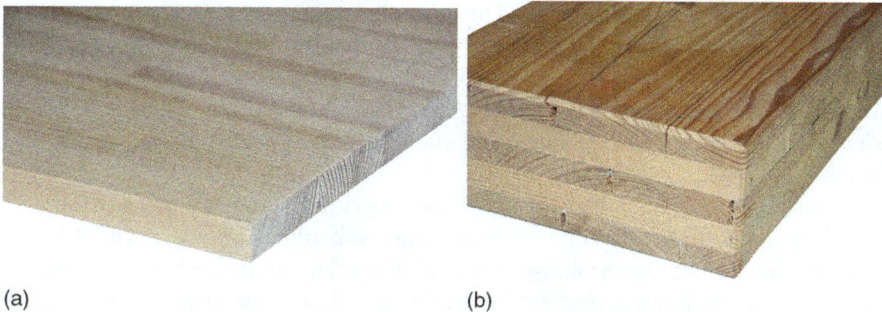

(a)   (b)

FIGURE 2.32   Panel-shaped EWPs based on sawn timber: (a) edge-glued panel and (b) cross-laminated timber.

(Courtesy of D. Sandberg.)

as cores for sandwich panels with outer sheets of either wood or non-wood material. A wide variety of species is used in EGPs.

The lamellae are glued with adhesives of different types depending on the properties asked for in the application (in the first-hand degree of moisture resistance). EGPs are therefore produced with three levels of adhesive-bond quality: (1) interior adhesives that are non-moisture resistant, (2) intermediate moisture-resistant adhesives, *i.e.* lower resistance to moisture than exterior adhesives, and (3) exterior adhesives that are moisture resistant. Minimum requirements for each type, developed from the results of long-term exposure, are detailed in product standards for EGPs. The most common adhesive in EGPs for interior usage is polyvinyl acetate (PVAc). The manufacture of EGPs ranges from manual carpenter-made panels to highly industrialised processes in specialised production units.

Cross-laminated timber (CLT), sometimes also called X-lam, is a flat, multi-layered timber member consisting of an odd number of layers, of which each second of three is orthogonally bonded to the others. CLT is commonly used as a structural system in multi-storey timber buildings and can be used as an alternative to concrete and steel or combined with these materials. CLT is also used for wooden bridges. The thickness is commonly about 100 mm, but up to 600 mm thicknesses are also available. CLT, often combined with glued-laminated timber and other materials, is the basis for a new class of building systems, usually called mass-timber construction. By the nature of its design, CLT has an inherent load-bearing capacity and can be used in both vertical and horizontal assemblies. The sawn timber in the outer layers of the CLT panels used as walls is usually oriented vertically, parallel to gravity loads, to maximise the wall's vertical load-bearing capacity. Likewise, the outer layers run parallel to the major span direction for flooring and roof systems. The on-site construction time is short since wall, floor and roof sections made of CLT are tailored off-site in a factory.

In multi-layer panels, the layers are connected by adhesives, steel nails or wooden dowels, sometimes using classic wood connections such as dovetail (Figure 2.30).

Softwood is mainly used, and hardwood is used very occasionally. When bonding hardwood to cross-laminated timber, the standard specifications of the delamination test according to EN 302-2 (CEN 2003b) are challenging due to the hardwood's more significant swelling and shrinkage. The strength properties are often higher compared to softwoods, which also places greater demands on the strength of the adhesive bond-line. In multi-layer panels of large thicknesses, the middle layers are often grooved, or the lamellae of the middle layers are inserted edge-wise unglued to reduce stresses caused by moisture variation, which reduces distortion and cracking (Figure 2.30).

"Brettstapel" is a plate-shaped EWP widely used in Central Europe. Sawn timber is flatwise joined by nails (aluminium or steel) or hardwood dowels. This EWP is also commonly referred to as nail-laminated timber. The Brettstapel system was successfully developed and promoted by Professor Julius Natterer as a sophisticated timber engineering method. However, using steel nails makes machining difficult, and panel production is often done manually. To address this challenge, nail-laminated technology has been further developed, replacing nails with wood dowels. The dowel may be diagonally oriented instead of those placed perpendicular to the

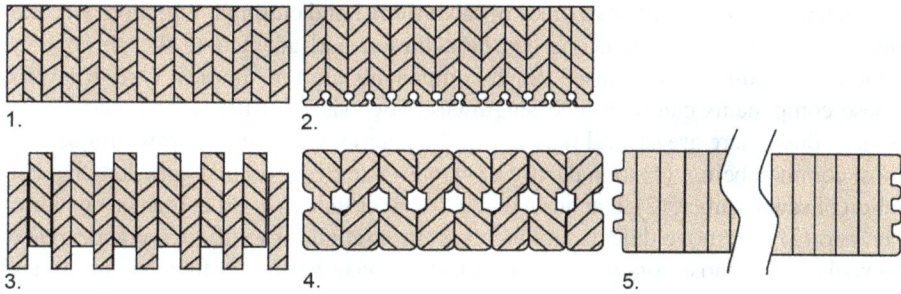

FIGURE 2.33 Adhesively bonded laminated floor system or Brettstapel floor system – different element structures with different joint formations: 1) standard version, 2) acoustic design, 3) shifted lamellae, 4) with cable ducts, and 5) tongue and groove edges for a tight connection of the elements for enhanced acoustic and fire-resistance properties.

(Courtesy of D. Sandberg.)

sawn timber surface. Wood-dowel-laminated timber is machinable using computer numerical control (CNC), enabling automated panel production. The wooden dowels create a stiffer and stronger connection between each sawn timber than nails, and the finger-jointed sawn timber eliminates the movement in the butt joints earlier commonly used in nail-laminated timber. The next step in the evolution of the laminated floor system involved bonding the sawn timber with adhesives, thus enabling longer spans (Figure 2.33).

2.4.3.2 Rod-shaped Components
Typical rood-shaped materials based on sawn timber are (Figure 2.34):

- components of sawn timber and
- glued-laminated timber.

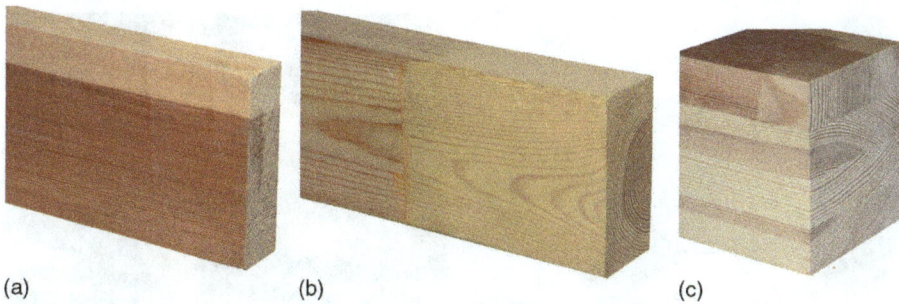

FIGURE 2.34 Rood-shaped components of saw timber: (a) component of sawn timber in specific dimensions and strength, (b) finger-jointed component and (c) flat-wise glued components.

(Courtesy of D. Sandberg.)

By selecting specific parts of a piece of sawn timber and cutting it out, more well-defined properties and dimensions can be achieved and adapted to the purpose for which the product is to be used. Such components are the simplest type of EWPs. These components can be joined lengthwise, edgewise and flatwise to increase the dimensions. There are several reasons for the lengthwise joining of sawn timber, the most common being: (1) to make use of wood waste from, for example, length adaptation of sawn timber, (2) to achieve a length longer than that of the logs or sawn timber, and (3) to remove defects and increase the strength or to affect the appearance. Nowadays, the most common method of lengthwise joining is finger-jointing, and it is used for structural members such as studs, glued-laminated timber (GLT) and non-structural purposes. A small-dimensional beam with good shape stability can be obtained by gluing together 2–5 boards by their face sides with the fibre direction running parallel (Figure 2.34c). The choice of adhesives depends on the final use of the product. For structural purposes, phenol-resorcinol-formaldehyde (PRF) or melamine-urea-formaldehyde (MUF) types are commonly used, and for non-structural products, polyvinylacetate (PVAc) or hot-melt adhesives are used.

Glued-laminated timber (GLT), commonly called glulam, is a structural element of sawn timber flat-wise glued with parallel fibre orientation to straight beams or members with curvature (Figure 2.35). Softwood and hardwood species are used in GLT, but the most common species for GLT production are Douglas fir, larch, Norway spruce, radiata pine, Scots pine, southern pine and yellow poplar. Preservative-treated sawn timber is used for structures that are expected to be exposed to the prolonged influence of moisture. The major advantage of GLT is its high strength and stiffness, which means that it can be used to manufacture structures for wide spans and enhanced bearing capacity. Stress-graded sawn timber is always used to manufacture GLT, distinguishing GLT from other flat-wise glued components. The sawn timber is graded (by machine) according to the mechanical characteristics into Strength Classes according to the EN 338 and EN 14080 standards (CEN 2013a, 2016a), which can significantly increase the load-bearing capacity of the GLT. To best utilise the timber's strength, high-strength sawn timber is

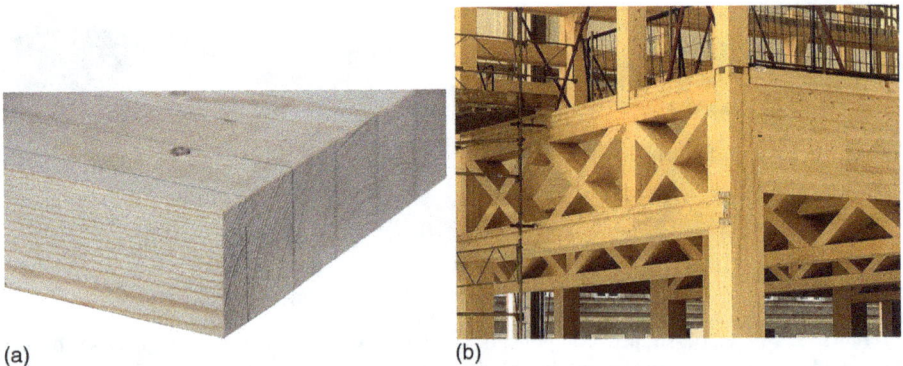

(a)   (b)

FIGURE 2.35   The use of glued-laminated timber (GLT): (a) the principle of GLT, (b) a load-bearing construction structure built up by GLT.

FIGURE 2.35 *(Continued)* (c) GLT tube-formed columns (columns with tubular cross-section) and a roof of conventional GLT beams, and (d) connection of horizontal and vertical GLT elements.

(Courtesy of D. Sandberg.)

often used for the outer parts of the GLT where the stresses are highest. On average, laminating increases strength by around 10% compared to native wood, while the variation of properties decreases considerably. In GLT manufacture, only adhesives with a documented high strength and durability under long-term loads are used and only those with which the producers have a long practical experience. The traditional and very common adhesives used in the manufacture of GLT are those of the synthetic two-component PRF type (phenol-resorcinol-formaldehyde), which give a dark reddish-brown bond line. MUF (melamine-urea-formaldehyde) adhesives are now being increasingly used, and these adhesives are initially light in colour but darken with time.

Laminated wood is also increasingly used to produce windows and doors (frames) because it has better shape stability than sawn timber. Glued-laminated timber can also be manufactured in curved shapes and, in exceptional cases, into three-dimensional free forms (Figure 2.36).

Glued-laminated timber is often made of Norway spruce but is sometimes covered with a thin (6–7 mm) layer of high-quality hardwood for aesthetic reasons. Large cross-sections of laminated timber, for example, for use in high-rise buildings, can be glued together from several smaller beams. This requires extensive knowledge to prevent the adhesive joints from damage during later use under variations in temperature and relative humidity.

FIGURE 2.36 A free-form timber structure. The Haesley Nine Bridges golf clubhouse in Yeoju, Republic of Korea. Shigeru Ban Architects.

(Courtesy of CJ Group, South Korea.)

2.4.3.3 Composite Elements

In the construction industry, cross-laminated timber, glued-laminated timber and strength-graded sawn timber are used for load-bearing purposes. In addition to these mass-timber construction components, lightweight hybrid frame-construction elements covered by panel material, such as oriented strand board (OSB), are used.

Typical rood-shaped materials based on sawn timber are (Figure 2.37):

- box-frame beams, and
- elements with heat and sound insulation.

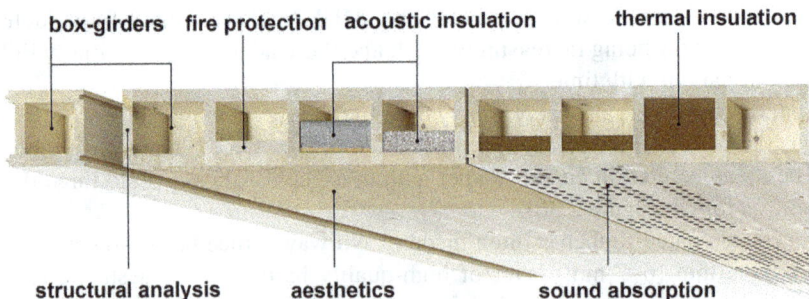

FIGURE 2.37 Construction element with heat and sound insulation. Box-girder element from the Lignatur™ company: elements combined into a large-scale surface element modified for acoustics (with acoustic holes or slits), fire protection and thermal insulation.

(Courtesy of Lignatur, Waldstatt, Switzerland.)

2.4.4   EWPs Based on Veneer

A veneer is a thin sheet of wood used as a surface covering on a core of wood-based or other material for decorative purposes or for building up laminates for strength. This means that veneers can be divided into decorative and structural categories. In laminated veneer products, these two categories are face veneers and core veneers. Typical EWPs based on veneer are:

- plywood,
- mass plywood panels,
- laminated veneer lumber and
- high-pressure laminated veneer.

This material group also includes modified veneers, such as synthetic resin-impregnated and densified veneers. The orientation of the veneer may also be varied to alter its strength properties (Figure 2.38).

2.4.4.1   Plywood and Mass Plywood Panels

Plywood is a rigid panel composed of an odd number of symmetrically stacked veneers glued together so that the fibre orientation of a veneer is perpendicular to the fibre orientation of the adjacent veneers (Figure 2.39). The maximum veneer thickness is about 3 mm. Blockboards (bar plywood, strip plywood) are classified as

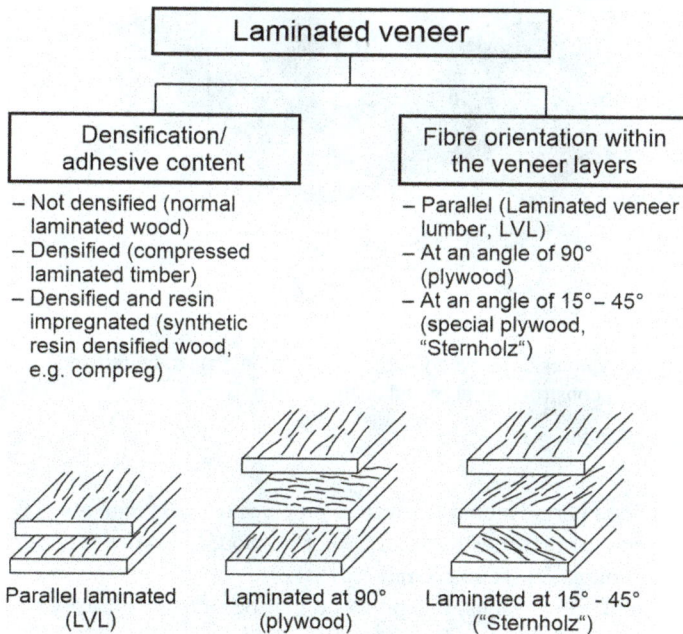

FIGURE 2.38   Principal variants of veneer-based materials due to modification and veneer orientation: (a) categorisation of veneer-based materials and (b) orientations of veneer layers.

(Courtesy of Hanser Verlag.)

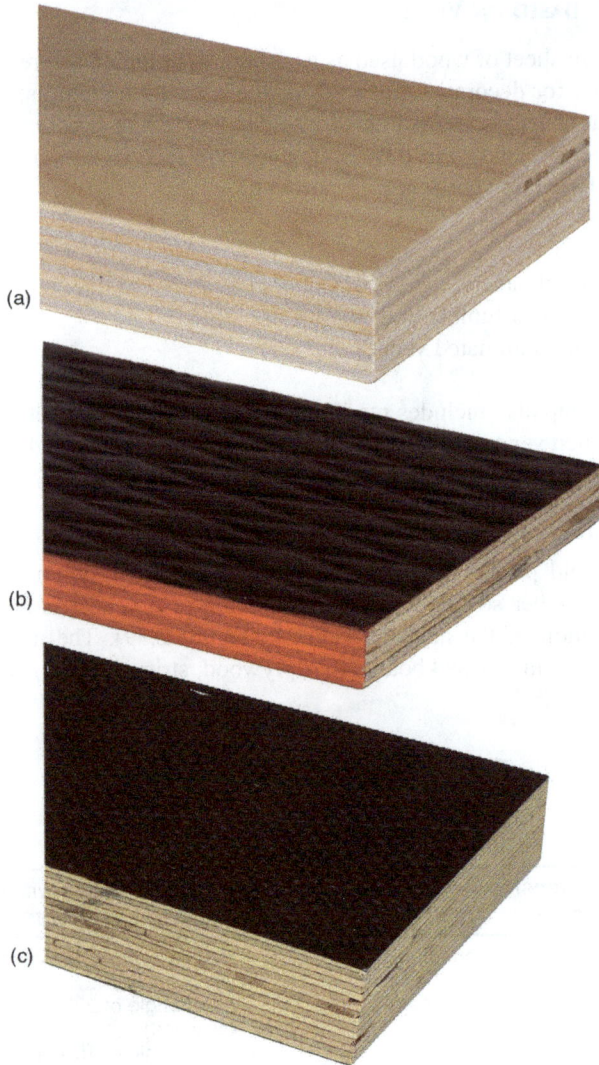

FIGURE 2.39 Examples of plywood: (a) birch plywood, (b) structured and thermally densified surface, and (c) construction plywood coated with polypropylene ISOPLYFORM.

(Courtesy of D. Sandberg.)

plywood. Moulded parts (such as furniture, pipes, and shells) can also be made from laminated veneers, sometimes reffered to as plywood, but more accurately as laminated veneer products (Navi and Sandberg 2012).

There are numerous grades, but plywood can be divided into plywood for constructional (exterior) purposes, interior use (joinery and decorative plywood), and special applications such as concrete shuttering, marine plywood, plywood with unique surface layers, *etc*. Panel types where veneer sheets are glued onto an edge-glued panel core, called core plywood or face-glued blockboard, are also considered plywood products.

Plywood is glued with thermosetting adhesives. Phenol-formaldehyde-based (PF) adhesives are used for exterior-type plywood, and urea-formaldehyde (UF), reinforced urea adhesives, and sometimes natural polyphenols (tannins) mixed with synthetic adhesives are used for interior-type plywood. Plywood may be surfaced with metals, plastics or other materials, or its veneers may be impregnated to achieve a superficial hardness or resistance to microorganisms, fire or other destructive agents.

Plywood and particleboard are the most widely used wood-based panel materials globally. It is manufactured and used, particularly in the Americas and many developing countries.

Mass plywood panels (MPPs) are a veneer-based product consisting of plywood bonded together with an adhesive under pressure to a thick structural element for use in, for example, walls, floors, beams or columns (Figure 2.40). Production, based on softwood plywood, started at the end of 2017 at Freres Lumber Co. situated between Lyons and Mill City, Oregon, USA. MPP can be used in construction packages for everything from skyscrapers to single-story houses. MPPs can be made to specifications, not only regarding length, width and thickness but also other custom qualities, depending on architects' and engineers' need for performance. In the first production plant, the maximum dimensions for an MPP are 3.6 m in width, 15 m in length and 600 mm in thickness. MPP columns with a height of up to 15 m can be manufactured, and these can, in addition to design purposes, be used to create temporary roads on construction sites. The panels can match the performance characteristics of a similar product, for example, cross-laminated timber (CLT), with the use of 15–20% less wood. MPP is bonded with a fire-resistant non-urea-formaldehyde-based adhesive.

2.4.4.2   Laminated Veneer Lumber

Laminated veneer lumber (LVL) is made of veneers with a thickness between 2 mm and 4 mm, generally bonded together with the same fibre orientation in the layers to make large format beams and panels (Figure 2.41). Some of the inner veneers of the

FIGURE 2.40   Mass plywood panels (MPPs).

(Courtesy of D. Sandberg.)

FIGURE 2.41 Laminated veneer lumber (LVL).

(Courtesy of D. Sandberg.)

LVL may be oriented perpendicular to the others, providing two-way structural capacity and reducing shrinkage/ and swelling perpendicular to the length direction. LVL is available on the market and is made from softwood, preferably Norway spruce, for example, Kerto LVL, and hardwoods, such as common beech, for example, BauBuche from Pollmeier. LVL is increasingly being used in construction as panel material and rod-shaped elements analogous to the use of glued-laminated timber. High rigidity and strength are achieved, especially when using high-density hardwoods.

The manufacturing procedure is very similar to that of plywood production, and the final product has a high strength, good dimensional stability and a homogeneity of physical and mechanical properties along the elements. The veneer is graded so that low-quality veneers are placed in the inner layers, and high-quality veneer is used in the outer layers, thus improving both flexural strength and modulus of elasticity. LVL can be utilised as large dimension panels for floors, roofs, and walls or as columns and beams. LVL can be combined with other EWPs or used in hollow box floor and roof cassettes to create a structural cross-section with good spanning characteristics.

In most cases, LVL is bonded with a phenol-formaldehyde (PF) or melamine-formaldehyde (MF) adhesive and has higher strength and stiffness than plywood. The number of veneers placed in LVL depends on the dimensions. LVL has a thickness of up to 275 mm, a width of usually about 2 m and a length of up to 25 m due to transport restrictions. LVL is cut to the final width and length according to customer requirements.

2.4.4.3 High-pressure Laminated Veneer

High-pressure laminated veneer (HPLV) is a material that has been compressed in the transverse direction to increase its density (Figure 2.42). This process is commonly called densification. Depending on the temperature, pressure and time used during pressing, the board develops different physical and mechanical properties. Under certain manufacturing conditions, bonding between the veneers is also possible without

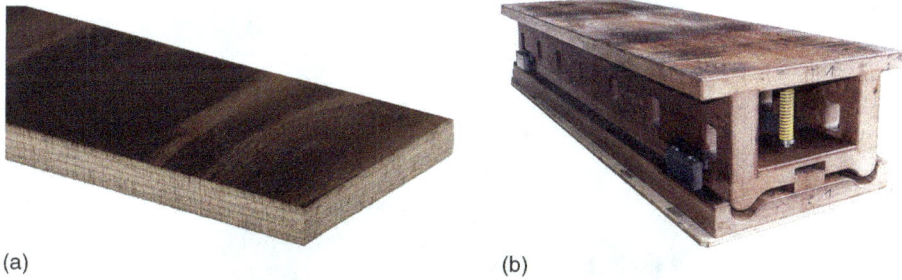

(a)   (b)

FIGURE 2.42   High-pressure laminated veneer (HPLV): (a) densified and heat-treated plywood, and (b) a mould of densified plywood.

(Courtesy of D. Sandberg and Otto Bosse GmbH & Co. KG, Stadthagen, Germany.)

adhesives, reducing the product's environmental impact. This product is, however, rare in use. The veneers are usually impregnated with a low molecular-weight resin (typically phenol-formaldehyde) and cured at an increased temperature for a dimensionally stable product. Electric transmission support components are typically resin-impregnated, laminated veneer. The billets are compressed in a heated press to a density of approximately 1300 kg/m³. Other applications for resin-impregnated veneer include wear plates for machinery and vehicles, machine pattern moulds, bullet-proof barriers, liquid natural gas (LNG) storage containers and associated support structures. HPLV should not be confused with high-pressure laminates (HPL), melamine-impregnated decorative surfaces backed with multiple layers of kraft paper impregnated with phenolic resins and consolidated into a solid sheet under high temperature and pressure.

2.4.5   EWPS BASED ON STRANDS, STRIPS, CHIPS OR PARTICLES

Using as many harvested trees as possible is important for sustainable use of forest resources. The low-value residues from forestry, in the form of low-quality trees or parts of trees, and from the sawmilling and joinery industry in the form of chips and sawdust, are used in various types of EWPs, such as:

- parallel strand lumber,
- waferboard and oriented strand board,
- laminated strand lumber,
- particleboard,
- inorganic bonded composites and
- wood-plastic composites.

2.4.5.1   Parallel Strand Lumber

Parallel strand lumber (PSL, also called Parallam) is manufactured by gluing flakes of veneer (strands) together with the fibre orientation of the flakes oriented parallel to the length direction of the PSL (Figure 2.43). The manufacture of PSL is based

FIGURE 2.43   Parallel strand lumber – PSL or Parallam.

(Courtesy of D. Sandberg.)

on a technology which makes it possible to convert small trees into elements with large cross-section dimensions (up to 0.3 × 0.5 m) and considerable lengths (up to 20 m). PSL can also be manufactured from rotary-cut veneer or veneer waste. These products are intended for construction as elements under compression, large trusses, beams or posts. After drying to a moisture content of about 6%, the strands are treated with an adhesive with hydrophobic properties and introduced in the longitudinal direction into a continuous press where microwave radiation polymerises the adhesive. After sanding, the PSL is cut into sections ready for use. PSL is very strong in its length direction (stronger than sawn timber). Additional strength is gained from the 10% densification relative to the original wood density. In tension, the strands fail because the overlap between strands is significant, and the shear resistance is higher than the tensile strength of the strand.

2.4.5.2   Waferboard, Oriented Strand Board and Laminated Strand Lumber

Waferboard (WB) and oriented strand board (OSB) belong to the subset of reconstituted wood panel products called flakeboards, made from wood wafers or strands (Figure 2.44). WB was developed in the 1950s in the USA, and OSB originated in

FIGURE 2.44   Oriented strand board – OSB.

(Courtesy of D. Sandberg.)

the early 1980s. OSB is produced from wood strands, typically 15–25 mm wide, 75–150 mm long and 0.3–0.7 mm thick, cut from small diameter logs. A water-resistant adhesive is used to bond the strands together, and the panels are fabricated under pressure at a high temperature with a high density. OSB is a multi-layer panel. The strands in the face layers are aligned parallel to the length direction of the board, whereas the internal strands are deposited randomly or perpendicular to the face layers. The panels are used in various structures or for decorative purposes such as under-floors, ceilings and sometimes walls, and in packaging and pallets, and they have replaced plywood and particleboard in many applications. Different species such as aspen, red maple, southern yellow pine, sweetgum, white birch and yellow poplar have been used as raw materials in North America, but in Europe, Norway spruce and Scots pine are more common as raw materials.

Laminated strand lumber (LSL) is made up of wood strands glued together, with the grain of each strand oriented parallel to the length of the finished product. LSL uses a manufacturing technology which makes it possible to convert small trees into elements with a large cross-section, but treetops and branches can also be used. The debarked wood is flattened and partially split in the longitudinal direction using crushing and coarse so-called scrimming rollers. The dried "scrims" are further processed using fine scrimming rollers to form mats (The product is therefore sometimes called laminated scrimmed lumber). These mats are, in general, layered to each other to achieve the thickness and properties required of the final component. The orientation of the mats in LSL is parallel (*i.e.* the fibre orientation of all the scrims run parallel to each other), perpendicular (*i.e.* the different layers are oriented at 90° to each other) and mixed before the mats are laminated together under pressure to a component of a beam or panel size. Aspen, yellow poplar and other hardwood species are used to manufacture LSL. A phenol-formaldehyde (PF) adhesive is generally used.

The following terminology is used for OSB, *i.e.* oriented strand boards (CEN 2006):

- OSB/1 – General purpose panels and panels for interior fitments (including furniture) for use in dry conditions,
- OSB/2 – Load-bearing panels for use in dry conditions,
- OSB/3 – Load-bearing panels for use in humid conditions and
- OSB/4 – Heavy-duty load-bearing panels for use in humid conditions.

2.4.5.3 Particleboard

Particleboards (PB) are panel-shaped materials manufactured from wood particles of small dimensions (length 1–10 mm and thickness 0.2–0.3 mm). The particles are adhesively bonded, usually under the influence of heat. The manufacture of PB is a dry process, and there are two different methods of production: flat hot pressing and extrusion, which give different types of boards with varying orientations of particles. In the first method, which is most common, the particles are oriented parallel to the panel surface, whereas in the second method, the particles are oriented perpendicular to the surface (Figure 2.45).

(a)   (b)

FIGURE 2.45   Particleboard – PB: (a) conventional flat hot pressed and (b) extruded.
(Courtesy of D. Sandberg.)

The following terminology is used for resin-bonded particleboard (CEN 2010):

- P1: general-purpose panels for use in dry conditions,
- P2: panels for interior fitments (including furniture) for use in dry conditions,
- P3: non-load-bearing panels for use in humid conditions,
- P4: load-bearing panels for use in dry conditions,
- P5: load-bearing panels for use in humid conditions,
- P6: heavy-duty load-bearing panels for use in dry conditions and
- P7: heavy-duty load-bearing panels for use in humid conditions.

There are many types of PBs. They are generally three-layered, with a lighter core of coarser particles and more dense surface layers with fine particles, making the surfaces smooth. Depending on the PB's layering, the panel achieve different density profiles throughout the thickness, for example, giving a lightweight panel with hard surfaces (Figure 2.46).

The mechanical properties of the panels depend on the dimensions, orientation and arrangement of the wood particles used in panel manufacturing. The PB industry

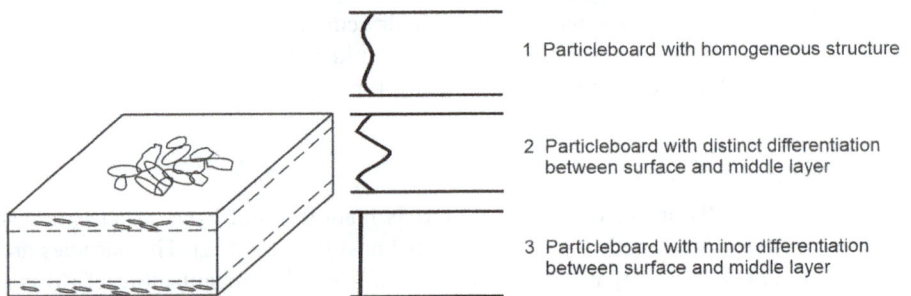

1 Particleboard with homogeneous structure

2 Particleboard with distinct differentiation
between surface and middle layer

3 Particleboard with minor differentiation
between surface and middle layer

FIGURE 2.46   Layering particleboard (PB) with particles of different sizes gives panels with a varying density (density profile) in the thickness direction: (a) the macrostructure of a layered PB with dense and hard surfaces, and (b) density profiles possible in a PB.

(Courtesy of Hanser Verlag.)

has been commercialised successfully worldwide because of the favourable conditions for raw-material supply and market demands. The PB process makes using wood trunks of small diameters and wood residues possible. PBs are very useful in the furnishing and construction industries. The production of PBs also leads to an effective use of wood with a very small percentage of waste, 10–25%, instead of 50% in the sawing of logs. Figure 2.47 shows possible ways of varying the structure and properties of PBs to achieve various properties.

Moulded wood particles are a unique feature. So-called moulded PB is manufactured with an increased adhesive content (10–30%) and increased density compared to conventional PBs and is usually covered on all surfaces with a polymer-based film for improved moisture resistance. In some cases, pallets are also manufactured by moulding (Anon. 2003, Niemz et al. 2023). Moulded parts are also used for facades, windowsills and garden furniture.

2.4.5.4 Inorganic Bonded Composites

Inorganic-bonded composites (IBCs) are panels or other types of constructional elements produced from wood wool, particles or flakes, or other types of fibre of vegetable biomass blended with cement, magnesite ($MgCO_3$) or gypsum (Figure 2.48). The most expedient binder that gives strength, durability and acoustic insulation properties is Portland cement. In this process, heat is not required. The wood-cement

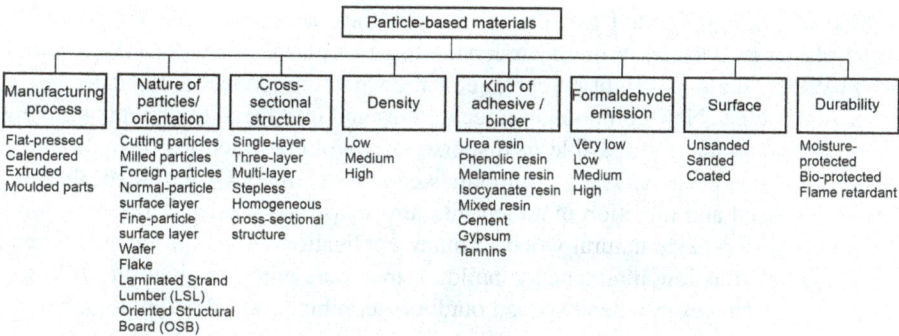

Manufacturing process	Nature of particles/orientation	Cross-sectional structure	Density	Kind of adhesive / binder	Formaldehyde emission	Surface	Durability
Flat-pressed	Cutting particles	Single-layer	Low	Urea resin	Very low	Unsanded	Moisture-protected
Calendered	Milled particles	Three-layer	Medium	Phenolic resin	Low	Sanded	Bio-protected
Extruded	Foreign particles	Multi-layer	High	Melamine resin	Medium	Coated	Flame retardant
Moulded parts	Normal-particle surface layer	Stepless		Isocyanate resin	High		
	Fine-particle surface layer	Homogeneous cross-sectional structure		Mixed resin			
	Wafer			Cement			
	Flake			Gypsum			
	Laminated Strand Lumber (LSL)			Tannins			
	Oriented Structural Board (OSB)						

FIGURE 2.47   Classification of particle-based materials production principles.

(Niemz and Sonderegger 2021, courtesy of Hanser Verlag.)

FIGURE 2.48   Cement-bonded particleboard.

(Courtesy of D. Sandberg.)

panels are used for specialised structural applications. They have outstanding properties concerning their resistance to fire, durability, sound insulation and stiffness, and this means that the product is most suitable for internal wall constructions in public places, the lining of lift shafts, the construction of cabling ducts, soffits, motorway acoustic fencing and the cladding of prefabricated house units. The cement-bonded panels have a harder surface and are more resistant than their components, with a lower cost and lower density than concrete.

Many manufacturers use additives like mica (silicate/phyllosilicate minerals), aluminium stearate and cenospheres (a lightweight, inert, hollow sphere made mainly of silica and alumina and filled with air or inert gas) to achieve certain panel qualities. A typical cement fibreboard comprises 40–60 weight% cement, 20–30% fillers, 8–10% cellulose and 10–15% mica. Additives such as aluminium stearate and polyvinyl alcohol (PVA) are typically used in quantities less than 1%. Cenospheres are used only in low-density boards with quantities between 10% and 15%.

2.4.5.5 Wood-plastic Composites

Although EWPs based on strands, strips, chips and particles have been made with thermosetting adhesives for many years, only in the last decades has a serious attempt been made to incorporate wood flour and chips into thermoplastics to produce wood-plastic composites (WPCs). WPC refers to any composite containing wood particles and a thermosetting or thermoplastic polymer (Figure 2.49). In contrast to wood-thermoset composites, wood-thermoplastic composites have seen phenomenal growth in the USA. For this reason, they are often referred to simply as wood-plastic composites, with the understanding that plastic is always a thermoplastic. In WPCs, the dry-weight percentage of the wood component is typically in the range of 50–60%. New compounding techniques and interfacial treatments utilising coupling agents make it feasible to disperse large-volume fractions of hydrophilic wood in various plastics. These compounds can be continuously extruded, thermoformed, pressed and injection moulded into any shape and size, and they thus have the potential to replace natural wood in many applications. WPC products are generally marketed as low-maintenance building materials with a high ability to resist fungal decay. However, when exposed outdoors, combined wood and polymers may have poor long-term durability.

FIGURE 2.49   An extruded wood-plastic composite – WPC.

(Courtesy of D. Sandberg.)

2.4.6 EWPS BASED ON FIBRES

Fibres for use in EWPs are taken from low-value residues from forestry, pulp production or sawmills in the form of low-quality trees, parts of trees, screening reject, chips and sawdust. The large-sized raw materials are heated in steam, and the fibres are separated from each other before forming the EWP. Typical EWPs based on fibres are:

- high-density fibreboard,
- medium-density fibreboard,
- low-density fibreboard, and
- bio-based insulation materials (ultra-low-density fibreboard).

Fibreboards are panel-shaped materials that consist of fibres, predominantly wood and have been produced with or without pressure, adhesives and additives under the influence of heat. The fibre length is 4–6 mm, and the fibre thickness is 0.04–0.15 mm. This results in a higher degree of slenderness (length/thickness ratio) than with chips. Therefore, the modulus of elasticity and strength are higher than those of, for example, particleboard. So far, only fibreboards can be produced industrially without adhesives using the wet process of high density. However, this production is on a sharp decline. The panels have largely been replaced by thin fibreboards produced using the dry process according to medium-density fibreboard (MDF) technology. The bulk density of fibreboard ranges from 50 kg/m^3 (insulating materials in the dry process) to over 1000 kg/m^3 for high-density fibreboard (HDF). Moulded parts, especially for vehicle construction (parcel shelves, side panels), are also manufactured based on fibres made from wood and other renewable raw materials. The variety of fibreboards is large, and Figure 2.50 shows a classification based on production parameters.

Fibreboards are used in the construction industry as insulating materials (in roofs for thermal insulation and sometimes in walls directly for plastering) and as a base material for flooring and other types of laminate.

High-density fibreboards with high adhesive content, so-called compact-density fibreboard (CDF), are often used as a substitute for plastic in interior design (see, *e.g.* Swiss Krono at https://www.swisskrono.com). Through-coloured MDF is also widely used in interior design.

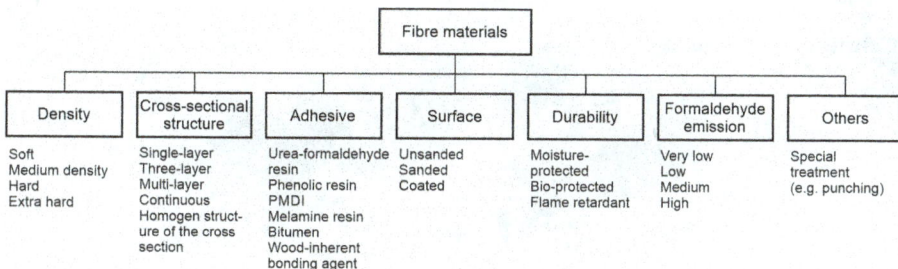

			Fibre materials			
Density	Cross-sectional structure	Adhesive	Surface	Durability	Formaldehyde emission	Others
Soft	Single-layer	Urea-formaldehyde resin	Unsanded	Moisture-protected	Very low	Special treatment (e.g. punching)
Medium density	Three-layer	Phenolic resin	Sanded	Bio-protected	Low	
Hard	Multi-layer	PMDI	Coated	Flame retardant	Medium	
Extra hard	Continuous	Melamine resin			High	
	Homogen structure of the cross section	Bitumen				
		Wood-inherent bonding agent				

FIGURE 2.50 Classification of fibre materials.

(Niemz and Sonderegger 2021, courtesy of Hanser Verlag.)

2.4.6.1 High-density Fibreboard

The high-density fibreboard (HDF, also called hardboard) is made of highly compressed randomly oriented wood fibres with a density of about 900–1200 kg/m³ that makes the panel denser, harder and stronger than, for example, particleboard and MDF (Figure 2.51). HDF with a lower density also occurs on the market. HDF is traditionally bonded without adhesives by hot-pressing, where the high-temperature-induced flow of lignin promotes the bonding. Today, 2–3% of PF adhesive can be added to improve the properties of the panel. HDF is ideal for high-quality furniture, cabinet-making, automobile dashboard panels, packaging and construction, and HDF is often used as a substrate for lamination. Today, HDF is mainly used as the core material for laminate flooring. A special grade of HDF is oil-tempered HDF that is impregnated with a special oil polymerised by heat treatment during manufacturing. This treatment gives the panel excellent moisture resistance properties and a higher bending strength. Typical end uses include internal wall or roof linings, signs and flooring underlay. Typical dimensions are a thickness between 2 mm and 8 mm and a width and length up to 2 and 4 m, respectively. Perforated HDF is smooth on one side with a mesh pattern on the reverse, uniformly perforated across the whole panel. The holes are between 3 and 10 mm in size. Perforated HDF can also be painted and is used, for example, for furniture, acoustic constructions and in the shopfitting industry.

2.4.6.2 Medium-density Fibreboard

The essential difference between HDF and medium-density fibreboard (MDF) is that adhesives are used as binders in MDF (Figure 2.52). The MDF manufacturing process began as a semi-dry process before being developed into a fully dry process. Since less

FIGURE 2.51　High-density fibreboard – HDF.

(Courtesy of D. Sandberg.)

FIGURE 2.52　Medium-density fibreboard – MDF.

(Courtesy of D. Sandberg.)

water is used than in the wet process, smaller amounts of polluted water are produced. In addition, this method makes it possible to fabricate panels with thicknesses from 2 mm up to 100 mm. A uniform distribution of fibres during manufacture ensures that the MDF has a homogeneous structure, and it is possible to manufacture MDF with different characteristics to suit particular applications. The MDF has normally a density of 600–800 kg/m³ but 400 kg/m³ and 1000 kg/m³ occur. MDF is easy to machine, and its regular surface is exceptionally well suited to painting or applying a decorative coating. This quality has given MDF the place it occupies in the furniture industry. The thick MDF is used in joinery and for door and window frames.

2.4.6.3   Low-density Fibreboard

Low-density fibreboard (LDF) is a structural and decorative lightweight panel or insulation material based on fibres, with a density ranging from ultra-low, *i.e.* less than 150 kg/m³ to low, which is about 150–300 kg/m³ (Figure 2.53). Bio-based insulation materials, *i.e.* ultra-low-density fibreboards (ULDF), are described in the next section. The processes for manufacturing fibreboards, in general, are divided into two types of processes, *i.e.* dry or wet. In the dry process, the fibres are bonded by an adhesive, and the product gets a relatively high density. In the wet process, the cellulose fibres bind together by natural forces (hydrogen bonds), achieved through drying and compression in a process which takes place in a wet condition, with no adhesives being used to bond the fibres together. LDF is a non-adhesive product. Additives may be added during manufacture to improve properties such as fire resistance. LDF has good heat and acoustic insulation properties, and it is therefore used as filling doors and as impact-sound insulation in floor construction. LDF is used in indoor ceilings because of its sound-absorbing properties. LDF can also be found in various uses such as packing, bulletin boards and hybrid EWPs. Special grades such as asphalt- or phenol-impregnated LDF are used for harsher environmental conditions in the external walls or for ceiling panelling and flooring. Typical dimensions are a thickness between 10 mm and 20 mm, width and length up to about 1.5–4 m, respectively.

2.4.6.4   Bio-based Insulation Materials

A bio-based insulation material is an ultra-low-density fibreboard (ULDF) with a density of less than 150 kg/m³ (Figure 2.54). ULDF is mainly used as insulation, with wood fibre dominating, but other bio-based fibre materials are also used.

FIGURE 2.53   Low-density fibreboard – LDF.

(Courtesy of D. Sandberg.)

FIGURE 2.54 Examples of bio-based insulation materials on the market: (a) wood-fibre, Bestwood Multitherm™, (b) jute, Thermojute™, (c) hemp, Thermohanf Premium™, (d) flax, Isolina™, (e) paperpulp, Thermocell™, (f) recycled newspaper, Icell™, (g) wood-fibre, Pavatherm combi™, (h) wood-fibre, Bestwood Wall™, (i) wood-fibre, STEICOflex™ and (j) wood-fibre, Hunton™.

(Courtesy of D. Sandberg.)

Bio-based insulation materials are made mainly from renewable natural resources such as agricultural by-products, animal fibres (*e.g.* sheep wool) and plant fibres, but recycled material such as newspapers is also used. Natural fibre insulations can be used alone as granulates or formed into flexible, semi-rigid or rigid panels using a binder. They usually provide significantly less thermal insulation than non-bio-based industrial products, but this can be compensated for by increasing the thickness of the insulation layer.

Bio-based insulation materials typically require fire-retardant or anti-insect/pest treatment. A clay coating is a non-toxic additive which often meets these requirements. The material can buffer heat and moisture but has high water uptake, and during, for example, a water leakage, this may be a disadvantage if not treated with moisture-repellent additives. It is suggested that bio-based insulation materials may have the potential to make a significant contribution to the potential reduction of global warming in the construction industry.

Due to their high specific heat capacity, bio-based insulation materials have advantages over foams or fibres based on glass or stone. The bulk density of insulating materials has a decisive influence on their thermal conductivity. Figure 2.55 compares the thermal conductivity coefficients of different materials.

2.4.7 HYBRID EWPS

By combining different materials, the properties of hybrid EWPs can be explicitly designed to give specific values such as low weight, high strength, low heat transmission or aesthetics. Hybrid-engineered wood products consist of different EWPs or of EWPs in combination with other materials, such as sandwich panels or members with box- or I-shaped cross-sections:

- composite or wood sandwich panels, and
- I-joists and box-beams.

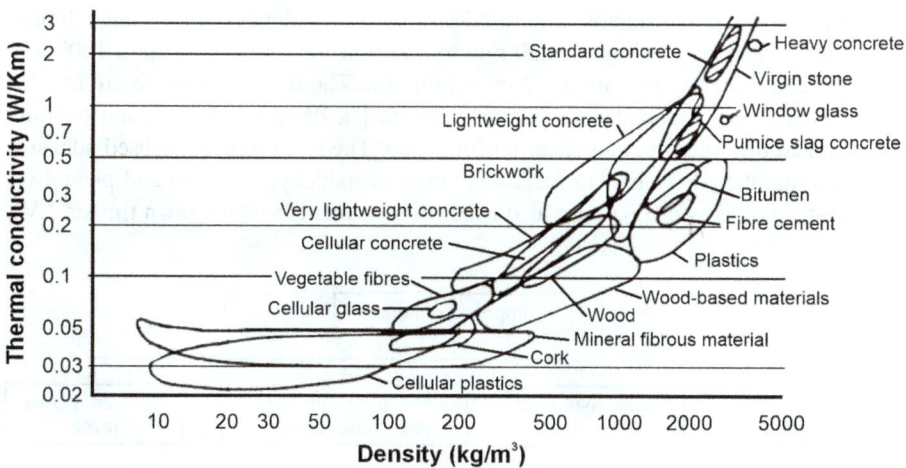

FIGURE 2.55 Thermal conductivity as a function of bulk density.

(Wesche 1996, courtesy of Bauverlag.)

Above the need for lightweight construction, the reuse of materials and improved functionality are of great importance for new materials (Engelhardt 2013). Werner Sobeck pioneered lightweight construction in the construction industry (Anon. 2018).

2.4.7.1 Composite Panels

Composite panels are panel-shaped materials symmetrically composed of several layers, predominantly with a thicker core layer than the surface layers. Figure 2.56 shows a classification of composite panels. The basic concept of sandwich panel construction is to use thin, dense, strong-facing materials bonded to a thick, lightweight core. Each component is relatively weak and flexible, but they provide an extremely stiff, strong, and lightweight structure when combined. Composite panels are used mainly for lightweight construction purposes.

A core of insulation materials, kraft paper honeycomb, polystyrene, polyester, lightweight wood such as balsa, wood wool *etc.* between outer sheets of a board material such as plywood, OSB, MDF, HDF or other types of wood-based laminates, exist in a wide variety with unique designs for different purposes such as high strength, fire resistance, low sound or heat transmission, or for low weight to give an aesthetic appearance to furniture or interior joinery (Figure 2.57). The sandwich is not a material with unique mechanical properties but rather a structure that must be designed for particular uses to which it will be subjected. If foam or honeycomb is used in the middle layer, the shear strength is low, and if solid wood, for example, balsa, is used, a high shear stiffness is achieved. Fibre insulation materials may also be used as a middle layer.

2.4.7.2 I-joists and Box-beams

I-joists are structurally timber joists with flanges made of strength-graded sawn timber or LVL of various widths, united with one or two webs made of OSB, HDF, plywood or particleboard of various depths (Figure 2.58). The section depth (flange-flange dimension) ranges from approximately 200 mm to 400 mm for I-joists and up to 1200 mm for box-beams. A length of up to 12 m is common. The flanges and web are bonded together to form an I-shaped cross-section member. The flanges resist everyday bending stresses, and the web provides shear performance. The most frequently used adhesive is urea-formaldehyde (UF), but melamine urea-formaldehyde (MUF) and phenol-resorcinol-formaldehyde (PRF) are also used. Box-beams consist of sawn timber, LVL

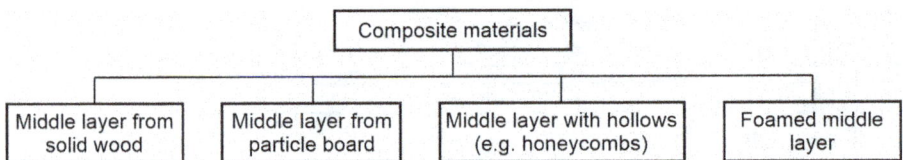

FIGURE 2.56 Classification of wood-based composite panels.

(Niemz and Sonderegger 2021, courtesy of Carl Hanser Verlag.)

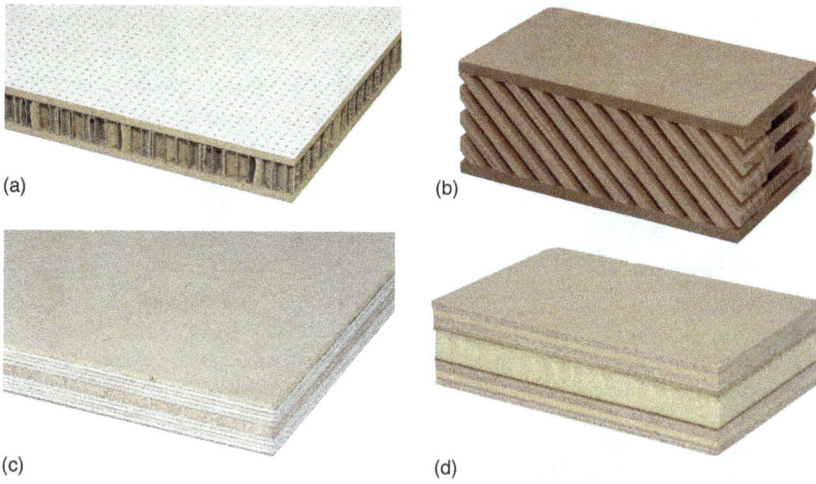

FIGURE 2.57 Examples of composite panels: (a) lightweight acoustic panel with a core of honeycomb paper and surface layers of perforated hardboard, (b) Dendrolight® with a core of solid wood and surfaces of hardboard, (c) a core of cork and surfaces layers of birch plywood, and (d) a core of polyurethane and spruce plywood as surfaces layers of birch plywood – Variotec.

(Courtesy of D. Sandberg.)

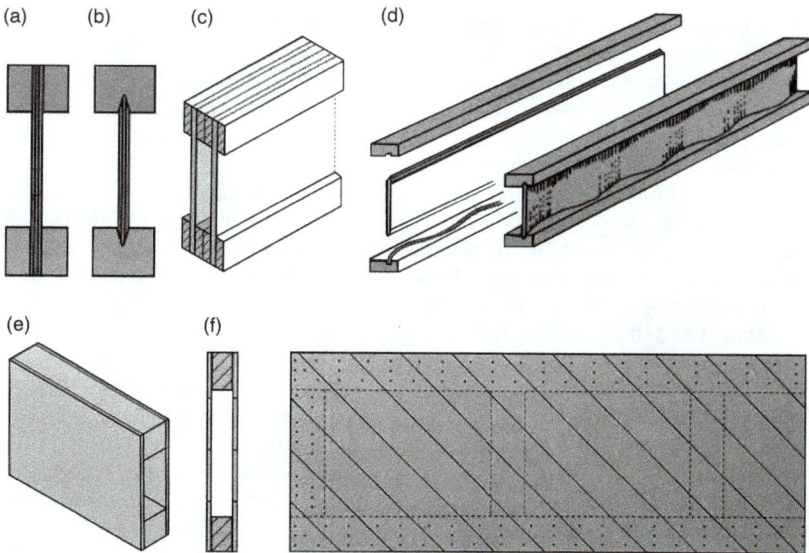

FIGURE 2.58 Examples of I-joist and box-beam design. An I-joist with a) flanges glued to a web of plywood (cross-section view), b) a V-shaped track between web and flanges to increase the gluing pressure, c) a twin-web beam and d) a sine-wave web to increase the gluing pressure. e) A box-beam with webs of plywood, OSB, hardboard etc., and flanges of sawn timber or LVL, and f) cross-section and side views of a box-beam with webs of nailed boards.

(Courtesy of D. Sandberg.)

or GLT flanges with, in most cases, OSB or plywood webs. The hollow cross-section of the beam also permits services to be run in the void inside the member. I-joists and box-beams are designed to carry heavy loads over long distances using less wood, a structurally efficient alternative to conventional sawn timber or small-dimension GLT. They are designed mainly for floor and roof constructions, but they can also be seen in other structural applications such as concrete forming.

REFERENCES

Anon. (2003). *Holz-Lexikon: Nachschlagewerk für die Holz- und Forstwirtschaft.* (4th Ed.), DRW-Verlag Weinbrenner, Germany, (1460 pp.).

Anon. (2018). Leichtbau im Bauwesen. *Ein Praxis-Leitfaden zur Entwicklung und Anwendung.* Leichtbau im Bauwesen, Stuttgart, Germany. mwk.baden-wuerttemberg.de, (27 pp.).

Burgert I. (2016). Werkstoffe I. *Teil 2 - Holz und Holzwerkstoffe.* ETH Zürich, Zurich, Switzerland, (35 pp.).

CEN (2001). EN 12369-1: Wood-based panels - Characteristic values for structural design - Part 1: OSB, particleboards and fibreboards. *The European Committee for Standardization (CEN)*, Brussels, Belgium.

CEN (2003a). EN 14251: Structural round timber - Test methods. *The European Committee for Standardization (CEN)*, Brussels, Belgium.

CEN (2003b). EN 302-2: Adhesives for load-bearing timber structures - Test methods - Part 2: Determination of resistance to delamination. *The European Committee for Standardization (CEN)*, Brussels, Belgium.

CEN (2004). EN 1995-1-1: 2004+A 1: Eurocode 5: Design of timber structures - Part 1-1: General Common rules and rules for buildings. *The European Committee for Standardization (CEN)*, Brussels, Belgium.

CEN (2006). EN 300: Oriented strand boards (OSB) - Definitions, classification and specifications. *The European Committee for Standardization (CEN)*, Brussels, Belgium.

CEN (2010). EN 312: Particleboards. *Specifications.* The European Committee for Standardization (CEN), Brussels, Belgium.

CEN (2013a). EN 14080: Timber structures - Glued laminated timber and glued solid timber – Requirements. *The European Committee for Standardization (CEN)*, Brussels, Belgium.

CEN (2013b). EN 335: Durability of wood and wood-based products - Use classes: definitions, application to solid wood and wood-based products. *The European Committee for Standardization (CEN)*, Brussels, Belgium.

CEN (2016a). EN 338: Structural timber - Strength classes. *The European Committee for Standardization (CEN)*, Brussels, Belgium.

CEN (2016b). EN 350: Durability of wood and wood-based products - Testing and classifying the durability to biological agents of wood and wood-based materials. *The European Committee for Standardization (CEN)*, Brussels, Belgium.

CEN (2019). EN 14081-1:2016+A1: Timber structures - Strength graded structural timber with rectangular cross section - Part 1: General requirements. *The European Committee for Standardization (CEN)*, Brussels, Belgium.

Cufar K., Liang E., Smith K.T., Ważny T., Wrobel S., Cherubini P., Schmitt U., Läänelaid A., Burgert I., Koch G., Pumijumnong N., Sander C., Seo J.-K., Sohar K., Yonenobu H. & Sass-Klaassen U. (2024). Dieter Eckstein's bibliography and legacy of connection to wood biology and tree-ring science. *Dendrochronologia*, 83(34), Article ID: 126165.

DIN (1979). DIN 7707-1: Resin impregnated and compressed laminated wood and insulating wood; test methods. *Deutsches Institut für Normung (DIN)*, Berlin, Germany.

Engelhardt O. (2013). Leichtbau 3.0: Material, Struktur, Energie. *Stahlbau*, 82(6), 421–427.

Esteves B., Nunes L. & Pereira H. (2009). Furfurylation of *Pinus pinaster* wood. In: Englund F., Hill C.A.S., Militz H & Segerholm B.K. (Eds.), *The Proceeding of the Fourth European Conference on Wood Modification*, Stockholm, Sweden. (pp. 415–418).

FAO, 2020. *Global Forest Resources Assessment 2020*. Main report. Rome: Food and Agriculture Organization of the United Nations (FAO), Rome, Italy, (12 pp.).

Fengel D. & Wegener G. (1984). *Wood: Chemistry, Ultrastructure, Reactions*. De Gruyter, Berlin, Germany, (613 pp.).

Gong M. (Ed.) (2021). *Engineered Wood Products for Construction*. IntechOpen. London, UK, (358 pp.).

Grosser D. (1977). *Die Hölzer Mitteleuropas: Ein mikrophotographischer Lehratlas*. Springer Verlag, Berlin, Germany, (217 pp.).

Ibach R.E. (2010). Specialty Treatments. Chapter 19. In: Ross R. (Ed.) *Wood Handbook. Wood as an Engineering Material. (Centenial Ed.)*, US Department of Agriculture, Forest Service, Forest Products Laboratory, Madison (WI), USA, (pp. 19.1–19.16).

International Thermowood Association (2021). *ThermoWood Handbook. International Thermowood Association*, Helsinki, Finland, (54 pp.).

ISO (2020) ISO 18775: Veneers — Terms and definitions, determination of physical characteristics and tolerances. *International Organization for Standardization (ISO)*, Geneva, Switzerland.

Jones D., Sandberg D., Goli G. & Todoro L. (2019). *Wood Modification in Europe: A state of the art about Processes, Products, Applications*. Firenze University Press, Florence, Italy, (113 pp.).

Kitek Kuzman M. & Sandberg D. (2023). Engineered wood products in contemporary architectural use – a concise overview. *Wood Material Science & Engineering*, 18(6), 1212–1215.

Krackler V., Keunecke D. & Niemz P. (2010). *Verarbeitung und Verwendungsmöglichkeiten von Laubholz und Laubholzresten*. ETH Zürich, IFB, Holzphysik, Zürich, Switzerland, (154 pp.).

Lanvermann C. (2014). *Sorption and swelling within growth rings of Norway spruce and implications on the macroscopic scale*. Doctoral Thesis, ETH Zürich, Zurich, Switzerland.

Mai C. & Militz H. (2023). Wood Modification. In: Niemz P., Teischinger A. & Sandberg D. (Eds.), *Springer Handbook of Wood Science and Technology*. Springer Nature, Cham, Switzerland, (pp. 873–910).

Mai C. & Zhang K. (2023). Wood Chemistry. In: Niemz P., Teischinger A. & Sandberg D. (Eds.), *Springer Handbook of Wood Science and Technology*. Springer Nature, Cham, Switzerland, (pp. 179–280).

Manuel W. (2012). Industrial Wood Preservation. *Specification and Practice*. UK Wood Protection Association, Castleford, UK, (57 pp.).

Navi P. & Sandberg D. (2012). Thermo-Hydro-Mechanical Processing of Wood. *Presses Polytechniques et Universitaires Romandes (EPFL Press)*, Lausanne, Switzerland. (376 pp.).

Niemz P. & Sonderegger W. (2021). Holzphysik. Physik des Holzes und der Holzwerkstoffe. (2nd Ed.) Carl Hanser Verlag, Leipzig, Germany, (580 pp.).

Niemz P., Teischinger A. & Sandberg D. (Eds.) (2023). *Springer Handbook of Wood Science and Technology*. Springer Nature, Cham, Switzerland, (XXV+2069 pp.).

Niemz P., Teischinger A. & Sandberg D. (Eds.) (2025). Wood Material and Processing Data: The Most Relevant Data, Tables, and Figures. Springer Nature, Cham, Switzerland, (XII+284 pp.).

Paulitsch M. (1989). *Moderne Holzwerkstoffe*. Springer, Berlin, Germany, (286 pp.).

Pitt W. (2010). *33 Farbtafeln Parkett*. Holzmann Buchverlag, Bobingen, Germany, (96 pp.).

Ross R. (2021). Wood Handbook. *Wood as an Engineering Material. US Department of Agriculture, Forest Service, Forest Products Laboratory*, Madison (WI), USA, (543 pp.).

Sandberg D., Haller P. & Navi P. (2013). Thermo-hydro-mechanical wood processing: A opportunity for future environmentally friendly wood products. *Wood Material Science & Engineering*, 6(1), 64–88.

Sandberg D., Kutnar A., Karlsson O. & Jones D. (2021). *Wood Modification Technologies. Principles, Sustainability, and the Need for Innovation.* CRC Press, Boca Raton (FL), USA, (432 pp.).

Schmitt U., Koch G., Hietz P. & Tholen D. (2023). Wood Biology. In: Niemz P., Teischinger A. & Sandberg D. (Eds.), *Springer Handbook of Wood Science and Technology.* Springer Nature, Cham, Switzerland, (pp. 41–148).

Schwarze F.W.M.R. (2008). *Diagnosis and Prognosis of the Development of Wood Decay in Urban Trees.* ENSPEC, Melbourne, Australia, (336 pp.).

Schweingruber F.H. (2007). *Wood Structure and Environment.* Springer, Berlin, Germany, (279 pp.).

Sell, J. (1997). *Eigenschaften und Kenngrössen von Holzarten.* (4th Ed.) Baufachverlag / Lignum, Dietikon–Zürich, Switzerland, (80 pp.).

Speck, T. & Burgert, I. (2011). Plant stems: Functional design. *Annual Review of Materials Research*, 41(1), 169–193.

Teischinger A., Isopp A. & Fellner J. (2023). *Holzarten Ansichten, Kennwerte und Beschreibungen.* Pro Holz Austria, Detail Verlag, Wien, Austria, (112 pp.).

Timofeeva G. (2017). *Elucidating the Drought Response of Scots pine (Pinus sylvestris L.) using stable Isotopes.* Doctoral Thesis, ETH Zürich, Zurich, Switzerland.

Tolf R.W. (1976). *Russian Rockefellers: The saga of the Nobel family and Russian oil.* Hoover Institution Press, Stanford (CA), USA, (15+169 pp.).

Trendelenburg R. & Mayer-Wegelin H. (1955). *Das Holz als Rohstoff.* (2nd Ed.), Carl Hanser Verlag, München, Germany, (15+141 pp.).

Wagenführ R. & Wagenführ A. (2022). *Holzatlas.* (7th Ed.), Carl Hanser Verlag, München, Germany, (928 pp.).

Wesche K. (1996). Band 1: Grundlagen. *Baustoffkenngrößen, Meß- und Prüftechnik, Statistik und Qualitätssicherung.* Bauverlag GmbH, Wiesbaden, Germany, (266 pp.).

Zobel B.J. & Sprauge J.R. (1998). *Juvenile Wood in Forest Trees.* Springer-Verlag, Berlin, Germany, (15+304 pp.).

3 Structure-Property Relationship of Wood and Wood-based Materials

P. Niemz and W. Sonderegger

3.1 INTRODUCTION

The properties of wood vary significantly, and there are two main reasons for this:

- Differences between species: There is considerable variation in density, which, in turn, affects all mechanical properties. For example, balsa, one of the lightest woods, has an average density of 130 kg/m³. In contrast, *Lignum vitæ* (wood of life) has a density of around 1230 kg/m³ – ten times greater.
- Environmental conditions, site factors, the position of the trees within the forest stand, and the soil type, altitude and local climate all influence the properties of the wood.

However, even within one wood species, the properties vary greatly. Therefore, wood used in construction is strength-graded visually or, more commonly, machine-graded. This enables homogenisation and classification into Strength Classes. The European EN 338 standard (CEN 2016a) specifies values for density, strength and modulus of elasticity. There is a separation into poplar and softwoods (Class C) and hardwoods (Class D).

Timber for interior use is often sorted and graded according to visual criteria (appearance grading) such as number of knots, colour and texture.

The variation of properties is divided into three main groups:

- within-tree variation,
- within-stand variation, *i.e.* variation within a group of trees with somewhat similar growing conditions, and
- inter-regional variations, *i.e.* due to regional influences such as altitude above sea level and climatic conditions (*e.g.*, long cold winters in northern regions or the mountains).

Figure 3.1 shows the variation in density for different wood species, with a clear overlap between the density ranges of individual species. This is why density and

DOI: 10.1201/9781003411994-3

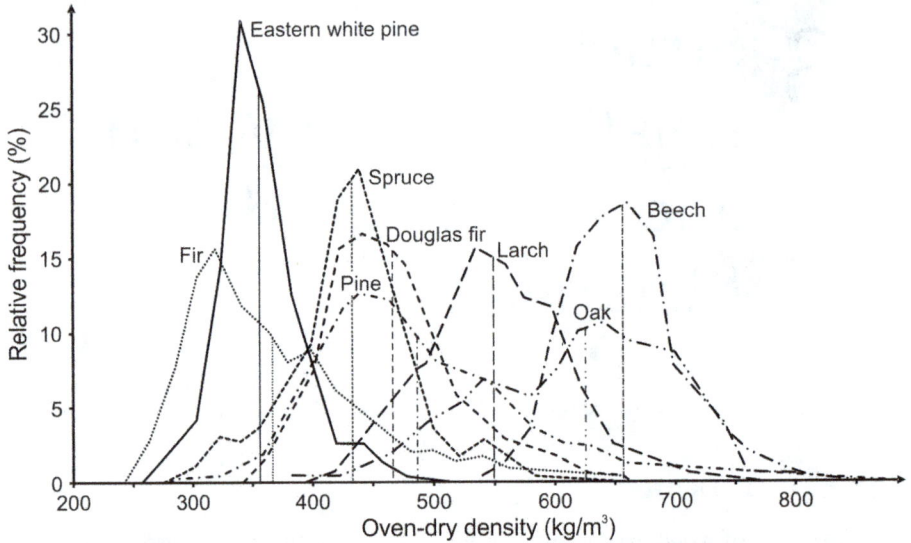

FIGURE 3.1 Frequency distribution of density of various species (Knigge and Schulz 1966). Beech – common beech, larch – European larch, oak – common oak, pine – Scots pine, spruce – Norway spruce.

(Courtesy of Carl Hanser Verlag.)

strength overlap for different wood species (Sell 1997, Wagenführ 2006). The phenomenon is very well known in practical use, for example, in glued-laminated timber production when sourcing timber from different regions. This is of great economic importance not only in glued-laminated timber production but also in other types of manufacturing, such as paper-pulp production.

Figure 3.2 shows the density distribution of Norway spruce from different regions in Central Europe, and Figure 3.3 shows the density distribution of radiata pine in the various areas of southern Chile. For Scots pine, the density is also significantly higher in, for example, south Sweden than in northern Sweden (Grönlund et al. 1991). The same applies to trees growing in the Alpine foothills and the Alps compared to lower heights (Niemz and Sonderegger 2021).

Strength correlates approximately linearly with density. This is also valid for other properties such as shrinkage/swelling, heat conduction and maximum water absorption. The region of origin of the wood is significant for its processing properties. This is particularly important when processing the trees into cross-laminated timber, glued-laminated timber and other wood-based materials. When gluing lamellae in different engineered wood products (EWPs), stresses will arise between the lamellae under moisture-level variation if there are large fluctuations in properties. Moisture variation can occur between lamellae due to sawmill drying and storage, water or other swelling agents in the adhesive, or fluctuations in the ambient climate. A certain degree of grading of the lamellae is therefore required. Permissible moisture variations between the lamellae are, thus, also subject to narrow limits. Suppose hardwood is glued or combined with softwoods, for example, to increase the

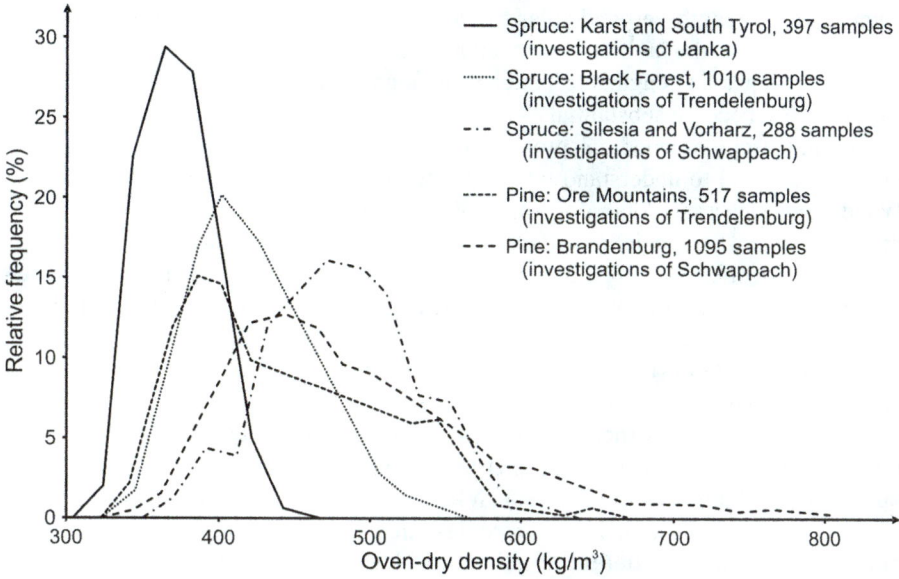

FIGURE 3.2 Oven-dry density of Norway spruce and Scots pine from different regions in Central Europe (Trendelenburg 1939).

(Courtesy of Carl Hanser Verlag.)

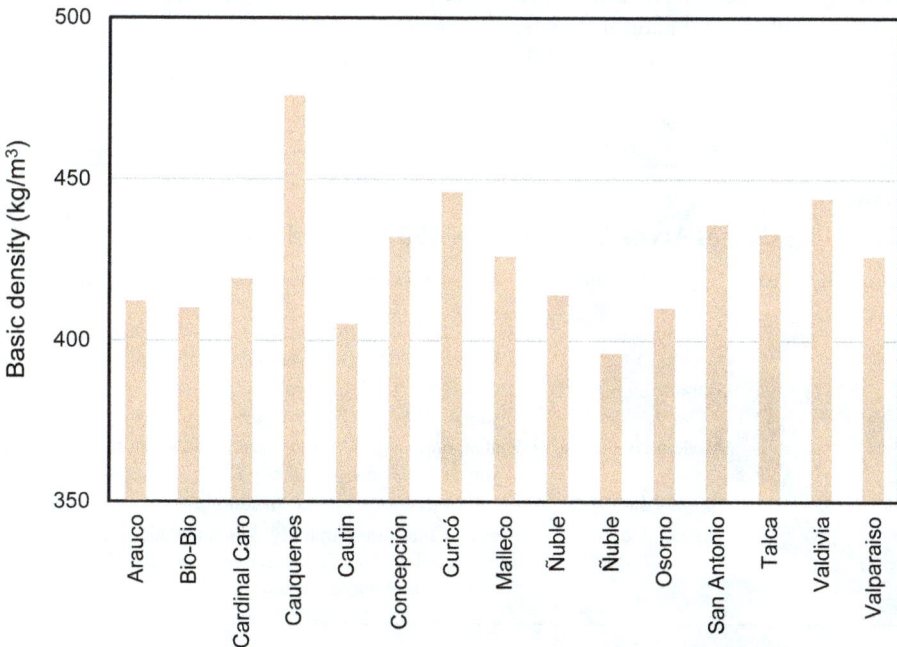

FIGURE 3.3 Basic density of radiata pine in different regions in Chile.

(Data from Walker 2006.)

compressive strength perpendicular to the length direction of a beam (grain direction). In that case, the problems are even more significant as hardwood and softwood properties (density, strength, etc.) often differ significantly, and the effect of moisture variation will become substantial.

The anisotropy of wood, *i.e.* property variation due to the different principal directions, is essential to understand to manage the proper use of wood in construction. Wood properties are similar along each of the three axes of the symmetry – longitudinal, radial and tangential (see Section 2.2.1), *i.e.* wood is an orthotropic material. Orthotropic properties gained importance with the use of wood in mechanical engineering, such as aircraft construction and vehicle construction. For further studies of orthotropic properties, the following literature is recommended: Hankinson (1921), Bodig and Jayne (1984), Niemz (1993), Pozgaj *et al.* (1997), Niemz and Sonderegger (2021), and Niemz *et al.* (2023).

Biological durability means how well a wood can resist degradation by fungus, insects, marine bores or mould, either as a living tree or as used as timber in various products (Sandberg *et al.* 2021). When growing in humid and warm environments such as the tropics or subtropics areas, trees often develop increased durability as a natural defence against fungi and insects by storing substances (extractives). These are particularly stored in the heartwood. However, the effect is also present in other non-tropical species, such as common oak, robinia and sweet chestnut.

The EN 350 standard (CEN 2016b) regulates Durability Classes for resistance to fungi and insects (Table 3.1). Tropical woods are particularly durable, but their use is severely restricted today. The durability of plantation timber is often reduced compared to timber from natural forests.

Chapter 10 describes the durability of timber in detail.

TABLE 3.1
Durability Classes According to the EN 350 Standard (CEN 2016b)

Durability Class	Description	Wood Species (Heartwood Only, Except Class 5)
1	Very durable	Asian teak (teak from plantations often only reaches Durability Class 3), bilinga, cumarú, maobi, massaranduba
2	Durable	Bankirai, bongossi, merbau, oak spp., sweet chestnut, wenge
3	Moderately durable	Bintangor, Douglas fir, European larch, kasai, mahogany spp. and high-density pine spp.
4	Slightly durable	Fir, Norway spruce, yellow meranti
5	Not durable	Less than one-year durability, which includes sapwood of all timbers, and lime, birch spp., common beech, common ash, white meranti

3.2 SOLID WOOD

In the last decades, numerous studies have been carried out in the sub-microscopic range down to the molecular range to determine the structure-property relationships. For example, Persson (2000) and Harrington (2002) developed multiscale softwood models. For hardwoods, which have a more complicated anatomical structure, there are only a few studies, for example Lavani *et al.* (2023). There are also approaches to studying the atomic structure of wood, for example Kulazinski (2015). Detailed descriptions of the anatomical and chemical composition of wood can be found in textbooks such as Bosshard (1982), Fengel and Wegener (1984), Wagenführ (1999) and Niemz *et al.* (2023).

Wood is a porous composite material. Depending on the bulk density of the wood, the proportion of pores is between 7% (*Lignum vitæ*) and 92% (balsa); on average, the porosity is approximately 60%. The wood substance can be regarded as a composite of long cellulose molecules as a fibre structure with hemicelluloses and lignin as a matrix. Figure 3.4 shows a structural model of wood as a macromolecular material at various structural levels, from molecular to a tree (Harrington 2002). A distinction is generally made between macro-, micro- and sub-micro-structural levels, each with its characteristics. On a macrostructural level, wood is determined by a cylindrical-conical shaped stem with the characteristic circular growth-ring pattern in the stem cross-section and a fibre orientation close to parallel to the length axis of the tree. This results in the anisotropic behaviour typical for wood in the three principal directions, longitudinal, radial and tangential. This also distinguishes between three planes or sections of the stem: the cross-section transverse to the tree axis, the radial section and the tangential sections (*cf.* Section 2.2.1). A distinction must also be made between sapwood and heartwood. Heartwood often contains a higher extractive content than sapwood, affecting sorption properties (Popper *et al.* 2006).

In the radial direction, the wood rays strongly influence the mechanical properties and shrinkage/swelling behaviour (Burgert 2000). Effects on the physical and mechanical properties occur in the sub-microscopic range mainly through variation of the microfibril orientation of the cell-wall S2 layer (Butterfield 1997) and the degree of lignification of the cell wall, for example, differences between juvenile and adult wood, reaction wood and normal wood.

The main structural features at the macroscopic level that influence the wood properties are:

- sapwood/heartwood and juvenile/mature wood (influenced by tree age),
- the fibre orientation (grain orientation),
- the fibre-to-load angle, *i.e.* the angle between (1) grain orientation in LR and LT sections or (2) the growth-ring orientation in RT direction, and the load,
- the growth-ring width and the proportion of latewood, and
- the presence of juvenile and reaction wood.

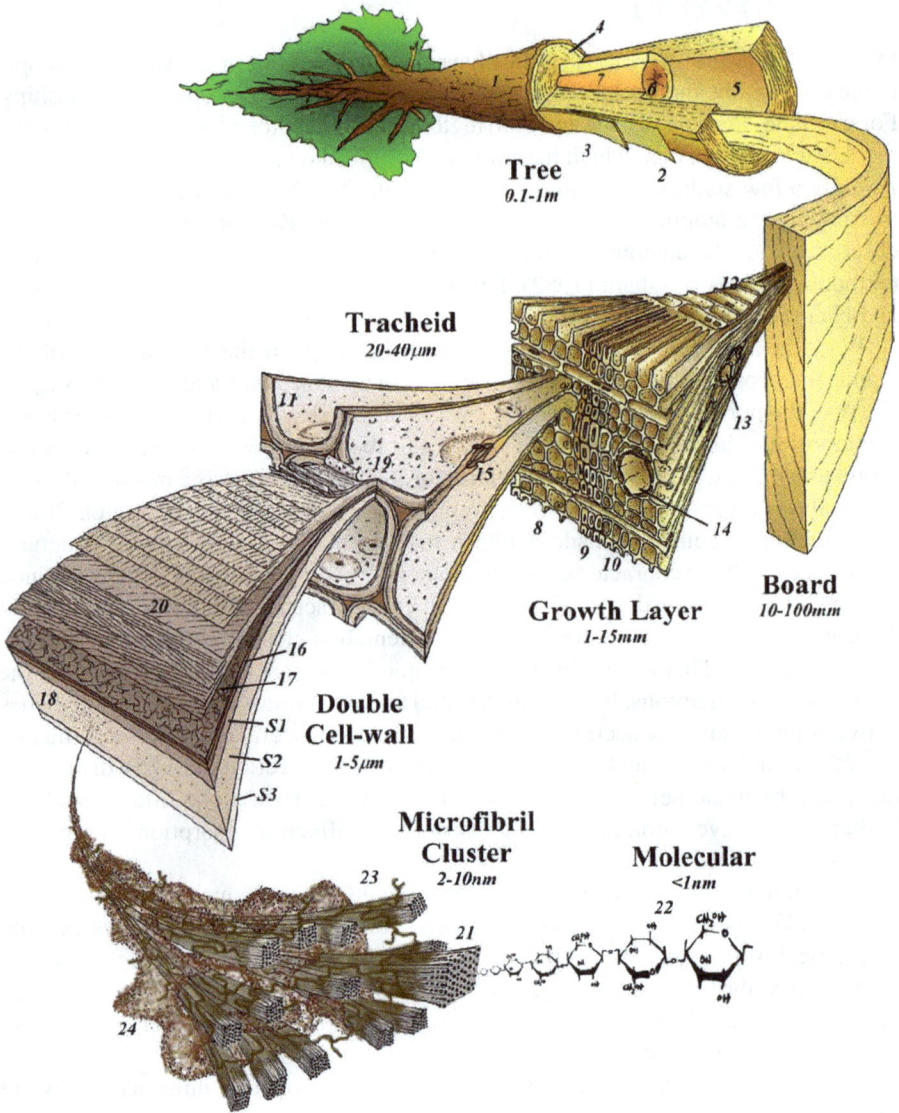

FIGURE 3.4   A structural model from chemical structure to tree level.

(Harrington 2002, courtesy of J. Harrington.)

At the microscopic level, the wood properties are affected by:

- the tissue proportions (proportion of vessels, fibres, rays, longitudinal parenchyma cells),
- the fibre lengths and fibre-wall thicknesses,
- the wood rays, especially on the mechanical properties and shrinkage/swelling, and
- the fibre orientation.

At the sub-microscopic level, the wood properties are affected by:

- the thickness of the cell-wall layers,
- the fibril orientation in the cell-wall S2 layer, and
- the degree of lignification of the cell wall.

Figure 3.5 shows the main structural parameters influencing the wood properties (Niemz and Sonderegger 2021, Arriaga *et al.* 2023)

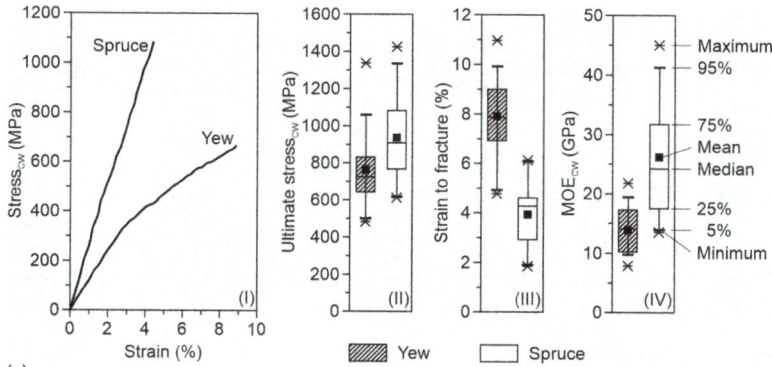

(a)

macroscopic:

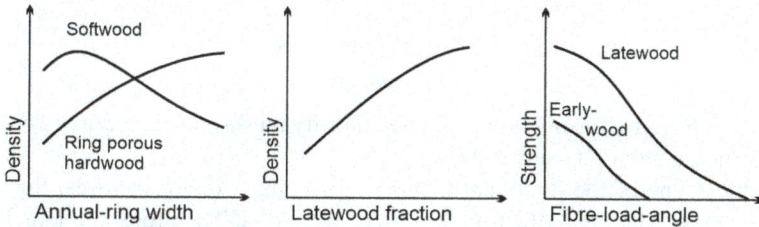

microscopic: submicroscopic:

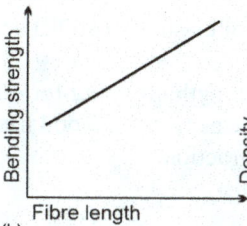 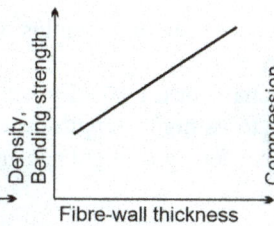 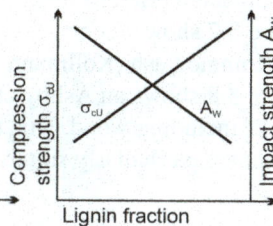

(b)

FIGURE 3.5 Schematically essential factors influencing the wood properties at different structural levels: (a) longitudinal tension of Norway spruce and yew fibres (the stress is based on the cell-wall cross-sectional area, cw). From the left: (I) the stress-strain diagram, (II) the ultimate stress, (III) the ultimate strain and (IV) the modulus of elasticity. The microfibril angles are for spruce 0–5° and yew 15–20° (Keunecke 2008). (b) Properties on macro- and microstructural levels, and _

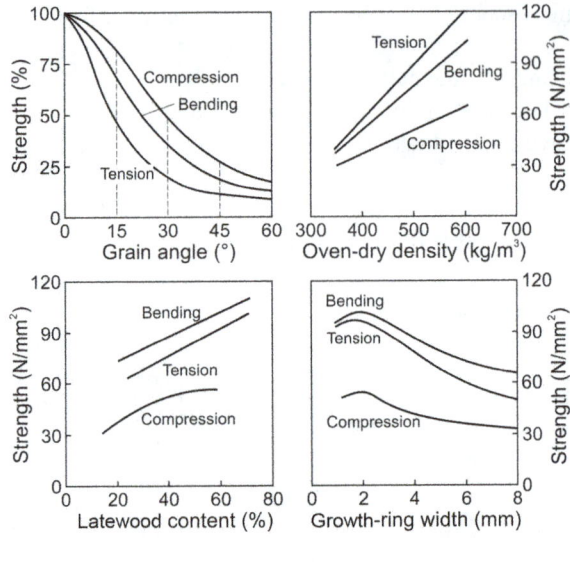

(c)

FIGURE 3.5 (*Continued*) (c) the strength as a function of latewood content, growth-ring orientation, grain angle, and oven-dry density 103 wood species from the literature.

(Courtesy of Carl Hanser Verlag.)

Figures 3.6 shows the influence of wood density on shrinkage, thermal conductivity and bending strength (Sell 1997).

Density is one of the dominant factors influencing wood properties. Significant variations in density also affect the gluing of the wood. Therefore, the lamellae are often sorted into density classes before gluing, for example, in the production of glued-laminated timber.

Figure 3.7 shows the relation between the modulus of elasticity (MOE) and density for common ash (Kollmann 1941).

Figure 3.8 shows an Ashby chart for the MOE-to-strength grouping for different materials, including wood. Due to its high strength-to-density ratio, wood has properties that are particularly suitable for lightweight construction.

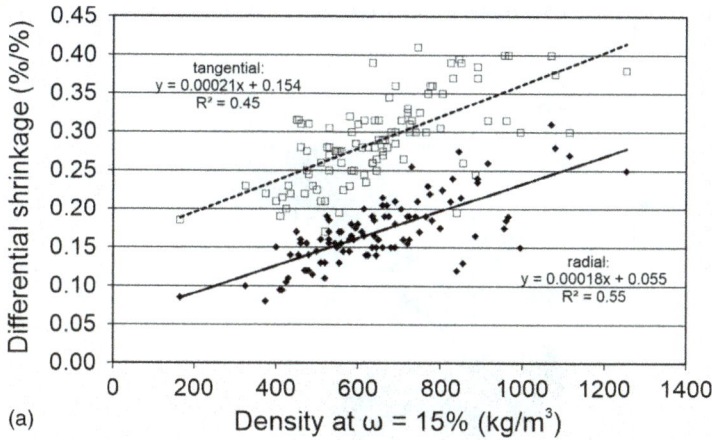

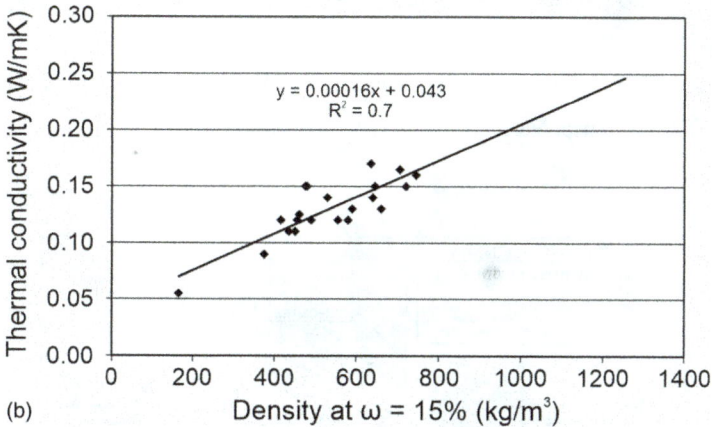

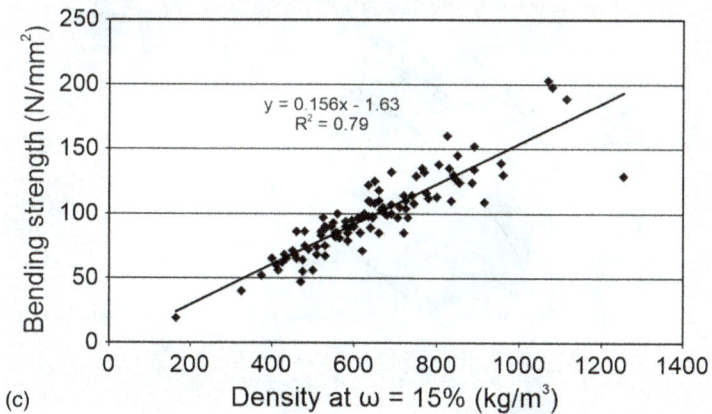

FIGURE 3.6 Influence of bulk density on (a) differential shrinkage, according to data from Sell (Sell 1997) for 103 different timber species, (b) thermal conductivity and c) density for selected wood species. ω – moisture content.

(Courtesy of Carl Hanser Verlag.)

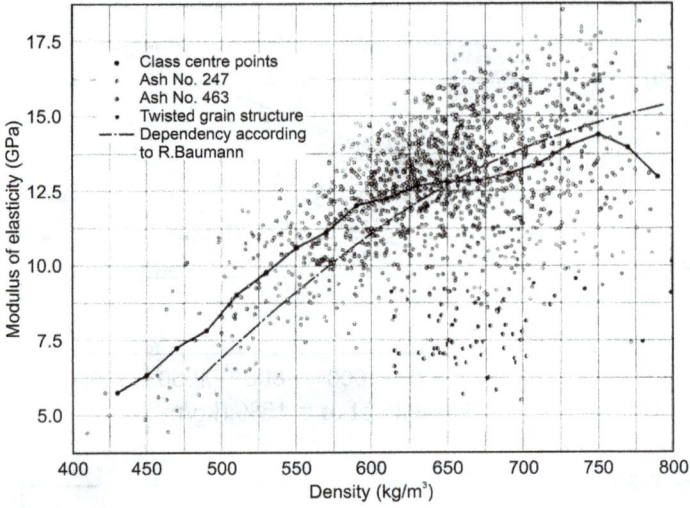

FIGURE 3.7 Modulus of elasticity for sawn timber of European ash as a function of density. **(Courtesy of Carl Hanser Verlag.)**

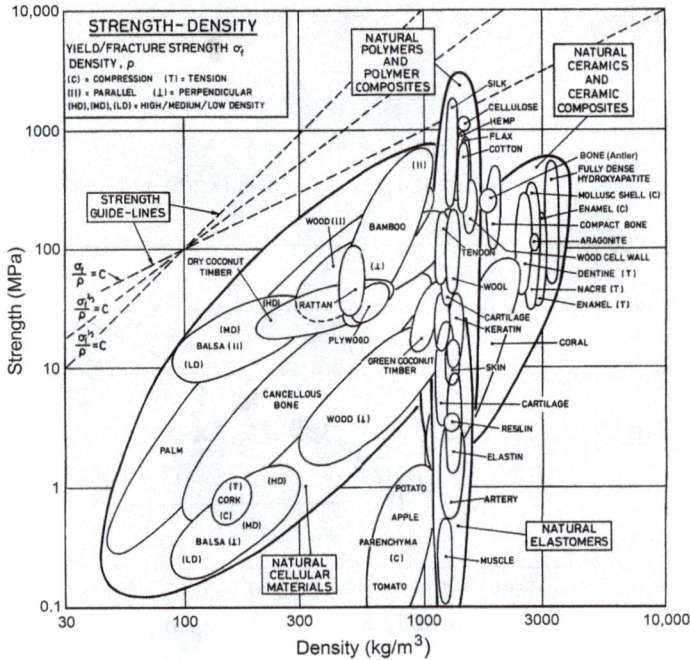

FIGURE 3.8 Material-property chart for engineered wood showing modulus of strength plotted against density (Wegst and Ashby 2004). Guidelines identify structurally efficient materials that are light and strong.

3.3 WOOD-BASED MATERIALS

The structure influences the properties of engineered wood products (EWPs). Smaller pieces of wood and lamination are used to improve the homogenisation of the material properties (*cf.* Section 2.4).

3.3.1 EWPs Made of Sawn Timber

In the case of EWPs made of sawn timber, such as cross-laminated timber (CLT) and glued-laminated timber (GLT), some factors are essential for its properties:

- the mechanical properties (MOE, MOR) of the lamellae and the layers in the case of CLT (Gereke 2009),
- the growth-ring orientation in the lamellae, which results in distortion (Gereke 2009),
- distribution from lamella properties over the EWP (higher stiffness or strength in the outer layers can be used for higher strength and stiffness of the whole beam or panel (Czaderski *et al.* 2007)),
- thickness of the lamellae and layers,
- bond-line properties such as adhesion, elastic properties, and strength (Sonderegger and Niemz 2009) and
- surface coating to reduce moisture exchange with ambient air, which reduces internal stresses and cracking.

3.3.2 EWPs Made of Veneer

Essential parameters for EWPs made of veneer are (Figure 3.9):

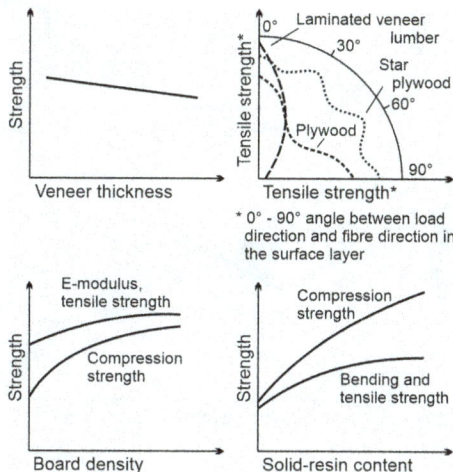

FIGURE 3.9 Important parameters on the strength of veneer-based materials.

(Courtesy of Carl Hanser Verlag.)

- veneer thickness,
- densification of the veneers increases the strength of the EWP, but when water-based adhesives are used, the degree of compression of the veneer may decrease due to set-recovery,
- veneer-layer orientation, and
- the adhesive solid content influences the strength of the EWP.

In laminated veneer lumber (LVL), some layers are sometimes oriented perpendicular to the length direction to improve the transverse strength and reduce shrinkage/swelling.

3.3.3 EWPs MADE OF PARTICLES AND STRANDS

For EWPs made of wood particles and strands, some essential parameters which influence the properties are shown in Figure 3.10, in particular:

- particle geometry (length and degree of slenderness, *i.e.* the ratio of length-to-thickness of the particle),
- the average density of the panel and the density profile through its thickness. The density profile gives a sandwich effect because of the higher density of the surface regions,
- the adhesive content and
- the proportion of hydrophobic agents, especially those types that reduce the thickness swelling.

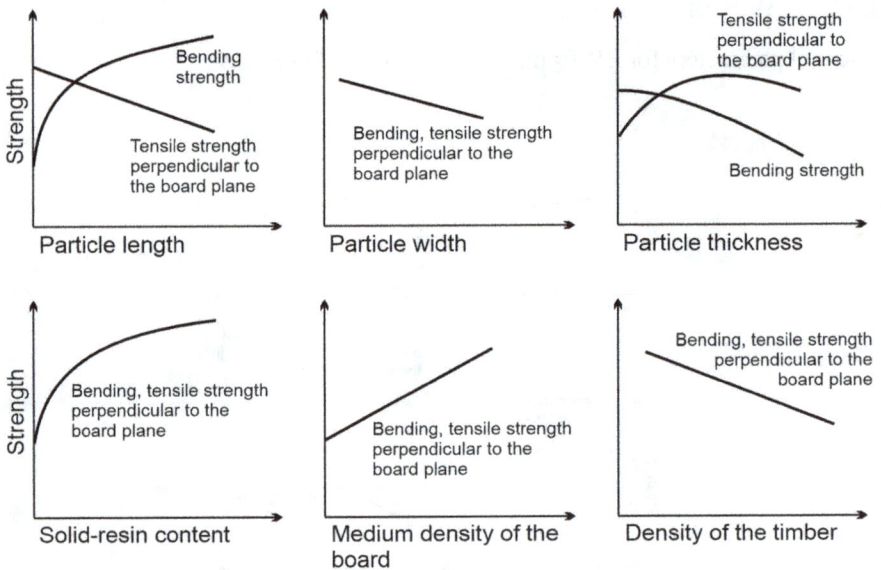

FIGURE 3.10   Important parameters for particleboard.

(Courtesy of Carl Hanser Verlag.)

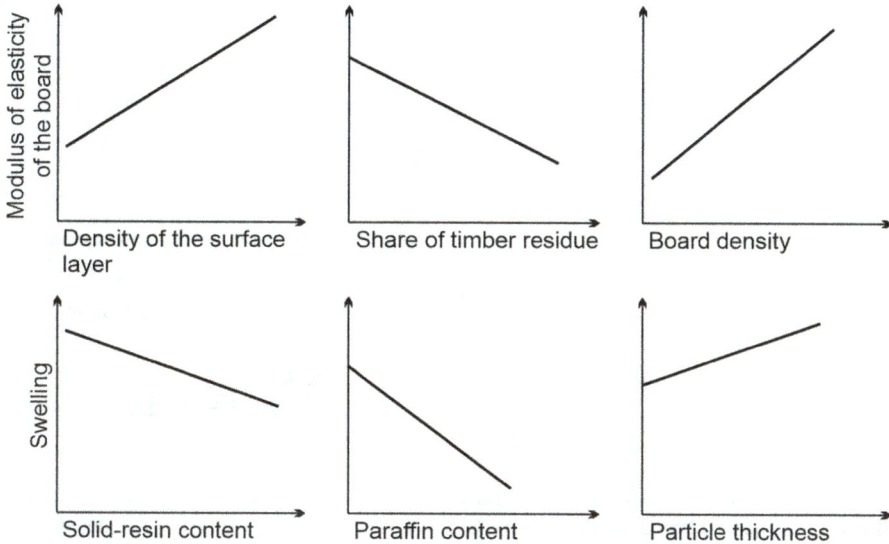

FIGURE 3.10   (*Continued*)

The size of the particles and strands influences the properties (Figure 3.10). Strands with a length of approximately 70 mm for oriented strand boards (OSB) and up to 300 mm for laminated strand lumber (LSL) result in a significantly higher strength for OSB and LSL in the strand direction than particleboards with small particles. OSB and LSL are often used in wall construction in the building industry.

3.4   PROPERTY VARIATIONS IN SOLID WOOD AND WOOD-BASED MATERIALS

Figure 3.11 schematically shows the variation in properties for some EWPs (Sandberg *et al.* 2018). The grading of sawn timber and raw materials in the production of wood-based panel materials can significantly limit the variability of the properties.

The properties of sawn timber vary considerably more than those of wood-based materials. The coefficients of variation for sawn timber in Table 3.2 serve as a rough guide. The coefficient of variation (CV) is the ratio of the standard deviation to the mean. The higher the CV, the greater the property dispersion around the mean.

Table 3.3 shows the variation coefficient used as guidelines for various wood-based materials.

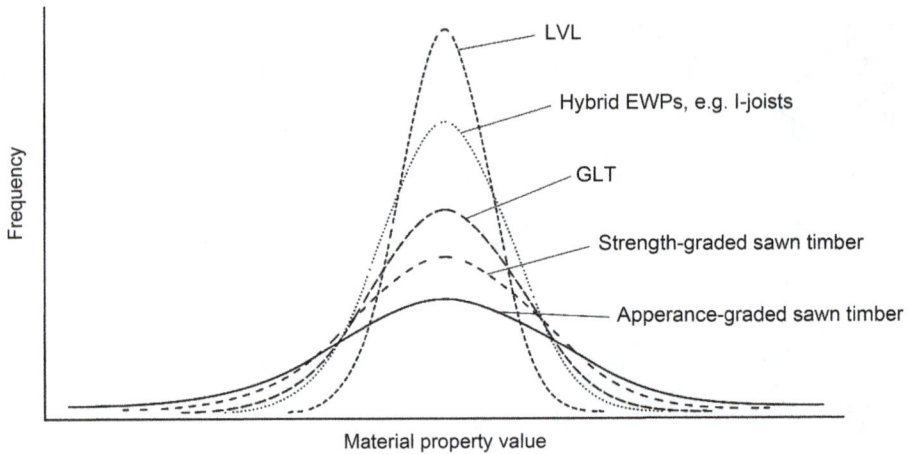

FIGURE 3.11   Frequency distribution and variability in mechanical properties of wood and different engineered wood products (EWPs). LVL – laminated veneer lumber and GLT – glued-laminated timber.

(Sandberg et al. 2018, courtesy of D. Sandberg.)

TABLE 3.2
Coefficient of Variation (CV) Used as Guidelines for Sawn Timber (Anon. 2003)

Sawn Timber	CV (%)
Density	10
Bending strength	16
Modulus of elasticity	22
Impact strength	30

TABLE 3.3
Coefficient of Variation (CV) That Can Be Used as Guidelines for Wood and Wood-based Materials

Type of Material	CV (%)
Clear (defect-free) wood	20
Visually graded sawn timber	40
Plywood	18
Particleboard (OSB)	12
Medium-density fibreboard (MDF)	8

REFERENCES

Anon. (2003). *Holz-Lexikon: Nachschlagewerk für die Holz- und Forstwirtschaft*. (4th Ed.), DRW-Verlag Weinbrenner, Germany, (1460 pp.).

Arriaga F., Wang X., Íñiguez-González G., Llana D., Esteban M. & Niemz P. (2023). Mechanical properties of wood: A review. *Forest*, 14(1202), 2–61.

Bodig J. & Jayne B.A. (1984). *Mechanics of Wood and Wood Composites*. Krieger Publishing Company, Malabar (FL), USA, (712 pp.).

Bosshard H.H. (1982–1984). *Holzkunde I-III*. (2nd Ed.), Birkhäuser, Basel, Germany, (224+312+286 pp.).

Burgert I. (2000). *Die mechanische Bedeutung der Holzstrahlen im lebenden Baum*. Doctoral Thesis, Universität Hamburg, Hamburg, Germany.

Butterfield B.G. (Ed.) (1997). Proceedings of International Workshop on the Significance of Microfibril Angle to Wood Quality. *International Association of Wood Anatomists, International Union of Forestry Research Organizations*. University of Canterbury, Westport, New Zealand, (410 pp.).

CEN (2016a). EN 338: Structural timber - Strength classes. *The European Committee for Standardization (CEN)*, Brussels, Belgium.

CEN (2016b). *EN 350: Durability of wood and wood-based products - Testing and classification of the durability to biological agents of wood and wood-based materials*. European Committee for Standardization, Brussels, Belgium.

Czaderski C., Steiger R., Howald M., Olia S., Gülzow A. & Niemz P. (2007). Versuche und Berechnungen an Allseitig Gelagerten 3-Schichtigen Massivholzplatten. *Holz als Roh- und Werkstoff*, 65(5), 383–402.

Fengel D. & Wegener G. (1984). *Wood: Chemistry, Ultrastructure, Reactions*. De Gruyter, Berlin, Germany, (613 pp.).

Gereke T. (2009). *Moisture-induced stresses in cross-laminated wood panels*. Doctoral Thesis, ETH Zürich, Zurich, Switzerland.

Grönlund A., Grönlund U. & Hagmann O. (1991). *Nordkalottenfura. [Scots Pine from Northern Sweden.]* Luleå University of Technology, Skellefteå, Sweden, (45 pp.).

Hankinson R.L. (1921). Investigation of crushing strength of spruce at varying angles of grain. *Air Service Information Circular*, 3(259), 3–15.

Harrington J. (2002). *Hierarchical modelling of softwood hygro-elastic properties*. Christchurch: Doctoral Thesis, Univeersity of Canterbury, Christchurch, New Zealand.

Keunecke D. (2008). *Elasto-mechanical characterisation of yew and spruce wood with regard to structure-property relationships*. Doctoral Thesis, ETH Zürich, Zurich, Switzerland.

Knigge W. & Schulz H. (1966). *Grundriss der Forstbenutzung*. Parey, Hamburg, Germany, (584 pp.).

Kollmann F. (1941). *Die Esche und ihr Holz*. Julius Springer, Berlin, Germany, (12+147 pp.).

Kulazinski K. (2015). *Physical and mechanical aspects of moisture adsorption in wood biopolymers investigated with atomistic simulations*. Doctoral Thesis, ETH Zürich, Zurich, Switzerland.

Lavani L., Suiker A., Crivellaro A. & Bosco E. (2023). A 3D multiscale hygromechanical model of oak wood. *Wood Science and Technology*, 57(6), 1215–1256.

Niemz P. (1993). *Physik des Holzes und der Holzwerkstoffe*. DRW-Verlag Weinbrenner GmbH & Co., Leinfelden-Echterdingen, Germany, (243 pp.).

Niemz P. & Sonderegger W. (2021). Holzphysik. *Physik des Holzes und der Holzwerkstoffe*. (2nd Ed.), Hanser Verlag, München, Germany, (580 pp.).

Niemz P., Teischinger A. & Sandberg D. (Eds.) (2023). *Springer Handbook of Wood Science and Technology*. Springer Nature, Cham, Switzerland, (XXV+2069 pp.).

Persson K. (2000). *Micromechanical modelling of wood and fibre properties*. Doctoral Thesis, Lund University, Lund, Sweden.

Popper R., Niemz P. & Torres M. (2006). Einfluss des Extraktstoffanteils ausgewählter fremd-ländischer Holzarten auf deren Gleichgewichtsfeuchte. *Holz als Roh- und Werkstoff*, 64(6), 491–496.

Pozgaj J., Chonavec D., Kurjatko S. & Babiak M. (1997). *Struktura a Vlasnosti Dreva. [Structure and Properties of Wood.]* Priroda, Bratislava, Zlovakia, (485 pp.).

Sandberg, D., Kitek Kuzman, M. & Gaff, M. (2018). *Kompozitní výrobky na bázi dřeva - Dřevo jako kompozitní a konstrukční materiál. [Engineered Wood Products: Wood as an Engineering and Architectural Material.]* Czech University of Life Sciences Prague (CULS), Faculty of Forestry and Wood Sciences, Prague, Czeck Republic, (185 pp.).

Sandberg D., Kutnar A., Karlsson O. & Jones D. (2021). *Wood Modification. Principles, Sustainability, and the Need for Innovation.* CRC Press, Boca Raton (FL), USA, (432 pp.).

Sell J. (1997). *Eigenschaften und Kenngrössen von Holzarten* (4th Ed.), Baufachverlag, Dietikon, Switzerland, (140 pp.).

Sonderegger W. & Niemz P. (2009). Thermal conductivity and water vapour transmission properties of wood-based materials. *European Journal of Wood and Wood Products*, 67(3), 313–321.

Trendelenburg R. (1939). *Das Holz als Rohstoff.* J.F. Lehmanns Verlag, München, Germany, (541 pp.).

Wagenführ R. (1999). *Anatomie des Holzes* (5th Ed.), DRW-Verlag, Leinfelden-Echterdingen, Germany, (188 pp.).

Wagenführ R. (2006). *Holzatlas.* (6th Ed.), Fachbuchverlag Leipzig im Carl Hanser Verlag, München, Germany, (816 pp.).

Walker J.C. (2006). *Primary Wood Processing: Principles and Practice* (2nd Ed.), Springer, Dordrecht, The Netherlands, (10+596 pp.).

Wegst U. & Ashby M. (2004). The mechanical efficiency of natural materials. *Philosophical Magazine*, 84(21), 2167–2181.

4 Overview of the Physical-Mechanical Properties of Wood and Wood-based Materials

P. Niemz and W. Sonderegger

4.1 INTRODUCTION

In line with classical physics, the properties of wood are classified into physical and mechanical categories. The mechanical behaviour pertains to these properties within the linear elastic range, where deformation is reversible upon unloading. Strength refers to the failure behaviour, describing the properties of wood in the inelastic range. It represents the stress at which the wood sustains damage or fracture (maximum load).

All of these properties are heavily influenced by environmental factors, particularly the relative humidity and temperature of the surrounding air, which can alter the moisture content of the wood. Additionally, the wood's structure plays a significant role, and in the case of elements exposed to challenging climates, coatings can mitigate moisture fluctuations within the material.

Corrosion behaviour (ageing) and the resistance of wood to fungi and insects are discussed in Chapter 10.

4.2 PHYSICAL PROPERTIES

The most essential physical properties are:

- density,
- shrinkage, swelling and sorption (wood-water relations),
- thermal behaviour, for example, thermal insulation and heat conduction.
- acoustic properties such as sound propagation, sound attenuation, and sound insulation,
- electrical properties such as electrical resistance and electrostatic charge,
- optical properties of the wood (colour, colour change due to exposure to light and weathering) and behaviour towards electromagnetic waves,
- friction properties, and
- natural durability in dry climates and resistance to aggressive media such as salts and acids.

DOI: 10.1201/9781003411994-4

4.3 MECHANICAL PROPERTIES

The mechanical properties are divided into:

- elastic properties, which describe the behaviour of the wood under mechanical stress in the linear-elastic region (including modulus of elasticity, shear modulus, and Poisson's ratio). The elastic properties are essential for the deformation calculations, for example, for deflection of flooring, roofs, and shelves under load)
- strength is the stress at which the wood is damaged or fractured in the inelastic range, and
- the hardness of the wood and its resistance to wear and tear are significant for parquet and other types of flooring and beams loaded in compression. Heavy wear occurs in areas such as high-traffic stairs and train floors and when bulk materials are transported through wooden pipes, which are rarely used today. The effect of abrasion on tabletops is a well-known example.

A timber structure is affected by various types of loads (compression, dynamic, static, and tension loads), the duration of the load (short-term or long-term load), and the load speed, such as sudden loads on baseball bats, ice hockey sticks, or wooden crash barriers.

The timber used in a building is exposed to the following:

- changes in humidity in the building (Figure 4.1),
- the mechanical loads (Figure 4.2),
- long-term deformation, *i.e.* creep, and relaxation (Figure 4.3), and
- ageing (Figures 4.5 to 4.8).

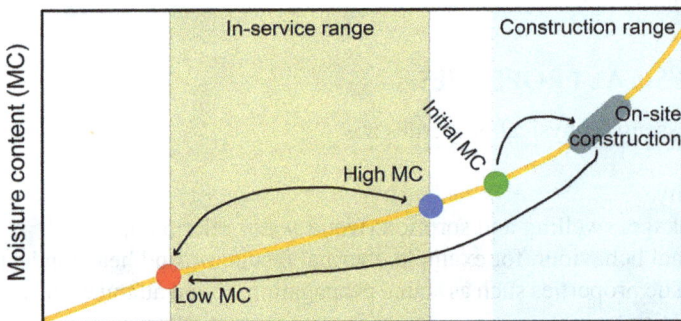

FIGURE 4.1 Principal moisture content variation in timber building under construction and in-service phases.

(Data from McLain and Steimle 2019, courtesy of D. Sandberg.)

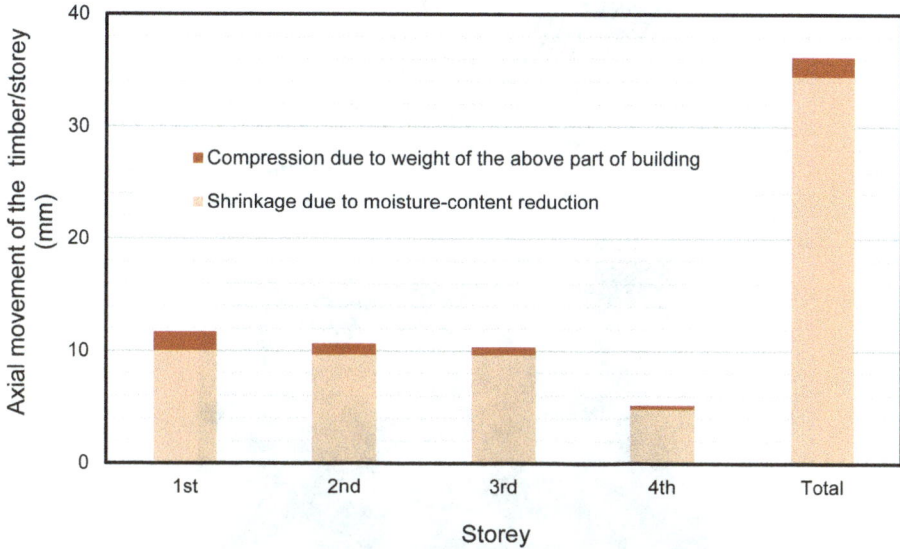

FIGURE 4.2   Estimated movement in a timber multi-storey building due to changes in moisture content and vertical compression loads.

(Data from Gong 2021, courtesy of D. Sandberg.)

FIGURE 4.3   Deformation of a barn roof in southern Sweden caused by creep under sustained loading.

(Courtesy of P. Niemz.)

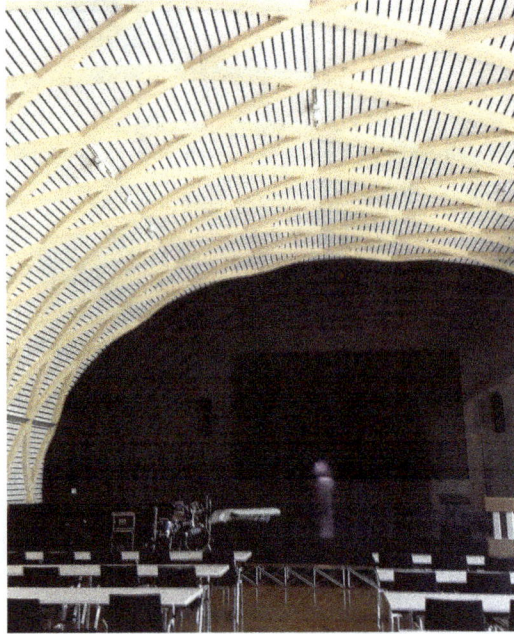

FIGURE 4.4   Wood with transparent coating and UV protection after five years of use at the IHD office in Dresden.

(Courtesy of P. Niemz.)

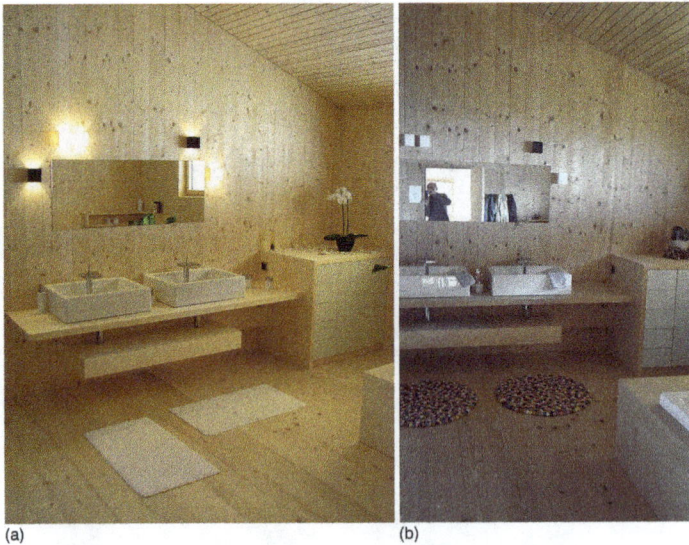

(a)   (b)

FIGURE 4.5   Natural colours change of untreated Norway spruce wood due to ageing: a) newly machined wood surfaces and b) after approximately five years.

(H. Nägeli, Gais, Switzerland, courtesy of P. Niemz.)

FIGURE 4.6 Colour change and water stains of untreated spruce glued-laminated timber. A multi-storey office building after around five years in use.

(Courtesy of P. Niemz.)

(a)   (b)

FIGURE 4.7 Façade made of untreated European larch timber: a) newly installed façade and b) after approximately three years of weathering (start of intense discolouration), Embrach, Switzerland.

(Courtesy of P. Niemz.)

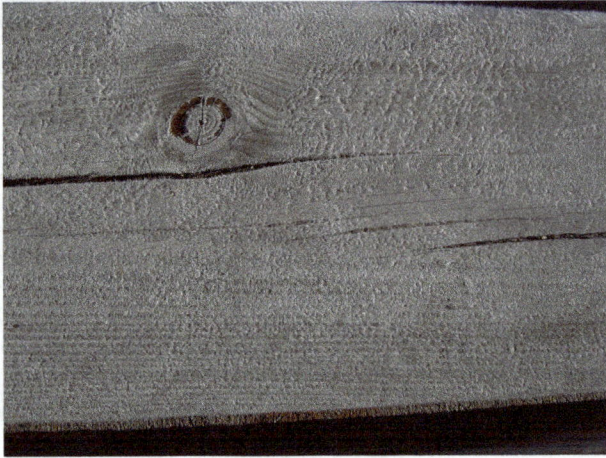

FIGURE 4.8   Wood surface after long-term exposure to salt water.
(Courtesy of P. Niemz.)

4.4   FACTORS INFLUENCING WOOD PROPERTIES

Properties of wood and wood-based materials are affected by:

- the structure of the wood itself, *i.e.* species, wood features, and fibre length and orientation, the type of wood-based material and its structure (density, orientation of the layers or particles, adhesive content), and
- the climate – particularly the relative humidity and temperature – as well as climate changes. In practical use, fluctuations in humidity may induce stresses in the material and typically have a far more significant effect than temperature variations.

Timber and wood-based materials (EWPs) in a building will exhibit:

- Signs of ageing due to UV radiation and locally increased temperature, for example, in conservatories (Li 2022).
- Wood itself ages to a low extent and not in an arid climate. Depending on the type of wood, colour changes and minor chemical changes may occur (Kránitz 2014, Kránitz *et al.* 2016, Niemz and Sonderegger 2021).
- Surface treatment can significantly reduce colour changes (such as darkening or lightening of untreated wood) and help mitigate moisture fluctuations that influence cracking or creep behaviour.
- The abrasion resistance (*e.g.*, in parquet) can sometimes be increased. Coatings (*e.g.*, UV blockers in paints) can also significantly reduce colour changes due to darkening (*e.g.*, spruce, larch) or lightening, for example, walnut, teak, thermally modified timber (Figure 4.4).

- Surface-treated wood is easier to clean, becoming increasingly important in timber construction. The visual appeal of wood is often a key selection criterion, and the 'furniture-like' character of buildings made from hardwood, for example, is gaining prominence. Untreated surfaces, commonly found in industrial buildings, can lead to problems in residential or office buildings after a certain period of use, with renovation being exceptionally costly.

Ageing can cause darkening (yellowing), as seen in maple, or lightening of the wood's colour (Pitt 2010). Figure 4.5 shows the colour change of untreated Norway spruce after five years of indoor use. Applying a transparent coating with UV blockers can significantly slow this process. Without such a coating, the darkening will persist over time (Figure 4.4).

Figure 4.6 shows the waiting area of a tram stop, where untreated glued-laminated timber was used. After a few years, the wood became heavily soiled and could only be restored through extensive sanding. This serves as a poor example and is not the best advertisement for the use of wood.

Renovations are highly complex and cost-intensive, a fact that customers are sometimes unaware of. In contrast, non-coated components can be easily reused or converted into energy. It is advisable to inform customers or users in advance about the ageing-related changes in the appearance of wood products. This is effectively done in the case of flooring, where ageing is typically visually documented (Pitt 2010).

Long-term exposure to salt water or chemicals can lead to degradation processes, resulting in a dissolved wooden surface. However, wood is generally quite resistant to chemicals, which is why it is often used for road salt storage (Figures 4.8 and 4.9).

Wood properties change very little when used in a dry indoor climate. The content of extractives is slightly reduced, but overall, the material remains relatively stable (Kránitz 2014).

FIGURE 4.9 Changes of wood surfaces by chemicals and high humidity in the Museum of Wool. **(Courtesy of P. Niemz.)**

(a)   (b)

FIGURE 4.10 Example of fracture in timber use: (a) tensile failure of a finger-jointed ash wood lamella due to the deviation in fibre orientation from the length direction of the sawn timber, which results in a substantial reduction in tensile strength (Photo: P. Niemz, Zurich, Switzerland), and (b) fracture of a wooden structure.

(P. J. Gustafsson, Lund University, Sweden, courtesy of P. Niemz.)

Wood is a viscoelastic material that exhibits creep and relaxation when subjected to loading and unloading. Figure 4.10 shows the fracture pattern of a broken finger-jointed connection of ash caused by a deviation in fibre orientation and the failure of the entire wooden structure at the connection point.

REFERENCES

Gong M. (2021). *Wood and Engineered Wood Products: Stress and Deformations*. IntechOpen, New Brunswick, Canada, (358 pp.).

Kránitz K. (2014). *Effect of Natural Ageing on Wood*. Doctoral Thesis, ETH Zürich, Zurich, Switzerland.

Kránitz K., Sonderegger W., Bues C-T. & Niemz P. (2016). Effects of ageing on wood: A literature review. *Wood Science and Technology*, 50(1), 7–22.

Li J. (2022). *Ageing of Wood as a Construction Material Measured by atomic force Microscopy*. Doctoral Thesis, TU Braunschweig, Braunschweig, Germany.

McLain R. & Steimle D. (2019). WW-WSP-10. *Accommodating Shrinkage in MultiStory Wood-Frame Structures*. WoodWorksTM, Wood Products Council, Washington DC, USA, (16 pp.).

Niemz P. & Sonderegger W. (2021). Holzphysik. *Physik des Holzes und der Holzwerkstoffe*. (2nd Ed.), Hanser Verlag, München, Germany, (580 pp.).

Pitt W. (2010). *33 Farbtafeln Parkett*. Holzmann Buchverlag, Bobingen, Germany, (96 pp.).

5 Understanding the Effects of Moisture in Wood and Wood-based Materials

P. Niemz, W. Sonderegger, and D. Sandberg

5.1 PARAMETERS OF WOOD AND MOISTURE

Wood is a porous substance with typical capillary properties. Its porosity (the ratio of the volume of voids to the total volume of the material) depends on its bulk density, and most wood species have a porosity between 40% and 80%. The higher the density, the lower the porosity.

Like all porous materials, the cavity system absorbs moisture from the air and can also absorb liquids, such as water, preservatives, and adhesives, through capillary transport. An applied pressure, for example, when impregnating or gluing wood, promotes the penetration of the liquid into the cavity system of the wood material (Suchsland 1958, Bellmann 1987, Hass *et al.* 2012).

Wood is a hygroscopic material, meaning it constantly seeks to reach equilibrium with the surrounding environment, particularly regarding relative humidity (RH) and temperature (T). The equilibrium moisture content (EMC) refers to the moisture content of wood when it has reached a stable balance with the ambient air. If the wood's moisture content exceeds the EMC at a given RH and T, it will lose moisture until equilibrium is restored. Conversely, if the wood's moisture content is lower than the EMC, it will absorb moisture from the surrounding air.

The amount of water in wood is measured as *moisture content*. The moisture content is defined as the ratio of the mass of water to the mass of the wood in its oven-dry state, using the same piece of wood. A reliable and straightforward method for determining moisture content is through gravimetric analysis, where the mass of the specimen is measured before and after drying in an oven at 103 °C until all water has been removed from the wood. The moisture content is then calculated as:

$$\omega = \frac{m_\omega - m_0}{m_0} \tag{5.1}$$

where

ω – moisture content
m_ω – the mass of wet wood
m_0 – the mass of oven-dry wood, *i.e.*, at $\omega = 0\%$

DOI: 10.1201/9781003411994-5

The difference $m_\omega - m_0$ in Equation 5.1 is the mass of the water in the wood. This is the "oven-dry method" described in the EN 13183-1 standard (CEN 2002a). The measured moisture content is, in fact, an average of the actual piece of wood. Moisture variation within the wood piece is impossible to detect with this method. Other methods to measure the moisture content are described in Section 5.7.

Depending on the moisture content, wood can be categorised into three distinct states (Figure 5.1):

1) *Absolutely dry or oven-dry wood: This state refers to wood that contains no water, meaning its moisture content is = 0%.*
2) *Fibre-saturation point (FSP)*: Below the FSP, water is present only within the cell walls, where it is chemically bound (bound water). Above the FSP, additional water not chemically bonded to the cell wall (free water) begins accumulating. Free water is typically found in various cavities within the wood, most notably in the lumina of the cells.
3) *Water-saturated wood*: In this state, the wood's micro and macrostructures (such as cell lumina and cavities in the cell walls) are saturated with water. The moisture content can reach up to 740% for low-density woods like balsa and around 31% for high-density woods like pockenholts (*Lignum vitae*).

Moisture content values for other wood species can be found in sources such as Trendelenburg (1939) and Trendelenburg and Mayer-Wegelin (1955). Freshly cut, never-dried wood is often referred to as "green" wood. In this condition, the cell walls are water-saturated, and water is found as a liquid, liquid–vapour mixture, or vapour in cell lumens.

The FSP is sometimes referred to as the fibre-saturation range because it is not a fixed value. At 20 °C, the moisture content at the FSP typically ranges from 28% to 32%, which will also vary with temperature. Despite this variability, the fibre-saturation point/range concept is crucial for understanding wood's behaviour and properties.

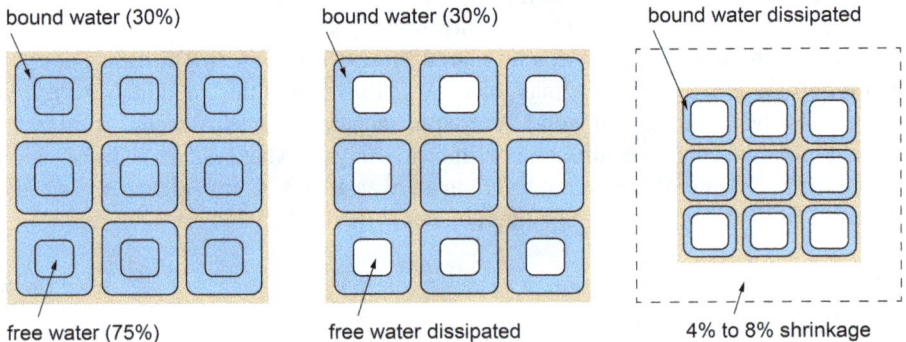

bound water (30%)   bound water (30%)   bound water dissipated

free water (75%)   free water dissipated   4% to 8% shrinkage

FIGURE 5.1 Diagram illustrating the drying process of green wood (left) to the oven-dry state.

(Courtesy of D. Sandberg and Taylor and Francis group.)

Dimensions, strength, and resistance to decay are key wood properties influenced by moisture content. When the moisture content changes below the FSP, the wood undergoes dimensional changes. Depending on whether moisture is absorbed or released, the wood either swells or shrinks in volume.

5.2 UPTAKE AND RELEASE OF WATER

5.2.1 WOOD AS A CAPILLARY POROUS MATERIAL

Wood is not a homogeneous material but a tissue of very different cell elements that fulfil the following functions:

- water conduction,
- strengthening, and
- storage of nutrients.

These structural elements are also decisive for moisture absorption and transport when working with, processing, and using wood.

The microsystem is responsible for water absorption from the air, *i.e.*, the hygroscopic behaviour. At the same time, the macrosystem is accountable for absorbing liquid water or other liquids, such as adhesives or coatings containing water or other solvents.

In the longitudinal direction, moisture is transported within the macrosystem via the cavity system of the cell elements (Figure 5.2). Moisture transport takes place predominantly via fibres or vessels. For example, an adhesive can penetrate through open fibres or vessels (open because of wood cutting) up to several millimetres beneath the surface (Suchsland 1958, Hass *et al.* 2012).

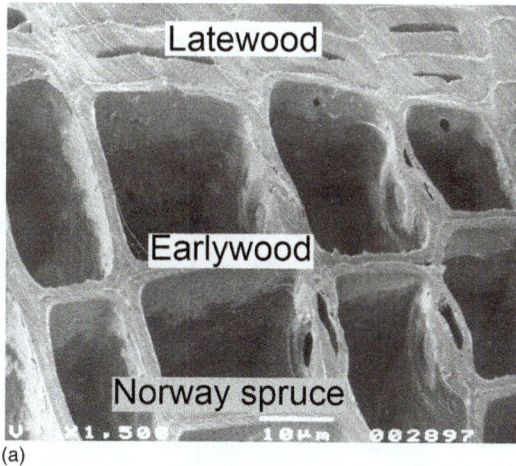

(a)

FIGURE 5.2   The porous system of the wood: a) scanning electron micrograph of the cross-section in Norway spruce and

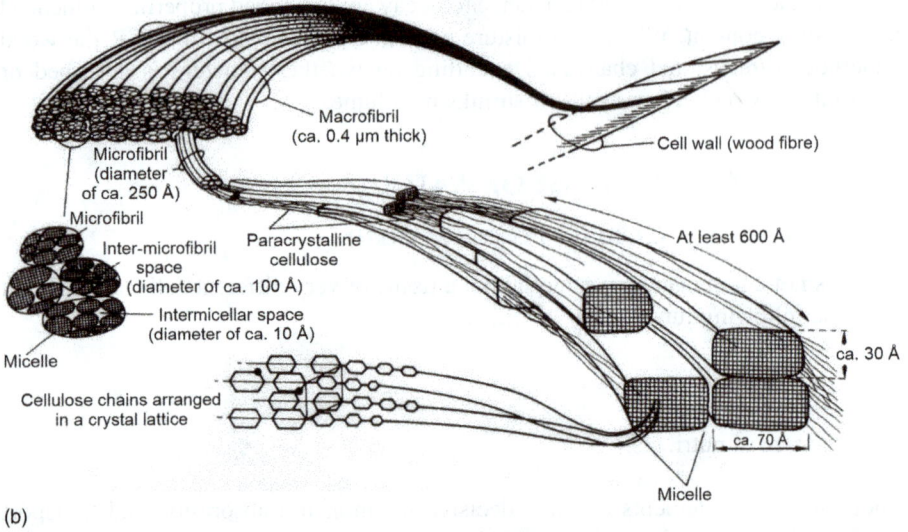

(b)

FIGURE 5.2 (*Continued*) b) the sub-microscopic structure.

(**Figures: U. Schmitt, Thünen Institute of Wood Research, Hamburg, courtesy of P. Niemz.**)

Liquid water is transported perpendicular to the fibre (transverse) direction via the pit openings and partly via radially oriented rays. Moisture is also transported within the cell-wall material via diffusion (Figures 5.3 and 5.4).

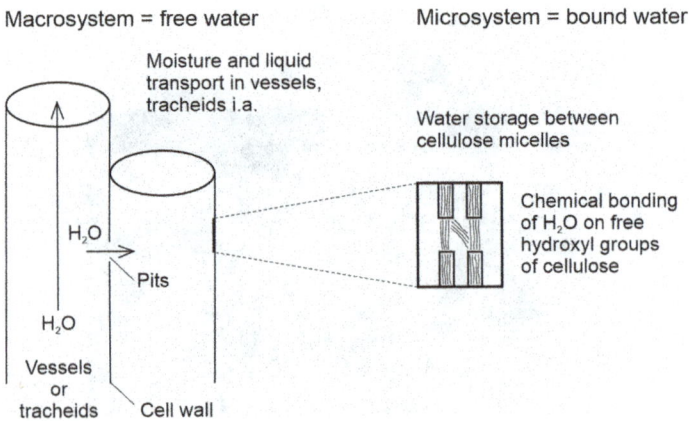

FIGURE 5.3 Schematic representation of moisture transport in wood and free and bound water distribution.

(**Courtesy of Carl Hanser Verlag.**)

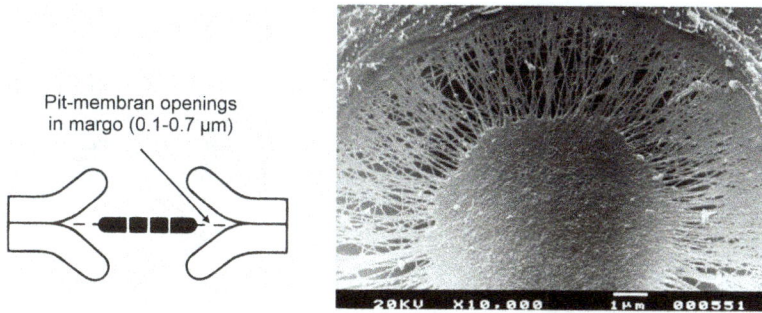

FIGURE 5.4 Openings for moisture transport through a pit: (a) schematic cross-section view of a cell wall with a pit and (b) SEM micrograph.

(G. Peschke, ETH Zürich, courtesy of P. Niemz.)

5.2.1.1 Changes in the Wood Structure

The pore system of the wood changes during the life of the living tree and during processing (*e.g.*, drying, pressing, thermal modification, gluing). For example, the heartwood formation of softwood leads to the closing of the pits. In certain hardwoods, such as black locus, common oak, and sweet chestnut, the vessels become gelatinised (a bubble-like protrusion of the cell walls), causing them to close (tylosis). This leads to a significant reduction in the absorption of liquids via capillary forces. Hardwoods with tyloses (*e.g.*, oak and chestnut spp.) are more difficult to dry and impregnate than species without tylosis (*e.g.*, common beech).

Up to the fibre saturation range, wood absorbs water bound in the cell wall; above this range, wood absorbs liquid water through capillary forces (Figure 5.3). Moisture transport in wood takes place according to the laws of capillary physics in wide to narrow capillaries. Below fibre saturation, diffusion predominates. The sorption behaviour can be described mathematically using the Hailwood-Horrobin or Brunauer-Emmett-Teller (BET) sorption theories.

The equilibrium moisture content (EMC) in wood can be reduced by thermal or hydro-thermal treatment (*e.g.*, high-temperature drying, thermal modification) and chemical treatments such as acetylation. Thermal treatment of the wood material reduces the content of hemicelluloses, reducing wood moisture under the same ambient conditions (RH and T) and improving dimensional stability (Burmester 1970, Navi and Sandberg 2012, Sandberg *et al.* 2021).

The extractive content in wood also changes moisture absorption (Popper *et al.* 2006). The proportion and type of extractives also influence the sorption behaviour and, in many cases, the gluing properties.

5.2.2 FLUID TRANSPORT IN CAPILLARY POROUS SYSTEMS (GAS AND WATER PERMEABILITY)

Wood and wood-based materials can be seen as a system of interconnected capillaries.

The following classification of capillary size should be used according to Lykow (1958) and the DIN 66131 standard (DIN 1993):

- macro-capillaries: radius > 1 µm,
- meso-capillaries: radius 0.1–1 µm, and
- micro-capillaries: radius < 0.1 µm.

While the macro-capillaries only absorb liquids under stationary conditions, micro-capillaries can also absorb liquids from the vapour phase through capillary condensation (Siau 1995).

5.2.2.1   Liquid Transport in Wood

In capillary porous systems like wood, which contain continuous and intercon-nected capillaries alongside air-filled spaces, pure capillary movement is impos-sible throughout the entire system. Moisture movement is caused by the water evaporating at the loaded menisci with the existing vapour pressure gradient, dif-fusing as steam into the air-filled pores, and condensing on the opposite, under-utilised menisci. Tensile differences in the menisci convey the condensed water. Above the fibre-saturation point, water is stored within the macroscopic system of the wood.

5.2.2.2   Diffusion

Bound water movement through wood is an example of diffusion, where the water molecules move from wetter wood to drier wood. The molecules jump from cellulose to cellulose molecules within and between adjacent cell walls.

The driving force of diffusion is the difference in humidity, moisture concentra-tion, or vapour pressure. The *diffusion number* K_D indicates the amount of liquid that flows per unit time with a moisture difference of 1% through a cross-section of 1 m^2 over a path length of 1 m.

The *vapour resistance* of a material is a measure of the material's reluctance (~resistance) to let water vapour pass through. The vapour resistance considers the material's thickness, so it can only be quoted for a particular thickness. It is measured in MNs/g. If the quantity is measured in MNs/gm (notice the "m" at the end), then it is *vapour resistivity*, which is also quoted as vapour resistance but is a bulk material property.

The *vapour resistance factor* (µ-value or mu-value) is often used as a parameter in the construction industry. It measures the material's relative reluctance to let water vapour pass through. It is measured in comparison to the properties of air, *i.e.*, the µ-value for air. This indicates that the diffusion resistance is higher than that of an air layer of the same thickness. The µ-value is a property of the bulk material and needs to be multiplied by the material's thickness when used in a particular construction. Because the µ-value is a relative quantity, it is just expressed as a number (it has no unit). The indicator depends on the moisture content of the wood, the bulk density, and the principal directions of the wood (Tables 5.1 and 5.2). The diffusion resistance increases as the moisture content in the wood decreases.

TABLE 5.1

The Vapour-resistence Factors for Wood-based Materials According to the EN ISO 10456 Standard (CEN 2007)

		Vapour-resistence	
	Density (kg/m³)	Dry (–)	Wet (–)
Wood	450–500	50	20
	700	200	50
CLT, LVL and plywood	300	150	50
	500	200	70
	700	220	90
	1000	250	110
Particleboard	300	50	10
	600	50	15
	900	50	20
OSB	650	50	30
Cement-bonded particleboard	1200	50	30
Insolation fibreboard	40–250	5	3
Fibreboard and MDF	400	10	5
	600	20	12
	800	30	20

Note: CLT – cross-laminated timber, LVL – laminated veneer lumber, OSB – oriented strand board, MDF – medium-density fibreboard

TABLE 5.2

The Water-vapour Resistance Factor of Wood in the Principal Directions According to Cammerer (1956), Vanek and Teischinger (1989), and Sonderegger *et al.* (2011)

		Water-vapour Resistance Factor, μ		
Species	MC (%)	Radial (–)	Tangential (–)	Longitudinal (–)
Norway spruce	4	230		
	6	160–240	170–300	
	8	110–150	120–50	
	10	80–100	80	5.7
	12	40–65	42–55	4.0
	16	18–28	15–26	2.0
	20	10–12	7–12	
Common beech	5	120–180	310–390	
	10	50–70	80–150	6.7
	15	11–22	27–60	3.4
	20	8.5–10.5	12–22	
	30	2.5		

Note: MC – moisture content.

TABLE 5.3

Water Absorption Coefficients of Wood and Wood-based Materials

Species		Water Absorption Coefficient $(kg/(m^2s^{0.5}))$
Common beech		
	Longitudinal	0.0440
	Radial	0.0050
	Tangential	0.0040
Common larch		
	Longitudinal	0.0470
	Radial	0.0020
	Tangential	0.0021
Norway spruce		
	Longitudinal	0.0170
	Radial	0.0030
	Tangential	0.0040
Particleboard (density 670 kg/m³)	Parallel to the surface	0.0250
	Perpendicular to the surface	0.0014

5.2.2.3 Measurement of Gas and Liquid Permeability

The permeability of gases and liquids is an essential property of wood. Wood containers such as wine barrels must have low permeability, but when wood is impregnated with, for example, preservatives, the permeability must be high. Measuring devices, such as those described in Ugolev (1986) and Acuña *et al.* (2014), measure the liquid transport through wood via a pressure gradient.

The measurement of the capillary water absorption coefficients for wood and wood-based materials is derived from building physics and is described in the EN ISO 15148 standard (CEN 2002b). The water absorption coefficient is given in kg/(m² · s⁰·⁵). Table 5.3 shows some examples of values for the capillary water absorption coefficient.

The water absorption coefficient of solid wood strongly depends on the principal direction of the wood and is significantly higher in the fibre direction (longitudinally) than radially or tangentially. In the case of wood-based materials, it is considerably higher in the plane of the panel than perpendicular, *i.e.*, in the thickness direction.

5.3 MOISTURE ABSORPTION AND RELEASE OF WOOD BY SORPTION

Due to its structure, wood is subject to the laws of capillary porous bodies. Both the micro and macrosystems can absorb water. Figure 5.5 schematically shows this process and how the water is bound to wood.

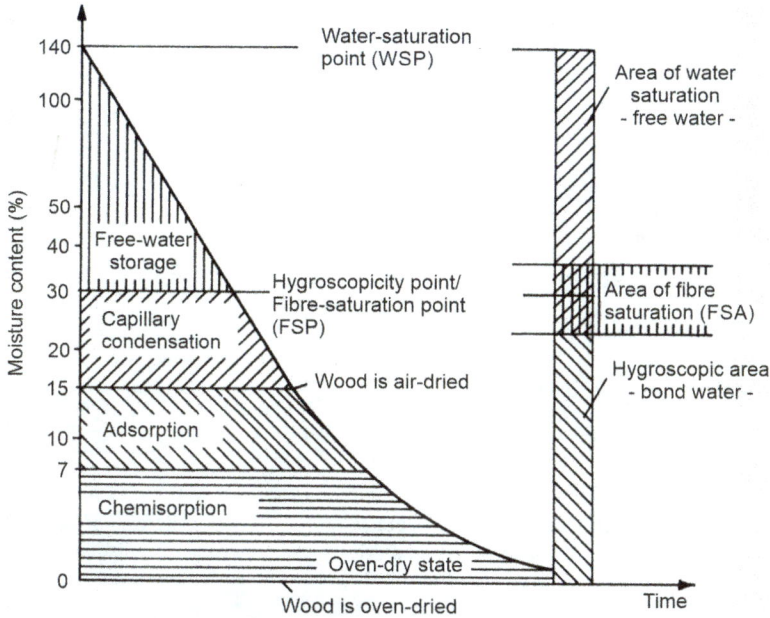

FIGURE 5.5   Schematic representation of the binding forms of water in wood.
(Courtesy of Carl Hanser Verlag.)

If dry wood is brought into contact with water, it absorbs moisture from the air (Figures 5.5 and 5.6). This phenomenon is called absorption. Sorption phenomena occur up to the fibre saturation. The water stored in this way is called "bound" water. There is a hysteresis difference of 1–2% between adsorption and desorption. When moist wood is dried for the first time, the equilibrium moisture is slightly higher than during subsequent cyclic adsorption and desorption. In practice, the hysteresis effect is mainly neglected. In most cases, only sorption isotherms from adsorption or desorption are used. During absorption, the following processes are superimposed:

- chemisorption,
- adsorption, and
- capillary condensation.

5.3.1   CHEMISORPTION

Chemisorption occurs at a moisture content between 0 and 6%, *i.e.*, at a relative humidity below 20%. Chemisorption is the first phase of sorption. Due to disordered molecular storage, the free secondary valences in the amorphous regions of the cellulose scaffold play an essential role. Water molecules are stored between adjacent cellulose chains via hydrogen bonds. The entire formation of a monomolecular layer of water molecules corresponds to a moisture content of 4–6%.

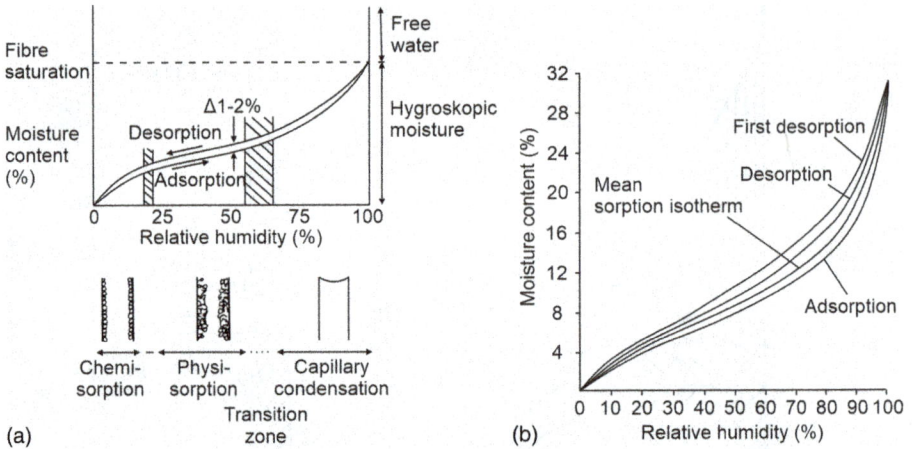

FIGURE 5.6 Sorption isotherms of wood and binding forms of water: (a) Phases of sorption isotherms including the principle of the hysteresis effect and (b) hysteresis effect in the case of first desorption of wood that has never been dried and in the case of multiple desorption.

(Courtesy of Carl Hanser Verlag.)

5.3.2 PHYSISORPTION OR ADSORPTION

Physisorption or adsorption occurs at a 6–15% moisture content (below 60% RH). Adsorption is caused by the accumulation of water in the pores of the microsystem due to molecular attraction forces. Van der Waals forces or electrostatic forces bind the water molecules, forming a polymolecular water layer.

5.3.3 CAPILLARY CONDENSATION

Capillary condensation occurs at a moisture content between 15% and the fibre-saturation point and at relative humidity from 60% to 100%.

For capillaries with a radius $r > 5 \cdot 10^{-10}$ to $1 \cdot 10^{-6}$ m, the saturation pressure above the capillaries is lower than above a flat liquid surface. As a result, some of the water vapour in these capillaries tends to condense and precipitate as a liquid on the wood surface of the cell wall system. As the moisture content increases, water is deposited in the intermicellar and interfibrillar cavities.

If the cell wall can no longer expand, the possibility of moisture absorption in the microsystem of the wood has reached its maximum. This moisture state is referred to as the fibre-saturation point (FSP). If this range is exceeded, water is stored in the macroscopic cavities (free water). Below the fibre saturation range, swelling and shrinkage phenomena occur. The fibre saturation range indicates the highest possible content of bound water. It is reached when the wood is surrounded by air saturated with water vapour (100% RH). This is not a point but a range between 22% and 35% moisture content, with an average of 28%. The differences between the individual types of wood are caused by differences in the fine cell-wall structure and the chemical structure.

5.3.4 SORPTION BEHAVIOUR OF WOOD

The wood moisture content's dependence on the surrounding air's relative humidity shows a typical S-shaped curve (Figure 5.6). These curves are called sorption isotherms. An equilibrium moisture content is established, which depends, among other things, on:

- the relative humidity,
- the temperature,
- the air pressure, and
- the chemical and structural composition of the wood material.

The isotherms for moisture absorption and release (desorption) are incongruent. Due to this hysteresis effect, the moisture content of the wood during desorption is 1–2% higher than the moisture content during absorption.

Figure 5.7 shows the influence of temperature and relative humidity on the equilibrium moisture content (EMC) of Sitka spruce. The relation in Figure 5.7 can be

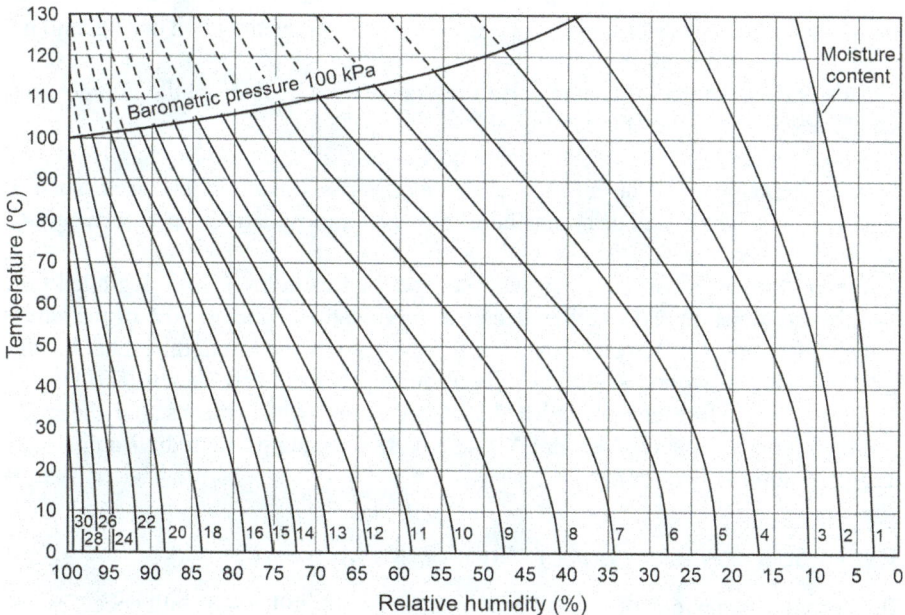

FIGURE 5.7 Hygroscopic equilibrium moisture content of Sitka spruce as a function of the relative humidity and temperature of ambient air. (Modified according to Kollmann (1951)).

(Courtesy of Carl Hanser Verlag.)

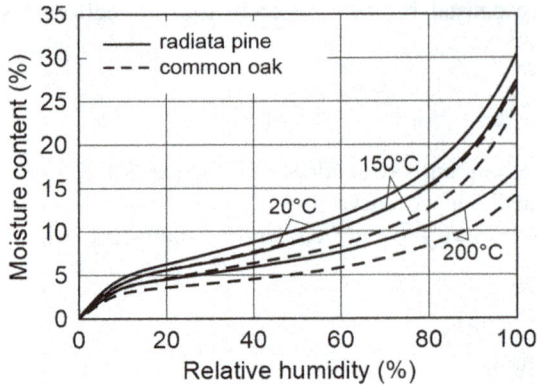

FIGURE 5.8 Influence of thermal pre-treatment on the sorption isotherms of common oak and radiata pine.

(Niemz and Sonderegger 2021, courtesy of Carl Hanser Verlag.)

used to approximate the EMC for most wood species, especially softwoods (Kollmann 1951).

Figure 5.8 shows that thermally pre-treated wood has a lower equilibrium moisture content than untreated wood (*e.g.*, after thermal treatment, wood drying, and hot pressing of wood-based materials).

This is due, among other things, to the degradation of hydrophilic components, which reduces moisture absorption.

Since the mid-1990s, the thermal modification of sawn timber has reached a specific market volume on an industrial scale. Various methods are used, such as treatment in a water vapour atmosphere, nitrogen atmosphere in an autoclave, and vacuum press drying.

When timber is thermally modified, the hemicelluloses are degraded, the equilibrium moisture and swelling decrease significantly (to up to 50% of the values of untreated timber), fungal resistance increases, and the strength, especially the breaking impact work, is significantly reduced (Niemz and Wetzig 2011). The equilibrium moisture content (EMC) also decreases, which means that thermally modified timber will have a lower EMC at a particular relative humidity-temperature stage than unmodified timber.

5.3.5   Sorption Behaviour of Wood-based Materials

The sorption behaviour of wood-based materials is significantly influenced by the type of adhesive or other binder used (Figure 5.9). For example, the alkali content of phenolic (PF) resin increases the moisture uptake, but urea-formaldehyde (UF) decreases the moisture uptake compared to the wood material itself. Adhesives themselves also absorb water when cured (Figure 5.10).

The structure of a particleboard and the temperature when the panel is pressed (*cf.* thermal modification) also influence hygroscopic behaviour.

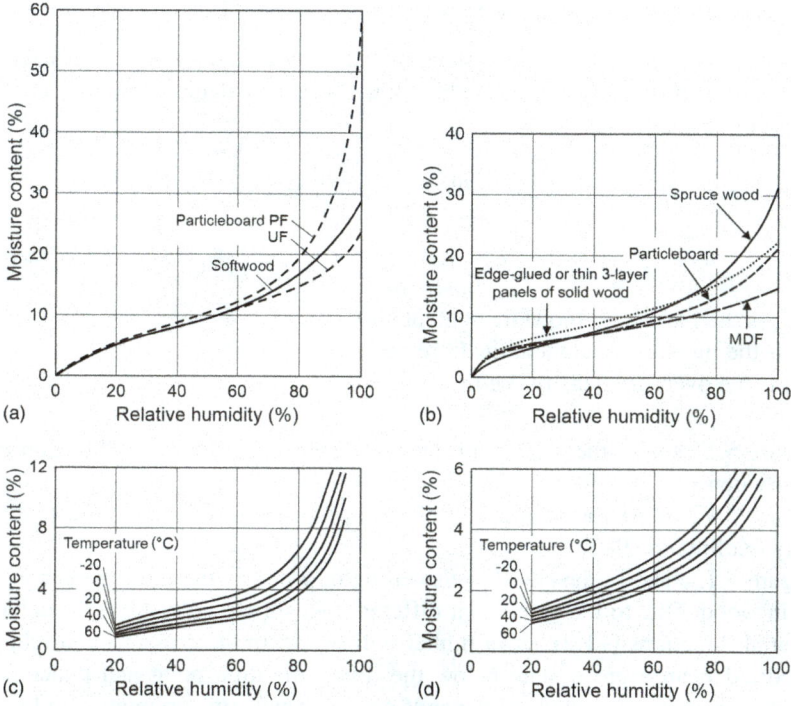

FIGURE 5.9   Sorption isotherms of various wood-based materials with a variable binder (Niemz and Sonderegger 2021): a) particleboard glued with urea-formaldehyde (UF), phenol-formalde-hyde (PF) adhesive, and non-glued softwood as a reference, b) Norway spruce solid wood and wood-based materials, c) gypsum-bonded particleboard, and d) cement-bonded particleboard.

(WKI Braunschweig, courtesy of Carl Hanser Verlag.)

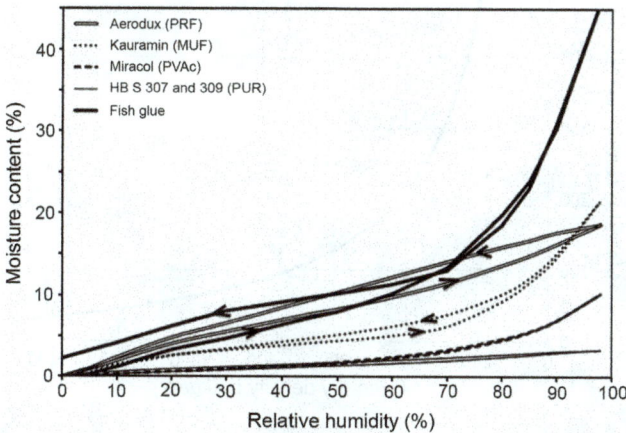

FIGURE 5.10   Sorption behaviour of adhesive films according to Wimmer *et al.* 2013. Adhesives: PRF - phenol-resorcin-formaldehyde, MUF - melamine-urea-formaldehyde, PVAc - polyvinyl-acetate, and PUR - polyurethane.

(Courtesy of Springer.)

5.4 MAXIMUM MOISTURE CONTENT OF WOOD

Free water is stored in the macrosystem when the fibre-saturation point is exceeded. The maximum possible moisture content of the wood is calculated by considering its porosity or bulk density:

$$\omega_{max} \approx \omega_F + \frac{1500 - \rho_{dtr}}{1.5 \cdot \rho_{dtr} \cdot 10^{-2}} \quad (5.2)$$

where

ω_{max} is the maximum moisture content (%)
ω_F is the moisture content at the fibre-saturation point (%)
ρ_{dtr} is the oven-dry density (kg/m³)

Figure 5.11 shows the maximum moisture content of wood as a function of the oven-dry density.

Figure 5.12 shows the volume shrinkage of various wood species depending on density (Kollmann 1982).

Figure 5.13a shows the shrinkage/swelling of wood in the three principal directions in wood. Due to the significant difference in shrinkage/swelling in radial and tangential directions (~1:2), sawn timber will distort in the cross-section (cupping) when dried from a green state below the fibre-saturation point and under further moisture variations. The cupping depends on the growth-ring orientation of the sawn timber. Figure 5.13b shows timber with various cross-section shapes and growth-ring orientations due to the location in the cross-section of a log, which results in no or multiple forms of cupping.

Table 5.4 shows examples of swelling and shrinkage values of some wood-based materials.

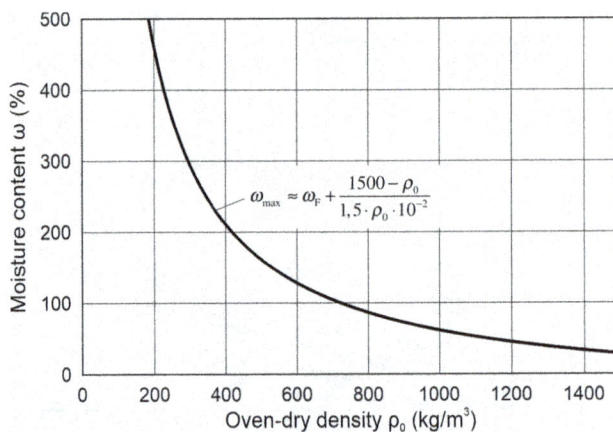

FIGURE 5.11 Maximum moisture content (ω_{max}) as a function of the oven-dry density (Kollmann 1951). ω_F – moisture content at the fibre-saturation point $\omega \approx 28\%$.

(Courtesy of Carl Hanser Verlag.)

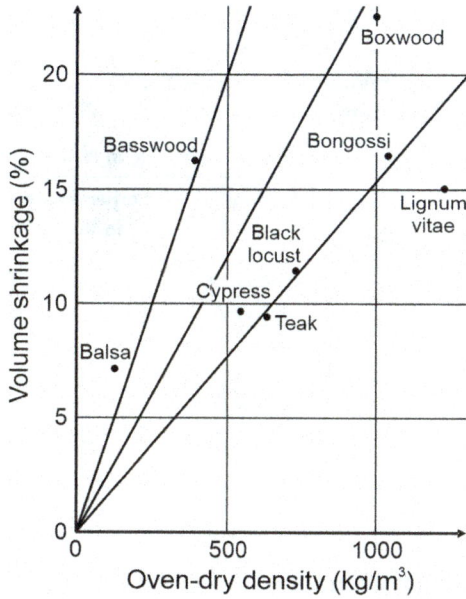

FIGURE 5.12   Volume shrinkage of wood species as a function of oven-dry density. (Courtesy of Carl Hanser Verlag.)

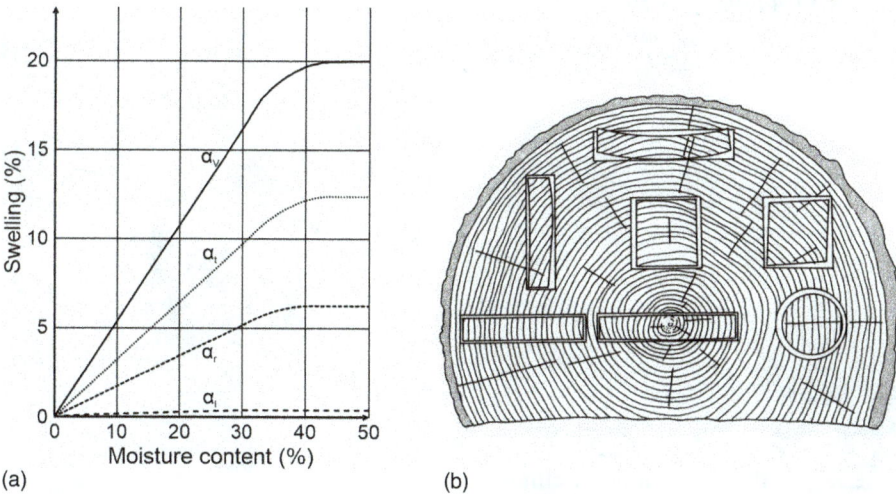

(a)   (b)

FIGURE 5.13   Swelling and shrinkage behaviour of wood: (a) swelling of common beech wood; A_l – longitudinal, α_r – radial, α_t – tangential, α_v – volume (Mörath 1931), and (b) distortion in the cross-section of wood due to anisotropic shrinkage.

(Forest Product Laboratory, Madison (WI), USA, cited in Kollmann 1951, courtesy of Carl Hanser Verlag.)

TABLE 5.4

Swelling and Shrinkage Values of Wood-based Materials (von Halász and Scheer 1986)

Type of Wood-based Material	Swelling or Shrinkage in % per 1% Change in EMC:	
	In-plane Direction	Perpendicular to the Plane, *i.e.*, in the Thickness Direction
Cement-bonded particleboard	0.030	0.3
Fibreboard	0.030	0.8
Particleboard	0.035	0.6
Plywood	0.020	0.3
Solid-wood panel, multilayer	0.015	0.3
Cement-bonded particleboard	0.030	0.3

Note: EMC – equilibrium moisture content.

5.5   SWELLING AND SHRINKAGE

The volume of the wood changes when moisture is released or absorbed in the range of 0% to the fibre-saturation point, *i.e.*, shrinkage and swelling. The shrinkage measure is related to the dimensions in the maximum swollen state, *i.e.*, dimensions at the fibre-saturation point and moisture content above. The swelling measure is related to the dimensions in the oven-dry state. The following linear relationships can be used to approximate the shrinkage and swelling:

$$\alpha = \frac{a_{\omega 2} - a_{\omega 1}}{a_{min}} \tag{5.3}$$

$$\beta = \frac{a_{\omega 2} - a_{\omega 1}}{a_{max}} \tag{5.4}$$

$$\alpha_{max} = \frac{a_{max} - a_{min}}{a_{min}} \tag{5.5}$$

$$\beta_{max} = \frac{a_{max} - a_{min}}{a_{max}} \tag{5.6}$$

where

α, α_{max} is the (maximum) swelling
β, β_{max} is the (maximum) shrinkage
a_{max} is the maximum specimen dimension, *i.e.*, at a moisture content above the fibre-saturation point
a_{min} is the minimum specimen dimensions, at oven-dry state
$a_{\omega 2}$, $a_{\omega 1}$ is the specimen dimensions at moisture contents ω_1, ω_2 for $\omega_2 > \omega_1$ and $a_{\omega 2} > a_{\omega 1}$

The (maximum) volume swelling α_v ($\alpha_{\text{max,v}}$) and (maximum) volume shrinkage β_v ($\beta_{\text{max,v}}$) are, when number is inserted in percentage, calculated as:

$$\alpha_v\left(\alpha_{\text{max,v}}\right) = \frac{\left(100+\alpha_r\right)\cdot\left(100+\alpha_t\right)\cdot\left(100+\alpha_l\right)}{10000} - 100(\%) \tag{5.7}$$

$$\beta_v\left(\beta_{\text{max,v}}\right) = \frac{\left(100+\beta_r\right)\cdot\left(100+\beta_t\right)\cdot\left(100+\beta_l\right)}{10000} - 100(\%) \tag{5.8}$$

An approximation is:

$$\alpha_v\left(\alpha_{\text{max,v}}\right) \approx \alpha_l + \alpha_r + \alpha_t \tag{5.9}$$

$$\beta_v\left(\beta_{\text{max,v}}\right) \approx \beta_l + \beta_r + \beta_t \tag{5.10}$$

where

l – longitudinal
r – radial
t – tangential

The maximum volume swelling can also be calculated roughly according to empirical experience from the moisture content at the fibre-saturation point and oven-dry density:

$$\alpha_{\text{max,v}} = \omega_F\cdot\rho_{\text{dtr}}\cdot10^{-3} \tag{5.11}$$

where

$\alpha_{\text{max,v}}$ is the maximum volume swelling (%)
ω_F is the moisture content at the fibre-saturation point (%)
ρ_{dtr} is the oven-dry density

The DIN 52184 standard (DIN 1979) describes how to determine swelling and shrinkage. Shrinkage values are calculated for a dimensional change in the principal directions of wood when the moisture content is reduced from the fibre-saturation point to 12% moisture content.

The swelling coefficient in wood's principal directions is the dimensional change per 1% unit change in relative humidity.

The swelling and shrinkage factor K (differential shrinkage) give the dimensional change in the principal directions in % per 1% unit change in moisture content.

According to Knigge and Schulz (1966), the longitudinal shrinkage of common timber species is 0.4% on average, the radial shrinkage is 4.3%, and the tangential shrinkage is 8.3%, i.e., the longitudinal shrinkage is ten to twenty times smaller than that in the transverse directions.

The considerably lower longitudinal swelling or shrinkage compared to the transverse directions is explained by the orientation of the fibrils in the longitudinal direction of the fibres.

Table 5.5 shows the differential swelling and shrinkage of some wood species.

TABLE 5.5

Maximum Swelling in % and Differential Swelling (in % / %) of Different Wood Species (von Halász and Scheer 1986)

Species	Maximum Swelling (%)			Swelling (%) at a Change of			
				RH of 1% Unit[a]		MC of 1% Unit[b]	
	longitudinal	Radial	Tangential	Radial	Tangential	Radial	Tangential
Softwoods							
Common larch	0.1–0.3	3.4	8.5	0.027	0.057	0.14	0.30
Douglas fir	0.1–0.3	5.0	8.0	0.025	0.046	0.15	0.27
Norway spruce	0.2–0.4	3.7	8.5	0.037	0.070	0.19	0.36
Scots pine	0.2–0.4	4.2	8.3	0.035	0.068	0.19	0.36
Western red cedar	0.2–0.6	2.5	5.3	0.015	0.030	0.10	0.20
Hardwoods							
Abachi	0.2–0.3	3.5	6.2	0.011	0.023	0.10	0.18
American mahogany	0.1–0.2	3.4	4.7	0.015	0.023	0.16	0.28
Azobe, bongossi	0.2–0.3	7.7	11.4	0.069	0.096	0.31	0.40
Common beech	0.2–0.6	6.2	13.4	0.032	0.065	0.20	0.41
Common oak	0.3–0.6	4.6	10.9	0.033	0.063	0.18	0.34
Dark red meranti	0.2–0.3	4.3	10.7	0.035	0.067	0.17	0.32
Iroko	0.2–0.7	3.5	5.5	0.031	0.045	0.19	0.28
Poplar spp.	0.2–0.4	3.4	8.9	0.012	0.029	0.10	0.28
Sipo mahogany	0.2–0.3	5.5	6.7	0.037	0.047	0.20	0.25
Teak	0.2–0.3	2.7	4.8	0.022	0.035	0.16	0.26

Note: MC = moisture content, RH = relative humidity.
[a] Swelling in the range from 35–80% RH
[b] Swelling in %/% unit change of MC

In practice, the transverse swelling (α_q) and shrinkage (β_q) are sometimes given only as average of the radial and tangential values:

$$\alpha_q = \frac{\alpha_t + \alpha_r}{2} \qquad (5.12)$$

$$\beta_q = \frac{\beta_t + \beta_r}{2} \qquad (5.13)$$

5.5.1 SWELLING AND SHRINKAGE BEHAVIOUR OF WOOD-BASED MATERIALS

The swelling and shrinkage behaviour of wood-based materials differs from sawn timber. Figure 5.14 shows the average length swelling of a particleboard in the plane of the panel and in the thickness direction perpendicular to the plane.

Figure 5.14 shows that:

- the in-plane swelling of particleboard is somewhat higher than that of sawn timber in the longitudinal direction. The particles are statistically randomly distributed in a particleboard and can swell in all three principal directions (longitudinal, radial, tangential) in the panel plane directions.
- The thickness swelling of particleboard is significantly higher than that of wood in the transverse directions. The wood particles are densified during the panel production and swell back partly when re-moistened.

Figure 5.15 shows the thickness swelling of various wood-based panels as a function of the increase in relative humidity.

There are some general reasons for the different hygroscopic properties of wood-based panels:

- The type of adhesive (in Figure 5.15 PF or UF) influences the hygroscopicity of the glued panels and, thereby, the swelling and shrinkage. For example, the alkali content of phenolic (PF) resin increases the moisture uptake.
- Paraffin is often added to particle- and fibre-based panels to reduce moisture uptake.
- The structural composition of the particleboard (particle size, bulk density, density profile, etc.) also influences its hygroscopic properties.

Similar relationships exist for plywood. Depending on the type and proportion of adhesive, a certain amount of swelling can occur in plywood production and use.

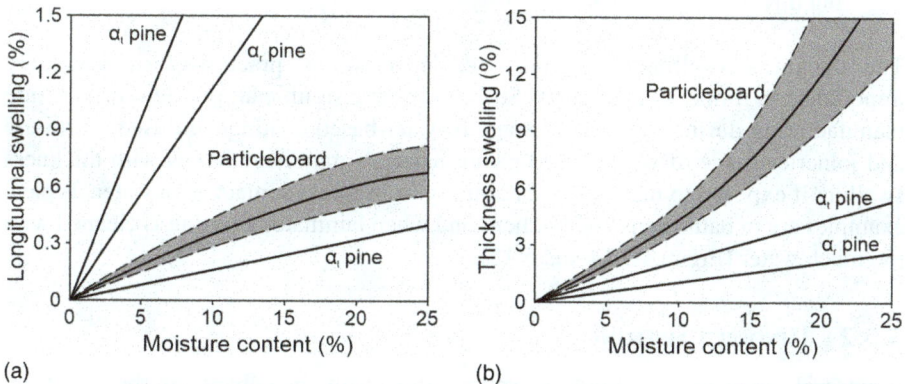

FIGURE 5.14 Swelling in particleboard: (a) in-plane and (b) in-thickness swelling of softwood particleboard. Sawn timber of Scots pine is given as a reference.

(Courtesy of Carl Hanser Verlag.)

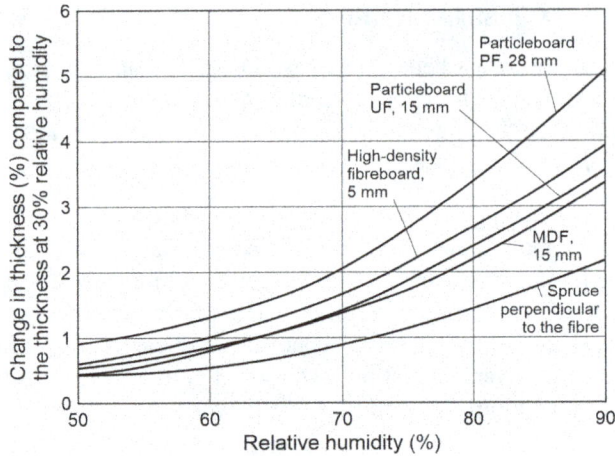

FIGURE 5.15 Changes in panel thickness in % as a function of relative humidity at 20 °C (Niemz and Sonderegger 2021). Norway spruce sawn timber is a reference. PF – phenol-formaldehyde-based and UF – urea-formaldehyde-based adhesives.

(Courtesy of Carl Hanser Verlag.)

The in-plane swelling of wood-based panels in the relative humidity range of 50% to 90% at 20 °C has been determined by Niemz and Sonderegger (2021) as follows:

- 0.2–0.3% for medium-density fibreboard (MDF),
- 0.4% for particleboard,
- 0.08% in the working direction and 0.40% perpendicular to the working direction for oriented strand board (OSB), and
- 0.12% in the working direction and 0.06–0.09% perpendicular to the working direction for plywood.

The thickness swelling of wood-based materials is practically of secondary importance in timber construction. Still, it can be a significant problem in furniture manufacturing during the coating stage if water-based coatings are used. Furniture and joineries made of wood-based panel materials may have issues with thickness swelling if exposed to high relative humidity or in direct contact with water. Typical examples are in bathrooms and kitchens and when laminated flooring is cleaned with too much water (Figures 5.14 and 5.15).

5.5.2 Hindered Swelling

If external forces restrict (hinder) swelling, this results in a change in the cell structure and plastic deformation may occur at re-drying (Burmester 1970). If the wood undergoes swelling multiple times in a restricted manner, the dimensions continuously reduce in the direction of the restrained swelling (Perkitny 1960). The swelling pressure increases as the re-strained swelling and initial oven-dry density of the wood (Figure 5.16a and b). The wood's density increases due to the compression/densification of the cells. For example, an axe handle tightly fitted into a steel axe head that swells will become looser when re-dried. Swelling pressure rises with increasing moisture content, and the stresses increase until the proportionality limit is exceeded, resulting in plastic deformation (Figure 5.16c).

The swelling pressure may be so high that it will destroy stone or concrete. Figure 5.16d shows a concrete failure caused by a wooden dowel cast into the concrete, which was probably forgotten to be removed after the concrete hardened. The swelling of the dowel then destroyed the concrete. In the past, stone blocks were split by drilling holes, wooden dowels were inserted, and water was poured onto the dowel to initiate swelling.

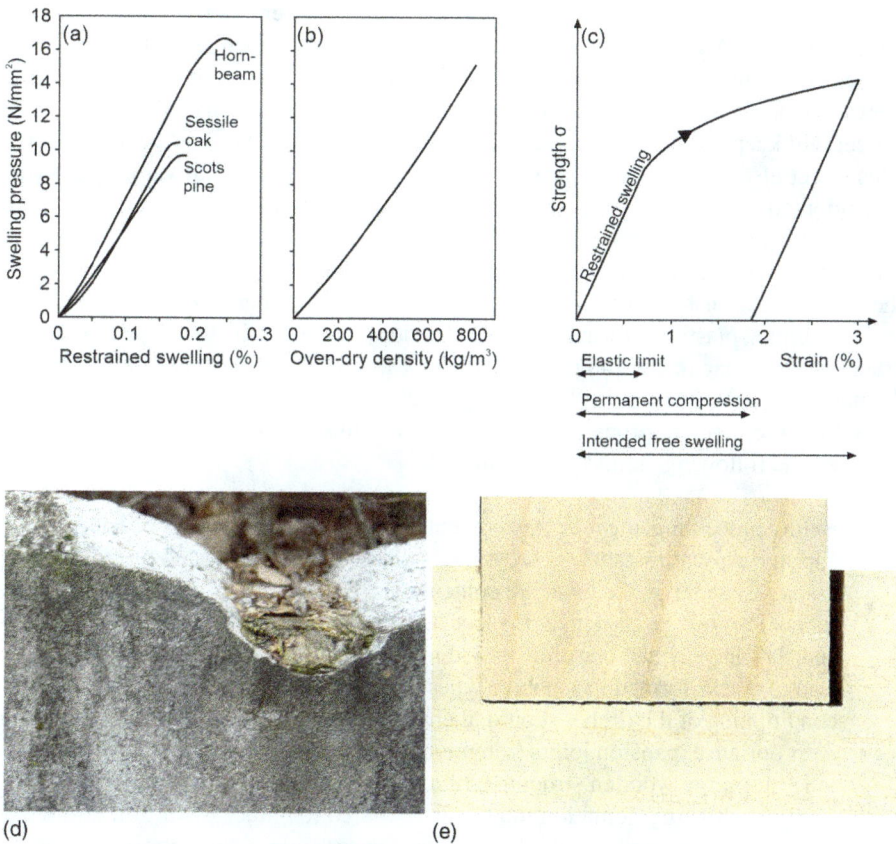

FIGURE 5.16   Hindered swelling of wood in the direction of the grain: (a) influence of the hindered swelling, (b) relationship between the oven-dry density and swelling pressure, (c) schematic representation of the stress-strain diagram in the case of hindered swelling (Krauss 1988), d) concrete failure due to swelling of wood, and e) opening of a wooden joint after cyclic moisture variation due to plastic deformation.

(Courtesy of Carl Hanser Verlag and P. Niemz.)

Considerable forces may also be initiated during wood drying. A moisture gradient through the thickness of sawn timber (high moisture content in the centre of the sawn timber) hinders the shrinkage of the surrounding wet core wood, which may cause cracking timber surfaces as the ultimate tensile strain in wood is very low.

5.5.3 EFFECTS OF THE SWELLING AND SHRINKAGE OF WOOD AND WOOD-BASED MATERIALS

A change in relative humidity inevitably leads to a change in the moisture content of wood and, thus, to swelling or shrinkage. The different swelling and shrinkage in the three principal directions of wood may result in significant distortion and internal stresses, causing cracks. For example, the cross-section of sawn timber changes its shape depending on the growth-ring orientation in the cross-section (*cf.* Figure 5.13b). This always occurs when the moisture content of the timber changes, *i.e.*, in the initial drying from a green state, and under relative-humidity changes in the use of the timber. A planing operation will temporarily give the cross-section a defined shape, but the shape will change when the moisture content changes. Only full quarter-sawn timber will keep the shape under moisture-content variation, but not the dimensions. This effect also influences the waviness of edge-glued panels with varying growth-ring orientation in the cross-sections of the lamellae after a longer ageing process.

Improper drying, for example, due to a pronounced moisture gradient between the outer and inner regions of the timber, may lead to significant cracking, particularly in the radial direction and for timber with a high density. If timber is dried too fast in the initial regime, plastic deformation will develop, and later in the drying process, cracks will appear in the interior of the sawn timber when the core begins to dry (Sonderegger 2011, Niemz and Sonderegger 2021).

Before a refinement process for sawn timber delivered from a sawmill or another supplier, the following should be considered:

- Before processing, the moisture content of the sawn timber must be controlled appropriately. If moisture content levels differ from the expected or moisture gradients are large, the moisture content must be adjusted by conditioning the timber to moisture content at the level expected in use. The different swelling and shrinkage behaviour of the wood in the three main cutting directions can lead to cracks formed in sensitive types of wood (hardwood). Fast drying and conditioning of the timber should, therefore, be avoided.
- Appropriate expansion joints (tongue-and-groove connections) must be provided for large wooden structures (*e.g.*, floors, ceilings, or wall cladding).
- The growth-ring orientation must be considered to reduce distortion of timber components and the final products, for example, frame constructions, doors, etc.
- The swelling and shrinkage of the wood must be considered for a proper fit, for example, clamping windows and drawers, during the entire annual cycle of temperature and relative humidity changes.
- When wooden surfaces are exposed to liquid moisture (water, stain, varnish), the surface will swell and thus increase the roughness. This may require regrinding after the first coating round after the surface has dried.

Due to their structural design, particleboard and fibreboard undergo significant swelling when exposed to moisture (Böhme 1980). Therefore, it is necessary to seal the edges (the narrow surfaces) of the panels. When coating a surface, the inevitable effect of humidity leads to roughening the panel surfaces (Figure 5.17). The particle thickness and the solid resin content significantly influence the surface roughness of the particleboard when exposed to water.

The roughness increases with increasing particle/thickness. Visually identical surfaces in the initial state can differ significantly after exposure to moisture.

The swelling of the individual particles in particleboard and interaction with variation in the bulk density of the panel results in an increase in surface roughness after

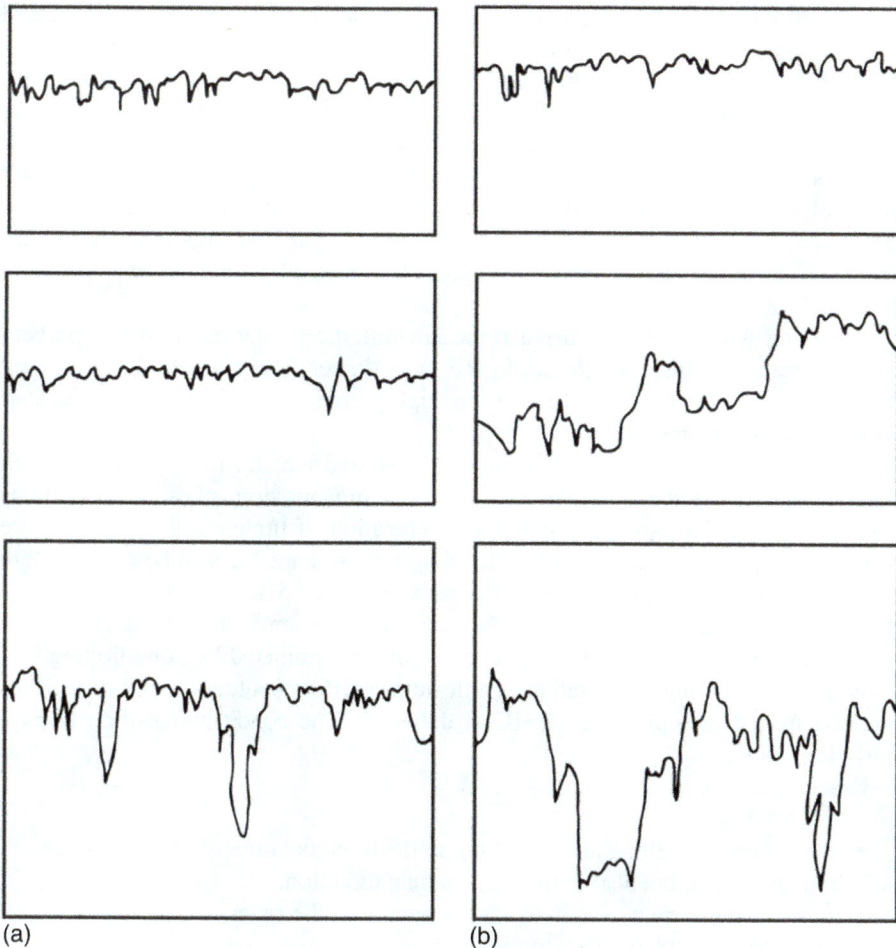

(a)   (b)

FIGURE 5.17 Principal roughness profiles of three particleboard surfaces (a) before wetting (a dry surface) and (b) after wetting of the surface.

(Courtesy of Carl Hanser Verlag.)

wetting, for example, coating with a water-based coating. Individual particles will be seen as minor marks on the surface. Cyclic swelling and shrinkage may also break the adhesive bond-lines between particles, which are brittle, particularly when urea-based adhesives are used. In the same way as for solid timber of large sizes or to cover large areas (*e.g.*, flooring), large-size panel elements must be provided with expansion joints.

5.6 DRYING OF WOOD AND WOOD-BASED MATERIALS

The drying process for wood begins immediately after the tree has been felled. Drying starts through the cross-sections, where the free water evaporates quickly from the log, while bound water releases much more slowly.

All cell cavities are filled with water at the beginning of timber drying, and evaporation begins on the wood surface. Industrial sawn-timber drying takes place in kilns at elevated temperatures, but usually below 100 °C. Drying in a kiln means a controlled process where it is possible to give the timber a low and well-defined moisture content. The large volumes of sawn timber that, in general, are dried at sawmills, however, make it hard to steer the drying process to fit all types of further use, especially the use in high-quality and advanced joinery, which may need further conditioning before processing. During drying, moisture gradients always develop to various degrees but should be kept as low as possible (Figure 5.18).

The faster wood is initially dried at the sawmill, the greater the moisture gradient and the greater the internal stresses in the sawn timber. Therefore, the conditioning regime in the drying kiln at the sawmill is highly important for getting sawn timber with low moisture gradients.

Moisture gradients cause stress (compressive and tensile) in the timber, which leads to distortion and cracking in most cases. Significant visco-plastic deformations occur due to mechano-sorptive creep and relaxation. If timber with high moisture gradients is split into smaller pieces (lamellae), these lamellae will be considerably distorted within some hours because the stresses are unbalanced in the lamella, and the timber moisture redistributes. The moisture gradients should be minimised already at the sawmill, but minor gradients can be eliminated by conditioning in a climate corresponding to the target equilibrium moisture content (EMC) of the timber (the from beginning target EMC at delivery). The conditioning often takes a fairly long time.

Crack formation is promoted by:

- the different swelling and shrinking in the principal directions, *i.e.*, tangential and radial, but also in the longitudinal direction,
- growth-rings orientation in the cross-section of the sawn timber, and
- the high modulus of elasticity in high-density hardwoods.

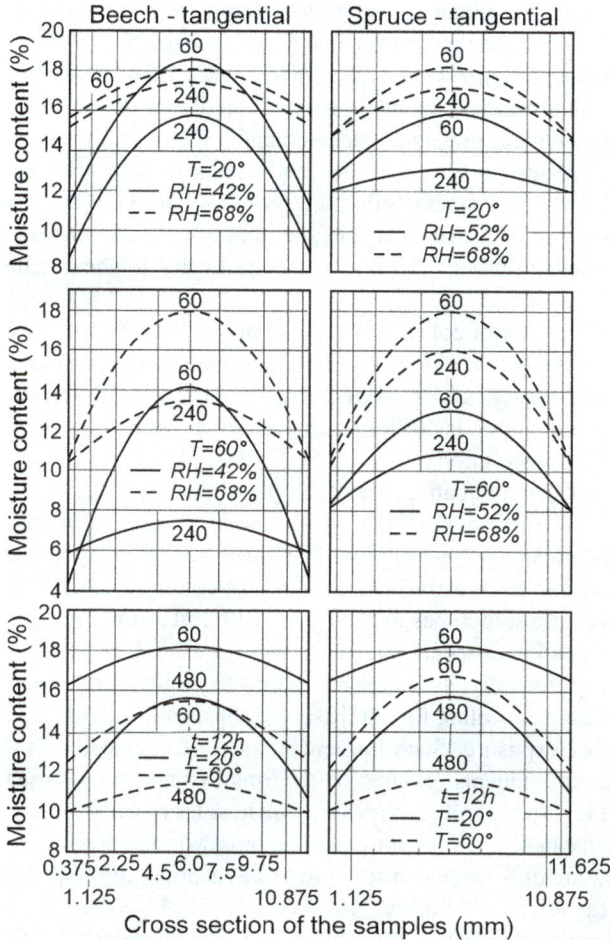

FIGURE 5.18 Moisture gradient in the cross-section in drying of flat-sawn (tangential) sawn timber. The numerical values in the curves indicate the drying time in minutes (Gerstetter 1976). RH – relative humidity, T – temperature, t – time.

(Courtesy of Carl Hanser Verlag.)

5.6.1 Stresses and Crack Formation When using Timber and Wood-based Materials in Dry Climate

Large cross-sections are required for load-bearing elements (CLT, glulam), particularly in high-rise buildings. Due to the moisture changes, these elements are more susceptible to cracking and bond-line delamination than timber elements with smaller cross-sections. An additional issue is the low relative humidity, often below 20%, in heated rooms during the winter period in, for example, Central and Northern Europe. The wood dries quickly on the surface, and a moisture gradient forms, which leads to tension and cracking (Niemz and Gereke 2009, Niemz 2016). Moisture profiles in the cross-sections of construction elements are built up due to drying surfaces and, at the same time, minor changes in the moisture content of the core of the element with large cross-sections. Stresses between the outer and the inner of the element and between lamellae, of which such elements often are produced, are built up. These cracks are rarely a problem for load-bearing capacity, but they are visually unattractive.

In cross-laminated timber (CLT), internal cracks due to shear behaviour (rolling shear) must be considered.

The moisture content of construction elements must be adjusted to the climate at later use to avoid severe cracking or delamination. An average moisture content of 12% is often too high for indoor use if the relative humidity in the ambient climate, for example, during winter, is 40%, i.e., an average moisture content of about 8% on the surface of the construction element.

Temporary transparent coatings, for example by wax, are applied to reduce moisture uptake in construction elements during transportation from the factory to the construction site and under construction. The Pollmeier company refers to this in their product information for "BauBuche", i.e., LVL of common beech (Pollmeier 2020). BauBuche trusses are sometimes also covered with foil during transport. Particularly severe damage (cracking) occurs after buildings have been converted from the moist construction stage to a dry use stage or in constructions under combined climatic exposure, for example, roofing for ski lifts, bus stops, etc., with parts of the truss non-cover for weather exposure. Such influences must be considered in the design and usage of timber components (Niemz 2016). Small or no roof overhangs on houses, combined with high-temperature exposure due to solar radiation, can also strengthen the effect, which can be seen on, for example, large window fronts.

The temperature difference can sometimes be the main driving force for moisture transport, not the relative humidity. Keylwerth (1966, 1969) tested timber elements with an initial constant moisture content of 12% and insulated on all surfaces under exposure to a differential climate in a double-sided climate chamber. After a few days with a temperature difference of 22°C (54°C on the warm side and 32°C on the cold side), the close-to-surface moisture contents on the warm and cold sides were 6% and 30%, respectively (Figure 5.19). For this reason, vapour barriers are installed in, for example, apartment entrance doors to prevent distortion caused by the difference in humidity.

Moisture-content variation in types of timber products may also cause problems, such as increased gaps between non-glued timber parts. A typical example is the joints between parquet slats that will increase with age and make the connection more visible in flooring. This is caused by continuous swelling and shrinking of the slats. Compressive stresses (restrained swelling) during swelling induce plastic deformation due to the low compressive strength perpendicular to the fibre direction, and the slats then become less wide. During the subsequent shrinkage and swelling cycles, the joints open up. The effect is usually noticeable a few years after the installation of the floor, especially with low-density wood and when dust is deposited between the slats.

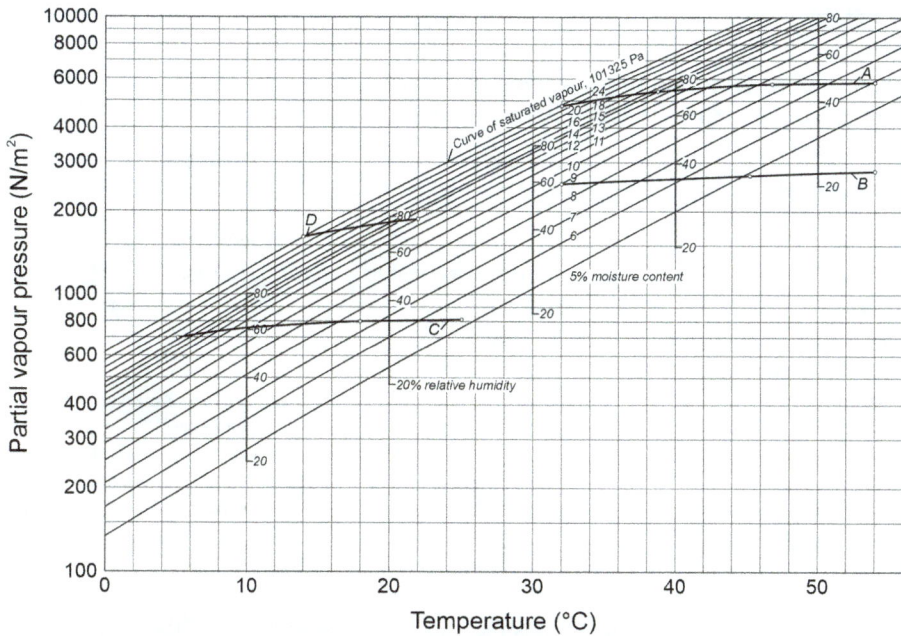

FIGURE 5.19 Moisture transfer in wood caused by temperature differences according to Keylwerth (1969). A, B, C, and D show examples.

5.7 MOISTURE MEASUREMENT PROCEDURES

Various methods can be used to determine the moisture content. The oven-dry method is considered the most accurate for solid wood and particle-based materials, but it is time-consuming compared to many other methods.

5.7.1 THE OVEN-DRY METHOD

In the oven-dry method, the specimens are dried at a temperature of 103±2 °C until the weight of the specimen between two subsequent (2 hours difference) weights is constant. The moisture content is then calculated according to Equation 5.1.

When the moisture content of sawn timber is measured, a piece more than 20 mm in the longitudinal direction is cut more than 300 mm from each end of the sawn timber to avoid a lower measured moisture content due to extreme drying through the end-grain (Table 5.6).

The oven-dry method can also be used for measuring other than the standardised dimensions of wooden specimens, for example, moisture distribution across the cross-section of a stem or a piece of sawn timber, as well as very small specimens. The stem measurements of a living tree are done with the help of an increment drill. A drill core can be removed from which the moisture content is determined, segment

TABLE 5.6

Overview of Determining the Moisture Content of Sawn Timber and Wood-based Panel Materials using the Oven-dry Method

Test Parameters	Sawn Timber EN 13183-1 (CEN 2002a)	Wood-based Materials EN 322 (CEN 1993)
Length	≥ 20 mm	No restrictions
Width	Width of the sawn timber	No restrictions
Thickness	Thickness of the sawn timber	Thickness of the panel
Mass	No restrictions	≥ 20 g
Accuracy of mass determination	0.1% of the mass	0.01 g
Mass (m) constancy during oven drying at 103±2 °C	Δm between 2 tests with a 2 h difference ≤ 0.1%	Δm between 2 measurements at an interval of 6 h ≤ 0.1%. Measurement must be performed after cooling the specimen in a desiccator with drying salt

by segment. The oven-dry method is then applied. The accuracy of the balance used must, of course, be relevant to the size of the specimens tested.

The basic principle of the oven-dry method is also used for various industrial devices to determine moisture more rapidly. For wooden chips, a device with a combined balance and heater is used to dry the chips (approximately 25 x 25 x 4 mm in size) and measure the mass until it is constant. The equipment has also been used with larger specimens (100 x 100 x 10 mm in size) with good results.

5.7.2 EXTRACTION OR DISTILLATION PROCESS

The oven-dry method becomes incorrect for woods that contain terpenes, essential oils, fats, and specific waterproofing agents because the substances mentioned evaporate. According to Kollmann (1951), the error that occurs can be up to 5–10%. Therefore, the extraction process is preferred for these materials.

The material specimens are chipped and heated in a flask with xylene, toluene, or another solvent. A condenser above the flask condenses the rising water vapour, collects the water, and determines the weight.

5.7.3 RESISTANCE MEASUREMENT METHOD

Another indirect method of determining the moisture content is measuring the wood specimen's electrical resistance. Typically, wood is an electrical insulator, but the moisture content means that electrical current can be conducted. The process is based on measuring wood's electrical resistance or conductivity, which depends on the moisture content. Readings are usually taken using stainless steel nails, screws, or cables attached to or inserted into the specimens to serve as electrodes.

The electrical resistance of wood increases as the moisture content decreases. It is influenced by the type of wood, the measurement orientation in the wood, and the temperature of the wood. The influence of species and temperature is significant, and the equipment must be calibrated due to these two parameters. However, moisture content is the factor that has the most considerable influence on resistance. When the moisture content drops below fibre saturation (approximately 30%) to 7% moisture content, resistance increases by a factor of one million. Above the fibre-saturation point, until wood cell lumens are filled with water, the resistance decreases by only a factor of 50 (James 1988).

Charge transport occurs with ions in the cell wall at moisture contents below fibre saturation. At moisture contents above fibre saturation, charge transport also occurs with ions in the free water. This charge transport is affected by the type of ions present in the water. Wood contains small amounts of water-soluble electrolytes that strongly affect charge transport in the free water. It is, therefore, difficult to measure the moisture content above fibre saturation through electrical resistance.

Small, portable moisture measuring devices are available to determine moisture content quickly. The equipment's software integrates data for species and temperature compensation. By designing the electrodes differently, the moisture content of sawn timber, veneer, and wood chips can be measured. Stationary devices can also measure moisture content well above the fibre-saturation point.

5.7.4 DIELECTRIC MEASURING METHOD

The dielectric moisture measurement method is based on the influence of the moisture content on the dielectric constant, which is 2–3 for absolutely dry wood and 80 for water. The measuring frequency is in the kHz or MHz range. The measurement frequency and the density of the wood influence the measurement result. The method is mainly used to measure the moisture content of wood chips for paper-pulp production.

5.7.5 RADIOMETRIC AND SPECTROMETRIC METHODS

The basis of the radiometric processes is that neutrons are decelerated significantly when they hit atoms, with the deceleration caused by the hydrogen (H) atoms of water being considerably stronger than that caused by the other elements in the wood. If you set the braking power of the H atoms equal to 1, then the braking power of the different atoms in wood (C, N, O) is 0.0044 and below. The number of decelerated neutrons is a measure of the wood moisture. The measurement result strongly depends on density (Mannes 2009, Lanvermann 2014). The method can be used successfully where conventional methods' spatial resolution is insufficient (moisture distribution between growth rings, moisture distribution under coatings, or in an adhesive bond-line). Magnetic resonance imaging enables high spatial resolution and the detection of bound water.

Light with wavelengths of 1930 nm and 1450 nm is absorbed by water so that the reflection of light of these wavelengths is reduced to a greater or lesser extent depending on the moisture content of the wood. Therefore, the reflected part of the light is a measure of the wood moisture. Due to the shallow penetration depth of the radiation (0.1 mm), this method only allows the moisture content of the surface to be measured. It is mainly used for bulk goods. However, it can also be used to measure wood moisture in areas close to the surface.

For further reading about methods to measure moisture content, see Sandberg *et al.* (2021).

5.8 MOISTURE DISTRIBUTION IN THE LIVING TREE

In the living tree, the moisture content varies within the stem, both in longitudinal and transverse directions and, to some extent, in the cardinal directions (north, east, south, and west). There may be significant differences in moisture content between trees of the same region or stand (Knigge and Schulz 1966).

In the radial direction of the trunk, the moisture content increases from the pith to the bark in almost all wood species, with a noticeable moisture difference between the heartwood and sapwood (Table 5.7). Some species, such as common beech, Douglas fir, fir, and poplar, do not have heartwood, which has a considerably lower moisture content than sapwood.

In the longitudinal direction of the trunk, the moisture content in softwoods generally increases from the root to the crown, while in heartwood, it often decreases after reaching a maximum. The situation is similar in ring-porous hardwoods, but for diffuse-porous and non-heartwood species, there is no clear trend in moisture variations within the length direction of the stem. Seasonal variations in moisture occur. A maximum moisture level is reached in spring, while a minimum occurs in late summer.

When the tree is felled, rapid drying may occur if the weather is warm, windy, and dry. At temperatures below zero degrees, drying is minimal. Newly felled trees contain a lot of water, and their weight is high, which must be considered in transportation. Green timber is sometimes used in construction, for example, for log houses, and the drying of the log after construction must be regarded as considerable shrinkage and cracking will occur.

TABLE 5.7

Examples of Moisture Content in the Living Tree (Trendelenburg 1939)

Wood Species	Moisture Content in the:	
	Sapwood (%)	Heartwood (%)
Common beech	70–100	50–80
Common oak	70–100	60–90
Norway spruce	130–160	30–42
Scots pine	120–150	30–50

5.9 PRACTICAL CONSEQUENCES OF WOOD AND MOISTURE RELATIONS

Sawn timber, veneer, and wood-based panel products are delivered dried to a moisture content well below 20%. Due to wood's hygroscopic nature and the associated swelling and shrinkage, the moisture content of the wood material may need to be adjusted before further processing into construction elements, carpentry products, furniture, or other wood products. This adjustment aims to reach a moisture content of the same level as in use and with low variation within and between pieces.

In winter, however, it may be an arid indoor climate in the production facilities where glued timber components are manufactured. This can result in an issue with the curing of, for example, one-component polyurethane (1C-PUR) adhesive if the humidity is very low and there is no equipment for humidifying the premises. Long-time timber storage in a dry climate may also lead to timber drying, followed by cracking, distortion, and the development of a moisture gradient within the timber.

There are rules of thumb for the moisture content of wood in different uses (Table 5.8). These rules are regional and depend mainly on climate conditions, *i.e.*, climate zones, the annual variation of the climate, and how the building is heated during the cold period of the year. However, other uses than those described in Table 5.8 may require other moisture-content levels. Figure 5.20 gives an example of how the average moisture content for timber in indoor use can vary over a continent due to variations in the outdoor climate. The difference in average moisture content over the USA seems to be lower than those expected in Central Europe (4–12% in Table 5.8).

TABLE 5.8

Example of Recommended Moisture Content (MC) of Wood and Wood-based Materials for Applications in Central Europe (Scheer *et al.* 2004)

Application	EMC (%)
Exterior:	
Wood exposed to weathering	> 20
Wood used in an open but roofed structure	12–20
Windows, entrance doors	10–15
Interior:	
Wood in a closed building without heating	9–15
Wood in a closed building with heating	4–12
Wood-based materials for kitchen furniture	10–12
Interior construction	6–10
Living room solid-wood furniture	8–10
Flooring, parquet	7–11
Radiator coverings	6–8

Note: EMC – equilibrium MC.

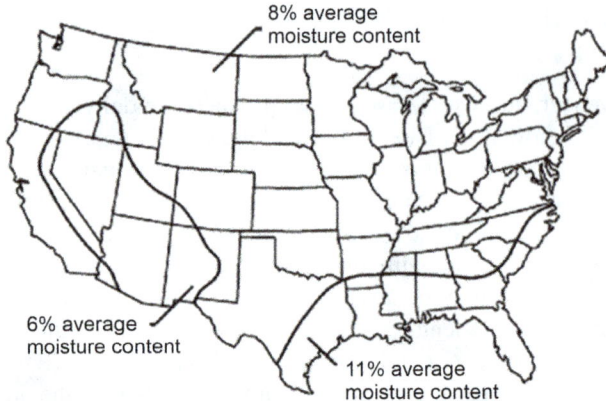

FIGURE 5.20   Average wood moisture content for interior used timber in various climatic regions of the USA.

(Ross 2010, courtesy of U.S. Department of Agriculture, USDA, Forest Products Laboratory.)

The moisture content of wood and wood-based materials is adjusted in production by conditioning under controlled temperature (T) and relative humidity (RH) (*cf.* Figure 5.7). At minor temperature variations, the variation in relative humidity is most important for the final moisture content and needs to be carefully controlled and regulated.

Non-modified solid wood, *i.e.*, logs, sawn timber, veneer, and non-glued or non-modified particles, typically reach about the same equilibrium moisture content when conditioned in the same temperature and relative humidity. Species with similar chemical properties also get about the same EMC. This is why the values for Sitka spruce presented in Figure 5.7 are commonly used to predict the EMC in a specific climate. However, depending on the species, there can be considerable variations, especially for tropical species that may differ considerably in chemical composition from species growing in the temperate and cold zones. Composite wood, *i.e.*, wood blended with other materials such as adhesives, cement, plastics, *etc.*, will have a different EMC level at a specific climate (RH, T) than non-modified solid wood. Table 5.9 shows the equilibrium moisture content of wood and wood-based materials conditioned at 20 °C and various relative humidities.

In regions with cold winters, indoor relative humidity varies considerably between summer and winter, with a dry to very dry climate during the winter and summers with indoor relative humidity more equal to that outdoors. The low indoor relative humidity in winter results from the very cold (below zero degrees) air being ventilated into the houses and heated to room temperature, which considerably lowers the relative humidity of that air. Figure 5.21 shows an example of such climate conditions from a typical farmhouse in the Midwest of the USA. The relative humidity in the living space is more than halved during the winter compared to summer; in the attic and the basement, the differences are even more significant.

TABLE 5.9

Expected Equilibrium Moisture Content of Wood and Wood-based Materials Conditioned at 20 °C and Relative Humidity (RH) of 30%, 65%, and 85%, Respectively (Based on von Halász and Scheer 1986)

Material	Moisture Content (%) at a RH of:		
	30%	65%	85%
Wood and veneer (based on Sitka spruce)	6	12	20
Plywood	4–6	8–12	12–18
Medium-density and hard fibreboards	3–5	6–8	10–14
Particleboard:			
With phenolic resin	4–6	10–12	15–23
With other resins	4–8	9–12	13–18
Cement-bonded		2–3	
Gypsum-bonded		2–3	

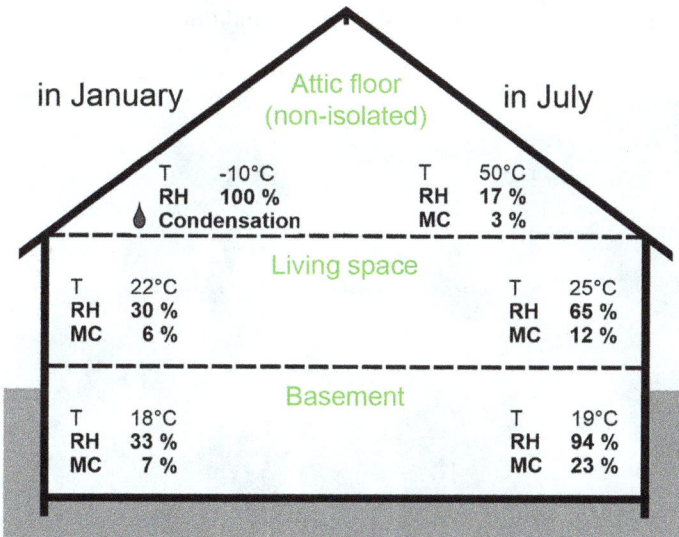

FIGURE 5.21 Indoor climate in a farmhouse in the Midwest region of the USA.

(Data from Wilcox *et al.* 1991, courtesy of D. Sandberg.)

In practice, the equilibrium moisture content of a timber-construction component with large cross-section dimensions, for example, glued-laminated timber, is never reached. With the typically seasonal variations in the indoor relative humidity (low relative humidity in winter and significantly higher in summer), the moisture transport to the centre of the component will never reach equilibrium with the climate of the ambient air. Only the surface-near regions of the element will achieve a balance in moisture content with ambient air. A moisture gradient will exist between the surface, with a low or high moisture content, and the component's core, with a high or low moisture content.

Suppose the moisture content of the timber-construction component at the time of installation is significantly higher than the expected average equilibrium moisture content in the space it is placed in. In that case, it may take years to reach "equilibrium". The component will be in a "drying regime" for the next years after installation, which may cause problems due to shrinkage distortion.

Green round timber typically leads to severe radial cracking when the logs dry after installation in construction. Grooves are introduced to reduce such cracks, which initiate cracking at a defined location in the log.

Internal stresses in timber components developed due to moisture variations may also affect a structure's load-bearing capacity (Gustafsson *et al.* 1998).

Cross-laminated timber and glued-laminated timber installed with a moisture content that is too high can develop drying cracks in the surface layers, and the adhesive joints may also partially delaminate (Figure 5.22). When timber components are used in premises with extreme climatic conditions, *i.e.*, very humid or dry indoor climates, extra consideration should be given to the choice of adhesive, and surface treatments are recommended to ensure that wood and wood-based materials will have properties that withstand the harsh climate conditions for long periods.

(a)   (b)

FIGURE 5.22 Cracks developed in timber components due to a dry climate: (a) drying cracks in glued-laminated timber and (b) in a log-house beam (Photo: P. Niemz, ETH Zurich).

(c)

FIGURE 5.22 (*Continued*) (c) Side-view of cracks and partial delamination in cross-laminated timber.

(Courtesy of D. Sandberg.)

Due to the structural composition of wood, almost all its properties are influenced by the moisture content. General relationships between wood and moisture are:

- the strength of wood decreases with increasing moisture content,
- the creep deformation of timber under long-term load increases with increasing moisture content,
- the thermal conductivity of wood increases with increased moisture content, and
- the susceptibility to fungi increases with a high moisture content, particularly above 20%.

For wood, adjusting the moisture content (at the production stage) to the climate where the wood product will be used is crucial to avoid undesirable shrinkage/swelling, surface cracking and distortion.

REFERENCES

Acuña L., Gonzales D., de La Fuente J. & Moya L. (2014). Influence of toasting treatment on the permeability of six wood species for ecological use. *Holzforschung*, 68(4), 447–454.
Bellmann H. (1987). Zur Bedeutung der Holzfeuchtigkeit bei der Kesseldrucktränkung von Nadelhölzern. *Holz-Zentralblatt*, 1857–1862, 2201–2203, 2312–2314.
Burmester A. (1970). Formbeständigkeit von Holz gegenüber Feuchtigkeit. *Grundlagen und Vergütungsverfahren*. Bundesanstalt für Materialforschung und -prüfung (BAM), Berlin, Germany, (179 pp.).

Böhme P. (1980). *Industrielle Oberflächenbehandlung von plattenförmigen Werkstoffen aus Holz*. Fachbuchverlag, Leipzig, Germany, (200 pp.).

Cammerer J.S. (1956). Bezeichnungen und Berechnungsverfahren für Diffusionsvorgänge im Bauwesen. *Kältetechnik*, 8(11), 339–343.

CEN (1993). *EN 322: Wood-based panels - Determination of moisture content*. The European Committee for Standardization (CEN). Brussels, Belgium.

CEN (2002a). *EN 13183-1: Moisture content of a piece of sawn timber - Part 1: Determination by oven dry method*. The European Committee for Standardization (CEN). Brussels, Belgium.

CEN (2002b). *EN ISO 15148: Hygrothermal performance of building materials and products — Determination of water absorption coefficient by partial immersion*. The European Committee for Standardization (CEN). Brussels, Belgium.

CEN (2007). *EN ISO 10456: Building materials and products - Hygrothermal properties -Tabulated design values and procedures for determining declared and design thermal values*. The European Committee for Standardization (CEN). Brussels.

DIN (1979). DIN 52184: Testing of wood – Determination of swelling and shrinkage. *Deutsches Institut für Normung (DIN)*, Berlin, Germany.

DIN (1993). DIN 66131: Determination of specific surface area of solids by gas adsorption using the method of Brunauer, Emmett and Teller (BET). *Deutsches Institut für Normung (DIN)*, Berlin, Germany.

Gerstetter E. (1976). *Untersuchungen über die Ausbildung von Schwind-Zugspannungen in Holz bei mechanischer Schwindungsbehinderung*. Doctoral Thesis, Universität Hamburg, Hamburg, Germany.

Gustafsson P.J., Hoffmeyer P. & Valentin G. (1998). DOL behavior of end-notched beams. *Holz als Roh- und Werkstoff*, 56(5), 307–317.

Hass P., Wittel F.K., Mendoza M, Herrmann H.J. & Niemz P. (2012). Adhesive penetration in beech wood: Experiments. *Wood Science and Technology*, 46(1/3), 243–256.

James W.L. (1988). Electric moisture meters for wood. *General Technical Report FPL-GTR-6*, U.S. Department of Agriculture, Forest Products Laboratory, Madison (WI), USA, (17 pp.).

Keylwerth R. (1966). Temperatur- und Feuchtigkeitsgefälle in Holzbauteilen. *Holz als Roh- und Werkstoff*, 24(10), 452–454.

Keylwerth R. (1969). Praktische Untersuchungen zum Holzfeuchtigkeits-Gleichgewicht. *Holz als Roh- und Werkstoff*, 27(8), 285–290.

Knigge W. & Schulz H. (1966). *Grundriss der Forstbenutzung*. Parey, Hamburg, Germany, (584 pp.).

Kollmann F. (1951). *Technologie des Holzes und der Holzwerkstoffe: Part 1: Anatomie und Pathologie, Chemie, Physik, Elastizität und Festigkeit*. (2nd Ed.), Springer Verlag, Berlin, Germany, (1051 pp.).

Kollmann F. (1982). Volumenschwindung von Holz und Rohdichteeinfluss. Ursachen von Ausreissern. *Holz als Roh- und Werkstoff*, 40(11), 429–432.

Krauss A. (1988). Untersuchungen über den Quelldruck des Holzes in Faserrichtung. *Holzforschung und Holzverwertung*, 40(4), 65–72.

Lanvermann C. (2014). *Sorption and swelling within growth rings of Norway spruce and implications on the macroscopic scale*. Doctoral Thesis, ETH Zürich, Zurich, Switzerland.

Lykow A.W. (1958). *Transporterscheinungen in kapillarporösen Körpern*. Akademie-Verlag, Berlin, Germany, (288 pp.).

Mannes D. (2009). *Non-destructive testing of wood by means of neutron imaging in comparison with similar methods*. Doctoral Thesis, ETH Zürich, Zurich, Switzerland.

Mörath E. (1931). Beiträge zur Kenntnis der Quellungserscheinungen des Buchenholzes. Doctoral Thesis, T.H. Darmstadt, Dresden, Steinkopf, Germany.

Navi P. & Sandberg D. (2012). *Thermo-hydromechanical Processing of Wood*. EPFL Press, Lausanne, Switzerland, (376 pp.).

Niemz P. (2016). Einfluss von Feuchteschwankungen bei Verklebungen. Formbeständigkeit, Rissbildung und Delaminierung von verklebten Holzwerkstoffen auf Vollholzbasis. *Holz-Zentralblatt*, 2016(50), 1219–1221.

Niemz P. & Gereke T. (2009). Auswirkungen kurz- und langzeitiger Luftfeuchteschwankungen auf die Holzfeuchte und die Eigenschaften von Holz. *Bauphysik*, 31(6), 380–385.

Niemz P. and Wetzig M. (2011). Auf spezielle Einsatzbereiche konzentriert. Einsatz von Thermoholz: Eigenschaften, Verarbeitung, Praxiserfahrungen. *Holz-Zentralblatt*, 137(1), 24–26.

Niemz P. & Sonderegger W. (2021). *Holzphysik. Physik des Holzes und der Holzwerkstoffe*. (2ⁿᵈ Ed.), Hanser Verlag, München, Germany, (580 pp.).

Perkitny T. (1960). Die Druckschwankungen in verschieden vorgepreßten und dann starr eingeklammerten Holzkörpern. *Holz als Roh- und Werkstoff*, 18(6), 200–210.

Pollmeier (2020). *BauBuche brochure 9: Wood Preservation and Surface Treatment (Ed. 12-20-EN)*. Pollmeier Massivholz GmbH & Co. KG, Creuzburg, Germany, (7 pp.).

Popper R., Niemz P. & Torres M. (2006). Einfluss des Extraktstoffanteils ausgewählter fremdländischer Holzarten auf deren Gleichgewichtsfeuchte. *Holz als Roh- und Werkstoff*, 64(6), 491–496.

Ross R. (2010). Wood Handbook. *Wood as an Engineering Material*. (Centennial Ed.), US Department of Agriculture, Forest Service, Forest Products Laboratory, Madison (WI), USA, (509 pp.).

Sandberg D., Kutnar A., Karlsson O. & Jones D. (2021). *Wood Modification. Principles, Sustainability, and the Need for Innovation*. CRC Press, Boca Raton (FL), USA, (432 pp.).

Scheer C., Peter M. & Stöhr S. (2004). *Holzbau-Taschenbuch. Bemessungsbeispiele nach DIN 1052, Ausgabe 2004*. (10ᵗʰ Ed.), Ernst und Sohn, Berlin, Germany, (396 pp.).

Siau, J.F. (1995). *Wood: Influence of Moisture on Physical Properties. Department of Wood Science and Forest Products*, Virginia Polytechnic Institute and State University, Blacksburg (VA), USA, (227 pp.).

Sonderegger W.U. (2011). *Experimental and theoretical investigations on the heat and water transport in wood and wood-based materials*. Doctoral Thesis, ETH Zürich, Zurich, Switzerland.

Sonderegger W., Vecellio M., Zwicker P. & Niemz P. (2011). Combined bound water and water vapour diffusion of Norway spruce and European beech in and between the principal anatomical directions. *Holzforschung* 65(6), 819–828.

Suchsland O. (1958). Über das Eindringen des Leimes bei der Holzverleimung und die Bedeutung der Eindringtiefe für die Fugenfestigkeit. *Holz als Roh- und Werkstoff*, 16(3), 101–108.

Trendelenburg R. (1939). *Das Holz als Rohstoff*. J.F. Lehmanns Verlag, München, Germany, (541 pp.).

Trendelenburg R. & Mayer-Wegelin H. (1955). *Das Holz als Rohstoff* (2ⁿᵈ Ed.), Carl Hanser Verlag, München, Germany, (541 pp.).

Ugolev B.N. (1986). *Wood science and bases of forest commodity science: Textbook for higher educational institution of forestry engineering*. [in Russian] Moscow State Forestry University (MGUL), Moscow, Soviet Union, (351 pp.).

Vanek M. & Teischinger A. (1989). Diffusionskoeffizienten und Diffusionswiderstandszahlen von verschiedenen Holzarten. *Holzforschung und Holzverwertung*, 41(1), 3–4.

von Halász R. & Scheer C. (1986). *Holzbau-Taschenbuch. Band 1: Grundlagen, Entwurf und Konstruktionen*. (2ⁿᵈ Ed.), Ernst & Sohn, Berlin, Germany, (712 pp.).

Wilcox W., Botsai E. & Kubler H. (1991). *Wood as a Building Material*. (2ⁿᵈ Ed.), John Wiley & Sons, New York, USA, (232 pp.).

Wimmer R., Kläusler O. & Niemz P. (2013). Water sorption mechanisms of commercial wood adhesive films. *Wood Science and Technology*, 47(4), 763–775.

6 Density of Wood and Wood-based Materials

P. Niemz and W. Sonderegger

6.1 DENSITY, POROSITY, AND SPECIFIC GRAVITY

Density, porosity, and specific gravity are measures of the same property, *i.e.* how much mass of the material is packed into a given volume of the same.

The density of wood is generally meant as "bulk density". Density, also called particle density, is a simple idea of how closely packed the molecules of substances are and how much a molecule weighs. Density is defined as the ratio of the mass of a substance to the volume occupied by that mass (Equation 6.1).

Bulk density is an extrinsic property of a material; it can change depending on how the material is handled or greatly vary depending on its condition. The bulk density of a substance varies with the state of the sample. Therefore, it is not an intrinsic property of a material, whereas density is an inherent property.

6.1.1 DENSITY

Density is the ratio of mass to volume:

$$\rho = \frac{m}{V} \tag{6.1}$$

where

ρ is density (kg/m³)
m is mass (kg)
V is volume (m³)

Wood is a capillary-porous system filled with water, water vapour, air, or sometimes a liquid such as paint or adhesive. Wood will swell when it absorbs moisture and shrink when it is released; the weight of wood also varies with moisture content, and density is a function of its moisture content (Figure 6.1). This means the mass and the volume will change with changing moisture content (ω). Density should, therefore, also be specified at a defined moisture content, *i.e.* the density at moisture content ω (ρ_ω). The volume change only occurs at moisture content changes below the so-called fibre-saturation point (FSP), which is approximately 30% at 20 °C.

The density of wood conditioned at 20 °C and 65% relative humidity (RH) is called normal density (ρ_{12}). The wood's equilibrium moisture content (EMC) is then approximately 12%.

DOI: 10.1201/9781003411994-6

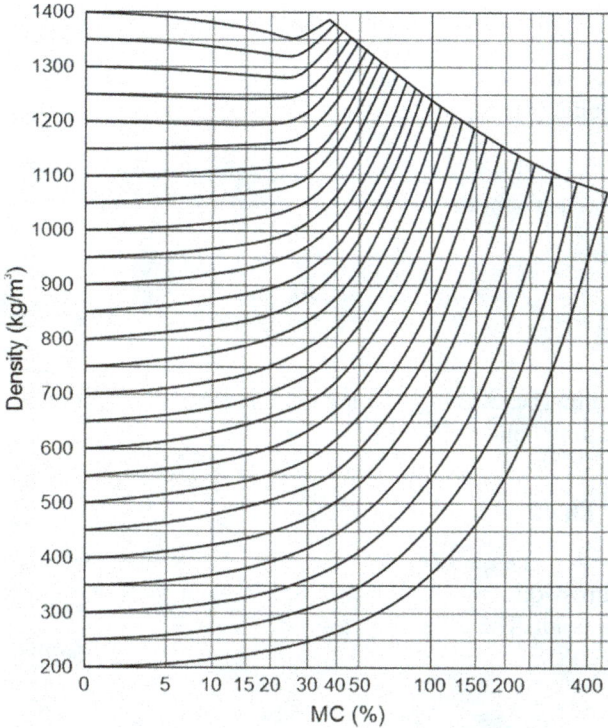

FIGURE 6.1   Density of wood as a function of the moisture content of wood.

(Kollmann 1951, courtesy of Carl Hanser Verlag.)

Suppose the mass and volume are determined at different moisture contents of the wood. In that case, the density should be given with two indexes, $\rho_{x \text{ and } y}$, where x indicates that the mass is determined at x moisture content and the volume is determined at y moisture content.

For particle-based materials, other measures for density are also used:

- spreading density or mat density,
- weight per unit area (the thickness is constant), and
- density profile.

6.1.2   OVEN-DRY DENSITY

The oven-dry density (ρ_0) is the quotient of the oven-dry mass and the oven-dry volume of the wood:

$$\rho_0 \text{ or } \rho_{0,0} = \frac{m_0}{V_0} \qquad (6.2)$$

where

m_0 is the mass of oven-dry wood
V_0 is the volume of oven-dry wood

For the conversion of the density $\rho\omega$ into the density ρ_0, the following applies to solid wood, taking into account the swelling and shrinkage:

$$\rho_0 = \rho_\omega \cdot \frac{100 + \alpha_V \omega}{100 + \omega} \tag{6.3}$$

where

α_V is the volumetric swelling (%) at moisture content ω (*cf.* Equation 5.7)
ω is the moisture content (%)

6.1.3 BASIC DENSITY

The basic density is the quotient of the mass of the oven-dry wood and the volume of the maximum swollen wood. The moisture content must, therefore, be above the fibre-saturation point. The following applies:

$$\rho_{\text{basic}} = \frac{m_0}{V_{\text{max}}} \tag{6.4}$$

where

m_0 is the mass of oven-dry wood
V_{max} is the volume of maximum swollen wood

The basic density is often used in forestry studies for practical reasons but not in the wood technology or civil engineering fields.

6.1.4 CELL-WALL DENSITY

The cell-wall density (ρ_{cw}) is the quotient of the mass of the oven-dry wood and the volume of the cell wall (without pores). It characterises the density of pure cell-wall substance:

$$\rho_{cw} = \frac{m_0}{V_{cw}} \tag{6.5}$$

The cell-wall density is almost the same for all wood species, on average 1500 kg/m³. According to Knigge and Schulz (1966), the value varies between 1440 kg/m³ and 1600 kg/m³.

These differences are caused by different test methods (Siau 1995) and due to the lignin and cellulose contents, which can vary between species. For lignin, a value of 1380 kg/m³ to 1460 kg/m³ is given, and for cellulose, a value of 1580 kg/m³ is given.

6.1.5 POROSITY CONTENT

Porosity (c) is the volume of all the voids (void fraction) in the wood at a dry state, based on the wood's volume. It results from the ratio of oven-dry density to cell-wall density.

$$c = 100 - \frac{(100 \times \rho_0)}{\rho_{cw}}$$

(6.6)

where,

ρ_0 – oven-dry density (kg/m³)
ρ_{cw} – cell-wall density (kg/m³)

for ρ_{cw} = 1500 kg/m³, it follows approximately:

$$c = 100 - 0.067 \cdot \rho_0$$

(6.7)

6.1.6 SPECIFIC GRAVITY

Specific gravity is the ratio between oven-dry mass and the mass of water displaced by the bulk specimen at a given moisture content (Siau 1995). It is dimensionless. Specific gravity is often used in the United States but not so usually in European countries.

$$G_m = \frac{m_0}{V_m \rho_\omega}$$

(6.8)

where

m_0 is the oven-dry mass (kg)
V_m is the moist volume (m³)
ρ_w is the density of water (1000 kg/m³) at 20 °C

The bound water causes swelling and when the amount of bound water increases, the specific gravity decreases. The maximum value is observed under oven-dry conditions, and the minimum (and a constant value) is achieved above the fibre-saturation point. This value is named *basic specific gravity*.

The specific gravity can also be calculated based on different moisture content (Siau 1995). Depending on the testing method, the dimensionless specific cell-wall gravity ranges from 1.53 tested in water to 1.44 tested with mercury porosimetry.

6.2   DENSITY CHARACTERISTICS FOR WOOD-BASED MATERIALS

Bulk density and mat density are production parameters used for particle-based panel materials, such as various particleboards and fibreboards (Niemz and Sandberg 2022). The bulk density or mat density is the quotient of mass and volume of wood particles deposited as a heap, particularly wood chips, particles, or fibres.

The mat density is the quotient of mass and volume of occasionally falling wood chips, particles, or fibres randomly deposited on the transport band before pressing the panel.

The density profile is the density variation in the thickness direction of the panel.

All high- and medium-density panel types have a density profile with a density maximum on the surfaces and a lower density in the core of the panels. The density profile is essential for the properties of the panel, for example, stiffness and dowels and screws pull-out strength.

Density is one of the dominant characteristics. It influences almost all wood properties, so it is often used as a guideline for strength. Figure 6.2 shows schematically the effect of bulk density on selected properties. Density variations of wood and wood-based materials, therefore, significantly impact strength. Data on the bulk density of wood and wood-based materials are given in Table 6.1. The bulk density alone is not sufficient for reliable strength grading.

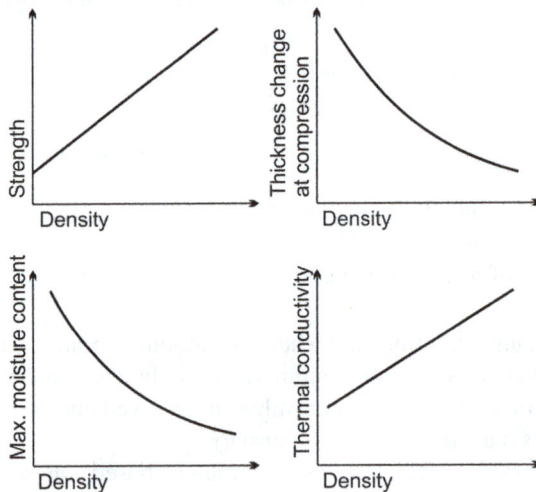

FIGURE 6.2   Schematic representation of the influence of bulk density on selected properties of wood and wood-based materials.

(Courtesy of Carl Hanser Verlag.)

TABLE 6.1

Density of Wood-based Materials

Type of Wood	Density (kg/m³)
Solid wood panels (softwood)	ca. 500
Plywood	400–850
Densified and impregnated plywood	800–1400
Particleboard	550–750
Oriented strand board (OSB)	600–700
Fibreboards (wet process)[a]	
High density	> 900
Medium density	400–900
Low density	230–400
Fibreboards (dry process)[b]	
High density (HDF)	> 800
Medium density (MDF)	650–800
Low density	550–650
Ultra-low density	< 550
Fibre-insulation materials (dry and wet process)	40–250

[a] EN 316 (CEN 2009).
[b] Marutzky and Schwab (2009).

6.3 DENSITY VARIATION OF WOOD AND WOOD-BASED MATERIALS

The ratio between cell wall to void fractions, *i.e.* the density, varies considerably between wood species. Table 6.2 shows the mean values of various wood species' oven-dry density and basic density.

Balsa, the lightest industrially usable species, has an average density of 130 kg/m³, while *Lignum vitæ* has an average density of 1230 kg/m³.

Significant differences in density exist within a single wood species due to growth (soil) and location-related factors (climatic conditions). Figure 6.3 shows the frequency distribution of density of different wood species.

The density varies approximately two to three times from the lowest to the highest density. There is generally no Gaussian distribution (Figure 6.3), but there is a tendency to asymmetry. The mean value is not identical to the peak value of the frequency distribution. Asymmetrical and uneven frequency curves are composed of symmetric normal frequency curves (Kollmann and Côté 1968).

There is also an influence of growing conditions and location and the sociological position of the tree in the forest. Walker (2006) divided the density variation in:

- within-tree variation,
- within-stand variation (*cf.* Chapter 3), and
- inter-regional variations.

TABLE 6.2

Mean Values ($\bar{x}$) and Range of Variation (x_{min}, x_{max}) of the Oven-dry Density and Basic Density (Knigge and Schulz 1966)

Species	Oven-dry Density x_{min} - $\bar{x}$ - x_{max} (kg/m³)	Basic Density $\bar{x}$ (kg/m³)
Softwoods		
Douglas fir	360–470–630	412
Eastern white pine	310–370–460	339
European larch	400–550–820	487
Grand fir	280–420–610	332
Norway spruce	370–430–540	377
Scots pine	300–490–860	431
Hardwoods		
Balsa	70–130–230	121
Black poplar	270–370–650	377
Common ash	410–650–820	564
Common oak	380–640–900	561
Common beech	540–660–840	554
Lignum vitæ	1200–1230–1320	1045
Scots elm	440–640–820	556
Sycamore maple	480–590–750	522

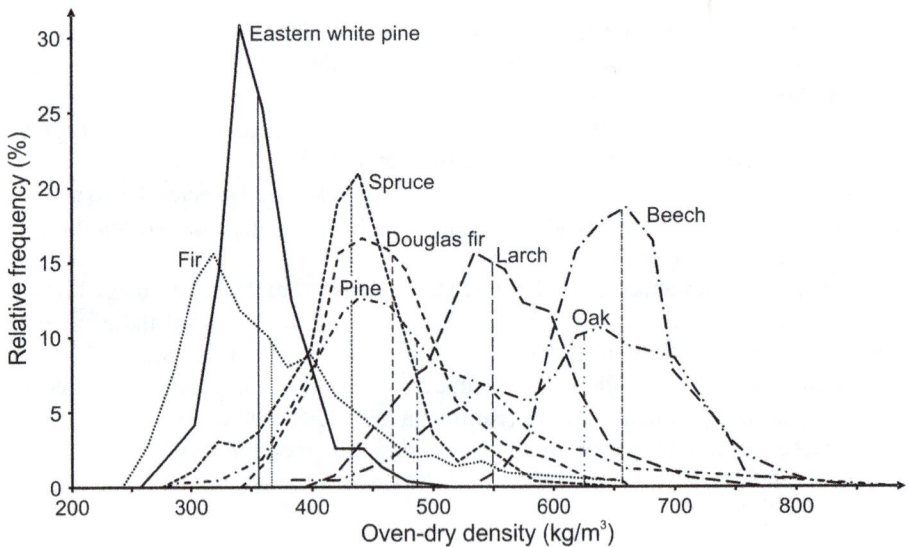

FIGURE 6.3   Frequency distribution of the density of various wood species.

(Knigge and Schulz 1966, courtesy of Carl Hanser Verlag.)

Soil and climate also have a decisive effect on the growth of the wood. The density decreases with the same soil conditions from the climatic optimum to colder locations (Mette 1984). It is lower at higher altitudes than at lower levels (Figure 6.4). In southern Sweden (Grönlund *et al.* 1992), the density of Norway spruce is higher than in northern Sweden.

Due to the growth conditions, the growth-ring width changes significantly in softwoods. It is, for example, higher in the Prealps than in the high altitudes of the Alps. Similar tendencies exist in other regions.

Even within a forest, there are differences in the trees' diameter, height, and stem shape.

The relationships have been studied less in hardwood. Schwappbach (quoted in Kollmann 1951) found that the density of European beech in the Northern Hemisphere decreases evenly from south to north and from lower to higher altitudes.

The density is lower in the predominant trees than in the suppressed trees.

In the case of Norway spruce, the basic density within a location increases with decreasing wood diameter.

Trendelenburg (1939) and Trendelenburg and Mayer-Wegelin (1955) also combine such trends and their causes. Likewise, Walker (2006) and Butterfield (1997) give an excellent overview of the variation in the density of plantation woods.

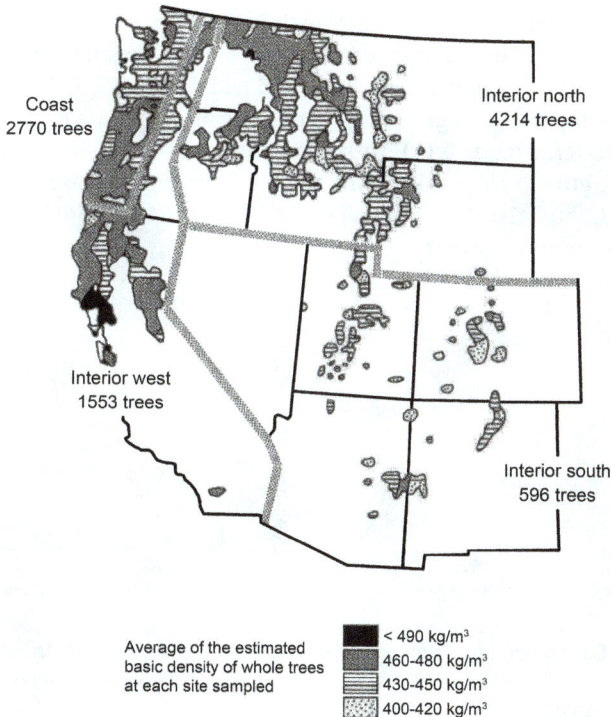

FIGURE 6.4   Basic density distribution of Douglas fir in the USA.

(USDA 1965, courtesy of U.S. Department of Agriculture, USDA, Forest Products Laboratory.)

6.4 INFLUENCE OF STRUCTURAL PARAMETERS ON DENSITY

6.4.1 EARLYWOOD AND LATEWOOD

Earlywood has a lower density than latewood, and the frequency distribution of the density may also differ (Figure 6.5).

Lanvermann (2014) determined minimum values of 200 kg/m^3 in earlywood and maximum values of about 1000 kg/m^3 in latewood of Norway spruce.

Yew is a slow-growing softwood with an exceptionally high density (590–670 kg/m^3). This results from the high density of the earlywood (500–600 kg/m^3) compared to the latewood, which has a density of around 1000 kg/m^3 – typical of softwoods (Keunecke, 2008). The latewood's maximum density values are similar for Norway spruce and yew.

There is a strong correlation between latewood content and mean density in softwoods (Figure 6.5). The relation between density and earlywood formation in hardwoods is more complicated; see, for example, Kollmann (1941) for ash wood.

As the growth rings widen, the proportion of latewood decreases, leading to a reduction in the density of softwood. When the growth-ring width increases, some softwoods exhibit a marked maximum density, often around 2 mm in ring width. Thereafter, the density drops (Figure 6.6).

Growing conditions may influence density. For example, Norway spruce from the Alpine foothills usually has a larger growth-ring width and a higher density than those from high altitudes (Trendelenburg and Mayer-Wegelin 1955). Thinning a forest stand significantly impacts growth speed and density.

The relationship between growth rings and density in hardwoods depends on the vessel structure (*cf.* Figure 2.11). In ring-porous hardwoods, the density increases with increasing growth-ring width as the latewood content increases with the growth-ring width and, thus, the density (Knigge and Schulz 1966). Such relations do not exist for diffuse-porous species.

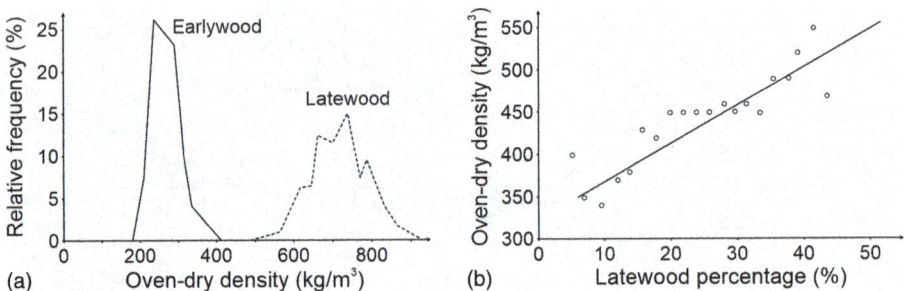

(a)   Oven-dry density (kg/m^3)   (b)   Latewood percentage (%)

FIGURE 6.5 Earlywood and latewood in softwoods: (a) frequency distribution of the density of earlywood and latewood in Douglas fir, and (b) density as a function of latewood content in Norway spruce.

(Knigge and Schulz 1966, courtesy of Carl Hanser Verlag.)

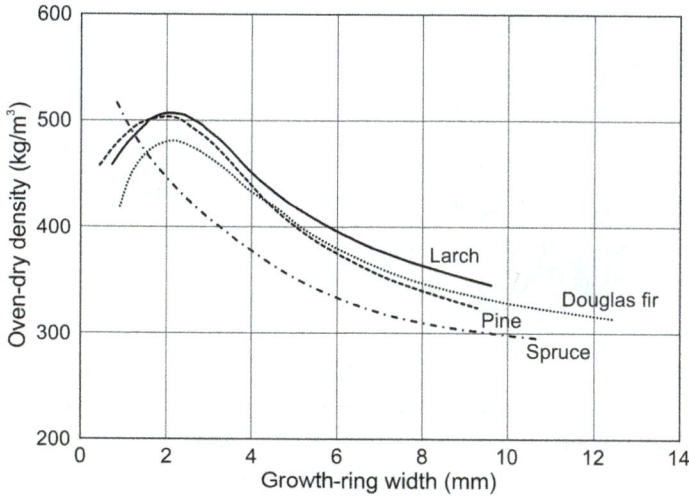

FIGURE 6.6 Oven-dry density as a function of growth-ring width for Douglas fir, European larch, Norway spruce and Scots pine.

(Knigge and Schulz 1966, courtesy of Carl Hanser Verlag.)

6.4.2 TREE AGE

Softwoods form denser wood with increasing age, independent of the growth-ring width. On the other hand, hardwoods' density decreases with tree age.

Growth rings, along with the width of earlywood and latewood and the density variations within the rings, are key factors in determining the age of wood in dendrochronology. The growth-ring pattern can also offer valuable insights into the climatic conditions of past periods (dendroclimatology).

For age determination, a continuous annual cycle is required up to the present. The longest European comparative chronology extends back to 5289 BCE. Dendrochronology and dendroclimatology, which focus on developing the Earth's environment over long periods, have become key fields in climate research. Age determination through growth-ring analysis is now an established method, as demonstrated in works such as Schweingruber (2007) and Günther (2013).

In the tropics, trees do not follow an annual growth cycle but grow in response to the varying timing of rainy seasons. As a result, there are no distinct yearly growth rings, although irregularities can be observed under a microscope (Anon. 2003).

Fast-growing wood from plantations, for example, radiata pine and various eucalyptus species, is often harvested after only 10–20 years as round timber for sawmilling (in Scandinavia 60–100 years). Practically, such timber contains only juvenile wood, the density is low, and the microfibril angle in the S2 cell-wall layer is more significant than that of slow-grown timber of the same species (Butterfield 1997).

6.4.3 HEARTWOOD AND SAPWOOD

Heartwood formation ends the physiologically active life phase of the tree's sap-wood. It increases density by incorporating extractives into the cell walls, making the heartwood darker.

6.4.4 WOOD FROM BRANCHES, ROOTS, AND REACTION WOOD

The wood from branches is commonly heavier than wood from the trunk, and wood from the roots is lighter. The branch wood of the Norway spruce can reach twice the density of the surrounding knotless wood. According to Mette (1984), the density of Norway spruce branch wood is approximately 900 kg/m³, and clear stem wood is, on average, 450 kg/m³. The density of the branch wood of Scots pine, however, is only about 600 kg/m³. In common beech, the differences between branch and trunk wood are smaller: the density of branches is about 750 kg/m³.

The density of compression wood (typical for softwood) is up to 40% higher than that of normal wood.

6.4.5 DENSITY DISTRIBUTION IN THE TRUNK

Figure 6.7 illustrates the variation in density within a tree stem. Density differences occur both along and across the trunk axis, and the wood species influences these variations. Mette (1984) provides the distribution of transverse density, shown in Figure 6.8, for different species. Overall, there are notable local density variations, meaning the density of the wood exhibits considerable differences within and between individual trees.

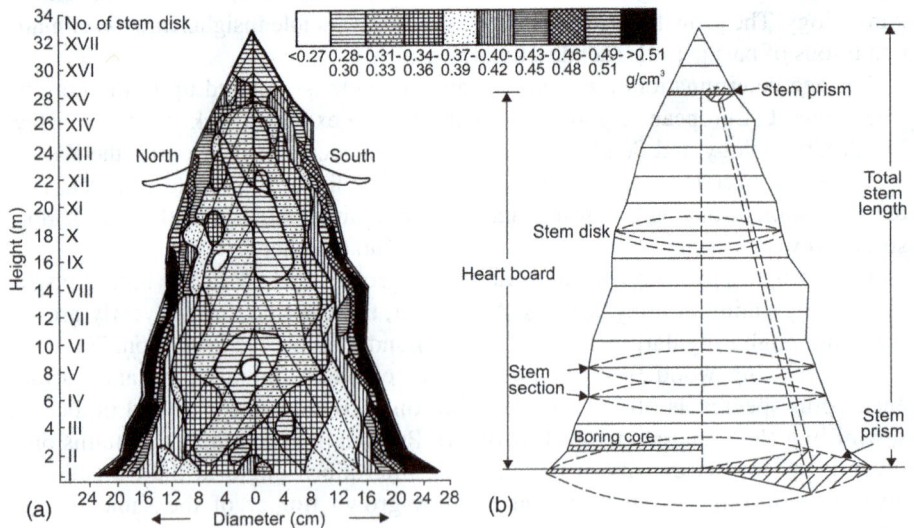

FIGURE 6.7 Basic density distribution in a fir stem: (a) density distribution and (b) sampling scheme for the density measurements.

(Knigge and Schulz 1966, courtesy of Carl Hanser Verlag.)

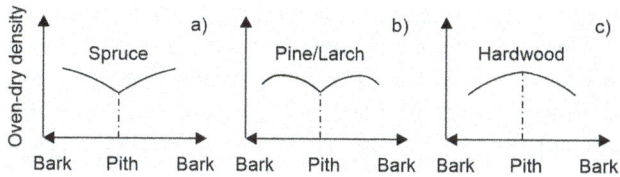

FIGURE 6.8 Principal distribution of oven-dry density across the stem cross-section: (a) spruce, (b) pine, and (c) hardwood models.

(Mette 1984, courtesy of Carl Hanser Verlag.)

6.4.6 Particleboard and Fibreboard

Due to manufacturing processes, particleboards and fibreboards exhibit significantly lower density variations than sawn timber and engineered wood products (EWPs) made from sawn timber. This is due to the properties of the particles (such as size and homogenisation) and the mat formation process. During mat formation, the density is carefully controlled and regulated.

The coefficient of variation for the average density of wood-based materials ranges from 3% to 5%, whereas for solid wood, it is between 15% and 20%.

Both spread density and bulk density are influenced by particle geometry and wood species.

6.5 METHODS FOR DENSITY AND POROSITY MEASUREMENTS

The determination of wood density is crucial for several reasons. Many properties of wood and wood-based materials, such as strength, elasticity, shrinkage, and swelling, are closely related to density. As a result, density or density profiles are used to grade wood and wood-based materials.

6.5.1 The Conventional Method for the Determination of Density

The density of wood and wood-based materials is usually determined using test specimens with constant dimensions. The density results from the mass and volume (Equation 6.1). Table 6.3 shows standard test specimen dimensions.

TABLE 6.3

Test Specimen Dimensions and Measurement Accuracy in Determining the Density of Solid Wood According to the DIN 52182 Standard (DIN 1976) and Wood-based Panels According to the EN 323 Standard (CEN 1993)

Parameter	Solid Wood	Wood-based Panels
Length	No restrictions	50 mm
Width	No restrictions	50 mm
Thickness	No restrictions	Board thickness
Accuracy of mass determination	0.01 g/cm³	0.01 g
Accuracy of the determination of dimensions	1% (volume)	0.10 mm in length and width; 0.05 mm in thickness

---

For the density determination of solid wood, for example, test specimens with dimensions of 20 mm × 20 mm × 30 mm are usual. Test specimens of wood-based materials usually have the dimensions 50 mm × 50 mm × board thickness. The density test is often also performed on specimens (or parts of this) intended for other tests (*e.g.*, on bending specimens). This allows a direct correlation of the density with the respective tested property.

6.5.1.1 Mass Determination
The mass is usually determined through a balance. Table 6.3 gives the required measurement accuracy.

6.5.1.2 Volume Determination
The volume of regularly shaped specimens is determined by measuring their length, width, and thickness. For irregularly shaped specimens, such as chips, displacement methods can be used to easily assess volume. This can be done using overflow or immersion vessels with a riser or a pycnometer (Figure 6.9). Before immersion in water, the specimens are coated with a thin layer of paraffin to prevent water absorption by the wood.

6.5.2 Other Methods for the Determination of Density
For the continuous determination of the bulk density, electromagnetic wave absorption is used to an increasing extent. X-rays are primarily used. The method is based on the effect that the electromagnetic waves are absorbed when passing through a material proportional to area-related mass.

The bulk density is thus considered the thickness:

$$\rho = \frac{ln\left(\dfrac{J_0 - J_N}{J - J_N}\right)}{\left(\dfrac{\mu}{\rho_m}\right)\cdot d} \tag{6.9}$$

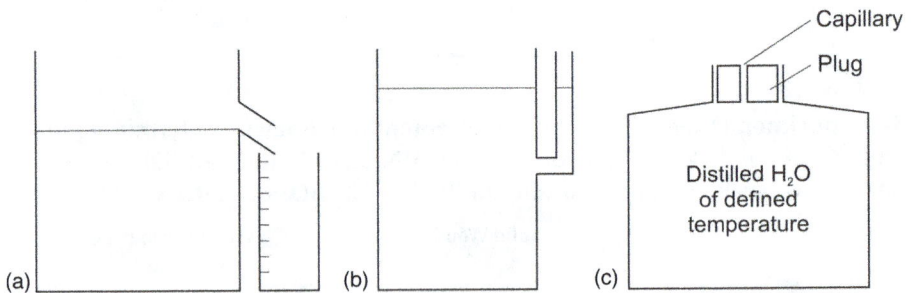

FIGURE 6.9 Volume determination of wood using displacement methods: (a) overflow vessel, (b) dip tank, and (c) pycnometer.

(Courtesy of Carl Hanser Verlag.)

where

J_0 is the count rate in air

J_N is zero effect

J is the count rate made by an absorber

$\dfrac{\mu}{\rho_m}$ is the mass attenuation coefficient (the attenuation coefficient normalised

by the density of the material)

d is the thickness of the sample

In this case, the radiation's absorption can be used to determine the average apparent density; the diffraction of the X-rays determines the density profile. An X-ray scattering system may also be used for the microfibril angle measurements (Butterfield 1997).

Figure 6.10 shows the density variation in spruce wood in dry and wet (green) conditions.

X-ray is used industrially to determine panel weight and the density profile in the thickness direction of the panels. It is also used to grade sawn timber, as it is easy to differentiate between clear wood and knots.

The wood moisture content and the mass attenuation coefficient influence the density measurement with X-rays. The mass attenuation coefficient for wood-based materials also depends on the adhesive content and the type of adhesives. The X-rays also determine local density differences and structural damage, such as insect holes.

Computed tomography (CT) is an X-ray method that gives 3D images of the wood structure describing voxel-wise density variation in the scanned volume. The industrial application of CT scanning for quality assessment of logs in sawmilling and veneer production is currently state-of-the-art. The methodological principle is creating a spatial image of the examined body, which allows the localisation of defects in the wood. In research, X-ray microtomography and synchrotron light are used, which give considerably higher resolutions than industrial CT scanners (Zauner 2014).

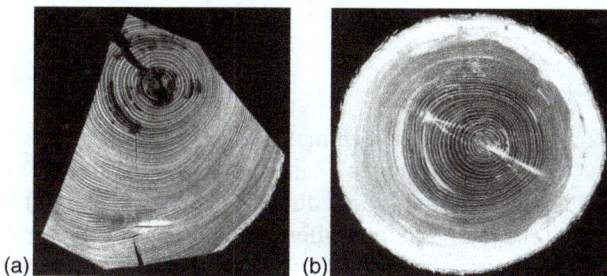

(a)   (b)

FIGURE 6.10 X-ray computed tomography (CT) of Norway spruce discs in (a) an air-dry condition with some regions of rot (black spots in the wood) and (b) in a green state (high moisture content over 100% in the sapwood (white regions).

(A. Flisch, EMPA, Dübendorf, Switzerland, courtesy of Carl Hanser Verlag.)

X-ray radiation is used for transmission measurements, which provide the average density over the measured thickness or can be used to determine a density profile in, for example, the thickness direction of a panel. For example, the panel specimen can be moved stepwise by 0.1 mm increments to measure its density profile.

The SilviScan™ is an X-ray-based equipment that measures the microfibril orientation in the cell-wall S2-layer. This method allows for determining the microfibril orientation distribution in the growth rings and estimating the modulus of elasticity (Keunecke 2008, Lanvermann 2014).

Due to its importance for processing wood-based materials, several other methods for determining density and density profiles have been developed.

6.5.3   METHODS FOR THE DETERMINATION OF POROSITY

6.5.3.1   Mercury Intrusion Porosimetry

The method is outlined in the ISO 15901-1 standard (ISO 2016) and is employed to experimentally determine the pore content and pore size distribution in wood and wood-based materials. This technique leverages the physical principle that a non-wetting liquid only penetrates capillaries when applied under pressure. The smaller the capillaries, the higher the pressure required for penetration. From the applied pressure, the capillary radius can be calculated.

The recorded plot can determine both the pore radius and pore volume. The Washburn equation calculates the pore radius based on the applied pressure (Washburn 1921). The method can measure pores ranging from 58 µm to 1.8 nm. Researchers such as Schneider (1979, 1982), Schweitzer and Niemz (1991), and Plötze and Niemz (2011) have conducted extensive measurements.

6.5.3.2   Gas Adsorption

Gas adsorption measurements can also be used to calculate the pore size distribution (Popper and Bariska 1972). According to Kelvin equation, the equilibrium pressure within a capillary is lower than the general saturation pressure, provided the contact angle is less than 90° (Popper 1985). Under these conditions, vapour will condense in all capillaries whose radius is smaller than the value calculated from this equation for the given pressure.

6.5.3.3   Other Methods

Other methods for measuring pore size include the pressure plate technique, which is commonly used in building physics (Zauer et al. 2016), and thermoporosimetry, which relies on the principle of dynamic differential calorimetry and heat flow measurements. Zauer's (2011) dissertation offers a comprehensive compilation of these methods and provides an overview of the pore sizes detectable by each technique. Additionally, partial X-ray or synchrotron tomography is utilised to detect larger pores (Hass 2012, Zauner 2014).

REFERENCES

Anon. (2003). *Holz-Lexikon: Nachschlagewerk für die Holz- und Forstwirtschaft.* (4th Ed.), DRW-Verlag Weinbrenner, Germany, (1460 pp.).

Butterfield B.G. (Ed.) (1997). Proceedings of International Workshop on the Significance of Microfibril Angle to Wood Quality. *International Association of Wood Anatomists, International Union of Forestry Research Organizations.* University of Canterbury, Westport, New Zealand, (410 pp.).

CEN (1993). EN 323: Wood-based panels - Determination of density. *The European Committee for Standardization (CEN)*, Brussels, Belgium.

CEN (2009). EN 316: Wood fibre boards - Definition, classification and symbols. *The European Committee for Standardization (CEN)*, Brussels, Belgium.

DIN (1976). DIN 52182: Testing of wood; determination of density. *Deutsches Institut für Normung (DIN)*, Berlin, Germany.

Grönlund A., Grönlund U. & Hagmann O. (1992). *Nordkalottenfura. [Scots Pine from Northern Sweden.]* Luleå University of Technology, Skellefteå, Sweden, (45 pp.).

Günther, B. (2013). *Erarbeitung einer Methode zur Erfassung von dendrochronologisch relevanten Jahrringmerkmalen der Trauben-Eiche (Quercus petraea [Matt.] Liebl.) auf Grundlage der Röntgendensitometrie.* Doctoral Thesis, TU Dresden, Dresden, Germany.

Hass P.F.S. (2012). *Penetration behavior of adhesives into solid wood and micromechanics of the bondline.* Doctoral Thesis, ETH Zürich, Zurich, Switzerland.

ISO (2016). ISO15901-1: Valuation of pore size distribution and porosity of solid materials by mercury porosimetry and gas adsorption. Part 1: Mercury porosimetry. *International Organization for Standardization (ISO)*, Geneva, Switzerland.

Keunecke D. (2008). *Elasto-mechanical characterisation of yew and spruce wood with regard to structure-property relationships.* Doctoral Thesis, ETH Zürich, Zurich, Switzerland.

Knigge W. & Schulz H. (1966). *Grundriss der Forstbenutzung.* Parey, Hamburg, Germany, (584 pp.).

Kollmann F. (1941). *Die Esche und ihr Holz.* Julius Springer, Berlin, Germany, (12+147 pp.).

Kollmann F. (1951). *Technologie des Holzes und der Holzwerkstoffe: Part 1: Anatomie und Pathologie, Chemie, Physik, Elastizität und Festigkeit.* Springer Verlag, Berlin, Germany, (14+1051 pp.).

Kollmann F. & Côté Jr. W. A. (1968). *Principles of Wood Science and Technology. Part 1: Solid Wood.* Springer, Berlin, Heidelberg, Germany. (12+592 pp.).

Lanvermann C. (2014). *Sorption and swelling within growth rings of Norway spruce and implications on the macroscopic scale.* Doctoral Thesis, ETH Zürich, Zurich, Switzerland.

Marutzky R. & Schwab H. (2009). Span- und Faserplatten, OSB. *Schriftenreihe: Informationsdienst Holz spezial.* Fraunhofer-Institut für Holzforschung (WKI), Verband der Deutschen Holzwerkstoffindustrie e.V. (VHI), Germany, (46 pp.).

Mette H.J. (1984). *Holzkundliche Grundlagen der Forstnutzung.* VEB Deutscher Landwirtschaftsverlag, Berlin, Germany, (148 pp.).

Niemz, P. & Sandberg, D. (2022). Critical wood-particle properties in the production of particleboard. *Wood Material Science & Engineering*, 17(5), 386–387.

Plötze M. & Niemz P. (2011). Porosity and pore size distribution of different wood types as determined by mercury intrusion porosimetry. *European Journal of Wood and Wood Products*, 69(4), 649–657.

Popper R. (1985). Das Holz-Sorbat-System mit Rücksicht auf die submikroskopische Betrachtungsweise. In: Kucera L.J. (Ed.), *Xylorama: Trends in Wood Research.* Birkhäuser Verlag Basel, Boston, Stuttgart, (pp. 155–163).

Popper R. & Bariska M. (1972). Die Acylierung des Holzes. 1. Mitteilung: Wasserdampf-Sorptionseigenschaften. *Holz als Roh- und Werkstoff* 30(8), 289–294.

Siau J.F. (1995). *Wood: Influence of Moisture on Physical Properties. Department of Wood Science and Forest Products*, Virginia Polytechnic Institute and State University, Blacksburg (VA), USA, (227 pp.).

Schneider A. (1979). Beitrag zur Porositätsanalyse von Holz mit dem Quecksilber-Porosimeter. *Holz als Roh- und Werkstoff*, 37(8), 295–302.

Schneider A. (1982). Untersuchungen über die Porenstruktur von Holzspanplatten mit Hilfe der Quecksilber-Porosimetrie. *Holz als Roh- und Werkstoff*, 40(10), 415–420.

Schweingruber F.H. (2007). Wood Structure and Environment. Springer, Berlin, Heidelberg, Germany, (279 pp.).

Schweitzer F. & Niemz P. (1991). Untersuchungen zum Einfluss ausgewählter Strukturparameter auf die Porosität von Spanplatten, *Holz als Roh- und Werkstoff*, 49(1), 27–29.

Trendelenburg R. (1939). *Das Holz als Rohstoff*. J.F. Lehmanns Verlag, München, Germany, (541 pp.).

Trendelenburg R. & Mayer-Wegelin H. (1955). *Das Holz als Rohstoff* (2nd Ed.) Carl Hanser Verlag, München, Germany, (541 pp.).

USDA (1965) *Western Wood Density Survey: Report Number 1*. USDA, Forest Products Laboratory, Research Paper FPL 27, Madison (WI), USA, (59 pp.).

Walker J.C. (2006). *Primary Wood Processing: Principles and Practice* (2nd Ed.), Springer, Dordrecht, The Netherlands, (10+596 pp.).

Washburn E.W. (1921). The dynamics of capillary flow. *Physical Review*, 17(3), 273–283.

Zauer M. (2011). *Untersuchungen zur Porenstruktur und zur kapillaren Wasserleitung in Holz und deren Änderung infolge einer thermischen Behandlung*. Doctoral Thesis, TU Dresden, Dresden, Germany.

Zauer M., Meissner F., Plagge R. & Wagenführ A. (2016). Capillary pore-size distribution and equilibrium moisture content of wood determined by means of pressure plate technique. *Holzforschung*, 70(2), 137–143.

Zauner M. (2014). *In-situ synchrotron based tomographic microscopy of uniaxially loaded wood: in-situ testing device, procedures and experimental investigations*. Doctoral Thesis, ETH Zürich, Zurich, Switzerland.

7 Other Important Physical Properties of Wood and Wood-based Materials

P. Niemz and W. Sonderegger

7.1 THERMAL PROPERTIES

7.1.1 THERMAL CONDUCTIVITY

Thermal conductivity (λ) is the quantity of heat which flows in one second through a cube of one cubic metre by a temperature difference of one Kelvin between the opposite sides. The unit of measurement is W/(m $\cdot$ K).

Wood is a poor heat conductor compared to other construction materials (Table 7.1). Their structure substantially determines the thermal conductivity of wood and wood-based materials, for example, the conductivity in the fibre direction is 1.50 to 2.75, as high as perpendicular to the grain (*cf.* Table 7.2). Values of the latter for many wood species are listed in the *USDA Wood Handbook* (Ross 2010) and *Springer Handbook in Wood Science and Technology* (Niemz *et al.* 2023). Maku in Kollmann and Malmquist (1956) calculated the thermal conductivity of the net wood substance across the fibre to 0.42 W/(m $\cdot$ K) and 0.65 W/(m $\cdot$ K) parallel to the fibre direction, which the latter is in the range of the thermal conductivity of water. Perpendicular to the fibre direction, there is only a slight difference between the radial and tangential directions (Table 7.2). Generally, thermal conductivity is about 10% higher in the radial direction than in the tangential direction. This difference is highly influenced by the rays (increasing with increasing ray cell volume) in hardwoods and by the latewood (increasing with increasing latewood volume) in softwoods (Steinhagen 1977).

The thermal conductivity increases with increasing density (Figure 7.1). Kollmann (1951) presents relationships for wood at 12% moisture content and 27 °C, perpendicular (Equation 7.1) and parallel to the fibre direction (Equation 7.2):

$$\lambda_{\perp} = 0.026 + 0.195 \cdot \rho \cdot 10^{-3} \left(\text{W}/\left(\text{m} \times \text{K} \right) \right) \tag{7.1}$$

$$\lambda_{\parallel} = 0.026 + 0.46 \cdot \rho \cdot 10^{-3} \left(\text{W}/\left(\text{m} \times \text{K} \right) \right) \tag{7.2}$$

where ρ is the wood density in (kg/m^3).

DOI: 10.1201/9781003411994-7

TABLE 7.1

Thermal Conductivity and Specific Heat of Various Materials According to the EN ISO 10456 Standard (CEN 2007)

Material	Density (kg/m³)	Thermal Conductivity (W/(m · K))	Specific Heat (J/(kg · K))
Air	1.23	0.025	1008
Water at 10 °C	1000	0.60	4190
Ice at 0 °C	900	2.2	2000
Ice at −10 °C	920	2.3	2000
Limestone, medium hard	2000	1.4	1000
Concrete of medium-density	1800	1.15	1000
Copper	8900	380	380
Cast iron	7500	50	450
Sawn timber (perpendicular to the fibre)	450	0.12	1600
	500	0.13	1600
	700	0.18	1600
Plywood, LVL, edge-glued panels	300	0.09	1600
	500	0.13	1600
	700	0.17	1600
	1000	0.24	1600
OSB	650	0.13	1700
Particleboard	300	0.10	1700
	600	0.14	1700
	900	0.18	1700
Fibreboard (wood-based)	250	0.07	1700
	400	0.10	1700
	600	0.14	1700
	800	0.18	1700
Cement-bonded particleboard	1200	0.23	1500

Thermal conductivity perpendicular to the grain increases with increasing moisture content due to water's higher thermal conductivity than oven-dry wood (Figure 7.2).

TABLE 7.2

Thermal Conductivity in the Principal Directions in Wood for Wood and Wood-based Materials

		Thermal Conductivity	
Material	Density (kg/m³)	Parallel to the Fibre or Board Plane (W/(m · K))	Perpendicular to the Fibre[a] or Board Plane (W/(m · K))
Ash[b]	740	0.31	0.16–0.18
Balsa[c]			0.049–0.077
Common beech[d, e]	640–680	0.25–0.49	0.13–0.15/0.15–0.17
Common oak[c]		0.24–0.35	0.16–0.18
Norway spruce[b, d, e]	410–450	0.22–0.33	0.09–0.12/0.10–0.12
Scots pine[d]	500	0.32	0.13–0.15
Particleboard[d, f]	600–760	0.26–0.29	0.10–0.15
OSB[f]	620–660		0.10–0.11
Extruded particleboard[d]	720	0.12[g]–0.24[h]	0.20
MDF[f]	740–840		0.11–0.12
Wood-fibre insulation board[i]	40–250		0.037–0.052

[a] Thermal conductivity in tangential (first value) and radial (second value) direction
[b] At 20 °C and 15–16% moisture content (Kollmann 1951)
[c] Oak at 10% moisture content (Vorreiter 1949)
[d] At 10 °C and 12% moisture content (0% moisture content for extruded particleboard) (Schneider and Engelhardt 1977)
[e] At 10 °C and 13–14% moisture content (Sonderegger et al. 2011)
[f] At 10 °C and 65% RH (Sonderegger and Niemz 2009)
[g] In the extrusion direction
[h] Perpendicular to the extrusion direction
[i] At 10 °C and 65% RH (Sonderegger and Niemz 2012)

To calculate the thermal conductivity from a given value to a value at different environmental conditions, the influence of moisture content in the range 5–35% is estimated according to Kollmann and Côté (1968):

$$\lambda_2 = \lambda_1 \left[1 - 0.0125 \left(\omega_1 - \omega_2 \right) \right] \qquad (7.3)$$

where

$\lambda_{1,2}$ is the thermal conductivity (W/(m · K)) at moisture content ω_1 and ω_2, and ω_1, ω_2 are the moisture content (%) at time/condition 1 and 2.

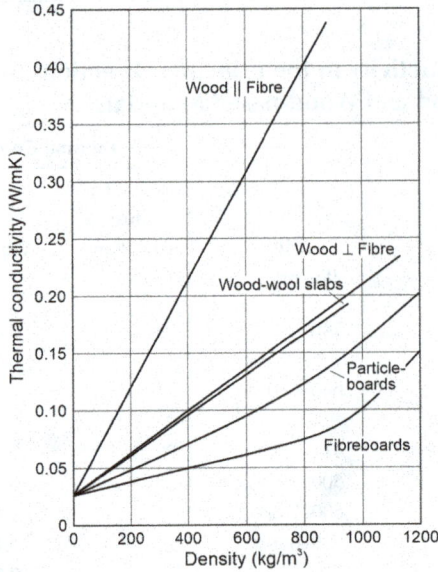

FIGURE 7.1 Thermal conductivity of wood-based materials perpendicular to the plane compared with wood.

(Data from Kollmann 1951, Kollmann and Malmquist 1956, courtesy of Carl Hanser Verlag.)

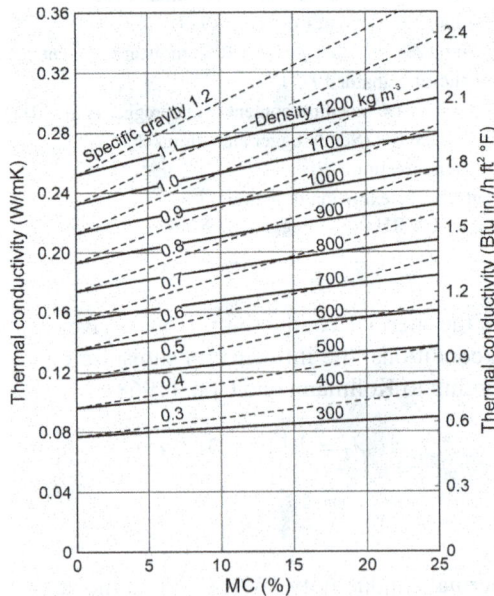

FIGURE 7.2 Thermal conductivity of wood perpendicular to the grain as a function of the moisture content and density or specific gravity, according to measurements in Kollmann and Côté (1968) and Ross (2010).

(Courtesy of Carl Hanser Verlag.)

7.1.2 Specific Heat

Specific heat (c) is defined as the heat quantity required to raise the temperature of one kilogram of a material by one Kelvin. The unit of measurement is J/(kg · K).

The specific heat of wood and wood-based materials highly depends on moisture content due to the high specific heat of water, which increases with increasing moisture content and temperature. In contrast, it is not influenced by density. The dependency of temperature and moisture content can be described as follows (Kollmann and Côté 1968):

$$c_0 = 1114 + 4.86 \cdot \theta \tag{7.4}$$

where

$$c_\omega = \frac{\omega \cdot c_w + c_0}{1 + \omega} \tag{7.5}$$

c_0 – heat capacity of oven-dry wood (J/(kg · K))
c_ω – heat capacity of wood at moisture content ω (J/(kg · K))
c_w – heat capacity of water (J/(kg · K))
θ – temperature (°C)
ω – moisture content (-)

A mean heat capacity value in the temperature range of 0 to 100 °C is suggested for drying technology (Kollmann and Côté 1968). Still, for application in building physics, a value of 20 °C or even lower is more appropriate (Sonderegger *et al.* 2011).

The specific heat of wood is about four times higher than that of iron and copper. Therefore, wood is well-suited for handles of cooking utensils and heating devices due to its high specific heat and low thermal conductivity. Table 7.3 shows the influence of moisture content on the specific heat of diverse wood species. In contrast to moisture content, the influence of wood species is low. Wood-based materials have a specific heat, like wood. Adhesives may influence it. Thus, according to Czajkowski *et al.* (2016), the specific heat of oriented strand board (OSB) at 6–7% moisture content is about 100 J/(kg · K) higher than that of low-density fibreboard and particle-board.

The high specific heat and the low thermal conductivity are advantageous for wood-fibre insulation boards compared to other insulating materials. These properties induce a clearly higher damping of the temperature amplitude (decrement factor) and a considerably higher phase shift (decrement delay) from the outdoor to the indoor climate compared to, for example, glass and stone/mineral wool insulation (Table 7.4). For example, wood-fibre insulation boards from the Pavatex company calculated the decrement delay with a high specific heat compared to "normal values" of mineral wool insulation (approximately 1000 J(kg·K)). The specific heat of wood-fibre insulation boards often shows lower values (1360–1630 J/(kg · K)) than those shown in Table 7.4 due to their lower moisture content (Ghazi *et al.* 2003, Czajkowski *et al.* 2016).

TABLE 7.3

Specific Heat Depending On Moisture Content (ω)

Species	Specific Heat (J/(kg · K))					
ω (%):	0	5	10	20	50	100
Common beech	1460	1600	1710	1920	2310	2830
Common oak	1450	1590	1670	1910	2370	2790
Norway spruce	1350	1510	1630	1800	2180	2800
Scots pine	1410	1540	1660	1870	2330	2800

TABLE 7.4

Specific Heat, Decrement Factor, and Decrement Delay of Wood-fibre Insulation Board (WFIB) Compared with Other Insulating Materials: Insulation Thickness 180 mm (or 160+20 mm), According to Pavatex SA, Switzerland[a]

Product	Decrement Factor (–)	Density (kg/m³)	Specific Heat (J/(kg · K)	Decrement Delay (h)
WFIB (Pavatex)[a]	0.09	140	2100	11.7
Cellulose (+ WFIB 20 mm)	0.16	45	1940	8.7
Flax	0.20	30	1550	7.4
Cotton	0.21	20	1900	7.1
Sheep wool (+ WFIB 20 mm)	0.22	25	1300	7.0
Rockwool	0.21	40	1000	6.7
Polystyrene	0.22	20	1500	6.3
Mineral wool	0.23	20	1000	5.9

Note: WFIB – wood-fibre insulation board.

[a] A significantly lower value is usually given in the literature. According to the EN ISO 10456 standard (CEN 2007), it is 2000 J/(kg · K); according to the withdrawn standards EN 12524 (CEN 2000), it is 1400 J/(kg · K), and according to Ghazi *et al.* (2003), specific heat for WFIB is 1361–1632 J/(kg · K).

7.1.3 THERMAL DIFFUSIVITY

Thermal diffusivity is a measure of how quickly heat diffuses through a material. It is defined as the ratio of thermal conductivity to the product of density and specific heat capacity, essentially indicating how rapidly a material's temperature changes in response to a temperature difference. It is influenced by thermal conductivity, specific heat, density, and, inversely, by moisture content. It is calculated as follows:

$$a = \frac{\lambda}{c \cdot \rho} \tag{7.6}$$

a – thermal diffusivity $\left(\text{m}^2 / \text{s}\right)$

c – specific heat $(\text{J}/(\text{kg} \cdot \text{K}))$

ρ – density (kg/m^3)

λ – thermal conductivity $(\text{W}/(\text{m} \cdot \text{K}))$

In practice, thermal diffusivity becomes important for wood, *i.e.*, for drying, bonding, and compression (*e.g.*, in the production of wood-based materials such as particle and fibreboards, laminated veneer, or solid wood), and for evaluating low- and high-temperature performance. Table 7.5 shows the influence of wood density and moisture content on thermal conductivity, specific heat, and thermal diffusivity.

7.1.4 THERMAL EXPANSION

Thermal expansion is characterised by the coefficient of thermal expansion, which is defined as the linear extension of a rod of 1 m at a temperature difference of one Kelvin:

$$\alpha_{th} = \frac{\Delta l}{l_0 \cdot \Delta t} \tag{7.7}$$

TABLE 7.5
Thermal Conductivity, Specific Heat, and Thermal Diffusivity of Wood at 27 °C Depending on Density and Moisture Content (ω) (Kollmann and Côté 1968)

Oven-dry Density (kg/m³)	ω (%)	Density (kg/m³)	Thermal Conductivity (W/(m·K))	Specific Heat (J/(kg·K))	Thermal Diffusivity (m²/s)
200	10	217	0.066	1610	1.89×10^{-7}
	20	233	0.074	1810	1.75×10^{-7}
	30	252	0.082	1990	1.64×10^{-7}
	50	287	0.098	2280	1.50×10^{-7}
	100	380	0.140	2750	1.44×10^{-7}
600	10	627	0.144	1610	1.42×10^{-7}
	20	657	0.162	1810	1.36×10^{-7}
	30	690	0.178	1990	1.31×10^{-7}
	50	776	0.216	2280	1.22×10^{-7}
	100	1030	0.306	2750	1.06×10^{-7}

where

α_{th} is the coefficient of thermal expansion (m/(m · K)) or (1/K)
Δl is the change in length (m)
l_0 is the initial length (m)
Δt is the temperature difference (K)

A linear correlation exists between thermal expansion and temperature, with a length increase by heating and a length decrease by cooling. The overall length l_2 after a temperature change from t_1 to t_2 and an initial length l_1 results in:

$$l_2 = l_1 \left[1 + \alpha_{th} \left(t_2 - t_1 \right) \right]$$ (7.8)

Wood species, density, and the principal directions in wood influence wood's thermal expansion coefficient (Table 7.6).

Thermal expansion has a lower impact compared to moisture-related swelling and shrinkage. In general, temperature change is combined with a moisture-content change; therefore, swelling and shrinkage appear. The swelling and shrinkage values are one order of magnitude higher than the thermal expansion values. The wooden coefficient of thermal expansion in the fibre direction accounts for $2.5 \cdot 10^{-6}$ to $11 \cdot 10^{-6}$ m/(m · K) in the temperature range of −60 °C to +50 °C. It increases by 2–3% in the fibre direction and 3–5% perpendicular to the fibre, with a decreasing moisture content of 1% unit.

TABLE 7.6

Linear Thermal Expansion of Dry Wood in the Range of −50 °C to +50 °C (Vorreiter 1949) and of Wood-based Materials in the Range of −40 °C to +60 °C (Sonderegger and Niemz 2009)

	Coefficient of Thermal Expansion (10^{-6}/K):		
	In the Fibre Direction or In-plane	Perpendicular to the Fibre Direction	
Wood species		Radial	Tangential
Balsa		16.3	24.1
Common ash	9.51		
Eastern white pine	3.65–4.00		63.6–72.7
Norway spruce	3.15–3.50	23.8–23.9	32.3–34.6
Red oak	3.43	28.3	42.3
Sugar maple	3.82–4.16	26.8–28.4	35.3–37.6
Yellow birch	3.36–3.57	30.7–32.2	38.3–39.4
MDF	6.80		
OSB	6.30		
Particleboard	6.20–7.20		
Plywood (larch)	4.20		

7.1.5 THERMAL INFLUENCE ON WOOD PROPERTIES

The force of attraction between the atoms in the wood material decreases as temperature and moisture content increase, resulting in greater elasticity. Typically, the wood softens, and its ductility increases significantly, allowing it to be permanently shaped, for example, through bending. The plasticising effect can be further enhanced by increasing the vapour pressure. These effects are utilised in applications such as wood bending in furniture making (*e.g.*, Thonét bentwood chairs) or thermo-mechanical pulping.

7.1.5.1 Thermal Influence of Short Duration

The modulus of elasticity (MOE) decreases with increasing temperature, and this behaviour is strengthened with increasing wood moisture. Figure 7.3 shows the relative change of MOE with increasing temperature as mean values of several wood species related to 20 °C (= 100%). Analogue tendencies exist for other elasto-mechanical properties or strengths (Figure 7.4).

Temperature change also influences other wood properties than elasticity, as can be seen in Table 7.7.

7.1.5.2 Long-term Effects of Temperature

The duration of thermal exposure influences the property change. The strength reduction with increasing duration is most substantial for impact bending strength. Further, sorption is influenced by long-term heating, which causes a decrease in the equilibrium moisture content (Hill 2006). Also, colour change occurs and is accelerated in a heated environment.

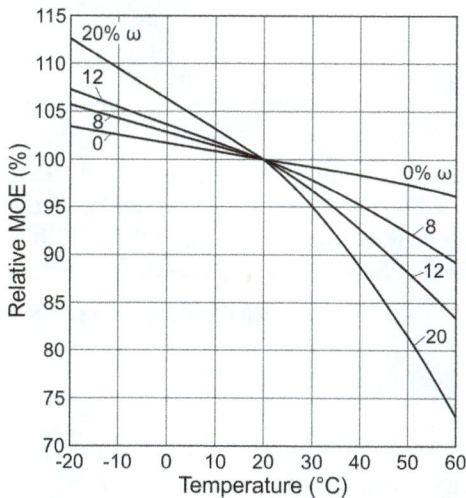

FIGURE 7.3 Relative change of wood's modulus of elasticity (MOE) depending on temperature. ω – moisture content.

(Data from Sulzberger in Kollmann 1951, courtesy of Carl Hanser Verlag.)

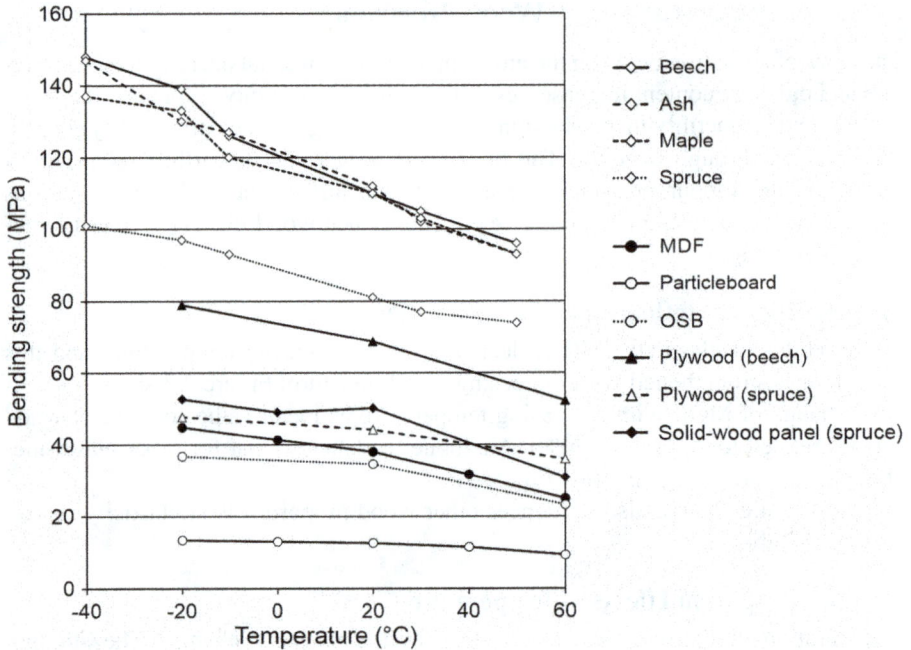

FIGURE 7.4 Bending strength of wood and wood-based panel materials as a function of temperature.

(Data from **Sonderegger and Niemz 2006, Niemz** *et al.* **2014, courtesy of Carl Hanser Verlag.**)

TABLE 7.7

Reduction of the Mechanical Properties at 100 °C Compared to at 20 °C for Clear Specimens of Norway Spruce at 7–10% Moisture Content (Kordina and Meyer Ottens 1983) and for Norway Spruce Sawn Timber (Glos and Henrici 1990)

Properties	Small Clear Specimens (%)	Sawn Timber (%)
Bending strength	55	28
Compression strength	51	44
Tensile strength	11	8
Modulus of elasticity in bending	27	12
Modulus of elasticity in tension		12
Modulus of elasticity in compression	35	25

Mönck and Erler (2004) recommend that, for the long-term effects of heat (35–50°C) on timber constructions, the allowable stress should be reduced to 80% of the standard value (and to 60–70% of the standard value at 50–80°C). This is particularly important in constructions in use at high temperatures, such as saunas and foundries.

7.1.5.3 Fire Properties

The fire properties of wood are described extensively in Chapter 11.

Wood is combustible and can, under fire, be described as a porous, carbon-forming solid. Its pyrolysis at fire is a highly complex process. Wood burns indirectly, which means combustion occurs as a reaction between oxygen and the gases released from wood. During heat exposure, wood efficiently produces substances that react eagerly with oxygen, leading to a high propensity of wood to ignite and burn. The reason is that the char layer reduces heat transfer and protects the underlying wood. The dimensions of the wood are also critical. Larger dimensions mean that the fire resistance of the wooden structure can be higher. The fire performance depends on the charring rate of the wood and on the reduced strength and stiffness that might occur during an extended fire exposure. The thickness of the layer with reduced strength is time-dependent; it is thin during the early fire stage and up to about 40 mm in the fully developed fire. The load-bearing capacity of a wooden structural element at fire exposure depends on its strength and rigidity and the joints between the components. Both the elements and the connections must fulfil their function during the intended time.

7.1.6 APPLICATION OF THERMAL PROPERTIES FOR QUALITY CONTROL

The thermography distinguishes differences in density and moisture content (only high differences), localisation knots, and other surface defects (López *et al.* 2014). When applied to laminated wood-based panels, different surface temperatures (cold and hot spots) indicate defects within the glued joint or the carrier material (Meinlschmidt 2005). Further, the method is widely implemented to control heat insulation in the building sector (Figure 7.5).

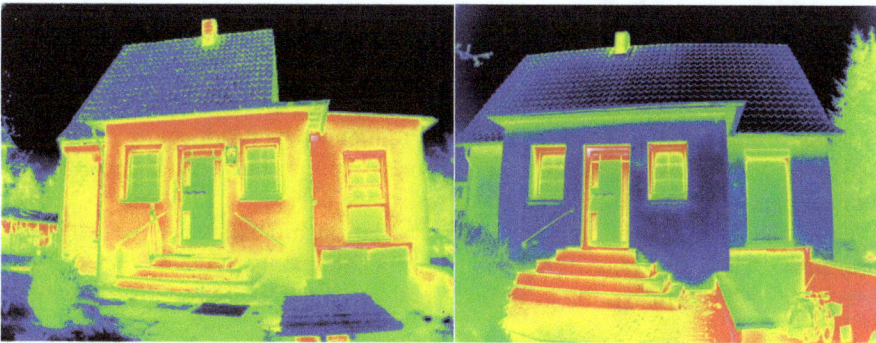

FIGURE 7.5 Thermograms of a building before (left) and after (right) restoration. Blue – low thermal radiation and red – high thermal radiation.

(P. Meinlschmidt, WKI, Braunschweig, courtesy of Carl Hanser Verlag.)

7.2   ELECTRICAL AND DIELECTRIC PROPERTIES

7.2.1   ELECTRICAL RESISTANCE AND CONDUCTIVITY

We use the following parameters to characterise the electrical properties of wood:

Electrical resistance (R) refers to the opposition that wood presents to the flow of an electrical current. This phenomenon occurs because the charge carriers must navigate between the atoms. In the process, they interact with the atoms, causing them to vibrate and converting electrical energy into heat energy.

Electrical resistivity (ρ) or specific electrical resistance is the resistance of a material with a cross-sectional area A and a length l:

$$\rho = \frac{R \cdot A}{l}\,(\Omega m)$$

(7.9)

Electrical conductance (G) is the ability of the substance to propagate electricity (*i.e.*, electrons and/or ions) from the point of origin throughout the body. G is the reciprocal of the electrical resistance ($G = 1/R$) with the unit Siemens (S).

Electrical conductivity (κ) or specific electrical conductance is analogous to the reciprocal of electrical resistivity:

$$\kappa = \frac{G \cdot l}{A}\left(\frac{S}{m}\right) \text{or} \left(\frac{1}{\Omega m}\right)$$

(7.10)

where

G – specific electrical conductance = $1/R$ (S), with the unit Siemens

Properties that influence the electrical resistance of wood are as follows.

7.2.1.1   Moisture Content

The moisture content of wood significantly influences its conductivity or electrical resistance. Very dry wood is a good insulator. If the moisture content increases, the conductivity increases, and the electrical resistance decreases.

Up to the fibre-saturation point, a linear relationship exists between the moisture content and the logarithm of the electrical conductivity or the electrical resistance. Above this point, the influence of moisture is significantly lower (Figure 7.6).

Determining the conductivity and electrical resistance of wood is important, as the relationship between these factors and moisture content forms the basis for the operation of wood moisture metres.

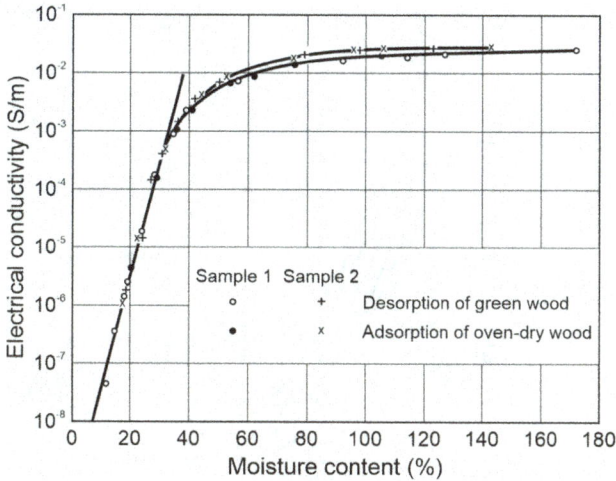

FIGURE 7.6   Dependence of electrical conductivity on moisture content in redwood.

(Data from Stamm 1964, courtesy of Carl Hanser Verlag.)

7.2.1.2   Temperature

The wood's temperature significantly influences the electrical resistance and, thus, the conductivity. As the temperature increases, the electrical resistance drops sharply (Figure 7.7).

Temperature influence must be considered when measuring electrical resistance, and therefore, it is also required when measuring the moisture content of wood and wood-based materials with electrical resistance-based equipment.

The reduced electrical resistance at high temperatures makes it possible to measure particle moisture content at low moisture levels, which is used in the wood industry to measure particle and fibre moisture content.

7.2.1.3   Wood Species, Structure, and Additives

Important parameters influencing the electrical properties of wood include the species, the three principal directions within the wood (longitudinal, radial, tangential), and its density. The influence of wood species is particularly significant due to variations in density and structure (Table 7.8). As a result, a species-specific characteristic curve is always used in electrical moisture measurement. Perpendicular to the fibre orientation, the electrical resistance of wood is approximately twice as high as in the fibre direction (Kollmann 1951).

When measuring the electrical resistance of particles or fibres, the density must be kept constant.

Particleboards used in data centres and clean rooms are manufactured by adding graphite to increase their conductivity, resulting in antistatic particleboards. The electrical resistance of these particleboards ranges from 10^5 to 10^9 Ω, compared to approximately 10^{11} Ω for solid wood. Special coating materials, such as high-pressure laminate (HPL) conductive panels made with carbon fibres, are also utilised.

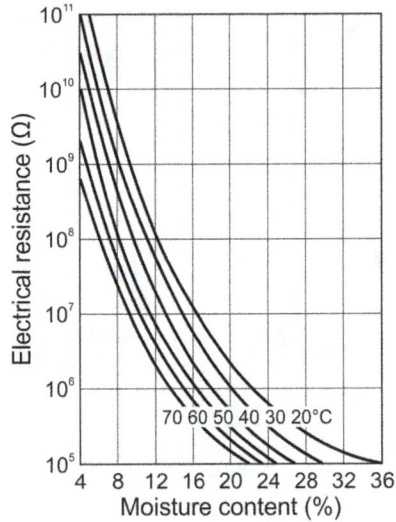

FIGURE 7.7 Influence of moisture content and temperature on the electrical resistance of wood.

(Data from Keylwerth and Noack 1956, courtesy of Carl Hanser Verlag.)

TABLE 7.8

Electrical Resistivity Perpendicular to the Grain for Various Species Depending on Moisture Content (Ugolev 1986)

Species	Electrical Resistivity (Ω m) at Different Moisture Content:		
	0%	**7%**	**20%**
Alder	$1.0 \cdot 10^{15}$	$9.0 \cdot 10^9$	$6.0 \cdot 10^6$
Birch spp.	$5.1 \cdot 10^{14}$	$9.0 \cdot 10^9$	$1.0 \cdot 10^6$
European oak	$1.5 \cdot 10^{14}$	$2.0 \cdot 10^9$	$7.0 \cdot 10^6$
Norway spruce	$7.6 \cdot 10^{14}$	$1.0 \cdot 10^{10}$	$3.0 \cdot 10^6$
Scots pine	$2.3 \cdot 10^{13}$	$5.0 \cdot 10^9$	$3.0 \cdot 10^6$

7.2.1.4 Dielectric Properties

Wood has dielectric properties. In its dry state, it can be considered a good insulator and a semiconductor in the air-dry state. However, no insulating ability is left when the fibre-saturation region is reached.

The relative permittivity (earlier called the dielectric constant) indicates how many times higher the dielectric conductivity of a material is than that of the vacuum ($\varepsilon_{r.vacuum} = 1$):

$$\varepsilon_r = \frac{\varepsilon}{\varepsilon_0} \qquad (7.11)$$

where

ε_r is the relative (index r) permittivity or dielectric constant

ε is permittivity of the material or dielectric conductivity, in farad per metre (F/m)

ε_0 is the vacuum permittivity, also called electric constant, is $8.854 \cdot 10^{-12}$ F/m

Influences on the dielectric constant of wood are as follows.

7.2.1.5 Frequency

The relative permittivity is the ratio of the capacitance formed by two plates with material between them to the capacitance of the same plates with air as the dielectric. The frequency of the applied voltage (over the plates) affects the dielectric constant. The relative permittivity of a material for a frequency of zero is known as its static relative permittivity.

7.2.1.6 Moisture Content

The moisture content of wood significantly influences the dielectric constant (Table 7.9). As the moisture content increases, the dielectric constant increases.

The wood species, the principal directions, and the grain angle significantly influence the parameters. Perpendicular to the fibre, the permittivity is lower than in the fibre direction. The differences between radial and tangential directions are minor. The temperature also influences the parameters (Ugolev 1986).

TABLE 7.9

Dielectric Constant of Wood Perpendicular to the Grain at 20 °C (Frequency: 2.4 GHz) Depending on the Type of Wood and the Moisture Content (Ugolev 1986, Torgovnikov 1993)

Species	Oven-dry Density (kg/m³)	Dielectric Constant Perpendicular to the Fibre Direction at Various Moisture Content:		
		0%	30%	80%
Aspen	470	1.7	3.7	7.8
Birch spp.	600	1.9	4.1	9.0
Norway spruce	420	1.7	3.5	7.0

7.3 ACOUSTIC PROPERTIES OF WOOD AND WOOD-BASED MATERIALS

7.3.1 OVERVIEW

Sound refers to mechanical vibrations or waves of an elastic medium. Depending on the type of medium in which the sound propagates, a distinction is made between airborne and structure-borne sound. Depending on the frequency, the sound differs in infrasound (not audible, frequency <16 Hz), audible sound (16 Hz to 20 kHz), and ultrasound (>20 kHz). Important acoustic properties are:

- the propagation velocity of sound waves in a material (sound velocity),
- acoustic emission (AE),
- sound absorption, and
- sound insulation.

7.3.2 PHYSICS OF SOUND

It is distinguished between two basic forms: longitudinal and transverse waves (Figure 7.8). For the longitudinal waves (P-waves), the propagation direction and the vibration (oscillation) direction are the same. For the transverse waves (referred to as shear waves or S-waves), the vibration direction is perpendicular to the propagation direction of the wave. In practice, both components usually occur, often making the evaluation difficult (*e.g.*, when measuring shear waves). The speed of transverse waves is significantly less than that of longitudinal waves. For example, Ozyhar *et al.* (2013) determined a sound velocity of 4682 m/s for longitudinal waves of beech at 20 °C/65% RH and 1485 m/s for transverse waves in the longitudinal-radial (LR) direction.

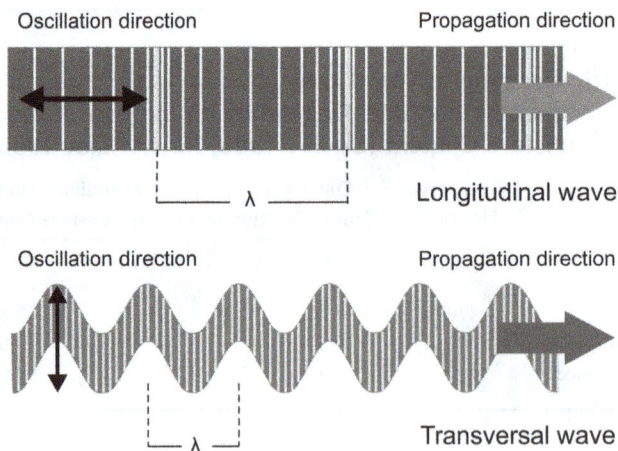

FIGURE 7.8   Different wave types.

(Courtesy of Carl Hanser Verlag.)

Testing of wood usually involves modified forms such as Lamb waves, Rayleigh waves, bending waves, or plate waves. Bending waves occur in rod-shaped or plate-shaped elements.

Dispersion is a property of certain waves, which means that the propagation velocity of the waves depends not only on external conditions such as temperature and medium but also on the wavelength or frequency of the wave. This effect must be considered for wood when working at different frequencies. Additionally, wave propagation is influenced by the tested component dimensions; see Schubert (2007) and Baensch (2015).

Further literature on ultrasound can be found, for example, in Krautkrämer and Krautkrämer (1986) and Bucur (2006).

7.3.2.1 Sound Velocity

Sound propagates as a longitudinal wave in a multi-directional broad medium (seawater, air). For longitudinal waves, the two-way impeded transverse contraction must be taken into account:

$$c_L = \sqrt{\frac{E}{\rho} \cdot \frac{1-\mu}{1-\mu-2\mu^2}} \qquad (7.12)$$

where

c is the sound velocity (m/s)
ρ is density (kg/m^3)
E is modulus of elasticity (Pa) (usually indicated in MPa or N/mm^2)
μ Poisson's ratio (-)

In a bar whose width and thickness are smaller than the wavelength, the sound propagates only as a strain wave or quasi-longitudinal wave. Therefore:

$$c_D = \sqrt{\frac{E}{\rho}} \text{ or } E = c_D^2 \cdot \rho \qquad (7.13)$$

Analogous to Equation 7.13, the shear modulus G is determined from the velocity of the transverse waves c_{ij} (first index: oscillation direction, second index: direction of wave propagation) and the density, as follows:

$$G = c_{ij}^2 \cdot \rho \qquad (7.14)$$

Equations 7.13 and 7.14 are often used to test specimen geometries other than long bars to calculate the modulus of elasticity. Depending on the frequencies used, this sometimes leads to significant inaccuracies.

Sound propagation is employed in the timber industry to grade sawn timber and standing trees. It is also used for quality control in the industry, including measurements of modulus of elasticity (MOE) and defect detection of sawn timber of glued-laminated timber (Chapters 19 and 20 in Niemz *et al.* 2023).

7.3.2.2 Wavelength

The wavelength λ is defined as:

$$\lambda = \frac{c}{f} \ (\mathrm{m})$$

(7.15)

where

  c is the sound velocity (m/s)
  f is frequency (1/s or Hz)

The wavelength, for example, provides information about the recognisability of failures. For instance, at a usual frequency for wood of 20 kHz and a sound velocity in fibre direction of 6000 m/s, a wavelength of 0.3 m results. Therefore, minor defects such as crushing and branches or the early stages of rot in trees are not recognisable. In contrast, when examining trees or beams for decay, large rots can be recognised due to the resulting longer path of the sound wave and, therefore, lower sound velocity (Schubert 2007, Niemz and Bächle 2002). The sound velocity reaches wood in grain direction values of 4000 to 6000 m/s and perpendicular to the grain values of 400 to 2000 m/s.

For the detection of defects in trees (extensive decay, hollows), the forced path change of the sound wave is used, as the wave propagates only in the solid state, not in air (Schubert 2007, Sanabria 2012).

7.3.3 Sound Velocity

The wood structure significantly influences sound velocity. All properties that correlate with the modulus of elasticity affect the speed of sound. Table 7.10 shows the sound velocity for different species and wood-based materials.

In solid wood, the sound velocity in the fibre direction is three to four times higher than that perpendicular to the fibre. In the tangential direction, it is always slightly smaller than in the radial direction. Particle-based materials have a significantly lower sound velocity than wood in the fibre direction due to the essentially random particle orientation. In oriented strand board (OSB), sound velocity is considerably higher in the direction of particle orientation than perpendicular to it. The sound velocity decreases when the moisture content increases (Figure 7.9).

The structural factors correlate with the modulus of elasticity and, thus, with sound velocity. Increasing the moisture content decreases sound velocity (associated with MOE).

7.3.3.1 Equipment for Sound Measurements

For sound measurements on wood, the following systems are used:

- Systems based on shock waves (also called stress-wave techniques). The waves are generated by impact at frequencies of some 100 Hz. Devices: for example, Fakopp and Picus Sonic Tomograph from Metrigard, USA (both used for tomography).

TABLE 7.10

Sound Velocity in Wood and Wood-based Materials at Normal Conditions (20 °C/65% Relative Humidity)

Material	Density (kg/m³)	Sound Velocity (m/s) in:		
		Fibre Direction	**Radial**	**Tangential**
Wood[a]				
Ash[b]	650	5330	2390	1680
Common beech[b]	680	5560	2180	1250
Common larch[b]	570	6660	2740	2100
Norway spruce[c]	400	5850	2130	1710
Red oak[b]	670	5350	2410	1780
Scots pine[b]	520	5690	2430	1900
Wood-based materials[d]		*In-board direction*	*Perpendicular to board direction*	
MDF (17 mm thickness)	770	2590		
OSB‖ (18 mm)	660	3290		
OSB⊥ (18 mm)	670	2990		
Particleboard (16 mm)	690	2090	600	
Particleboard (31 mm)	630	1110	470	

[a] Tested on cubes (1 cm³), frequency 2.25–10 MHz
[b] Test by Keunecke *et al.* (2011)
[c] Test by Keunecke *et al.* (2007)
[d] Bekhta *et al.* (2002) tested on bending specimens according to the DIN 52362 standard (DIN 1965) (recent: EN 310 standard (CEN 1993)) at 50 kHz; ‖ in the direction of particle orientation; ⊥ perpendicular to the particle orientation

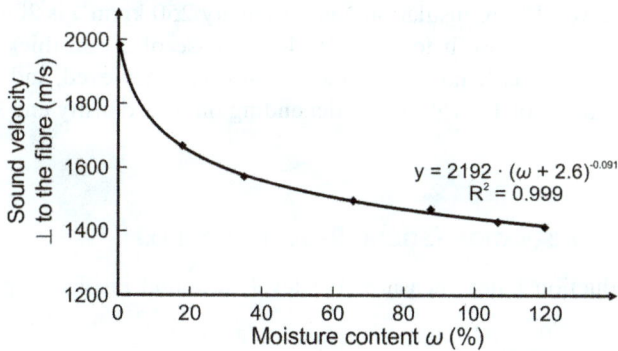

FIGURE 7.9 The sound velocity perpendicular to the fibre direction of roble wood as a function of the moisture content.

(Data from Niemz 1996, courtesy of Carl Hanser Verlag.)

- Devices based on piezoelectrical generated waves in the range of 20–50 kHz. Devices: for example, ultrasonic tester BP-700 from Steinkamp, Sylvatest. These devices usually show the transit time.
- Non-contact measuring devices (air-coupled ultrasound, frequency approximately 50 kHz marketed by for example, GreCon). Sanabria (2012) describes the principles of contactless coupled ultrasound and its use for delamination detection in glued-laminated timber.

Scientific studies for measuring elastic constants often use high-frequency devices of several MHz for longitudinal and transverse waves. Thus, transversal waves have considerably lower speeds than longitudinal waves.

A coupling to the wood with constant contact pressure is essential for reasonable measures, for example, special coupling gels or honey are used. Hasenstab (2006) also successfully used the pulse-echo method, *i.e.*, the transmitter and receiver are identical.

7.3.4 SOUND ATTENUATION OR SOUND ABSORPTION

The sound absorption (*S*) is the ratio of the non-reflected to the incident sound power:

$$S = \frac{K_s}{K_a} \tag{7.16}$$

where

K_s is non-reflected sound power (W)
K_a is incident sound power (W)

Sound absorption is caused by the dissipation of sound energy in the material (*e.g.*, conversion into heat) and by transmission. It is frequency-dependent and increases in particle materials with the material's porosity. The sound absorption of a 12-mm-thick wood-fibre insulation board (density 250 kg/m³) is 20–30%; punching or slitting can increase it to 60–80%. In the case of 5-mm-thick high-density fibreboards, on the other hand, only values of 5–8% are achieved, and in the case of particleboard, values of 10–15% result depending on bulk density and surface finish (Table 7.11).

7.3.5 SOUND INSULATION (SOUND REDUCTION INDEX)

The sound reduction index measures the level of sound insolation provided by a structure:

$$R = 10 \cdot \lg \frac{P_1}{P_2} \tag{7.17}$$

TABLE 7.11

Sound Absorption of Wood and Wood-based Materials (Anon. 1990)

	Sound Absorption (%)
Scots pine (sawn timber)	10–11
Particleboard	20–50
Sound absorption panels (*e.g.*, perforated)	60–80
Low-density fibreboard	20–30
High-density fibreboard	5–8

where

R is the sound reduction index (dB)
P_1 is the sound power striking the component (W/m^2)
P_2 is the sound power radiated on the back (W/m^2)

Good sound insulation is achieved through reflection and dissipation. The sound reduction index increases with the frequency of the sound waves and the surface dimensions of the material. Lightweight wooden structures are, therefore, only slightly suitable for sound insulation. A double-shell design, on the other hand, effectively insulates the sound. For particleboard with a surface density of 15–20 kg/m^2, the sound insulation is 24–26 dB. With each doubling of the surface density, the sound power is reduced by 5 dB (Anon. 1975).

7.3.6 ACOUSTIC EMISSION

Acoustic emission is the emission of sound waves in the audible and ultrasonic range (frequency > 20 kHz) caused by microscopic fractures, friction of fracture surfaces, outflow of liquids, transport processes in capillaries, or other effects. Therefore, the process is not non-destructive.

7.3.6.1 Practical Use of Acoustic Emission

The structure (density, fibre length, particle geometry), moisture content, type and percentage of adhesive, and history, for example, fungal or insect damage, mechanical or climatic pre-stresses, of wood or wood-based materials, all influence their properties. Changes in the wood induced by climatic conditions can be acoustically detected.

Acoustic emission analysis can identify fracture processes occurring during the creep or relaxation of wood and wood-based materials. There is a strong correlation between deflection and emission counts over time. Sound emission begins at a load factor of 30%. Under dynamic loading, a significant increase in emissions was observed with the increasing number of load cycles.

Comparative studies on bonded elements and wood-based materials demonstrated that acoustic emission analysis is particularly effective in detecting the brittleness of adhesive joints. The relatively brittle urea resins emitted significantly more pulses than other adhesives, such as polyvinyl acetate (PVAc) or one-component polyure-thane (1C-PUR) adhesives (Baensch 2015).

7.3.7 NATURAL FREQUENCY (EIGENFREQUENCY) AND MODAL ANALYSIS

The natural frequency (eigenfrequency) of a material excited to vibrate can be used as a criterion for determining the elastic constants. The process is increasingly used industrially for quality control, *i.e.*, for grading sawn timber and wood-based panels (see further Niemz *et al.* 2023).

The impact generates longitudinal, bending or torsional vibrations, and the elastic constants are then calculated from the determined natural frequency. The corresponding device systems are based on the fundamental principles described below. A rod, which is excited to oscillate, is placed on two supports (Görlacher 1984, 1990).

7.3.7.1 Determination of Tensile and Compressive Dynamic Modulus of Elasticity

The dynamic modulus of elasticity differs from the static modulus of elasticity, as static modulus is obtained from stress-strain data, and the dynamic modulus is deter-mined from wave velocity.

In the case of longitudinal (axial) waves in the sawn timber activated by an impact in the longitudinal direction of the timber, the dynamic modulus of elasticity (E) can be calculated from the natural frequency as:

$$E = \frac{4 \cdot l^2 \cdot f^2 \cdot \rho}{n^2} \qquad (7.18)$$

where

l is the length in the longitudinal (the wave) direction (m)
f is the natural frequency (Hz)
n is the order of vibration

Equation (7.18) applies to all cross-sectional shapes when the rod length is multiplied larger than the rod cross-sectional area. With known rod length and density, the tensile or compressive modulus of elasticity can thus be determined from natural frequency.

7.3.7.2 Determination of the Elastic Modulus in Bending (Flexural Vibration)

According to Görlacher (1984), the calculation of the dynamic modulus of elasticity, considering the transverse contraction is:

$$E = \frac{4\pi^2 \cdot l^4 \cdot f^2 \cdot \rho}{m_n^4 \cdot i^2} \cdot \left(1 + \frac{i^2}{l^2}\left(K_1 + K_2 \cdot s \cdot \frac{E}{G_{xy}}\right)\right) \qquad (7.19)$$

where

E is the modulus of elasticity (Pa)

G_{xy} is the shear modulus in bending plane (Pa)

I is the moment of inertia (m⁴)

A is the cross-section (m²)

ρ is the density (kg/m³)

l is the rod length (m)

i is the radius of inertia in the direction of bending vibration ($i^2 = I/A$); for rectangular cross-sections: $i^2 = h^2/12$ (h = rod height (m))

f is frequency (Hz)

K_1, K_2, m_n are constants (depending on the order of vibration)

s is the form factor (for isotropic rectangular cross-sections 1.20, for wood 1.06)

K_1, K_2, and m_n are constants that depend on the order of vibration. The form factor (s) assumes a value of 1.20 for isotropic rectangular cross-sections (wood is expected to have a shape factor of 1.06 as determined by Hearmon 1966). Görlacher (1984, 1990) and Görlacher and Hättich (1990) describe the method in detail and emphasise first-order vibrations: $K_1 = 49.84$, $K_2 = 12.3$, and $m_n^4 = 500.6$.

7.3.7.3 Dynamic Torsion Modulus

The determination of the torsional dynamic modulus (G_t) applies as follows:

$$G_t = 4 \cdot f^2 \cdot l^2 \cdot \rho \qquad (7.20)$$

where

f is frequency (Hz)

l is the rod length (m)

ρ is density (kg/m³)

Natural frequency measurement is often used to determine elastic constants, especially under industrial conditions, like grading from solid wood. Modal analysis is often used, which makes it possible to determine all elastic constants of wood or wood-based materials (see Gülzow (2008) for cross-laminated timber).

Several measuring principles are used to perform natural-frequency measurements, and they can be combined with other methods, such as X-ray for detecting knots or laser to detect fibre orientation (tracheid effect in softwoods). Figure 7.10 shows the correlation of the modulus of elasticity determined from eigenfrequency with that specified in the bending test for particleboard. This method also carried out experiments to measure the stiffness of adhesive joints (Hass 2012). To a certain extent, the penetration of the adhesive into the wood is integrated into the measurement result.

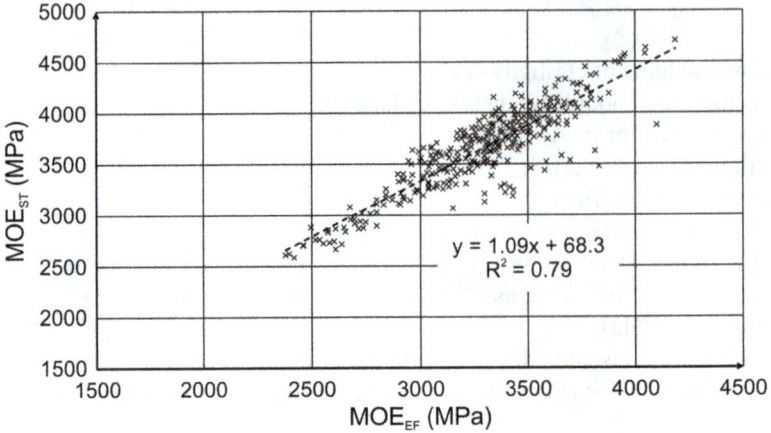

FIGURE 7.10   Correlation between static (MOE$_{ST}$) and dynamic (MOE$_{EF}$) modulus of elasticity in bending of particleboard.

(Data from Grundström *et al.* 1999, courtesy of Carl Hanser Verlag.)

7.4   FRICTION PROPERTIES OF WOOD AND WOOD-BASED MATERIALS

Friction is the resistance a surface or object encounters when moving over another. The friction decreases with increasing speed, increasing the smoothness and hardness of the surfaces. The parameter used is the coefficient of static (μ_s) or kinetic (dynamic or sliding) friction (μ_k). According to Figure 7.11, the following balance of forces is required to overcome the static (F$_s$) or kinetic (F$_k$) friction:

$$F_s \geq \mu_s \cdot F_N \tag{7.21}$$

$$F_k \geq \mu_k \cdot F_N \tag{7.22}$$

where

μ_s, μ_k – static and kinetic friction coefficients
F – forces, see Figure 7.11

Static friction (F$_s$) is the resistance encountered by two contacting surfaces when they attempt to move relative to each other. This resistance increases as the moisture content rises and when making the wood surfaces softer and rougher. Regarding the influence of the principal direction of the wood on friction, the following order of precedence is observed (with the coefficient of static friction decreasing from top to bottom):

- end-grain to end-grain,
- radial section, and
- tangential section.

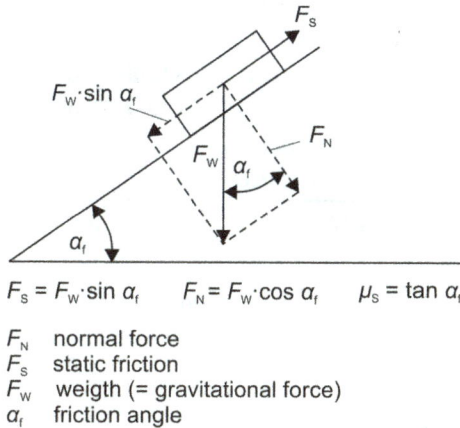

$F_s = F_w \cdot \sin \alpha_f \qquad F_N = F_w \cdot \cos \alpha_f \qquad \mu_s = \tan \alpha_f$

F_N normal force
F_s static friction
F_w weigth (= gravitational force)
α_f friction angle

FIGURE 7.11 Forces balance for determining the friction coefficients. **(Courtesy of Carl Hanser Verlag.)**

Kinetic friction (F_k) is the resistance to movement when one surface slides over another. This resistance decreases with increasing the smoothness and hardness of the surfaces. Kinetic friction is most pronounced in wood when the movement is perpendicular to the grain (cross-sections) and in its dry state. Applying oil, water, or grease lubrication reduces sliding friction (Vorreiter 1949, Xu *et al.* 2014).

Tables 7.12–7.13 present the friction coefficients for selected woods, derived timber products, and material combinations.

TABLE 7.12

Friction Coefficients for Norway Spruce[a] Depending on the Friction Interface (Möhler and Herröder 1979)

Material Pairs	Static Friction Coefficient μ_s	Kinetic (Sliding) Friction Coefficient μ_k
Spruce ‖, spruce ‖	0.65–0.80	0.40–0.45
Spruce ‖, spruce ⊥	0.70–0.80	0.45–0.55
Spruce ⊥, spruce ⊥	0.85	0.50–0.55
Spruce ‖, cross-section spruce	0.40–0.85	0.25–0.55
Spruce ⊥, cross-section spruce	0.90	0.60
Cross-section spruce, cross-section spruce	0.45–0.90	0.30–0.50
Spruce-to-concrete	0.90	0.55–0.65

‖ Parallel to the grain; ⊥ perpendicular to the grain
[a] Sawn timber, 10–25% moisture content (for spruce-to-concrete: 20–25% moisture content)

TABLE 7.13

Static Friction Coefficients for Various Solid Wood Surfaces. Design Values According to the EN 1995-2 Standard (CEN 2004)

Type of Meeting Surfaces	Perpendicular to the Grain Direction		In the Grain Direction	
	$\omega \leq 12\%$	$\omega \geq 16\%$	$\omega \leq 12\%$	$\omega \geq 16\%$
Sawn-to-sawn	0.30	0.45	0.23	0.35
Planed-to-planed	0.20	0.40	0.17	0.30
Sawn-to-planed	0.30	0.45	0.23	0.35
Wood-to-concrete	0.40	0.40	0.40	0.40

Note: ω – moisture content

The design values used (Table 7.13) are significantly lower than those listed in Table 7.12. Möhler and Maier (1969) and Möhler and Herröder (1979) recommend a maximum value of 0.4 (rough-sawn) or 0.25 (planed) for fibre-parallel contact surfaces. Meisel *et al.* (2015) propose a design value of 0.18 (characteristic value 0.25) for parallel to the grain (LR, LT) surface and a value of 0.25 (characteristic value 0.35) for cross-section (RT or TR) to parallel to the grain (LR, LT) surfaces.

Wood friction wood-to-steel also strongly depends on moisture content and surface conditions (Möhler and Herröder 1979, Niemz and Sonderegger 2021). Friction is important in securing loads against slipping (*e.g.*, transporting particleboards or furniture parts), but it is also essential when dealing with other wood products. Chapter 6 by Niemz *et al.* (2023) provides more extensive data for other material combinations, for example, wood and steel.

7.5 OPTICAL PROPERTIES OF WOOD AND WOOD-BASED MATERIALS

7.5.1 COLOUR

7.5.1.1 Parameters

Due to its natural characteristics, wood exhibits significant colour variations within and between different species. The colour can range from almost white to yellowish (*e.g.*, Norway spruce), dark red (*e.g.*, rauli from southern South America), brown (*e.g.*, walnut, teak), or even black (*e.g.*, African ebony). These colour differences are primarily due to the extractives (their proportion and type). Heartwood and sapwood typically differ considerably in colour. For instance, the heartwood is dark red in yew, while the sapwood is yellowish-white. In Europe, only a few species with a dark-coloured heartwood are found, such as walnut, oak, elm, and yew. Timber from tropical and subtropical regions is more frequently darker than European or North American species (Wagenführ 2006).

The colour of wood changes significantly during thermal modification (*cf.* Section 2.3) or through boiling or steaming. This darkening is a key criterion for specific applications, such as furniture and parquet, and is often preferred over other treatment effects, such as reduced sorption and swelling. In addition to colour, the texture, including colour variations within the wood, such as between earlywood and latewood, is an important feature when selecting wood for furniture, interior joineries and, to some degree, also for timber construction.

The colour characteristics are measured according to the CIE-Lab system, which follows the colour space according to the ISO 11664-4 standard (ISO 2019), as shown in Figure 7.12. Characteristic values are the lightness $L*$ (white-black) and the colour components green-red $a*$ and blue-yellow $b*$. The system can be used to quantify the colour characteristics or to determine age-related colour changes.

7.5.1.2   Colour Change

7.5.1.2.1   *Effect of Transparent Coatings*

Surface treatments using coatings (such as paints), oils, or waxes significantly impact the visual colour of the wood (Teischinger *et al.* 2012). These treatments can raise the surface grain to varying degrees, intensifying the colour effect. Sometimes, this effect can be observed by lightly moistening the wood surface.

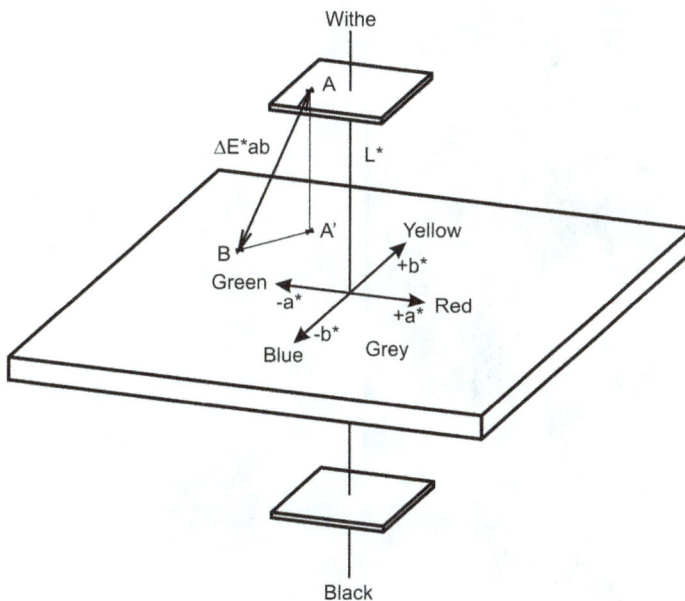

FIGURE 7.12   CIELab colour space.

(Courtesy of Carl Hanser Verlag.)

7.5.1.3 Ageing of Wood Used Indoors

The colour of wood in interior spaces changes over time due to oxidation. The room's environmental conditions and specific wood species strongly influence this colour change. Many types of wood darken when used indoors (Figures 7.13 and 7.14), while a few species may lighten. The most noticeable colour changes occur when wood is exposed to UV light, such as in areas near windows and balcony doors. This should be taken into account when choosing materials for flooring and furniture. A comprehensive overview of colour changes in parquet woods can be found in Pitt (2010). UV-induced colour change can be significantly reduced by applying coatings with UV blockers (Lukowsky 2013). Holz-Lexicon (Anon. 2003) and other sources provide the following general tendencies regarding natural colour changes:

- oak spp. become greyish brown by age,
- changes in mahogany and cherry depend on their natural original colour,
- softwoods such as Scots pine or common larch will darken and get more red-brown colour of the heartwood; Norway spruce turns slightly yellowish brown, and
- maple and birch spp. turn yellow.

Ammonia-fumed oak and beech wood are more colour-stable than, for example, thermally modified timber. Thermally modified timber and untreated rosewood become brighter due to UV radiation (Volkmer *et al.* 2014).

FIGURE 7.13 Natural indoor ageing of Norway spruce, resulting in brownish colour (left), compared to the same piece of spruce wood before ageing.

(Courtesy of P. Niemz.)

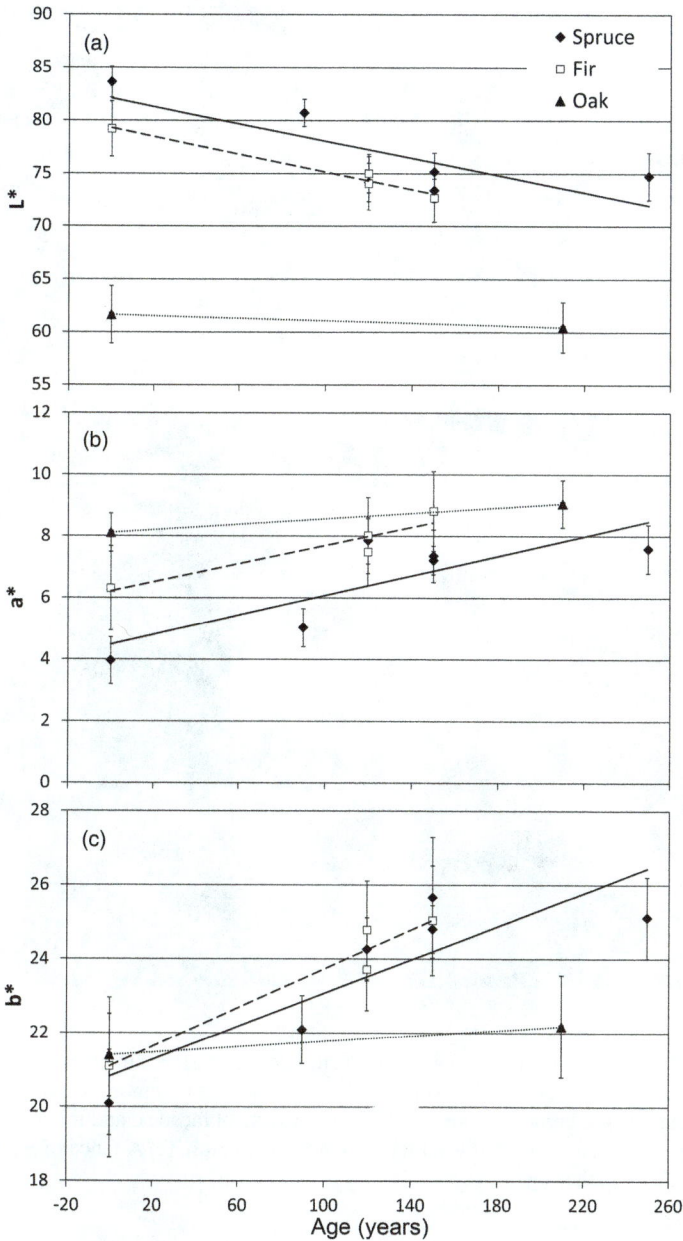

FIGURE 7.14   Colour change of Norway spruce wood related to age after the first installation. **(Data from Sonderegger *et al.* 2015, courtesy of Carl Hanser Verlag.)**

Colour will change significantly due to natural weathering (Figure 7.15). This is the result of chemical degradation of wood constituents close beneath the exposed surface and colonisation by staining fungi on the surfaces (Lukowsky 2013).

(a)   (b)

(c)

FIGURE 7.15   Weathering changes of wood surfaces after approximately one year of exterior exposure: (a) common beech and (b) thermally modified common ash, both without any surface coating: (1) reference before weathering, (2) vertical façade cladding protected by roof construction, and (3) unprotected weathering at 45° inclination. (c) A fence of European larch before and after weathering.

(Courtesy of P. Niemz.)

7.6   SPECTROMETRIC PROPERTIES

Near-infrared spectroscopy (NIR) is a method that uses the near-infrared region of the electromagnetic spectrum from about 700–2500 nanometres (nm), commonly used for quantitative detection under industrial conditions. By measuring light scattered off and through a sample, NIR reflectance spectra can quickly determine a material's

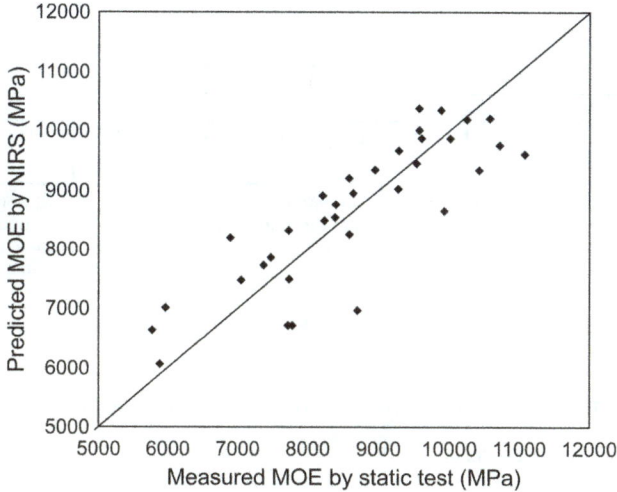

FIGURE 7.16   Correlation between the modulus of elasticity (MOE) determined by near-infrared spectroscopy (NIR) and a static bending test.

(Data from Meder *et al.* 2002, courtesy of Carl Hanser Verlag.)

properties without altering the sample. For example, the wavelengths of 1930 nm or 1430 nm are used to determine the moisture content of wooden particles, sometimes also for solid wood and veneer. The method is based on the complex interaction between the wood's chemical structure and physico-mechanical properties.

For many years, spectroscopy, particularly NIR spectroscopy, has been used to assess wood quality, particle adhesive content, and properties such as density, strength, and fungal resistance.

A wide range of characteristics is determined from the spectra and then correlated using multivariate statistics with parameters specified in standard experiments, for example, in bending, tension, and compression (Figure 7.16). Also, density, microfibril orientation, fungal resistance and wood ageing, can be studied. Achieving reliable correlations requires a substantial number of experiments. Research on this subject has been conducted by various scholars, including Wienhaus *et al.* (1988), Meder *et al.* (2002), Sandak *et al.* (2011), Tsuchikawa and Schwanninger (2013), and Sandak *et al.* (2021).

7.7   AGEING AND CORROSION

7.7.1   OVERVIEW

Wood and wood-based materials' ageing and corrosion properties have numerous practical applications. Wood generally remains stable under regular use, provided the climate remains consistent and there is no attack by fungi or insects. However, as described in Section 7.5, colour changes can be noticeable. Several factors can influence ageing and corrosion (Figure 7.17):

```
                        ┌──────────────────┐
                        │      Aging/       │
                        │     corrosion     │
                        └──────────────────┘
          ┌──────────────────────┼──────────────────────┐
┌──────────────────┐  ┌──────────────────┐  ┌──────────────────┐
│     Climate       │  │   Aggressive     │  │   Mechanical     │
│                   │  │      media       │  │     stress       │
└──────────────────┘  └──────────────────┘  └──────────────────┘
```

– Temperature/ relative humidity/   – Weakening of the cross   – Strength reduction
  UV-radiation   section through destruction   through microcracks
– Damage of wood substance   of the wood substance   by extremely high
  and/or glue joints by alternate   static strain
  swelling and shrinkage and
  associated crack formation
– Discoloration and decomposition
  of the near-surface region

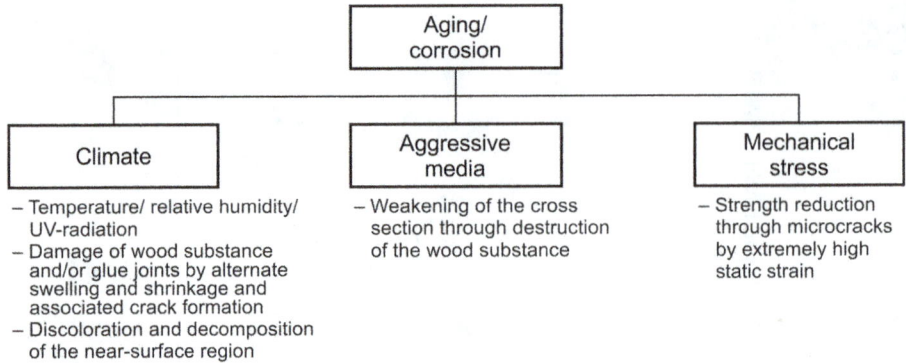

FIGURE 7.17   Factors influencing wood's corrosion/ageing behaviour.

(Courtesy of Carl Hanser Verlag.)

- climate,
- type of aggressive media,
- mechanical stresses, and
- natural ageing and wood preservation.

7.7.2 INFLUENCE OF CLIMATE AND METHODS FOR DETERMINATION OF CLIMATIC RESISTANCE

The ambient climate is crucial in determining the lifespan of wood and wood-based materials. Key parameters include moisture content, UV radiation, visible solar radiation, temperature fluctuations, and changes in rainfall and humidity. Variations in moisture content cause wood to swell or shrink.

UV radiation results in a brownish discolouration in the near-surface regions, followed by the degradation and leaching of degraded lignin, which causes the wood's surface to turn grey (with the uncoloured cellulose structure remaining intact). The degraded lignin and hemicelluloses are washed out by water (rain), leaving a rough, porous surface. The extent of weathering-induced erosion largely depends on the structural composition of the wood. Factors such as the density and thickness of the cell walls, as well as the presence of wood extractives, affect the wood's resistance to weathering. As timber density increases, weathering resistance improves. Consequently, depending on the weathering process and microclimate, various discolouration often occurs on timber façades, ranging from grey over brown to black. The surface typically turns uniform brown in high-altitude regions with dry climates and proper timber-construction techniques.

Significant variations in moisture content lead to the swelling and shrinkage of wood, often resulting in cracking. This occurs under dry conditions in heated interiors, particularly in winter, when cold outdoor air is brought indoors and heated. In central and northern European countries, indoor relative humidity can be below 20%. This impact is typically limited to the surface regions of the timber and the effect will be more pronounced in small-dimension timber (with thicknesses of only a few centimetres), such as façade boards. However, in timber with larger cross-sectional dimensions, such as cross-laminated timber and glued-laminated timber, moisture fluctuations can cause deep and severe cracks, which may allow moisture penetration and, consequently, lead to fungal attacks.

The weathering resistance of wood species varies very strongly due to differences in the wood structure and constituents. The durability of the wood is regulated in the European standard EN 350 (CEN 2016), and the corresponding humidity range in use according to the EN 335 standard (CEN 2013).

If wood is stored for a long time at elevated temperatures considerably above 20 °C, its strength is reduced (*cf.* Section 7.1.5). Mönck (1999) recommends the following reduction factors K_t:

$$35 \text{ °C to } 50 \text{ °C} \quad K_t = 0.8$$

$$> 50 \text{ °C to } 80 \text{ °C} \quad - K_t = 0.6 \text{ to } 0.7$$

7.7.2.1 Wood-based Materials

As with solid wood, the structure of wood-based materials is also affected by changes in climatic conditions. A significant factor is the type of adhesive used. Urea-formaldehyde (UF) adhesives are relatively susceptible to degradation through hydrolysis, swelling, and shrinkage, reducing strength (*e.g.*, weakened internal bond) and increasing thickness swelling. In contrast, phenol-resorcinol-formaldehyde (PRF), polyurethane (*e.g.*, 1C-PUR), and urea-melamine (MUF) adhesives are considerably more stable than UF. Additionally, the adhesives themselves are hygroscopic (Wimmer *et al.* 2013). For example, hide, bone, and fish adhesives exhibit high moisture content in the glued wood ambient environment at high relative humidity.

Using phenolic resin, melamine resin, isocyanate, or PUR, some weather resistance can be achieved for wood-based materials (such as cross-laminated timber, glued-laminated timber, particleboards, and plywood). However, particle-based panels are generally unsuitable for long-term use in high-humidity conditions, such as exterior façades. In contrast, solid wood, plywood, and cross-laminated timber are more moisture resistant and, therefore, suitable for such use. Surface coatings can significantly enhance weathering resistance. However, it is essential to note that if the surface coating is damaged, moisture can rapidly penetrate the material, but it cannot escape to the same extent.

7.7.3 AGEING OF WOOD AND WOOD-BASED MATERIALS

Ageing refers to the change in properties of materials under natural or artificial conditions over an extended period. It is well known that many plastics undergo partial ageing and embrittlement due to UV radiation or the migration of plasticisers. In contrast, the effects of ageing on solid wood have been comparatively under-researched. An overview of the current state of knowledge can be found in the works of Holz (1981), Obataya (2007), Oltean *et al.* (2008), Yokoyama *et al.* (2009), Kránitz (2014), and Kránitz *et al.* (2016).

For wood-based materials, the influence of adhesives and the material structure must also be considered. This is due to the differing orientations of the particles or layers (as in cross-laminated timber, CLT), which restrict the wood's ability to shrink and swell. Additionally, the moisture resistance of the adhesives must be carefully observed.

7.7.3.1 Sawn Timber

Various reports have described the ageing of wood stored in dry interior conditions. In very dry climates (*e.g.*, in the catacombs of the Egyptian pyramids), wood can survive for thousands of years without considerable degradation due to only age. However, according to Holz-Lexikon (Anon. 2003), it is well-established that the coefficient of variation for swelling and shrinkage values of aged wood is reduced. Measurements by Sonderegger *et al.* (2015) indicate that the strength properties show only minimal or no changes, with the most noticeable reduction observed in impact bending strength, as shown in Figure 7.18.

In contrast, the chemical properties of wood undergo more significant changes (Kránitz 2014). Notable differences by age were observed in oak wood, particularly conserning the changes in lignin and polysaccharide contents and the degree of cellulose crystallinity. Wood extractives from both aged and non-aged wood will reveal degradation processes due to oxidation and slight hydrolysis. The study demonstrated that the relative content of the structural wood components depends on the wood's age and the conditions under which it was stored.

7.7.3.2 Wood-based Materials

Extensive work has been carried out on wood-based panels, especially for particleboard and fibreboard, by Anon. (1975), Deppe and Ernst (2000), Dunky and Niemz (2002), and Paulitsch and Barbu (2015). Cross-laminated timber and glued-laminated timber are less experienced. Moisture resistance is mostly tested using tensile-shear tests according to the EN 302-1 standard (CEN 2023a) or delamination tests according to EN 302-2 (CEN 2023b) after various storage series: A1 – dry storage, A4 – storage in hot and cold water, and A5 – redried to the A1 condition.

Glued-laminated timber constructions over 100 years old are still in use today. Remarkably, even these older constructions show no reduction in load-bearing capacity. In contrast, research by Konnerth *et al.* (2012) demonstrated that urea-based adhesives used in aircraft wood constructions undergo ageing (*cf.* also ageing of adhesives in Chapter 12). The ageing effect of urea has been well-documented. Various methods are employed to simulate ageing, *e.g.*, wet-dry cycles, temperature fluctuations, exposure to UV radiation, water storage (*e.g.*, cooking or drying), or the greenhouse test suggested by the Materials Testing Institute (MPA) in Stuttgart, Germany. Mechanical loading and climate change can also simulate ageing (Aicher 2006).

UV radiation typically has little effect on bond-lines for lamellae with a thickness of 30–40 mm (except on the edge surfaces where the bond-lines are visible). However, in large cross-section timber, cracks and, in some cases, delamination may occur. These problems are most evident in buildings exposed to very dry winter conditions with extremely low relative humidity.

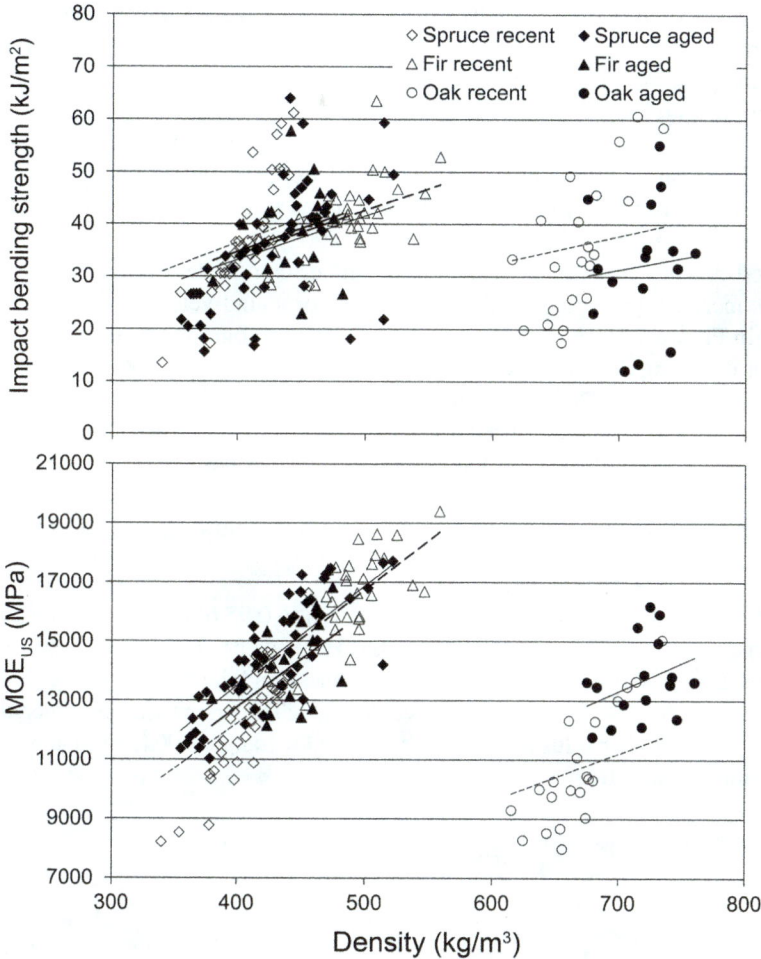

FIGURE 7.18 Impact bending strength and MOE tested with ultrasound, depending on density for recent and aged wood.

(Data from Sonderegger *et al.* 2015, courtesy of Carl Hanser Verlag.)

In cross-laminated timber and glued-laminated timber, the properties of the elements, the growth-ring orientation of the lamellae, the lamella thickness, and moisture differences between the lamellae during bonding are crucial factors for increased cracking and lamination. The stresses caused by humidity changes are typically greater for cross-laminated timber than those in glued-laminated timber.

7.7.4 INFLUENCE OF MECHANICAL LOAD HISTORY AND CRACKING

Wood in construction use is often subjected to considerable mechanical stress. However, the load is typically only 10–30% of the material's ultimate load capacity. As a result, damage from exceeding the permissible stresses occurs relatively infrequently. Such damage usually arises with shrinkage cracks or when the tensile strength perpendicular to the grain of curved glued-laminated timber beams is too high. Creep deformation and the relaxation of stresses play a key role in selecting appropriate wood and wood-based materials.

Particleboard and solid wood do not experience a permanent reduction in strength under normal stress levels; this only happens under extremely high loads.

The often-discussed influence of lunar phases on wood properties is yet to be conclusively proven (Fellner and Teischinger 2001).

Wood damage in trees or construction timber can be detected through methods such as incremental coring, drilling resistance, or sound-velocity measurements, as shown in Figure 7.19 (Niemz *et al.* 2002). Sound tomography is also used, particularly for detecting defects in wood, especially in standing trees.

Mönck and Erler (2004) recommend that for the static testing of older timber structures, reduction factors should be applied for continuous use after more than 20 years: particularly reduction in load capacity between 0.6 and 0.9 of the values for non-aged timber (design values).

When cracks are caused by climate variation, the following is recommended:

1. At moderate weakening:
   Reduction of the load-bearing cross-section to 0.95 of the modulus of resistance or 0.91 of the moment of inertia. Here, determining the critical stress for tear propagation using fracture toughness and the fracture energy would be an alternative for a more exact calculation. The first works have been done on sawn timber (Gustafsson *et al.* 1998, Dill-Langer 2004, Ammann 2015, Titner *et al.* 2016).

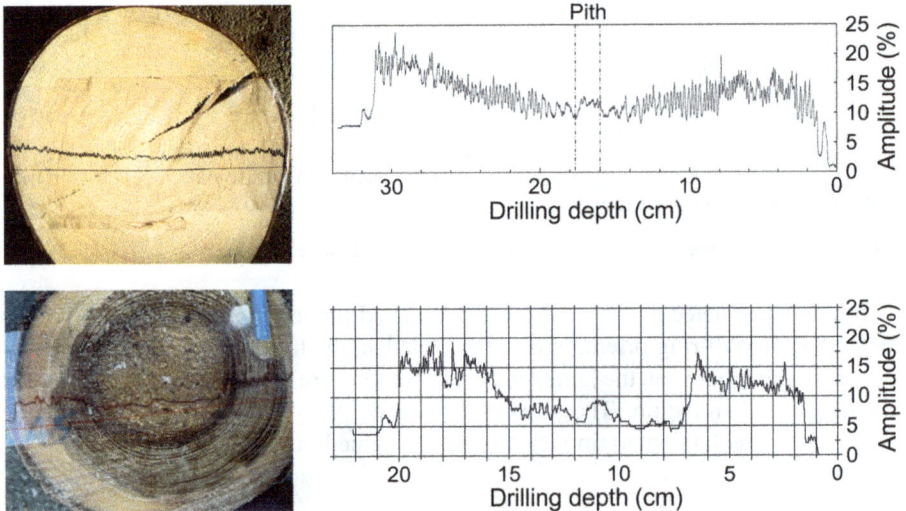

FIGURE 7.19   Drilling resistance of non-decayed (top) and decayed (below) Norway spruce log.

(S. Herrmann and F. Bächle, ETH Zürich, courtesy of Carl Hanser Verlag.)

2. Substantial weakening:
   The exact determination of the load-bearing cross-section is relevant for the moment of inertia or the modulus of resistance, and the calculation is made with the determined values.

7.7.5 INFLUENCE OF AGGRESSIVE MEDIA

Wood corrosion is defined as damage or destruction that progresses from the surface into the wood, caused by chemical and/or physicochemical reactions influenced by the surrounding environment (Erler 1998). Compared to other materials, wood has a high level of corrosion resistance. As a result, timber is often used in applications such as roof trusses in fertiliser storage halls and for the storage of de-icing salts (Titner *et al.* 2016). However, it is essential to note that aggressive substances primarily attack hemicelluloses and lignin, with cellulose being less affected. Therefore, softwoods are generally more resistant to corrosion than hardwoods. Unger *et al.* (2001) provide a comprehensive overview of wood corrosion.

7.7.5.1 Water

At normal temperatures, water does not cause any chemical change in the wood. However, storing wood in water for a very long time (*e.g.*, during rafting of round timber) causes certain wood constituents to leach out, which increases the wood's resistance to insects. If the water is not recycled or exchanged, bacteria may degrade the timber. Hydrolytic degradation can occur if water acts on the wood at high temperatures and pressures (Kollmann 1951, Sonderegger and Niemz 2006). This can be observed, for example, in timber, which has previously been used to construct cooling towers in timber.

7.7.5.2 Chemical Substances

According to Erler (1998), the aggressive effect of chemicals on wood depends on the physical state of the medium:

- gaseous media: type, concentration, humidity,
- liquid media: pH, concentration, degree of dissociation, and
- solid media: type, solubility in water, hygroscopicity, pH, humidity.

Certain chemicals, such as dimethylformamide (DMF), cause substantial swelling of the wood and will probably also cause mechanical damage to the wood. When priming wood for bonding, damage to the surface regions adjacent to the bond-line by the primer is often discussed. Primers also improve wetting.

The corrosion phenomena are associated with most media (Erler 1990):

- a brown or yellowish-brown colouring of the wood, which penetrates from the surface zones inwards,
- an increased moisture content of the wood surface area,
- a storage of salt crystals or acid radicals in the wood,

- a fibrous structure at the surface, for example, a wool-like surface after salt influence (Rug and Lissner 2011),
- delamination along the growth rings,
- strength reduction of wood in the surface regions, and
- flat breaks in the case of acid exposure.

Figure 7.20 shows the increase of the corrosion layer thickness on wood as a function of the service life for different moisture classes. A reduction in strength occurs only in the relatively narrow margins of the wood, while the strength in 10 to 15 mm depth corresponds already mainly to the strength of the native wood. In load-bearing capacity calculations, it is, therefore, customary to reduce the load-bearing cross-section of the timber.

7.7.5.3 Metals

Wood's contact with metals partially causes corrosion. This depends on the type of wood and its moisture. Iron, for example, induces discolouration in tannin-rich wood species such as oak. Even thermally treated wood causes increased corrosion due to its lower pH value (Niemz and Wetzig 2011). Therefore, stainless steel is used for fasteners.

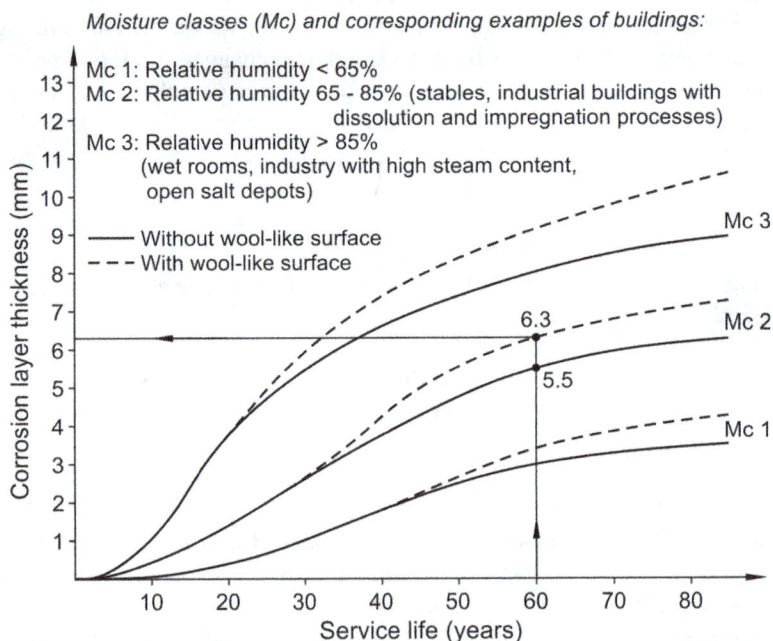

FIGURE 7.20 Corrosion layer thickness of wood used in the fertiliser salt industry as a function of service life at different humidity conditions.

(Data from Erler 1990, courtesy of Carl Hanser Verlag.)

Unger (1988) gives the following specifications concerning the influences of metals:

- Wood at a moisture content of <10% is less affected by metals. At a moisture content of 20% to 30%, the influence is more pronounced regarding interactions between metal and wood. This is because weakly acidic extracts are often formed at these moisture contents, which promote corrosion.
- Chestnut and oak wood are highly corrosive, and poplar and mahogany are corrosive to a low degree.
- Iron stains oak and other tannin-rich species (oak changes to black, linden to green).
- The tensile strength of wood decreases after prolonged contact with iron due to degradation of polyoses, (a class of carbohydrates composed of chains of monosaccharide molecules) caused by iron corrosion products; the compression strength, however, remains unaffected.

REFERENCES

Aicher S. (2006). *Bestandsaufnahme: Schadensfälle, Dauerstandfestigkeit, Prüfmethoden.* Materials Testing Institute (MPA), University of Stuttgart, Stuttgart, Germany.

Ammann S.D. (2015). *Mechanical Performance of Glue Joints in Structural Hardwood Elements.* Doctoral Thesis, ETH Zürich, Zurich, Switzerland.

Anon. (1990). Lexikon der Holztechnik (4th Ed.). Fachbuchverlag, Leipzig, Germany.

Anon. (2003). *Holz-Lexikon: Nachschlagewerk für die Holz- und Forstwirtschaft.* (4th Ed.), DRW-Verlag Weinbrenner, Germany, (1460 pp.).

Anon. (1975). *Werkstoffe aus Holz und andere Werkstoffe der Holzindustrie.* Fachbuchverlag, Leipzig, Germany, (928 pp.).

Baensch F. (2015). *Damage Evolution in Wood and Layered Wood Composites monitored in situ by Acoustic Emission, Digital Image Correlation and Synchrotron based Tomographic Microscopy.* Doctoral Thesis, ETH Zürich, Zurich, Switzerland.

Bekhta P., Niemz P. & Kucera L.J. (2002). Untersuchungen einiger Einflussfaktoren auf die Schallausbreitung in Holzwerkstoffen. *Holz als Roh- und Werkstoff*, 60(1), 41–45.

Bucur V. (2006). *Acoustics of Wood.* (2nd Ed.), Springer, Berlin, Germany, (394 pp.).

CEN (2000). EN 12524: Building materials and products - Hygrothermal properties - Tabulated design values. *The European Committee for Standardization (CEN)*, Brussels, Belgium.

CEN (2004). EN 1995-2: Eurocode 5: Design of timber structures - Part 2: Bridges. *The European Committee for Standardization (CEN)*, Brussels, Belgium.

CEN (2007). EN ISO 10456: Building materials and products — Hygrothermal properties — Tabulated design values and procedures for determining declared and designed thermal values. *The European Committee for Standardization (CEN)*, Brussels, Belgium.

CEN (2013). EN 335: Durability of wood and wood-based products - Use classes: definitions, application to solid wood and wood-based products. *The European Committee for Standardization (CEN)*, Brussels, Belgium.

CEN (2016). EN 350: Durability of wood and wood-based products - Testing and classification of the durability to biological agents of wood and wood-based materials. *The European Committee for Standardization (CEN)*, Brussels, Belgium.

CEN (2023a). EN 302-1: Adhesives for load-bearing timber structures - Test methods - Part 1: Determination of longitudinal tensile shear strength. *The European Committee for Standardization (CEN)*, Brussels, Belgium.

CEN (2023b). EN 302-2: Adhesives for load-bearing timber structures. *Test methods Determination of resistance to delamination.* The European Committee for Standardization (CEN), Brussels, Belgium.

Czajkowski L., Olek W., Weres J. & Guzenda R. (2016). Thermal properties of wood-based panels: Specific heat determination. *Wood Science and Technology*, 50(3), 537–545.

Deppe H.-J. & Ernst K. (2000). *Taschenbuch der Spanplattentechnik.* (4th Ed.), DRW-Verlag, Leinfelden-Echterdingen, Germany, (552 pp.).

Dill-Langer G. (2004). *Schädigung von Brettschichtholz bei Zugbeanspruchung rechtwinklig zur Faserrichtung.* Doctoral Thesis, Universität Stuttgart, Stuttgart, Germany.

DIN (1965). DIN 52362-1: Testing of wood chipboards; bending test, determination of bending strength. *Deutsches Institut für Normung (DIN)*, Berlin, Germany.

Dunky M. & Niemz P. (2002). *Holzwerkstoffe und Leime: Technologie und Einflussfaktoren.* Springer, Berlin, Germany, (978 pp.).

Erler K. (1990). Korrosion und Anpassungsfaktoren für chemisch aggressive Medien bei Holzkonstruktionen. *Holztechnologie*, 30(5), 228–233.

Erler K. (1998). Korrosion von Vollholz und Brettschichtholz. *Bautechnik*, 75(8), 530–538.

Fellner J. & Teischinger A. (2001). Alte Holzregeln. *Von Mythen und Brauchbarem über Fehlinterpretationen zu neuen Erkenntnissen.* Österreichischer Kunst- u. Kulturverlag, Wien, Austria, (160 pp.).

Ghazi Wakili K., Binder B. & Vonbank R. (2003). A simple method to determine the specific heat capacity of thermal insulation used in building construction. *Energy and Buildings*, 35(4), 413–415.

Glos P. & Henrici D. (1990). *Festigkeit von Bauholz bei hohen Temperaturen.* Forschungsbericht, Institut für Holzforschung der Universität München, München, Germany, (83 pp.).

Grundström F., Kucera L.J. & Niemz P. (1999). Schalluntersuchungen an Spanplatten. Bestimmung der Platteneigenschaften durch eine Kombination aus Schallgeschwindigkeit und Eigenfrequenz. *Holz-Zentralblatt*, 125(127), 1734–1736.

Gustafsson P.J., Hoffmeyer P. & Valentin G. (1998). DOL behaviour of end-notched beams. *Holz als Roh- und Werkstoff*, 56(5), 307–317.

Gülzow A. (2008). *Zerstörungsfreie Bestimmung der Biegesteifigkeiten von Brettsperrholzplatten.* Doctoral Thesis, ETH Zürich, Zurich, Switzerland.

Görlacher R. (1984). Ein neues Messverfahren zur Bestimmung des Elastizitätsmoduls von Holz. *Holz als Roh- und Werkstoff*, 42(6), 219–221.

Görlacher R. (1990). *Klassifizierung von Brettschichtholzlamellen durch Messung von Longitudinalschwingungen.* Doctoral Thesis, Universität Karlsruhe, Karlsruhe, Germany.

Görlacher R. & Hättich R. (1990). Untersuchung von altem Konstruktionsholz. Die Bohrwiderstandsmessung. *Bauen mit Holz*, 92(6), 455–459.

Hass P.F. (2012). *Penetration behavior of Adhesives into Solid Wood and Micromechanics of the Bondline.* Doctoral Thesis, ETH Zürich, Zurich, Switzerland.

Hasenstab A.G. (2006). *Integritätsprüfung von Holz mit dem zerstörungsfreien Ultraschallechoverfahren.* Doctoral Thesis, TU Berlin, Berlin, Germany.

Hearmon R.F.S. (1966). Vibration testing of wood. *Forest Products Journal*, 16(8), 29–40.

Hill C.A. (2006). Wood Modification: Chemical, *Thermal and Other Processes.* Wiley, Chichester, UK, (264 pp.).

Holz D. (1981). Zum Alterungsverhalten des Werkstoffes Holz - einige Ansichten, Untersuchungen, Ergebnisse. *Holztechnologie*, 22(2), 80–85.

ISO (2019). ISO/CIE 11664-4: Colorimetry - Part 4: CIE 1976 L*a*b* colour space. *International Organization for Standardization (ISO)*, Geneva, Switzerland.

Keunecke D., Sonderegger W., Pereteanu K., Lüthi T. & Niemz P. (2007). Determination of Young's and shear moduli of common yew and Norway spruce by means of ultrasonic waves. *Wood Science and Technololgy*, 41(4), 309–327.

Keunecke D., Merz T., Sonderegger W., Schnider T. & Niemz P. (2011). Stiffness moduli of various softwood and hardwood species determined with ultrasound. *Wood Material Science & Engineering*, 6(3), 91–94.

Keylwerth R. & Noack D. (1956). Über den Einfluß höherer Temperaturen auf die elektrische Holzfeuchtigkeitsmessung nach dem Widerstandsprinzip. *Holz als Roh- und Werkstoff*, 14(5), 162–172.

Kollmann F. (1951). *Technologie des Holzes und der Holzwerkstoffe: Part 1: Anatomie und Pathologie, Chemie, Physik, Elastizität und Festigkeit.* (2nd Ed.), Springer Verlag, Berlin, Germany, (1051 pp.).

Kollmann F. & Malmquist L. (1956). Über die Wärmeleitzahl von Holz und Holzwerkstoffen. *Holz als Roh- und Werkstoff*, 14(6), 201–204.

Kollmann F. & Côté Jr. W.A. (1968). Principles of Wood Science and Technology. *Part 1: Solid Wood.* Springer, Berlin, Heidelberg, Germany, (592 pp.).

Konnerth J., Müller U., Gindl W. & Buksnowitz C. (2012). Reliability of wood adhesive bonds in a 50 year old glider construction. *European Journal of Wood and Wood Products*, 70(1), 381–384.

Kordina K. & Meyer Ottens C. (1983). *Holz-Brandschutz-Handbuch.* Deutsche Gesellschaft für Holzforschung, München, Germany, (290 pp.).

Kránitz K. (2014). *Effect of Natural Aging on Wood.* Doctoral Thesis, ETH Zürich, Zurich, Switzerland.

Kránitz K., Sonderegger W., Bues C.-T. & Niemz P. (2016). Effects of aging on wood: A literature review. *Wood Science and Technology*, 50(1), 7–22.

Krautkrämer J. & Krautkrämer H. (1986). Werkstoffprüfung mit Ultraschall. (5th Ed.), Springer, Berlin, Germany, (723 pp.).

López G., Basterra L.-A., Ramón-Cueto G. & de Diego A. (2014). Detection of singularities and subsurface defects in wood by infrared thermography. *International Journal of Architectural Heritage*, 8(4), 517–536.

Lukowsky D. (2013). *Schadensanalyse Holz und Holzwerkstoffe.* Fraunhofer IRB Verlag, Stuttgart, Germany, (240 pp.).

Meder R., Thumm A. & Bier H. (2002). Veneer stiffness predicted by NIR spectroscopy calibrated using mini-LVL test panels. *Holz als Roh- und Werkstoff*, 62(3), 159–164.

Meinlschmidt P. (2005). Thermographic detection of defects in wood and wood-based materials. In: *Proceedings of the 14th International Symposium on Nondestructive Testing of Wood. Shaker Verlag, Aachen*, Germany, (pp. 41–45).

Meisel A., Wallner B. & Schickhofer G. (2015). Tragfähigkeit und Verformungsverhalten von Kammverbindungen. *Bautechnik*, 92(6), 412–423.

Möhler K. & Maier G. (1969). Der Reibbeiwert bei Fichtenholz im Hinblick auf die Wirksamkeit reibschlüssiger Holzverbindungen. *Holz als Roh- und Werkstoff*, 27(8), 303–307.

Möhler K. & Herröder W. (1979). Obere und untere Reibbeiwerte von sägerauhem Fichtenholz. *Holz als Roh- und Werkstoff*, 37(1), 27–32.

Mönck W. (1999). *Schäden an Holzkonstruktionen: Analyse und Behebung.* (3rd Ed.), Huss-Medien Verlag Bauwesen, Berlin, Germany, (318 pp.).

Mönck W. & Erler K. (2004). *Schäden an Holzkonstruktionen: Analyse und Behebung.* (4th Ed.), Huss-Medien Verlag Bauwesen, Berlin, Germany, (308 pp.).

Niemz P. (1996). Untersuchungen zum Einfluss der Holzfeuchte auf die Schallausbreitungsgeschwindigkeit in Roble. *Holz als Roh- und Werkstoff*, 54(1), 60.

Niemz P. & Wetzig M. (2011). Auf spezielle Einsatzbereiche konzentriert. Einsatz von Thermoholz: Eigenschaften, Verarbeitung, Praxiserfahrungen. *Holz-Zentralblatt*, 137(1), 24–26.

Niemz P. & Bächle F. (2002). Bohrwiderstandsmessung und Schallgeschwindigkeitsmessung - Untersuchungen zur Erkennung der Fäule in Fichtenholz. *Stadt und Grün*, 51(10), 52–55.

Niemz P., Hug S. & Schnider T. (2014). Einfluss der Temperatur auf ausgewählte mechanische Eigenschaften von Esche, Buche, Ahorn und Fichte. *Forstarchiv*, 85(5), 163–168.

Niemz P. & Sonderegger W. (2021). *Holzphysik. Physik des Holzes und der Holzwerkstoffe.* (2nd Ed.), Hanser Verlag, München, Germany, (580 p.).

Niemz P., Teischinger A. & Sandberg D. (Eds.) (2023). *Springer Handbook of Wood Science and Technology.* Springer Nature, Cham, Switzerland, (XXV+2069 pp.).

Obataya E. (2007). Characteristics of aged wood and Japanese traditional coating technology for wood protection. *Cité de la Musique - Conserver aujourd'hui: les "vieillissements" du bois, journée d'étude du 2 février*, Philharmonie de Paris, Paris, France, (pp. 26–43).

Oltean L., Teischinger A. & Hansmann C. (2008). Wood surface discolouration due to simulated indoor sunlight exposure. *Holz als Roh- und Werkstoff*, 66(1), 51–56.

Ozyhar T., Hering S., Sanabria S.J. & Niemz P. (2013). Determining moisture dependent elastic characteristics of beech wood by mean of ultrasonic waves. *Wood Science and Technology*, 47(2), 329–341.

Paulitsch M. & Barbu M.C. (2015). *Holzwerkstoffe der Moderne.* DRW-Verlag, Leinfelden-Echterdingen, Germany, (528 pp.).

Pitt W. (2010). *33 Farbtafeln Parkett.* Holzmann Buchverlag, Bobingen, Germany.

Ross R. (2010). *Wood Handbook. Wood as an Engineering Material. (Centenial Ed.)*, US Department of Agriculture, Forest Service, Forest Products Laboratory, Madison (WI), USA, (509 pp.).

Rug W. & Lissner A. (2011). Studies on the strength and ultimate limit design of timber under the influence of chemically aggressive media. *Bautechnik*, 88(3), 177–188.

Sanabria S.J. (2012). *Air-Coupled Ultrasound Propagation and novel Non-Destructive Bonding Quality Assessment of Timber Composites.* Doctoral Thesis, ETH Zürich, Zurich, Switzerland.

Sandak A., Sandak J. & Negri M. (2011). Relationship between near-infrared (NIR) spectra and the geographical provenance of timber. *Wood Science and Technology*, 45(1), 35–48.

Sandak J., Niemz P., Hänsel A., Mai J. & Sandak A. (2021). Feasibility of portable NIR spectrometer for quality assurance in glue-laminated timber production. *Construction and Building Materials*, 308(188), Article ID: 125026.

Schneider A. & Engelhardt F. (1977). Vergleichende Untersuchungen über die Wärmeleitfähigkeit von Holzspan- und Rindenplatten. *Holz als Roh- und Werkstoff*, 35(7), 273–278

Schubert S. (2007). *Acousto-Ultrasound Assessment of inner Wood-Decay in standing Trees: Possibilities and Limitations.* Doctoral Thesis, ETH Zürich, Zurich, Switzerland.

Sonderegger W. & Niemz P. (2006). Untersuchungen zur Quellung und Wärmedehnung von Faser-, Span- und Sperrholzplatten. *Holz als Roh- und Werkstoff*, 64(1), 11–20.

Sonderegger W. & Niemz P. (2009). Thermal conductivity and water vapour transmission properties of wood-based materials. *European Journal of Wood and Wood Products*, 67(3), 313–321.

Sonderegger W. & Niemz P. (2012). Thermal and moisture flux in soft fibreboards. *European Journal of Wood and Wood Products*, 70(1/3), 25–35.

Sonderegger W., Hering S. & Niemz P. (2011). Thermal behavior of Norway spruce and European beech in and between the principal anatomical directions. *Holzforschung*, 65(3), 369–375.

Sonderegger W., Kránitz K., Bues C.-T. & Niemz P. (2015). Aging effects on physical and mechanical properties of spruce, fir and oak wood. *Journal of Cultural Heritage*, 16(6), 883–889.

Stamm A.J. (1964). *Wood and Cellulose Science.* Ronald Press, New York, USA, (549 pp.).

Steinhagen H.P. (1977). Thermal conductive properties of wood, green or dry, from −40° to +100°C: A literature review. *Technical Report FPL-9, Forest Products Laboratory*, U.S. Department of Agriculture, Forest Service, Madison (WI), USA, (10 pp.).

Teischinger A., Zukal M.L., Meints T., Hansmann C. & Stingl L. (2012). Colour characterization of various hardwoods. In: Nemeth R. & Teischinger A. (Eds.), *The 5th Conference of Hardwood Research and Utilization in Europe: Hardwood Science and Technology.* University of West Hungary Press, Sopron, Hungary, (pp. 180–188).

Titner J., Smid E., Tieben J., Reschreiter H., Kovarik K. & Grabner M. (2016). Aging of wood under long-term storage in a salt environment. *Wood Science and Technology*, 50(5), 953–961.

Torgovnikov G.I. (1993). *Dielectric Properties of Wood and Wood-based Materials.* Springer-Verlag, Berlin, Heidelberg, Germany, (196 pp.).

Tsuchikawa S. & Schwanninger M. (2013). A review of recent near-infrared research in wood and paper (Part 2). *Applied Spectroscopy Reviews*, 48(7), 560–587.

Ugolev B.N. (1986). *Wood Science and Fundamentals of Wood Products Science.* [in Russian] Lesnaja Prom., Moskau, Soviet Union.

Unger A. (1988). *Holzkonservierung.* VEB Fachbuchverlag, Leipzig, Germany, (220 pp.).

Unger A., Schniewind A.P. & Unger W. (2001). *Conservation of Wood Artifacts.* Springer-Verlag Berlin and Heidelberg, Germany, (578 pp.).

Volkmer T., Lorenz T., Hass P. & Niemz P. (2014). Influence of heat pressure steaming (HPS) on the mechanical and physical properties of common oak wood. *European Journal of Wood and Wood Products*, 72(2), 249–259.

Vorreiter L. (1949). *Holztechnologisches Handbuch: Band 1.* Fromme, Wien, Austria, (548 pp.).

Wagenführ R. (2006). *Holzatlas.* (6th Ed.), Fachbuchverlag Leipzig im Carl Hanser Verlag, München, Germany, (816 pp.).

Wienhaus O., Niemz P. & Fabian S. (1988). Untersuchungen zur Holzartendifferenzierung mittels IR-Spektroskopie. *Holzforschung und Holzverwertung*, 32(6), 120–125.

Wimmer R. Kläusler O. & Niemz P. (2013). Water sorption mechanisms of commercial wood adhesive films. *Wood Science and Technology*, 47(4), 763–775.

Xu M., Li L., Wang M. & Luo B. (2014). Effects of surface roughness and wood grain on the friction coefficient of wooden materials for wood-wood frictional pair. *Tribology Transactions*, 57(5), 871–878.

Yokoyama M., Gril J., Matsuo M., Yano H., Suigiyama J., Clair B., Kubodera S., Mistutani T., Sakamoto M., Ozaki H., Imamura M. & Kawai S. (2009). Mechanical characteristics of aged Hinoki wood from Japanese historical buildings. *Comptes Rendus Physique*, 10(7), 601–611.

8 Elastic and Inelastic Properties of Wood and Wood-based Materials

P. Niemz and W. Sonderegger

8.1 OVERVIEW

The properties of wood and wood-based materials can be divided into two groups:

- elastic properties (modulus of elasticity, shear modulus, Poisson's ratio), and
- inelastic properties, *i.e.*,
  - viscoelastic properties (creep and relaxation),
  - mechano-sorptive properties, which reflect the behaviour under load and humidity change, and
  - plastic properties manifest in permanent deformation (*e.g.*, plastic components of creep deformation, plastic deformation under load above the limit of proportionality).

In addition, all properties are highly dependent on the moisture content of the wood. All properties are, therefore, functions of characteristic values.

8.2 ELASTIC PROPERTIES

8.2.1 The Elastic Law and Stress-Strain Diagram

A linear relationship exists between stress and strain in ideal elastic bodies (Hooke's Law). A solid body becomes more extensive in the case of tensile load and more minor under compression. After release, the deformation of an ideal elastic body is completely regressed (Figure 8.1a). This applies if the mechanical stress is below the proportional limit. The proportionality limit varies with the moisture content and the load mode of the wood. It is, for example, a tensile load of about 50–60% of the maximum stress under normal climatic conditions. The modulus of elasticity (Young's modulus) is calculated from the slope of the straight line in the stress-strain diagram. Above the proportional limit, a plastic deformation occurs. In the direction of the fibre and perpendicular to the fibre, the ultimate strain in tensile load is low (0.7–1.0%). Wood-based materials exhibit stress-strain behaviour analogous to that of solid wood. Perpendicular to the fibre direction, compressive loading above

DOI: 10.1201/9781003411994-8

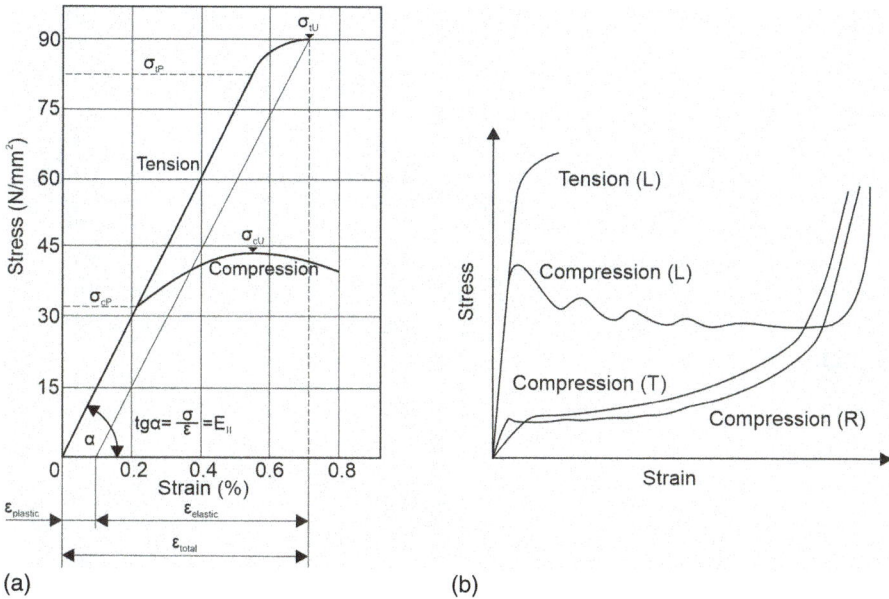

FIGURE 8.1 Stress-strain diagrams of solid wood and wood-based materials: (a) wood under tensile and compressive stress in the fibre direction, and (b) tension and compression parallel and perpendicular to the fibre direction.

(Courtesy of Carl Hanser Verlag.)

the proportional limit leads to considerable plastic deformation. The wood can be strongly densified, particularly in the radial direction (Figure 8.1b). After densification (especially of the earlywood), stress and strain also have an almost linear relationship. Therefore, an ultimate strain at a break of 2% or 5% is commonly defined for calculating strength under pressure perpendicular to the fibre.

For wood-based particle panel materials, the particles are strongly compressed perpendicular to the plane, *i.e.*, in the thickness direction of the panel, where most of the particles have their transverse direction (perpendicular to the wood's fibre or grain direction). Depending on panel density, the densification is approximately 50% in particleboard and 80% in medium-density fibreboard (MDF) and high-density fibreboard (HDF).

The elongation is calculated in the uniaxially stressed, isotropic material to:

$$\varepsilon = \frac{\Delta l}{l} \tag{8.1}$$

where
   ε is the strain
   Δl is the change of length loaded-unloaded
   l is the initial length (unloaded)

Applying Hooke's law, the stress can then be calculated in the uniaxial case:

$$\sigma = E \cdot \varepsilon \tag{8.2}$$

where

σ is the stress (normal stress) in a material under load
E is the modulus of elasticity (MOE)
ε is the strain

For the MOE in the three principal directions for wood, *i.e.* longitudinal (L), radial (R), and tangential (T), the transverse contraction (Poisson's effect) is neglected, and:

$$E_L = \frac{\sigma_L}{\varepsilon_L}$$

$$E_R = \frac{\sigma_R}{\varepsilon_R} \tag{8.3}$$

$$E_T = \frac{\sigma_T}{\varepsilon_T}$$

For shear stresses:

$$\tau = G \cdot \gamma \tag{8.4}$$

τ is the stress (shear stress)
G is the modulus of rigidity (shear modulus)
γ is the shear strain

8.2.2 GENERALISED HOOKE'S LAW FOR ORTHOTROPIC MATERIALS

Wood is an orthotropic material with strong property differentiation in the three principal axes. Figures 8.2 and 8.3 show the coordinate system of solid wood and wood-based materials, as well as the assignment of the coordinate axes. Different names are used for the axes, such as L, R, and T for solid wood or x_1, x_2, x_3 or x, y, z.

The properties of wood-based materials, particularly particleboard and fibreboard, are often more or less isotropic in the plane. In contrast, due to the production method, a strong differentiation is usually present between the properties in plane and perpendicular to the panel plane.

As is usual in solid mechanics, the coordinate system is classified according to its properties' significance, *i.e.*, for wood, 1 – in longitudinal (fibre), 2 – radial, and 3 – in the tangential direction.

For orthotropic materials such as wood or wood-based materials, the following applies to the three-dimensional, orthotropic state:

$$[\varepsilon] = [S] \cdot [\sigma] \tag{8.5}$$

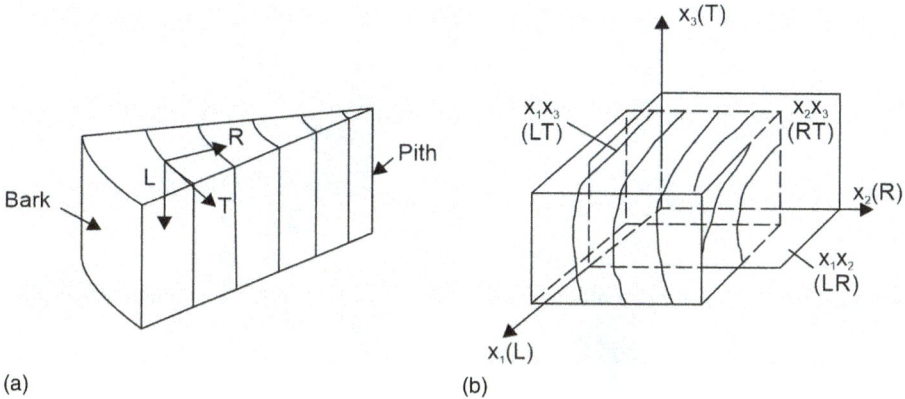

FIGURE 8.2 Orientations in wood: (a) the orthotropic coordinate system and (b) principal axes for solid wood.

(Courtesy of Carl Hanser Verlag.)

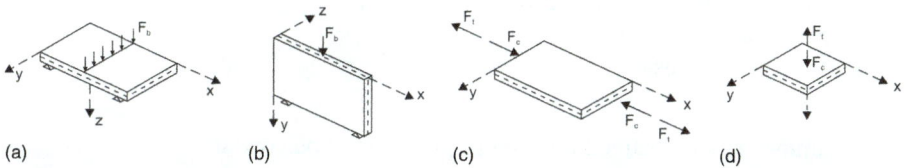

FIGURE 8.3 Loading options of wood-based materials: a) bending perpendicular to the plane of the panel (flat), b) bending parallel to the plane of the panel (upright), c) tension/pressure in the panel plane, and d) tension perpendicular to the plane of the panel. F_b, F_t, F_c = bending, tensile, or compression load; x, y, z = coordinate axes.

(Courtesy of Carl Hanser Verlag.)

[S] is the compliance matrix, where the S components are defined as:

$$S = \frac{\Delta l}{l \cdot \sigma} \tag{8.6}$$

where

S is the compliance (n.b. the unit m²/N)
Δl is the change in length due to load
l is the unloaded length
σ is the stress

For an orthotropic material such as wood, using the compliance matrix [S] in Voigt's notation yields:

$$
\begin{bmatrix} \varepsilon_{11} \\ \varepsilon_{22} \\ \varepsilon_{33} \\ \gamma_{23} \\ \gamma_{13} \\ \gamma_{12} \end{bmatrix} = \begin{bmatrix} S_{11} & S_{12} & S_{13} & 0 & 0 & 0 \\ S_{21} & S_{22} & S_{23} & 0 & 0 & 0 \\ S_{31} & S_{32} & S_{33} & 0 & 0 & 0 \\ 0 & 0 & 0 & S_{44} & 0 & 0 \\ 0 & 0 & 0 & 0 & S_{55} & 0 \\ 0 & 0 & 0 & 0 & 0 & S_{66} \end{bmatrix} \cdot \begin{bmatrix} \sigma_{11} \\ \sigma_{22} \\ \sigma_{33} \\ \tau_{23} \\ \tau_{13} \\ \tau_{12} \end{bmatrix}
\tag{8.7}
$$

In principle, the representation as stiffness matrix [C] in analogous form is also possible:

$$
[\sigma] = [C] \cdot [\varepsilon]
\tag{8.8}
$$

The following applies:

$$
C = S^{-1} \text{ and } S = C^{-1}
\tag{8.9}
$$

where

$\varepsilon_{11}, \varepsilon_{22}, \varepsilon_{33}$ are the normal strains
$\gamma_{23}, \gamma_{13}, \gamma_{12}$ are the shear strains
σ is the normal stress
τ is the shear stress

Nine parameters are assumed for orthotropic material behaviour:

- three moduli of elasticity,
- three shear moduli, and
- six Poisson's ratios (three can be derived from the other parameters using Equations in 8.10).

For calculations, only three Poisson's ratios are used, and for solid wood, the following applies for the Poisson's ratios:

$$
\frac{\mu_{RL}}{E_R} = \frac{\mu_{LR}}{E_L}
$$

$$
\frac{\mu_{TL}}{E_T} = \frac{\mu_{LT}}{E_L}
\tag{8.10}
$$

$$
\frac{\mu_{TR}}{E_T} = \frac{\mu_{RT}}{E_R}
$$

Practical measurements usually involve some deviations from symmetry, so the mean value is used in calculations to maintain the necessary symmetry conditions (Bodig and Jayne 1993). This also applies to the shear modulus.

The first index indicates the direction of the load, and the second index indicates the direction of elongation. In the literature, reverse notation is often used. The index system used here refers to Bodig and Jayne (1993) and Altenbach et al. (1996).

The distortion-stress relationships can be replaced by the engineering constants E and G. In the distortion-stress state, the engineering constants can be summarised as follows:

$$
\begin{bmatrix} \varepsilon_{11} \\ \varepsilon_{22} \\ \varepsilon_{33} \\ \gamma_{23} \\ \gamma_{13} \\ \gamma_{12} \end{bmatrix} = \begin{bmatrix} \dfrac{1}{E_1} & \dfrac{\mu_{21}}{E_2} & \dfrac{\mu_{31}}{E_3} & 0 & 0 & 0 \\[2ex] -\dfrac{\mu_{12}}{E_1} & \dfrac{1}{E_2} & \dfrac{\mu_{32}}{E_3} & 0 & 0 & 0 \\[2ex] -\dfrac{\mu_{13}}{E_1} & -\dfrac{\mu_{23}}{E_2} & \dfrac{1}{E_3} & 0 & 0 & 0 \\[2ex] 0 & 0 & 0 & \dfrac{1}{G_{23}} & 0 & 0 \\[2ex] 0 & 0 & 0 & 0 & \dfrac{1}{G_{13}} & 0 \\[2ex] 0 & 0 & 0 & 0 & 0 & \dfrac{1}{G_{12}} \end{bmatrix} \cdot \begin{bmatrix} \sigma_{11} \\ \sigma_{22} \\ \sigma_{33} \\ \tau_{23} \\ \tau_{13} \\ \tau_{12} \end{bmatrix} \qquad (8.11)
$$

σ is the normal stress
τ is the shear stress
γ is the shear strain or torsion
ε is the normal strain

8.2.3 ORTHOTROPIC PROPERTIES OF WOOD AND WOOD-BASED MATERIALS

Elastic and strength properties differ significantly in the three main cutting directions. Noack and Schwab (in von Halász and Scheer 1986) provide the proportional values shown in Table 8.1.

The shear modulus G_{RT} is very low in softwood (caused by the low density in the earlywood). For softwood, G_{RT} is about 10% of G_{LT}, and for hardwood, about 40% of G_{LT}. This can lead to shear failure in the transverse layers of multilayer boards (*e.g.*, cross-laminated timber).

TABLE 8.1

Ratios of the Modulus of Elasticity and Shear Modulus in the Significant Axes According to Noack and Schwab in von Halász and Scheer (1986)

Properties	Softwoods	Hardwoods
$MOE_T : MOE_R : MOE_L$	1 : 1.7 : 20	1 : 1.7 : 13
$G_{LR} : G_{LT} : G_{RT}$	1 : 1 : 0.1[a]	1.3 : 1 : 0.4

[a] Lower value than hardwood due to the continuous earlywood zone with low density and rigidity.

In addition, the influence of the load-to-fibre angle and the growth-ring orientation must be considered. According to Hankinson (1921), it applies:

$$E_\varphi = \frac{E_{//} \cdot E_\perp}{E_\perp \cos^n \varphi + E_{//} \sin^n \varphi} \qquad (8.12)$$

where

E_φ is the MOE in the direction of load
$E_{//}$ is the MOE parallel to the fibre direction
$E_\perp$ is the MOE perpendicular to the fibre direction
φ is the angle between the direction of the load and the fibre orientation
n – is a constant that can take values between 1.5 and 2.

Figure 8.4 shows the influence of the fibre-to-load angle and the growth-ring inclination on the modulus of elasticity under compressive loading for ash wood.

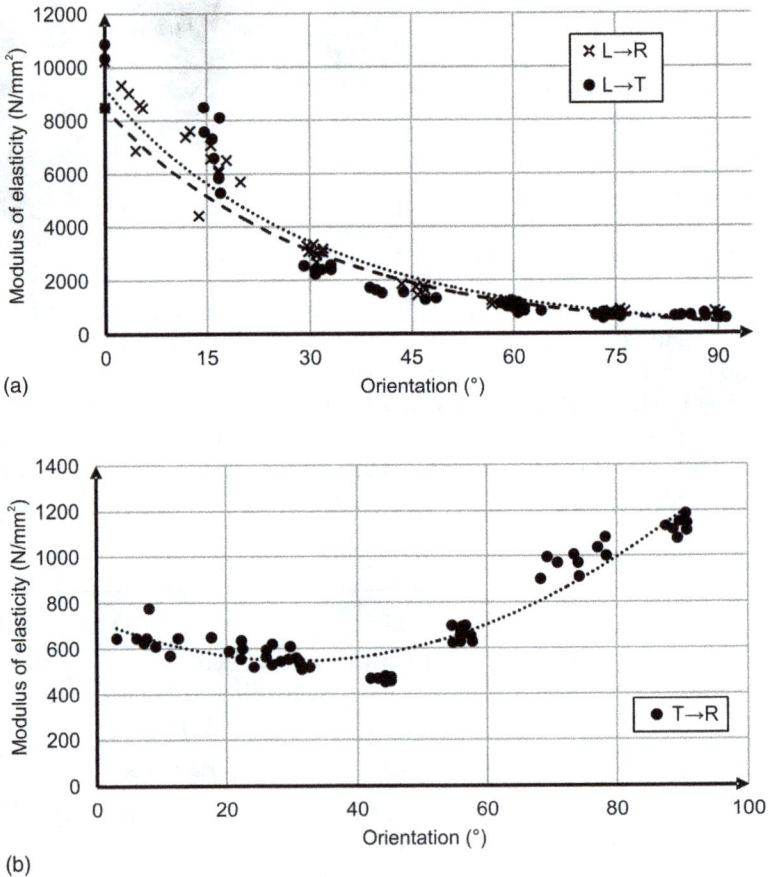

(a)

(b)

FIGURE 8.4 Influence of the fibre-to-load angle in (a) the LR and LT planes and (b) of the growth-ring orientation in the RT plane on the modulus of elasticity of common ash timber.

(Data from Clauss et al. 2014, courtesy of Carl Hanser Verlag.)

Perpendicular to the fibre direction, the modulus of elasticity is significantly lower than parallel; even a low fibre load angle causes a significant reduction.

Radially, the modulus of elasticity is, on average, considerably higher than tangentially, which is due, among other things, to the honeycomb structure but also the stiffening effect of the rays (Bodig and Jayne 1993, Burgert 2000, Sjölund 2015). At the RT section, a minimum can be seen at about 45 degrees (Figure 8.4b). Analogue dependencies can be found in Bodig and Jayne (1993) for Sitka spruce.

For plywood, cross-laminated timber, and wood particle-based panel materials, there is a specific directional dependence in the plane of the panels. The modulus of elasticity and strength in bending in the plane of the panels is in the production direction about 75% higher than perpendicular to it for oriented strand boards (OSB), for particleboards 15–20% and for MDF 2–5%. This difference is generally neglected in practice for particleboard and medium-density fibreboard (MDF). The tensile strength in plane is about 12 to 14 times higher than that perpendicular to it.

8.3 DETERMINATION OF THE ELASTIC PARAMETERS

8.3.1 Modulus of Elasticity

8.3.1.1 Static Methods

Table 8.2 provides an overview of the static methods used to determine the modulus of elasticity. The commonly used testing method is the bending test.

8.3.1.2 Loading Due to Normal Stresses in Compression and Tension

The modulus of elasticity is determined at normal stresses (tension, pressure) from Equation 8.13 and Table 8.2 using Hooke's law. Either the increase (Figure 8.1) of the straight line in the stress-strain diagram or the strain in the area below the proportional limit is determined, *e.g.*, with optical or incremental strain measuring

---

TABLE 8.2

Overview of the Static Methods for Determining the Moduli of Elasticity

Load Type	Graphic	Equations
Bending		
a) Three-point bending (static central-load bending) *e.g.*, DIN 52186 (DIN 1978), EN 310 (CEN 1993) (rectangular cross-section)		$E = \dfrac{l_s^{3}}{4 \cdot b \cdot h^{3}} \cdot \dfrac{\Delta F}{\Delta f}$
b) Four-point bending *e.g.*, DIN 52186 (DIN 1978), EN 408 (CEN 2012) (rectangular cross-section)		Global MOE (whole support width): $E = \dfrac{2 \cdot l_s^{3} - 3 \cdot l_s \cdot l'^{2} + l'^{3}}{8 \cdot b \cdot h^{3}} \cdot \dfrac{\Delta F}{\Delta f}$ Local MOE (between both loads F/2): $E = \dfrac{3 \cdot l_s \cdot l'^{2} - 3 \cdot l'^{3}}{8 \cdot b \cdot h^{3}} \cdot \dfrac{\Delta F}{\Delta f'}$

(Continued)

TABLE 8.2   (Continued)

Load Type	Graphic	Equations
Tensile load		$E = \dfrac{\Delta\sigma}{\Delta\varepsilon} = \dfrac{\sigma_2 - \sigma_1}{\Delta\varepsilon}$ $\Delta\varepsilon = \dfrac{\Delta l}{l_0} = \dfrac{l_2 - l_1}{l_0}$
Compression load *e.g.*, EN 408 (CEN 2012)		$E = \dfrac{\Delta\sigma}{\Delta\varepsilon} = \dfrac{\sigma_2 - \sigma_1}{\Delta\varepsilon}$ $\Delta\varepsilon = \dfrac{\Delta l}{l_0} = \dfrac{l_1 - l_2}{l_0}$

(Courtesy of Carl Hanser Verlag.)

E is the modulus of elasticity (N/mm²)

l_S is the span length between the supports (mm)

l' is the distance of the force application points at four-point loading (mm)

b is the specimen width (mm)

h is the specimen height (mm)

ΔF is the difference of load in the elastic range (N)

Δf is the deflection corresponding to the force difference in the middle of the specimen (mm)

$\Delta f'$ is the deflection between the force application points at four-point loading corresponding to the force difference (mm)

$\Delta\sigma$ is the stress difference in the elastic deformation range (N/mm²)

$\Delta\varepsilon$ is the strain change in the range of $\Delta\sigma$ (-)

Δl is the change in length (mm)

l_0 is the initial length (unloaded) (mm)

l_1, l_2 is the measured length at stress σ_1 and σ_2 (mm)

systems. Frequently, a measurement is made between preload (5–10% of the ultimate load) and a value below the proportional limit (*e.g.*, 30% of the ultimate load).

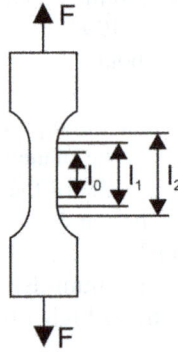

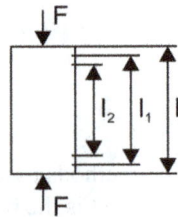

$$E = \frac{\Delta\sigma}{\Delta\varepsilon} = \frac{\sigma_2 - \sigma_1}{\varepsilon_2 - \varepsilon_1} \qquad (8.13)$$

8.3.1.3   Bending

Bending is the most used test method (three- or four-point loading).

For three-point bending, the modulus of elasticity depends on the span-to-thickness ratio. It increases with an increasing span-to-thickness ratio up to about a ratio of 15 to 20. In addition to normal stresses, shear stresses also occur. Deflection is composed of the components of pure bending and shear deformation. For the total deflection f applies, considering the shear deformation according to Timoshenko:

$$f = \frac{F \cdot l_s^{\,3}}{48 \cdot E \cdot I} + \frac{3}{10} \cdot \frac{F \cdot l_s}{G \cdot A} \qquad (8.14)$$

where

F is the load
l_s is the span length
E is the modulus of elasticity in pure bending
I is the moment of inertia
G is the shear modulus
A is the cross-sectional area of the specimen

The proportion of shear deformation or the shearing loss is thus influenced by the ratios h/l_s and G/E. The smaller the G/E, the lower the shearing influencing factor at the same ratio of l_s/h and the greater the shearing loss. The loss of shear increases as the ratio of l_s/h decreases.

This shearing loss is not included in calculating the modulus of elasticity in the three-point bending (according to the equation in Table 8.2). For wood-based panel materials, therefore, a span-to-thickness ratio of 20 is required (l_s at least 100 mm, EN 310 (CEN 1993)). The shearing loss is still 10–15% (strongly dependent on the wood material). Therefore, the modulus of elasticity determined by this method is always smaller than when determined in bending.

Of course, the same applies to the test of solid wood. Usually, a span/thickness ratio of at least 15 is used, for example, in DIN 52186 (DIN 1978).

In four-point bending, a modulus of elasticity independent of the span-to-sample thickness ratio is obtained when bending in the lateral shear stress-free area is measured (Table 8.2). There is no influence of shearing. Therefore, the modulus of elasticity in three-point bending is lower than that determined by pure bending (four-point bending) due to the impact of shearing.

8.3.1.4 Dynamic Modulus of Elasticity, Natural Frequency Measurement (Modal Analysis)

The modulus of elasticity test can also be carried out by dynamic methods, for example, measurement of the sound transit time and natural frequency measurement, including the logarithmic damping decrement (Görlacher 1987). Indeed, the orthotropic characteristics in the L, R, and T directions determined on small specimens with high ultrasound frequencies in the range of a few MHz are often clearly too high if the transverse contraction is neglected (Ozyhar 2013, Bachtiar 2017). At frequencies of about 30 kHz, however, the dynamic moduli of elasticity are only 10–20% higher than those determined in the static test. See also Section 7.3. However, the congruence between the natural frequency measurement and the static test is considerably better. The most frequent method used is the determination of the modulus of elasticity by bending load. This measurement is easy to perform.

Modal analysis is a very suitable method for identifying the material parameters of wood and wood-based materials. It can determine the modulus of elasticity, shear modulus, and Poisson's ratios (Grimsel 1999, Berner *et al.* 2007). The method offers the advantage of deciding several characteristic values on one specimen.

8.3.2 SHEAR MODULUS

8.3.2.1 Parameters

The shear modulus or modulus of rigidity is defined as the ratio of shear stress to the shear (Figure 8.5). It applies:

$$\gamma = \frac{1}{G} \cdot \tau \tag{8.15}$$

or

$$G = \frac{\tau}{\gamma} \tag{8.16}$$

where

G is the shear modulus
τ is the shear stress
γ is the shear strain

Solid wood has three shear moduli G_{LR}, G_{LT}, and G_{RT}. For wood-based panel materials such as particleboard, there are two shear moduli for bending perpendicular to the plane of the panel G_{zx} and parallel to the panel plane G_{yx} (z is the thickness direction of the panel). A differentiation in and perpendicular to the "production direction" usually does not occur.

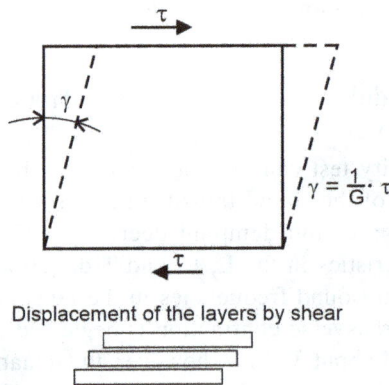

Displacement of the layers by shear

FIGURE 8.5   Shearing of wood and wood-based materials.

(Courtesy of Carl Hanser Verlag.)

TABLE 8.3

Static shear moduli determined by Arcan-test (Bachtiar *et al.* 2017)

Species	Method	Density (kg/m³)	G_{LR} (N/mm²)	G_{LT} (N/mm²)	G_{RT} (N/mm²)
Walnut	Arcan test	564	1020	868	194
Cherry	Arcan test	551	1188	782	218

Even with dynamic methods, the shear modulus can be determined:

- measure the sound-propagation time with transverse waves and calculate the shear modulus from density and sound velocity (Ozyhar 2013, Bachtiar 2017), or
- eigenfrequency measurement or modal analysis (Grimsel 1999, Berner *et al.* 2007, Gülzow 2008).

However, analogous to the modulus of elasticity, specific differences exist between the shear modulus determined in the static and dynamic experiments. As Table 8.3 shows, the values determined in the static test are mostly lower than those determined by ultrasound.

8.3.3 Poisson's Ratio

If a body is subjected to compression or tension stresses, the shape changes in and perpendicular to the load direction (Figure 8.6). Under tensile loading, the specimen becomes more prolonged and narrower, shorter and broader under compression.

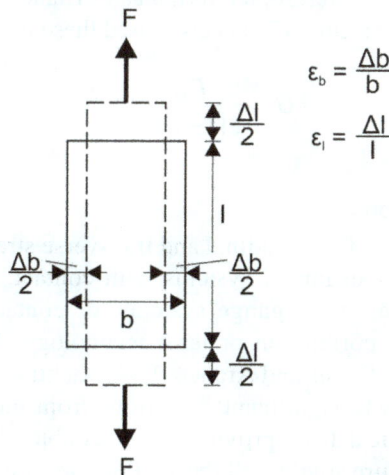

$$\varepsilon_b = \frac{\Delta b}{b}$$

$$\varepsilon_l = \frac{\Delta l}{l}$$

FIGURE 8.6 Elongation and transverse contraction of a body under tensile load.

(Courtesy of Carl Hanser Verlag.)

The Poisson's ratio is the quotient of the change in width to the change in length.

$$\frac{\Delta b}{b} = -\mu \cdot \frac{\Delta l}{l}$$

$$\mu = -\frac{\varepsilon_{\text{transverse}}}{\varepsilon_{\text{length}}} \tag{8.17}$$

where

b is the width
Δb is the change in width
l is the specimen length
Δl is the change in length
μ is the Poisson's ratio
$\varepsilon_{\text{transverse}}$ is the transverse strain
$\varepsilon_{\text{length}}$ is the longitudinal strain

There are six Poisson ratios (Equation 8.10). Due to the matrix's symmetry, only three must be determined; the other three can be calculated. However, in practice, symmetry is not always present in measurements. This is caused, for example, by the asymmetry of the growth rings.

The first index of the Poisson's ratio indicates the direction of the load, and the second is the direction of the transverse contraction. The reverse version is also often used in the literature.

According to Ugolev (1986), however, the three shear moduli G_{LR}, G_{LT}, and G_{RT} can be estimated by measuring the modulus of elasticity and the Poisson's ratio at 45 degrees (Equation 8.18). Therefore, the formulas of Hankinson and the relationship of Kon (1948) are used. Bachtiar (2017) also fused these relations.

$$G = \frac{E_{45}}{2\left(1 + \mu_{45}\right)} \tag{8.18}$$

8.3.3.1 Testing Methods

A standard measurement of longitudinal and transverse strain is required. Suitable for this purpose are, for example, systems with contact, such as strain gauges, incremental or electrical strain gauge systems, or contact-free optical systems based on digital image correlation or laser technology. Also, measuring sound propagation is possible (Bucur and Archer 1984, Bachtiar et al. 2017). The sound propagation often leads to significant deviations from the values determined in the static test. Mechanical tests provide more reliable characteristics. In many wood technology literature studies, all the six Poisson's ratios are listed; for finite-element (FE) calculations, only the three Poisson's ratios of the symmetry axis are used.

8.3.4 BUCKLING

When a structure is subjected to compressive axial stress, buckling may occur. A sudden sideways deflection of a structural member characterises buckling. This may occur even though the stresses that develop in the structure are well below those needed to cause failure of the material the structure is composed of. There is a risk of buckling in pillars with a sizeable length-to-cross-sectional dimension ratio.

8.3.4.1 Euler's Solutions for Buckling with Various End Conditions

Figure 8.7 shows the four load cases according to Euler. If the load exceeds the Euler limit, the column will suddenly bend or buckle sideways. If the critical buckling load F_k exceeds, the system changes from stable to unstable, and the specimen deviates laterally.

For case 2 in Figure 8.7 and a significant degree of slenderness (for wood: $\lambda >$ 100), the buckling theory developed by Euler applies. The buckling behaviour is also time and moisture-dependent.

8.3.4.2 Short Columns (Inelastic Buckling after Temajer)

For shorter columns, below a limit of slenderness, there is a region of buckling which is no longer characterised generally by the material's elasticity. The limit of slenderness can then be calculated:

$$\lambda_g = \pi \cdot \sqrt{\frac{E}{\sigma_p}} \tag{8.19}$$

where σ_P is the proportionality limit of the material in the compression test.

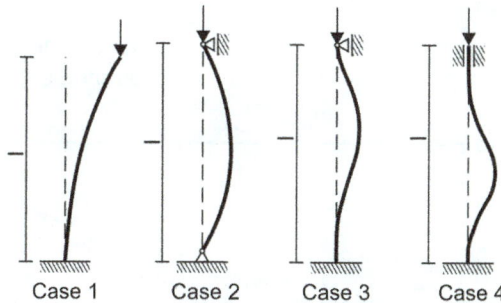

	Case 1	Case 2	Case 3	Case 4
Effective length factor	$\beta = 2$	$\beta = 1$	$\beta = \frac{1}{2} \cdot \sqrt{2} \approx 0.7$	$\beta = 0.5$
Buckling length $s = \beta \cdot L$	$s = 2 \cdot L$	$s = L$	$s \approx 0.7 \cdot L$	$s = 0.5 \cdot L$
Critical buckling load	$F_{crit} = \pi^2/4L^2 \cdot EI$	$F_{crit} = \pi^2/L^2 \cdot EI$	$F_{crit} = 2\pi^2/L^2 \cdot EI$	$F_{crit} = 4\pi^2/L^2 \cdot EI$

FIGURE 8.7 Euler's critical load of columns with constant stiffness and various attachment modes. L = column length, s = buckling length.

(Courtesy of Carl Hanser Verlag.)

At a value below this limit of slenderness, the Tetmajer equations are valid. These are equations which have the slenderness as an independent variable in the function and have the following structure:

$$\sigma_k = a + b \cdot \lambda + c \cdot \lambda^2 \tag{8.20}$$

For softwoods: $a = 29.3$, $b = -0.194$, and $c = 0$.

Further experimental work on sawn timber and wood-based materials is described in Kollmann (1951), Pozgaj *et al.* (1997), and Ross (2010) and related to timber construction literature such as Becker (2002), Werner and Zimmer (2009), and Neuhaus (2011).

8.4   MATERIAL CHARACTERISTICS AND INFLUENCING FACTORS

8.4.1   OVERVIEW

Figure 8.8 shows the correlation between the modulus of elasticity and the bending strength of spruce.

The essential factors that influence the elastic properties are:

- solid wood: species, bulk density, fibre orientation, microfibril orientation, moisture content, temperature,
- particleboard: density, particle dimensions, particle orientation, solid resin content, moisture content,
- fibreboards: fibre geometry, density, solid resin content, moisture content, and
- plywood and cross-laminated timber: the layer thickness ratio, the quality of the timber used in the layers, the narrow-surface bonding of the veneers, and the size of eventual gaps between lamellae in the middle layer.

The type of stress can affect the results, but this influence is usually not considered. Table 8.4 shows the impact of moisture content on common beech.

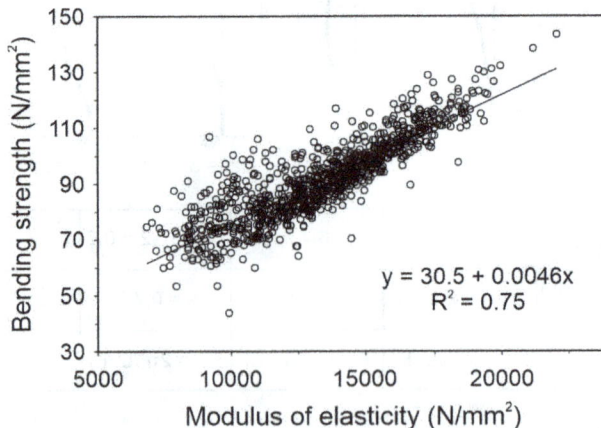

FIGURE 8.8   Correlation between modulus of elasticity in bending and bending strength for Norway spruce.

(Data from Sonderegger *et al.* 2008, courtesy of Carl Hanser Verlag.)

TABLE 8.4

Relative Change of the Modulus of Elasticity (MOE) for Common Beech within the Hygroscopic Range as a Function of the Load Type (Ozyhar 2013)

Load	ΔMOE (%) per 1% unit change of EMC		
	Longitudinal	Radial	Tangential
Tension	3.1	2.6	2.5
Compression	2.1	5.5	5.2
Bending	1.1		

EMC = equilibrium moisture content

The influence of the load speed is regulated by the standards (*e.g.*, 90 seconds for bending test according to DIN 52186 (DIN 1978)). The influence of the sample geometry is also present. Therefore, DIN 52186 requires a minimum of five growth rings per specimen in the cross-section.

8.4.2 MODULUS OF ELASTICITY AND SHEAR MODULUS

Tables 8.5 and 8.6 show values for the modulus of elasticity and the shear modulus. The modulus of elasticity of solid wood is the highest in the fibre direction, and in the radial direction about twice as large as in the tangential direction due to the effect of the rays (Sjölund 2015).

For torsional calculations, which are made according to the theory of elasticity for isotropic materials, the torsion modulus ($G_{torsion}$) in the LR and LT sections for sawn timber (solid wood) is 2/3 of the shear modulus (G_{LR} or G_{LT}). For glued-laminated timber, $G_{torsion}$ is equal to the shear modulus (Table 8.5).

8.4.2.1 Important Parameters

Density (Figure 8.9), the grain angle (Figure 8.4a), and the growth-ring orientation (Figure 8.4b) influence the modulus of elasticity. The minimum is found at approximately 45 degrees between radial and tangential directions. The modulus of elasticity in the radial direction is up to two times higher than in the tangential direction.

In solid wood, the modulus of elasticity generally increases linearly between species with density. The same applies to wood particle-based panel materials. However, even within a wood species, the influence of density on the modulus of elasticity is high. Other factors, such as microfibril angle, are also important, for example, if plantation wood (mostly juvenile wood with a high microfibril angle) is used (Butterfield 1997, Walker 2006).

Wood's shear modulus increases with increasing density. Structural factors such as the microfibril angle and density differences in the growth ring have a significant impact. Thus, Norway spruce, with high differences between earlywood and latewood, has a much lower shear modulus in the RT section than for example, yew.

For particleboard, for example, the particle geometry, the proportion of adhesive, and the bulk density significantly influence the modulus of elasticity and the shear modulus.

TABLE 8.5

Modulus of Elasticity and Shear Modulus of Wood in the Principal Directions According to the DIN 68364-11 Standard (DIN 1979)

Solid wood	Density (kg/m³)	E_L (N/mm²)	E_T (N/mm²)	E_R (N/mm²)	G_{LR} (N/mm²)	G_{LT} (N/mm²)	G_{RT} (N/mm²)
Softwoods							
Douglas fir	540	12000	700	900	800	900	80
Norway spruce	470	10000	450	800	600	650	40
Pitch pine	520	11000	500	1000		680	70
Hardwoods							
African teak	780	12900	1470	2250	1280	980	520
Bangkirai	940	20000	1500	2740			
Birch spp.	650	14000	630	1130	1200	930	
Common ash	690	13000	820	1500	880	620	
Common beech	690	14000	1160	2280	1640	1080	470
Common oak	670	13000	920	1580	1150	800	400
Doussie	790	13500	1450	1840	1330	980	420
Mahogany spp.	540	9500	570	990	770	590	
Sycamore	610	9400	890	1550	1240	1120	

TABLE 8.6

Modulus of Elasticity (Stiffness Values) for Wood-based Materials According to EN 12369-2 (CEN 2011). Values in ( ) are Perpendicular to the Surface Layers' Main Direction, *i.e.* in the Fibre Direction of the Surface Layers

Wood-based materials	Density (kg/m³)	Bending E_m (E_p) (N/mm²)	Tension, Compression E_t, E_c (N/mm²)	Shear G_v (G_r) (N/mm²)
Cross-laminated timber$_\parallel$	410	7100–10,000 (1800–4700)	2400–4700[1]	470 (41)
Cross-laminated timber$_\perp$	410	550–1500 (3500–4700)	2900[a]	470 (41)
Plywood$_{\parallel,\perp}$	350...750	500–14,000	250–11,200	220–550 (7.3–110)
OSB/2$_\parallel$, OSB/3$_\parallel$	550	4930	3800	1080 (50)
OSB/2$_\perp$, OSB/3$_\perp$	550	1980	3000	1080 (50)
OSB/4$_\parallel$	550	6780	4300	1090 (60)

(Continued)

TABLE 8.6   (Continued)

Wood-based materials	Density	Bending	Tension, Compression	Shear
		E_m (E_p)	E_t, E_c	G_v (G_r)
	(kg/m³)	(N/mm²)	(N/mm²)	(N/mm²)
OSB/4$_\perp$	550	2680	3200	1090 (60)
Particleboard, Type P4	500–650	1800–3200	1100–1800	550–860
Particleboard, Type P5	500–650	2100–3500	1300–2000	660–960
Particleboard, Type P6	500–650	2800–4400	1700–2500	880–1200
Particleboard, Type P7	500–650	3200–4600	2000–2600	1000–1250
Fibreboard, HB.HLA2	800–900	4600–5000	4600–5000	1900–2100
Fibreboard, MBH.LA2	600–650	2900–3100	2900–3100	1200–1300
MDF.LA	500–650	2700–3700	1600–2900	600–800
MDF.HLS	500–650	2800–3700	2400–3100	800–1000

‖ – in the direction of the main axis or fibre direction of the surface layer; ⊥ – in the direction of the minor axis or perpendicular to the fibre direction of the surface layer; E_m = modulus of elasticity when bending transversely to the plane of the panel; E_p – modulus of elasticity when bending in panel plane; E_t, E_c – modulus of elasticity in tension and compression; G_v – shear modulus transverse to the plane of the panel; G_r – shear modulus in plane of the panel
[a] Values for the modulus of elasticity under tensile load.

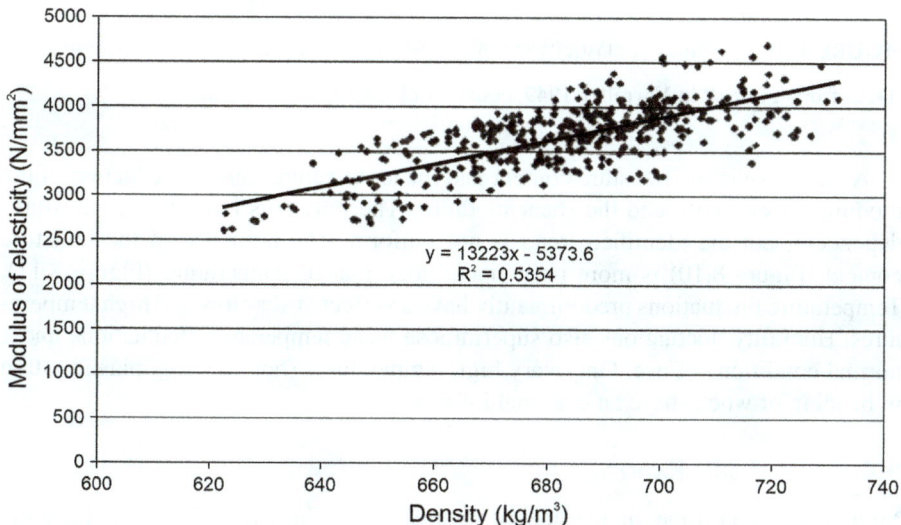

FIGURE 8.9 Modulus of elasticity as a function of particleboard density.

(Measurements ETH Zürich, courtesy of Carl Hanser Verlag.)

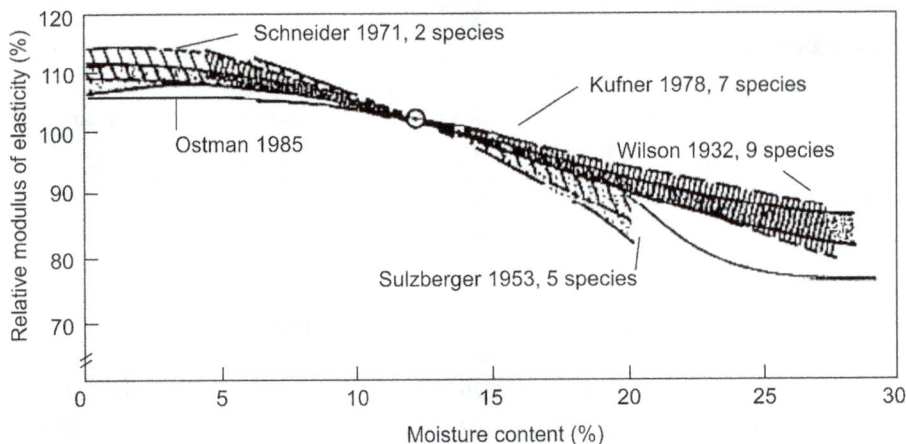

FIGURE 8.10 Modulus of elasticity as a function of the moisture content, normalised at 12% moisture content.

(Data from Ross 2010, courtesy of Carl Hanser Verlag.)

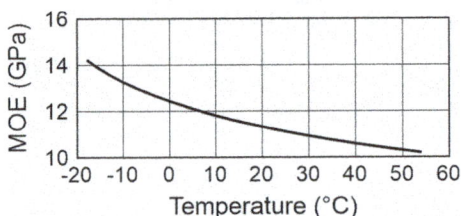

FIGURE 8.11 Modulus of elasticity (MOE) of Scots pine as a function of temperature.

(Data from Thunell in Vorreiter 1949, courtesy of Carl Hanser Verlag.)

An increase in the moisture content and the temperature causes a reduction of the modulus of elasticity and the shear modulus. The Poisson's ratio is also moisture-dependent, but the identified trend is not uniform. The influence of the moisture content (Figure 8.10) is more pronounced than that of temperature (Figure 8.11). Temperature fluctuations predominantly have an effect at very low and high temperatures. Humidity fluctuations also superimpose usual temperature fluctuations under normal conditions of use. Only very high temperatures (hot pressing, plasticisation in bending of wood) have an essential influence.

8.4.3 POISSON'S RATIO

Moisture content, time, and structural parameters are important factors influencing the Poisson's ratio (Neuhaus 1981, Ozyhar 2013). According to Ozyhar (2013), the influence of moisture content on the Poisson's ratio under tensile and compressive loading shows a not consistently uniform tendency. Carrington (1922) found similar

TABLE 8.7
Poisson's Ratios for Solid Wood and Wood-based Panel Materials

Material	Type of Load	μ_{12}	μ_{13}	μ_{21}	μ_{23}	μ_{31}	μ_{32}	
Common ash	Compression	0.30	0.21	0.04	0.58	0.05	0.36	Claus *et al.* (2014)
	Tension	0.41	0.27	0.05	0.66	0.07	0.37	Claus *et al.* (2014)
European beech	Compression	0.27	0.24	0.07	0.64	0.09	0.27	Hering *et al.* (2012)
	Tension	0.43	0.58	0.04	0.61	0.04	0.31	Ozyhar *et al.* (2012)
Maple	Compression	0.34	0.47	0.16	0.59	0.05	0.40	Sonderegger *et al.* (2013)
	Tension	0.49		0.06	0.65	0.04	0.38	Sonderegger et al. (2013)
Norway spruce	-	0.46	0.45	0.041	0.50	0.01	0.20	Keunecke *et al.* (2008)
Particleboard	Compression	0.26	0.21	0.26	0.35			Schreiber *et al.* (2007)
	Tension	0.23	0.33	0.29	0.29			Schreiber *et al.* (2007)
	-	0.23	0.27	0.25	0.24	0.02	0.02	Albers (1970)
OSB	Compression	0.24	0.25	0.33	0.24			Schreiber *et al.* (2007)
MDF	Compression	0.24	0.25	0.28	0.19			Schreiber *et al.* (2007)
	Tension	0.17	0.22	0.16	0.26			Schreiber *et al.* (2007)
Plywood	Compression	0.04	0.30	0.10	0.44	0.08	0.05	Albers (1970)

Poisson's ratios with j – direction of transverse strain and i – load direction; Index: wood: 1 – longitudinal (in fibre direction), 2 – radial, 3 – tangential; wood-based materials: 1 – in the production direction or parallel to the grain direction of the surface layer, 2 – perpendicular to the production direction or the grain direction of the surface layer, 3 – perpendicular to the plane of the panel.

trends for Norway spruce. Table 8.7 shows Poisson's ratios for different species and wood-based panel materials. The Poisson's ratios are also time-dependent (Ozyhar *et al.* 2013, Huč and Svensson 2018).

8.5   RHEOLOGICAL PROPERTIES

8.5.1   OVERVIEW

Rheology is the science that deals with materials' deformation and flow behaviour. The term "Rheo" comes from the Greek word for flow. Solid wood and wood-based materials are viscoelastic materials. So, they have elastic behaviour (defined by Hooke's Law) and partly viscous behaviour. Viscoelastic materials combine the properties of solids and liquids.

Roth (1935) carried out the first studies on the rheological behaviour of wood. Figure 8.12 shows an overview of rheological properties. The elastic properties, *i.e.*, modulus of elasticity, shear modulus, Poisson's ratio, and strength, are time-dependent. This time dependency affects the influence of the load speed on mechanical properties-testing results. The rheological properties define test speed in static short-term tests, *i.e.* the time until break (*e.g.*, 60 or 90 ± 30 seconds). Time dependence is also detectable when comparing static and dynamic experiments.

The viscoelastic behaviour is detectable under static and dynamic loads. Dynamic load leads to fatigue. Temperature, time, frequency, and moisture content determine the storage modulus E', the loss modulus E'', and the loss factor (tan $\delta = E''/E'$), indicating how viscous a material is relative to its elasticity.

Important rheological properties are:

- creep,
- stress relaxation, and
- long-term strength (duration of load).

Thus, strengths (σ) and deformations (ε) of wood and wood-based materials are time-dependent, *i.e.* σ and $\varepsilon = f$(time), as shown in Figure 8.12.

8.5.1.1   Creep
Creep means increased deformation under a constant load over time. Creep phenomena occur in all types of loads. They are, for example, recognisable in the time-dependent deflection of shelves in furniture or roofs of old houses or bridges. Creep deformation must be considered in timber construction.

8.5.1.2   Stress Relaxation
Stress relaxation is the decrease in stress under constant strain as a function of time. It occurs, for example, in adhesive joints or prestressed timber structures.

8.5.1.3   Long-term Strength
Long-term strength is the stress with which a material can be loaded infinitely long (duration of load, DOL). It must be considered when dimensioning timber structures.

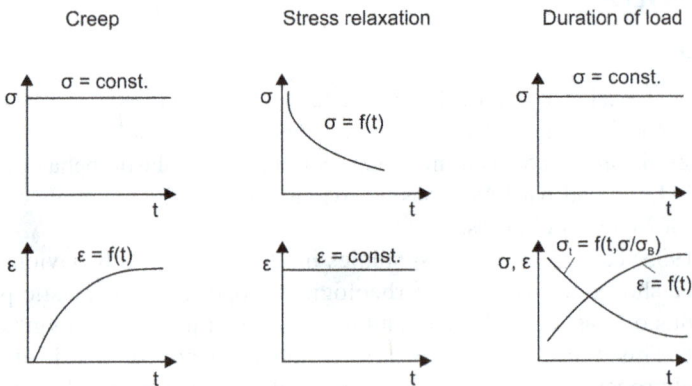

FIGURE 8.12   Rheological behaviour of wood and wood-based materials under static load. t – time, ε – strain, σ – stress, and σ_B – stress.

(Courtesy of Carl Hanser Verlag.)

8.5.2 CREEP

8.5.2.1 Sawn Timber

Depending on load, three creep phases are distinguished over time: primary, secondary, and tertiary. The creep is influenced by the main constituents of wood: cellulose, hemicelluloses, and lignin. In addition, the moisture content and the extractives are essential. The total deformation of a viscoelastic material can be divided into three deformation phases (Figure 8.13):

- instantaneous (elastic),
- flow, and
- permanent (plastic).

The proportion of the three phases in the total deformation depends on the load level and the moisture content (Figure 8.13).

At loading, the deformation is instantaneous and elastic, *i.e.*, this part of the deformation will instantly recover upon unloading. This is followed by delayed elastic (viscoelastic) deformation, which also is reversible but delayed (non-instantaneous), *i.e.* the wood flows. Depending on load level, a permanent "plastic" deformation may occur, irreversible even after an indefinitely long unloaded time. The

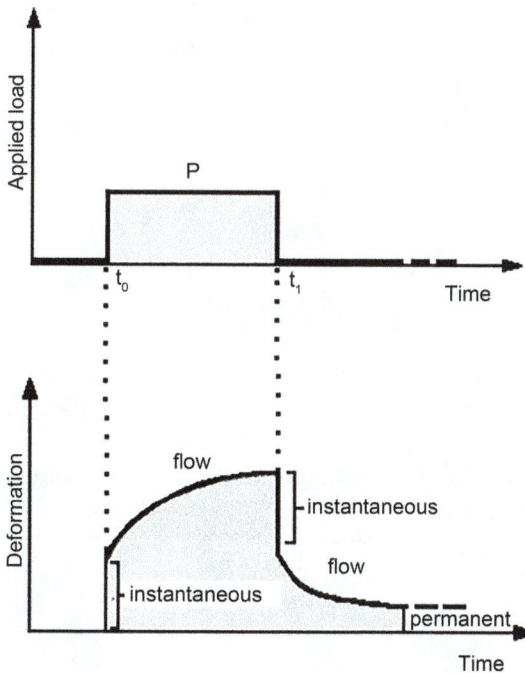

FIGURE 8.13 Creep curve under a constant load.

(Based on Bodig and Jayne 1993, courtesy of Carl Hanser Verlag.)

total creep after a load/un-load cycle is thus the sum of the partial deformations. The proportion of the individual sections depends strongly on the load level and climate conditions.

8.5.2.2 Wood-based Materials

The creep deformation of wood-based materials leads to an interaction of the following creep components (Niemz 1982).

- creep of the structural elements (*e.g.*, veneers, strands, particles),
- creep of the adhesive, and
- movements between the structural elements.

The creep of the adhesive joints is relatively low for urea, melamine, or polymeric methyl diphenyl diisocyanate (PMDI) resins, which are the most used adhesive types in particle-based panel materials. In the case of phenol-formaldehyde resins, however, the alkali component significantly influences the creep. An increase in alkali content will increase wood moisture content and creep deformation.

8.5.2.3 Parameters and Testing Methods

The bending test is often used to determine the creep properties (*e.g.*, for wood-based materials and glued-laminated timber). However, tension and compression tests may also be applied. Parameters for creep evaluation are the creep factor (φ) and relative creep (F):

$$\varphi = \frac{f_t - f_o}{f_o} \tag{8.21}$$

$$F = \frac{f_t}{f_o} \tag{8.22}$$

where

f_t is the deflection/strain at time t
f_o is the elastic deflection/strain at first loading (initial deflection)

For static calculations, the time-dependent modulus of elasticity (E_t) is used to consider creep deformation:

$$E_t = \frac{E_o}{1+\varphi} = \frac{E_o}{F} \tag{8.23}$$

where

E_o is the modulus of elasticity at short-term loading
E_t is the time-dependent (reduced) modulus of elasticity

The European standard EN 1995-1-1 (CEN 2004) applies analogically to the deformation parameter (k_{def}) in timber construction.

The creep deformation can be determined for all load types (tensile, fracture, bending, shearing). In this case, a constant load is applied, and the deformation is defined as a function of time, mostly in bending.

Gressel (1984) developed a test method for testing the components in timber construction due to creep deformation (Svensson 2011).

The test load is usually between 10% and 30% of the ultimate load. To obtain reproducible results, a constant climate (*e.g.*, 23 °C and 50% relative humidity or 20 °C and 65% relative humidity) is important.

Various approaches are applied to describe the creep behaviour and extrapolate the measurement results mathematically. Some examples are:

The tested load duration is at least 140 days. Attempts have been made for over ten years (Gressel 1986, Dinwoodie *et al.* 1990, CEN 2013).

According to EN 1156 (CEN 2013), wood-based materials in four-point bending, creep deformation, and creep rupture strength are standardised today. The specimens are loaded with 25% of the ultimate load. The material is assumed to behave linearly viscoelastic up to 40% of the breaking strength. The deflection is measured after 5, 10, 50, and 500 minutes and then at 24-hour intervals. The test duration should be at least 26 weeks (preferably 52 weeks). The creep factor is plotted in a logarithm over time diagram (but the first five minutes are ignored), and deformation is estimated at ten years linear extrapolation.

After about one year, only about 2/5 to 2/3 and after three years, about 2/3 to 4/5 of the assumed final creep deformation are reached. Therefore, the extrapolation to ten years is reasonable. In addition, long-term exposure under actual conditions leads to a superposition with the mechano-sorptive effect due to the change in relative humidity.

In practice, testing is often only six months up to one year, sometimes even considerably shorter. In many cases, only these values are listed in literature without extrapolation. Generally, it is to be noted that only a few full-scale creep tests have been carried out in recent decades.

Deformation factors for long-term loads (k_{def}) in the timber construction standards (Eurocode 5) for solid timber and wood-based materials indicate very high creep deformations, especially for particle-based panels (Table 8.8). Leps (2012) found very high creep factors for newly produced particleboards, *i.e.* creep factors of 3 after extrapolation to ten years.

The structure, the load duration, and the climate determine the creep behaviour. The dominant factor influencing creep is the relative humidity.

8.5.2.4 Structure

The creep increases with grain angle. Wood loaded in the radial direction creeps less than wood, which is loaded in the tangential direction. Sapwood has a higher creep deformation than heartwood, and earlywood has a higher creep deformation than latewood. As density increases, creep deformation generally decreases. The influence of the density on the creep deformation is not as pronounced as the influence on the elasto-mechanical properties of the wood (Niemz 1980). The higher creep perpendicular to the fibre is mainly responsible for the higher creep deformation of particleboard compared to solid wood in the fibre direction. Also, the lower creep deformation of OSB in the production direction is due to the influence

TABLE 8.8

Deformation Factor (k$_{def}$) for Deflection Correction, According to Eurocode 5, *i.e.* EN 1995-1-1:2010-12 (CEN 2004)

Material	Type	Service Class 1	2	3
Sawn timber				
Glued-laminated timber		0.60	0.80	2.00
Laminated veneer lumber				
Plywood	EN 636-1	0.80		
	EN 636-2	0.80	1.00	
	EN 636-3	0.80	1.00	2.50
Oriented strand board	OSB/2	2.25		
	OSB/3, OSB/4	1.50	2.25	
Particleboard	P4	2.25		
	P5	2.25	3.00	
	P6	1.50		
	P7	1.50	2.25	
High-density fibreboard	HB.LA	2.25		
	HB.HLA1/2	2.25	3.00	
Medium-density fibreboard	MBH.HLA1/2	3.00		
	MBH.HLS 1/2	3.00	4.00	
	MDF.LA	2.25		
	MDF.HLS	2.25	3.00	

of the grain angle of the wood. Wood modification at the cell wall or molecular level (for swelling reduction) reduces creep. The creep decreases with increasing length of the structural elements. Thus, the creep deformation increases in the following order:

- solid wood,
- laminated wood such as laminated veneer lumber (LVL) and Parallam©,
- plywood, edge-glued panels, and cross-laminated timber (CLT),
- oriented strand board (OSB),
- particleboard, and
- fibreboards.

Perkitny and Perkitny (1966) determined the ratio of creep deformation for different wood-based materials such as sawn timber: particleboard: fibreboard as 1:4:5.

The creep deformation increases with decreasing particle length/size, even within particle panel materials. The type of chip production also has considerable influence. Particleboards made from chips from a hammer mill with the same chip geometry have a much higher creep deformation than particleboards made from a disk flaker. An apparent influence of the density on the creep deformation of particleboard is not detectable. Surface veneer on particleboard significantly reduces the creep deformation, but a coating with decorative (overlay) foil does not considerably reduce creep deformation. The type of adhesive is essential for creep. Urea-glued panels creep the least, and polyurethane (1C-PUR)-bonded panels creep the most. For phenolic resins, creep deformation increases with the increase in the alkali content of the adhesive.

8.5.2.5   Climatic Conditions

With increasing relative humidity of the ambient air, creep deformation generally increases in a constant climate; more pronounced for particle-based materials than for solid wood. In alternating climates, the creep deformation of particleboard increases in the wet period and decreases in the dry period. Surface protection can reduce the effect of moisture. With increasing temperature, the creep deformation increases (Niemz 1980, Hunt 1984).

8.5.2.6   Load Mode

Figure 8.14 shows the effect of the load mode on creep. The lowest creep deformation occurs in tension. In the cases of bending and torsion, shear also affects the results. Further information can be found in Niemz (1982) and Dinwoodie (2000). The bending mode combines three types of stresses, *i.e.*, compression, shear, and tension.

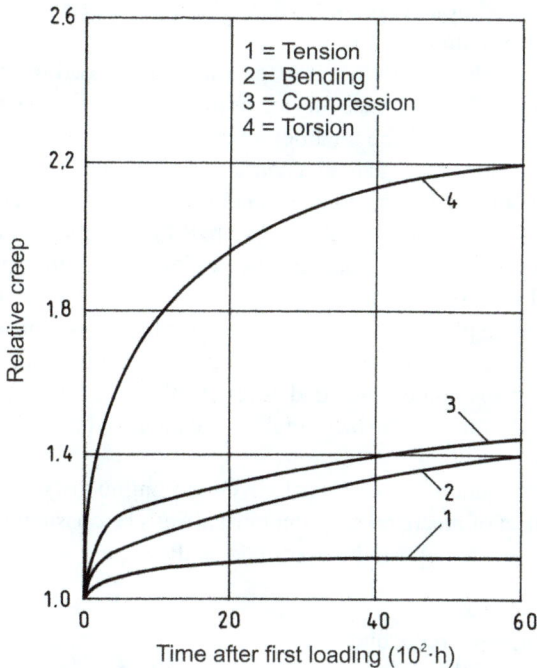

FIGURE 8.14   Relative creep as a function of the time after loading for Norway spruce.

(Data from Gressel 1984, courtesy of Carl Hanser Verlag.)

Experimental data on viscoelastic creep in different directions of various wood species under constant tensile, compression, or shear loading are reported by Ozyhar 2013, and Huč and Svensson 2018. Time dependence (creep) can also be detected when determining Poisson's ratios (Ozyhar 2013).

8.5.2.7 Material Parameters

Table 8.8 shows creep deformation characteristics. Thus, the creep deformation increases with increasing load (Figure 8.16a). The deformation factors in timber construction influence long-term loading, and the elastic deformation will be multiplied by the factor $1 + k_{def}$. Otherwise, the modulus of elasticity can be reduced by a factor of $1 + k_{def}$. Gressel (1983) summarised various studies in the field that have entered today's standardisation.

The influence of temperature on creep deformation of wood and wood-based materials is negligible, up to approximately 50 °C. A surface coating significantly reduces moisture absorption and desorption, thereby reducing creep.

8.5.3 MECHANO-SORPTION

As the wood moisture increases, the creep deformation in the constant climate increases significantly (Figure 8.15a). In the alternating climate (changing humidity, *e.g.*, dry/wet cycles), the swelling behaviour and resulting stresses/strains and creep are superimposed. The creep under changing climatic conditions is higher than at constant climatic conditions (Figure 8.16b). This phenomenon is called mechano-sorption (Figures 8.15b and 8.16b).

Mechano-sorption also occurs when a specimen under load changes the moisture content in a constant ambient climate (drying or moistening). It should be considered in addition to viscoelastic creep. The total deformation is higher in a cycling climate and under moisture-content changes than in a constant climate with the same sorption maxima. In the case of solid wood, creep deformation increases in the drying phase under bending and decreases in the wetting phase (Figure 8.17). Thereby, creep deformation in solid wood increases in the drying phase and decreases in the wetting phase. In medium-density fibreboard (MDF) and particleboard, on the other hand, creep deformation increases in the wetting phase and decreases in the drying phase (Dinwoodie 2000).

Creep is significantly influenced by the climate's exposure time, history, the cross-section area of the specimen, and load level. With large cross-sections, moisture changes occur only in the surface regions, which reduces creep deformation (Hanhijärvi 1998).

Total creep deformation due to climate cycling continuously increases with time. Therefore, the impact of mechano-sorption must always be considered in construction.

Simplified, the total creep can be described as the superposition of:

- viscoelastic creep,
- mechano-sorptive creep, and
- swelling and shrinkage.

The mechano-sorptive effect occurs both under creep and stress relaxation. Mechano-sorption was first described by Armstrong and Kingston (1960), later

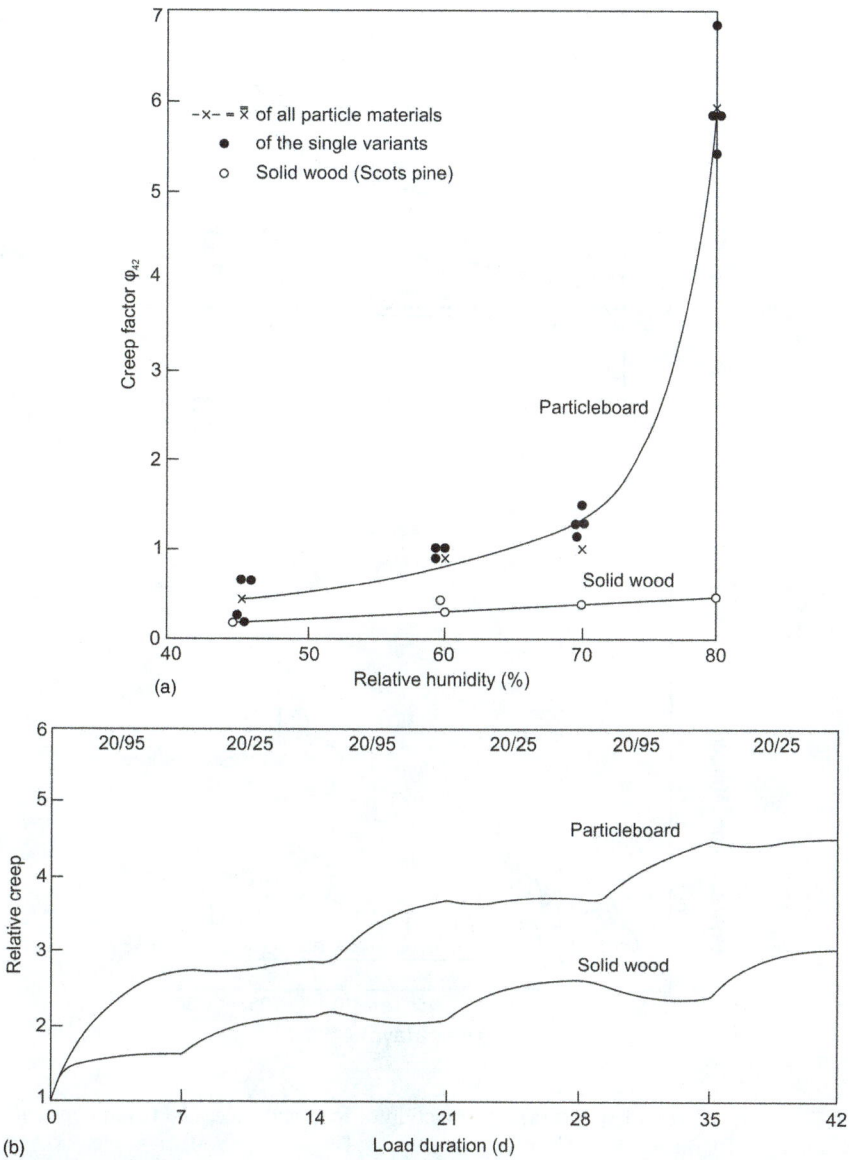

FIGURE 8.15 Creep of wood and particleboard as a function of relative humidity: (a) in a constant climate (Gressel 1971) and (b) at alternating climate.

(Data from Gressel 1971, courtesy of Carl Hanser Verlag.)

followed by numerous works dealing with the influence of chemical and hygro-mechanical modification of wood (Montero *et al.* 2012, Norimoto *et al.* 2012). For further reading, see, Smith *et al.* (2003), Navi and Sandberg (2012), and Huč and Svensson (2018).

$$\sigma_4 > \sigma_3 > \sigma_2 > \sigma_1$$

(a)

(b)

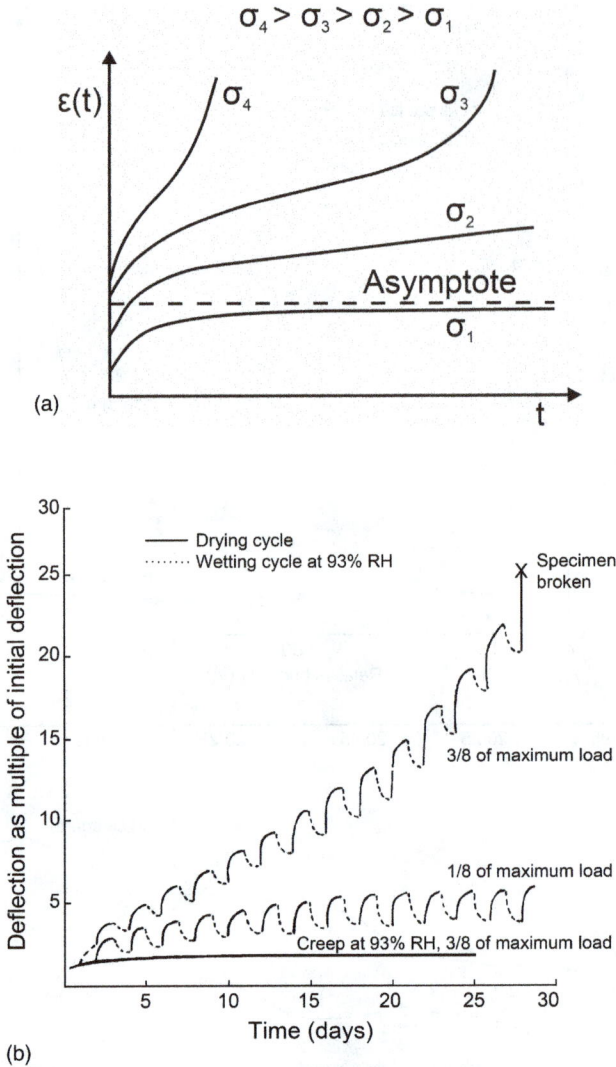

FIGURE 8.16 Creep of wood: (a) as a function of load and time, and (b) in constant and cyclic climate for different load levels.

(Data from Dinwoodie 2000, courtesy of _Carl_ Hanser Verlag.)

8.5.4 STRESS RELAXATION

If a wood specimen is constantly deformed, the stress required to maintain the constant strain decreases with time. This is called stress relaxation. In stress relaxation, the same processes occur inside the material as when it creeps, and stress relaxation is affected by the same factors as in creep. The ambient climate and material structure are essential for the level of creep.

Relaxation

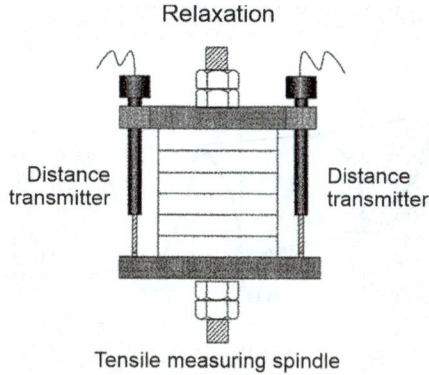

Tensile measuring spindle

FIGURE 8.17 Principle of the relaxation test in compression perpendicular to the grain. (Based on Popper *et al.* 1999, courtesy of Carl Hanser Verlag.)

Figure 8.17 schematically shows a device for stress-relaxation tests of wood and wood-based materials.

The relaxation is calculated as:

$$R = \frac{\sigma_t}{\sigma_0} \qquad (8.24)$$

where

σ_t is the stress at time t

σ_0 is the maximum stress at the beginning of the test (time = 0, $\sigma_0 > \sigma_t$)

The stress relaxation at a temperature of 20 °C is approximately 60%.

Stress relaxation occurs, for example, in prestressed wooden construction elements and is approximately of the magnitude of creep deformation. Relaxations may occur in adhesive joints after two lamellae with different moisture contents have been glued together and thereafter are conditioned.

An overview of stress relaxation and related modelling can be found, for example, in Navi and Stanzl-Tschegg (2009) and Huč *et al.* (2018).

Figure 8.18 shows the stress relaxation at a compressive load perpendicular to the grain in glued-laminated timber in a cyclic climate. In the drying phase, the stress decreases (caused by the shrinkage), and in the wetting phase, it increases. The mechano-sorptive effect occurs analogously to creep during climate change. As the number of cycles increases, the stress decreases significantly.

There is a difference between constant and alternating climate stress relaxation. The stress relaxation in prestressed glued-laminated timber after 70 days is according to Popper *et al.* (1999):

- 10% in a constant climate of 20 °C and 65% RH,
- 48% in a constant humid climate at 88% RH,
- 25% when the relative humidity increases from 65% to 88%, and
- 60% when the relative humidity decreases from 88% to 65%.

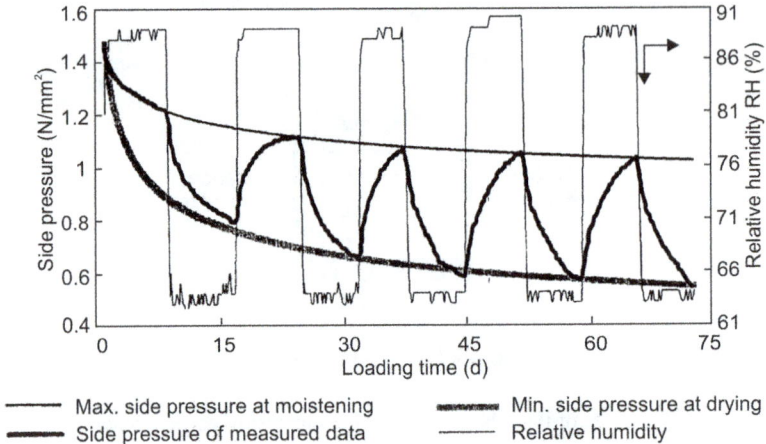

FIGURE 8.18 Stress (side pressure) relaxation of glued-laminated timber under prestress at cyclic changing climatic conditions.

(Data from Popper et al. 1999, courtesy of Carl Hanser Verlag.)

8.5.5 DURATION OF LOAD TEST

The duration-of-load (DOL) test is a creep strength or a stress-to-rupture test. The stress-to-rupture test is a material testing method for determining the material's behaviour at a constant temperature and after prolonged exposure to a constant force in bending, compression, or tension.

The creep deformation is primarily determined by the stress level. If threshold stress is exceeded, a progressive increase in deformation occurs due to the increasing microfractures in the material, which finally results in a break.

8.5.5.1 Parameters and Testing Methods

When testing the duration of the load, specimens are loaded in steps, for example, 80%, 60%, and 50% of the ultimate bending strength, and the time until the break is determined. At creep tests, 30% of the ultimate bending strength is applied to avoid failure. The EN 1156 standard (CEN 2013) is followed when testing wood-based panel materials. The test is performed at 20°C and 65% RH, and the specimens are loaded to 55%, 60%, 70%, and 75% of the ultimate load in bending. The time to break is determined by extrapolation. The bending stress is reduced when the test is made at higher relative humidities (50–75% of static strength, due to the lower strength at higher moisture content).

According to the *Wood Handbook* (Ross 2010), "time under intermittent loading has a cumulative effect. In tests where a constant load was periodically placed on a timber beam and then removed, the cumulative time the load was applied to the beam before failure was equal to the time to failure for a similar beam under the same load applied continuously".

8.5.5.2 Material Parameters

Figure 8.19 shows the duration-of-load (DOL) test for Norway spruce in bending under static load with variable wood moisture content. The commonly used "Madison curve" invented by the Forest Products Laboratory (FPL) in Madison, USA, is used as reference (Hoffmeyer 1995).

However, as it is a probabilistic estimation, the same behaviour as in creep is expected. In a normal climate (20 °C/65% RH), the fatigue strength of solid wood and wood-based panel materials is 40–60% of the static short-term strength. At an elevated moisture content of the material, the fatigue strength decreases. It can be expected that the type of load also influences fatigue strength, but this is not adequately studied.

Studies on the influence of structural parameters on the fatigue strength of wood and wood-based panel materials are not available. However, creep-rupture models linking the accumulated damage to creep deformation exist.

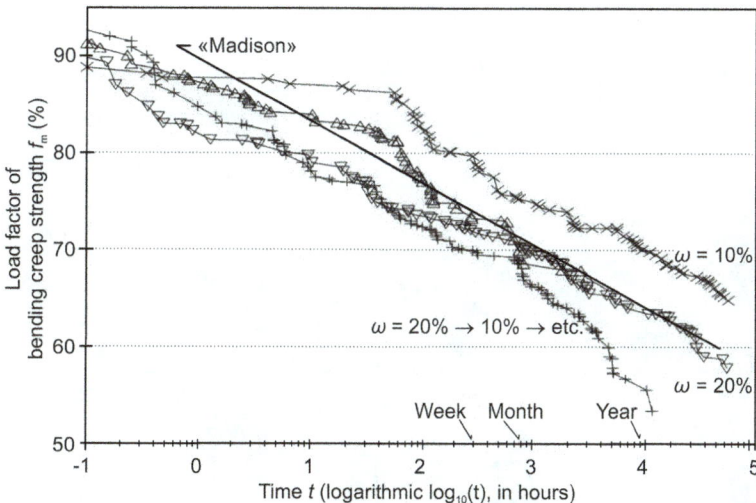

FIGURE 8.19 Creep strength in bending of Norway spruce (50 mm x 100 mm in cross-section) with 10%, 20%, and varying moisture content (ω). The straight line shows the so-called Madison curve.

(Data from Hoffmeyer 1995, courtesy of Carl Hanser Verlag.)

8.5.6 RHEOLOGICAL MODELS

Both elastic and inelastic behaviour can be represented for isotropic materials by rheological models. These are often used also for wood and wood-based panel materials. In that case, a spring represents the elastic component (Hooke's element) and a dashpot (damper) the viscous component (Newton's element). A dashpot, a damper, is a mechanical device that resists motion via viscous friction. When these two elements are in series, a Maxwell element is achieved, which can be used to describe creep and relaxation in a material.

Extensive work on this topic can be found, for example, in Molier (1992), Hassani *et al.* (2015), Reichel and Kaliske (2015), and Huč *et al.* (2018) for creep and Huč and Svensson (2018) for relaxation. The diffusion behaviour can be simulated in combination with the other parameters (Reichel and Kaliske 2015). Based on experimental investigations, the entire approach is applied to investigate the stress-level-dependent time to failure of timber beams subjected to a constant or a varying climate. Figure 8.20 shows the so-called Burger model for creep (Bodig and Jayne 1993).

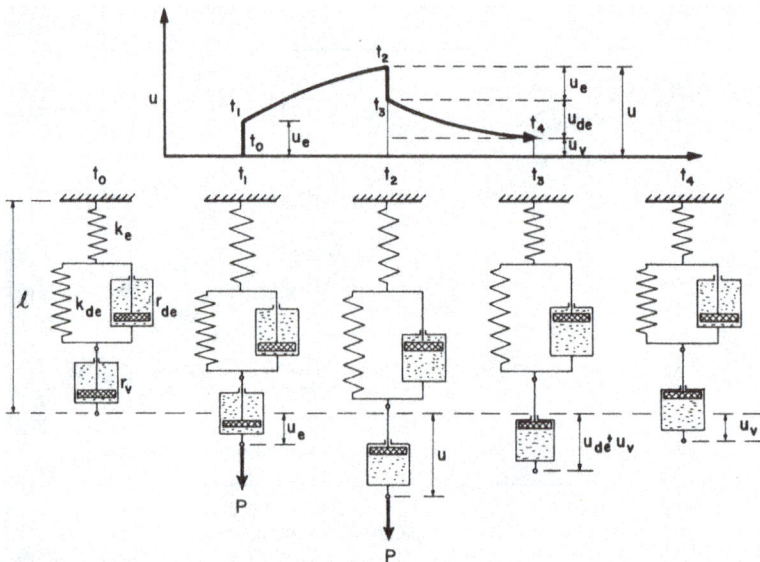

FIGURE 8.20 Burger model for the visco-elastic behaviour of wood (Bodig and Jayne 1993). P – load, t_i – time at i, u_e – instantaneous elastic deformation, u – total deformation after loading at time t_2, u_{de} – creep after unloading, and u_v – residual (permanent) deformation.

(Courtesy of Carl Hanser Verlag.)

REFERENCES

Albers K. (1970). *Querdehnungs- und Gleitzahlen sowie Schub- und Scherfestigkeiten von Holzwerkstoffen*. Doctoral Thesis, Universität Hamburg, Hamburg, Germany.

Altenbach H., Altenbach J. & Rikards R. (1996). *Einführung in die Mechanik der Laminat- und Sandwichtragwerke*. Deutscher Verlag für Grundstoffindustrie, Stuttgart, Germany, (15+410 pp.).

Armstrong L.D. & Kingston R.S.T. (1960). Effect of moisture changes on creep in wood. *Nature*, 185(4716), 862–863.

Bachtiar E. (2017). *Material Charcterization of Wood, Adhesive and Coating under various climatic Conditions*. Doctoral Thesis, ETH Zürich, Zurich, Switzerland.

Bachtiar E.V., Sanabria S.J., Mittig J.P. & Niemz P. (2017). Moisture-dependent elastic characteristics of walnut and cherry wood by means of mechanical and ultrasonic test incorporating three different ultrasound data evaluation techniques. *Wood Science and Technology*, 51(1), 47–67.

Becker P. (2002). *Modellierung des zeit- und feuchteabhängigen Materialverhaltens zur Untersuchung des Langzeitverhaltens von Druckstäben aus Holz*. Doctoral Thesis, Bauhaus-Universität Weimar, Weimar, Germany.

Berner M., Gier J., Scheffler M. & Hardtke H.-J. (2007). Identifikation von Werkstoffparametern an Platten aus Holz und Holzwerkstoffen mittels Modalanalyse. *Holz als Roh- und Werkstoff*, 65(5), 367–375.

Bodig J. & Jayne B.A. (1993). *Mechanics of Wood and Wood Composites*. (2nd Ed.) Krieger Publishing Company, Malabar (FL), USA, (21+712 pp.).

Bucur V. & Archer R.R. (1984). Elastic constants for wood by an ultrasonic method. *Wood Science and Technology*, 18(4), 255–265.

Burgert I. (2000). *Die Mechanische Bedeutung der Holzstrahlen im lebenden Baum*. Doctoral Thesis, Universität Hamburg, Hamburg, Germany.

Butterfield B.G. (1997). Proceedings of International Workshop on the Significance of Microfibril Angle to Wood Quality. *International Association of Wood Anatomists, International Union of Forestry Research Organizations*. University of Canterbury, Westport, New Zealand, (410 pp.).

Carrington H. (1922). The elastic constants of spruce as affected by moisture content. *Aeronautical Journal*, 26(144), 462–471.

CEN (1993). EN 310: Wood-based panels - Determination of modulus of elasticity in bending and of bending strength. *The European Committee for Standardization (CEN)*, Brussels, Belgium.

CEN (2004). EN 1995-1-1: 2004+A 1: Eurocode 5: Design of timber structures - Part 1-1: General Common rules and rules for buildings. *The European Committee for Standardization (CEN)*, Brussels, Belgium.

CEN (2011). EN 12369-2: Wood-based panels. Characteristic values for structural design – Plywood. *The European Committee for Standardization (CEN)*, Brussels, Belgium.

CEN (2012). EN 408: Timber structures - Structural timber and glued laminated timber - Determination of some physical and mechanical properties (includes Amendment A1:2012). *The European Committee for Standardization (CEN)*, Brussels, Belgium.

CEN (2013). EN 1156: Wood-based panels - Determination of duration of load and creep factors. *The European Committee for Standardization (CEN)*, Brussels, Belgium.

Clauss S., Pescatore C. & Niemz P. (2014). Anisotropic elastic properties of common ash (*Fraxinus excelsior* L.). *Holzforschung*, 68(8), 941–949.

DIN (1978). DIN 52186: Testing of wood; bending test. *Deutsches Institut für Normung (DIN)*, Berlin, Germany.

DIN (1979). DIN 68364-1979-11: Characteristic values for wood species; strength, elasticity, resistance. *Deutsches Institut für Normung (DIN)*, Berlin, Germany.

Dinwoodie J.M. (2000). *Timber: Its Nature and Behaviour*. E and FN Spon, London, UK, (272 pp.).

Dinwoodie J.M., Higgins J.-A., Robson D.J. & Paxton H.B. (1990). Creep in chipboard. Part 7: Testing the efficacy of models on 7-10 years data and evaluating optimum period of prediction. *Wood Science and Technology*, 24(2), 181–189.

Gressel P. (1971). *Untersuchungen über das Zeitstandbiegeverhalten von Holzwerkstoffen in Abhängigkeit von Klima und Belastung*. Doctoral Thesis, Universität Hamburg, Hamburg, Germany.

Gressel P. (1983). Erfassung, Systematische Auswertung und Ergänzung bisheriger Untersuchungen über das rheologische Verhalten von Holz und Holzwerkstoffen. *Ein Beitrag zur Verbesserung des Formänderungsnachweises nach DIN 1052 "Holzbauwerke". Forschungsbericht zu den AIF Vorhaben 4298 und 5348*, Universität Karlsruhe, Karlsruhe, Germany.

Gressel P. (1984). Kriechzahlen von Holz und Holzwerkstoffen. *Bauen mit Holz*, 86(4), 216–223.

Gressel P. (1986). Vorschlag einheitlicher Prüfgrundsätze zur Durchführung und Bewertung von Kriechversuchen. *Holz als Roh- und Werkstoff*, 44(4), 133–138.

Grimsel M. (1999). *Mechanisches Verhalten von Holz: Struktur- und Parameteridentifikation eines anisotropen Werkstoffes*. Doctoral Thesis, TU Dresden, Dresden, Germany.

Gülzow A. (2008). *Zerstörungsfreie Bestimmung der Biegesteifigkeiten von Brettsperrholzplatten*. Doctoral Thesis, ETH Zürich, Zurich, Switzerland.

Görlacher R. (1987). Zerstörungsfreie Prüfung von Holz: Ein "in Situ"-Verfahren zur Bestimmung der Rohdichte. *Holz als Roh- und Werkstoff*, 45(7), 273–278.

Hanhijärvi A. & Hunt D. (1998). Experimental indication of interaction between viscoelastic and mechano-sorptive creep. *Wood Science and Technology*, 32(1), 57–70.

Hankinson R.L. (1921). Investigation of Crushing Strength of Spruce at Varying Angles of Grain. *Air Service Information Circular*, 3(259), 3–15.

Hassani M.M., Wittel F.K., Hering S. & Herrmann H.J. (2015). Rheological model for wood. *Computer Methods in Applied Mechanics and Engineering*, 283(1), 1032–1060.

Hering S., Keunecke D. & Niemz P. (2012). Moisture-dependent orthotropic elasticity of beech wood. *Wood Science and Technology*, 46(2), 927–938.

Hoffmeyer P. (1995). Holz als Baustoff. In: Blaß H.J., Görlacher R. & Steck G. (Eds.). *Step 1: Holzbauwerke Nach Eurocode 5. Bemessung und Baustoffe. Arbeitsgemeinschaft Holz*, Düsseldorf Germany, (pp. 1–22).

Huč S. & Svensson S. (2018). Coupled two-dimensional modeling of viscoelastic creep of wood. *Wood Science and Technology*, 52(1), 29–43.

Huč S., Hozjan T. & Svensson S. (2018). Rheological behavior of wood in stress relaxation under compression. *Wood Science and Technology*, 52(1), 793–808.

Hunt D.G. (1984). Creep trajectories for beech during moisture changes under load. *Journal of Materials Science*, 19(5), 1456–1467.

Keunecke D., Hering S. & Niemz P. (2008). Three-dimensional elastic behaviour of common yew and Norway spruce. *Wood Science and Technology*, 42(8), 633–647.

Kollmann F. (1951). *Technologie des Holzes und der Holzwerkstoffe: Part 1: Anatomie und Pathologie, Chemie, Physik, Elastizität und Festigkeit*. (2nd Ed.), Springer Verlag, Berlin, Germany, (1051 pp.).

Kon T. (1948). On the law of variation of the modulus of elasticity for bending in wooden beams. *Bulletin of Hakaido University, Department of Engineering*, 1, 157–166.

Leps T. (2012). Das optimale Material für jedes Möbelteil. *Holz-Zentralblatt*, 138, 1271.

Molier P. (1992). Creep in Timber Structures. *Rilem Report No. 8*, E & FN SPON, London, UK, (157 pp.).

Montero C., Gril J., Legeas C., Hunt D.G. & Clair B. (2012). Influence of hygromechanical history on the longitudinal mechanosorptive creep of wood. *Holzforschung*, 66(6), 757–764.

Navi P. & Stanzl-Tschegg S. (2009). Micromechanics of creep and relaxation of wood. *A review. Holzforschung*, 63(2), 186–195.

Navi P. & Sandberg D. (2012). *Thermo-hydromechanical Processing of Wood*. EPFL Press, Lausanne, Switzerland, (7+360 pp.).

Neuhaus F. (1981). *Elastizitätszahlen von Fichtenholz in Abhängigkeit von der Holzfeuchtigkeit*. Doctoral Thesis, Universität Bochum, Bochum, Germany.

Neuhaus H. (2011). *Ingenieurholzbau*. Vieweg +Teubner, Wiesbaden, Germany, (1111 pp.).

Niemz P. (1980). Über einige Erkenntnisse zum Kriechverhalten von Vollholz. *Holztechnologie*, 21(4), 195–199.

Niemz P. (1982). *Untersuchungen zum Kriechverhalten von Spanplatten unter besonderer Berücksichtigung des Einflusses der Werkstoffstruktur*. Doctoral Thesis, TU Dresden, Dresden, Germany.

Norimoto M., Gril J. & Rowell R. (2012). Rheological properties of chemically modified wood: Relationship between dimensional and creep stability. *Wood and Fibre Science*, 24(1), 25–35.

Ozyhar T. (2013). *Moisture and time dependent Orthotropic Mechanical Characterization of Beech Wood*. Doctoral Thesis, ETH Zürich, Zurich, Switzerland.

Ozyhar T., Hering S. & Niemz P. (2012). Moisture-dependent elastic and strength anisotropy of European beech wood in tension. *Journal of Materials Science*, 47(16), 6141–6150.

Ozyhar T., Hering S., Sanabria J. & Niemz P. (2013). Determining moisture dependent elastic characteristics of beech wood by mean of ultrasonic waves. *Wood Science and Technology*, 47(2), 329–341.

Perkitny T. & Perkitny J. (1966). Vergleichende Untersuchungen über die Verformung von Holz, Span- und Faserplatten. *Holztechnologie*, 7(4), 265–271.

Popper R., Gehri E. & Eberle G. (1999). Mechanosorptive Eigenschaften von bewehrtem Brettschichtholz: Relaxation bei zyklischer Klimabelastung. *Drevarsky Vyskum*, 44(1), 1–11.

Pozgaj J., Chonavec D., Kurjatko S. & Babiak M. (1997). *Struktura a Vlasnosti Dreva. [Structure and Properties of Wood.]* Priroda, Bratislava, Slovakia, (485 pp.)

Reichel S. & Kaliske M. (2015). Hygro-mechanically coupled modelling of creep in wooden structures, Part I: Mechanics. *International Journal of Solids and Structures*, 77(1), 28–44.

Ross R. (2010). Wood Handbook. Wood as an Engineering Material. *(Centenial Ed.), US Department of Agriculture, Forest Service, Forest Products Laboratory*, Madison (WI), USA, (509 pp.).

Roth P. (1935). *Dauerbeanspruchung von Eichenholz- und von Tannenholz-Prismen in Faserrichtung durch konstante und durch wechselnde Druckkräfte und Dauerbiegebeanspruchung von Tannenholzbalken*. Doctoral Thesis, Universität Karlsruhe, Karlsruhe, Germany.

Schreiber J., Niemz P. & Mannes D. (2007). Vergleichende Untersuchungen zu ausgewählten Eigenschaften von Holzpartikelwerkstoffen bei unterschiedlicher Belastungsart. *Holztechnologie*, 48(1), 1–10.

Sjölund J. (2015). *Effect of Cell Structure Geometric and Elastic Parameters on Wood Rigidity*. Doctoral Thesis, Aalto University, Espoo, Finland.

Smith I., Landis E. & Gong M. (2003). *Fracture and Fatigue in Wood*. Wiley, Chichester, UK, (248 pp.).

Sonderegger W., Mandallaz D. & Niemz P. (2008). An investigation of the influence of selected factors on the properties of spruce wood. *Wood Science and Technology*, 42(4), 281–298.

Sonderegger W., Martienssen A., Nitsche C., Ozyhar T., Kaliske M. & Niemz P. (2013). Investigations on the physical and mechanical behaviour of sycamore maple (*Acer pseudoplatanus* L.). *European Journal of Wood and Wood Products*, 71(1), 91–99.

Svensson S. (2011). Duration of load effect: Experimental research in the past, present and future. In: *Proceeding of the Conference in Advances in Physics and Reliability of Wood*, 16–17 December, ETH Zürich, Zurich, Switzerland.

Ugolev B.N. (1986). *Holzkunde und Grundlagen der Holzwarenkunde [Properties of wood and wood based materials]*. [in Russian] Lesnaja Prom., Moskow, Soviet Union.

von Halász R. & Scheer C. (1986). Holzbau-Taschenbuch. *Band 1: Grundlagen, Entwurf und Konstruktionen*. Ernst & Sohn, Berlin, Germany, (712 pp.).

Vorreiter L. (1949). *Holztechnologisches Handbuch, Band 1-3 (1949-1963)*. Fromme, Wien, Switzerland.

Walker J.C.F. (2006). *Primary Wood Processing: Principles and Practice*. Springer, Dordrecht, The Netherlands.

Werner G. & Zimmer K.-H. (2009). *Holzbau 1: Grundlagen DIN 1052 (neu 2008) und Eurocode 5*. (4th Ed.), Springer-Verlag, Berlin, Germany, (20+370 pp.).

9 Strength Properties of Wood and Wood-based Materials

P. Niemz and W. Sonderegger

9.1 OVERVIEW

Strength is the stress calculated from the maximum force at failure or a defined strain level, for example, at 2% strain in compression perpendicular to the fibre direction of the wood. Depending on the load speed, a distinction is made between (1) static strength, *i.e.* the failure of the test specimen is caused by a slowly rising load, and (2) dynamic strength, *i.e.* the fracture is induced by a transient load (*e.g.* impact bending test) or an alternating load (*e.g.* Wöhler test).

Depending on the type of load, strength properties are defined as:

- tensile strength,
- compressive strength,
- bending strength,
- shear strength, and
- splitting strength (cleavage).

The strength properties are also defined in relation to the principal directions of the wood material (longitudinal, radial, tangential). The load type and direction significantly affect the strength properties. The strength properties further include nail and screw pull-out resistance, which is often applied in timber construction, as well as hardness and abrasion resistance. Nördlinger (1860) published one of the first studies about wood properties.

The properties of wood and wood-based materials are subject to significant variability, for example, because of the influence of the growth conditions of the tree. Safety factors, particularly characteristic properties, are used in practical timber construction, for example, according to Eurocode 5 (CEN 2004a). For safety factors, the following applies:

$$\text{Safety factor} = \frac{\text{ultimate stress}}{\text{existing stress}} \leq \frac{\text{ultimate stress}}{\text{allowed max. stress}} \tag{9.1}$$

DOI: 10.1201/9781003411994-9

TABLE 9.1
Distribution Functions for Wood Properties (Steiger 1996), Supplemented for Vibration Tests and Fracture Toughness by (Scheffler 2000)

Attribute	Distribution Function
Moisture content	Normal (Gaussian)
Density	Normal (Gaussian)
Modulus of elasticity	Normal (Gaussian)
Modulus of elasticity (vibration test)	Pearson
Sound velocity	Normal (Gaussian)
Strength	Log-normal (2-dimensional)
Fracture toughness	Weibull

Generally, a safety factor between 2 and 10 is expected, but a safety factor of 4 is sufficient for visually or mechanically strength-graded timber. In the literature of wood physics, only mean values are given, including standard deviation and/or the coefficient of variation when appropriate. In the literature regarding timber construction, characteristic values are used where a normal distribution is commonly assumed. However, other distributions, such as the Pearson distribution, may be present under special load conditions (Scheffler 2000). Table 9.1 shows different distributions that are used.

In timber construction, characteristic values determined on sawn timber are calculated according to the EN 408 standard (CEN 2012). Due to knots, fibre orientation, density fluctuations, etc., the properties are not comparable to those determined on small, clear (defect-free) specimens. The evaluation of characteristics of clear specimens is a specific field of research in civil engineering, see for example, Glos (1982), Bodig and Jayne (1993), Niemz and Sonderegger (2003), Denzler (2007), Neuhaus (2011). An extensive overview of the mechanical and physical properties of various wood species worldwide and bamboo can be found in Niemz *et al.* (2023, 2025).

9.1.1 Characteristic Values

In construction, the so-called 5% fractile (the characteristic value) is used (Figure 9.1), and factors are added to the characteristic value to calculate the so-called design values used in the Eurocode 5 standard (CEN 2004a, *cf.* Tables 9.4 and 9.5).

In furniture-making, however, the elements are first-hand dimensioned according to their permissible deformation. Furniture construction will not be further discussed in this book.

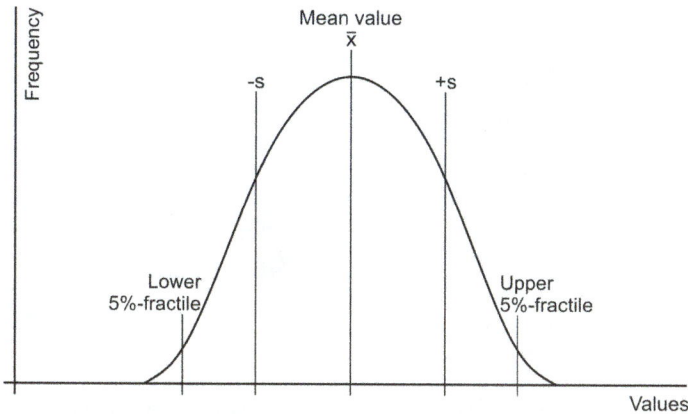

FIGURE 9.1   Normal distribution with mean value and 5% fractiles.

(Courtesy of Carl Hanser Verlag.)

9.2   FACTORS AFFECTING STRENGTH

The main factors affecting the strength of wood and wood-based materials are analogous to those of the elastic properties (*cf.* Chapter 8):

- wood structure,
- moisture content,
- ageing,
- mechanical and climatic history, and
- testing method.

The influence of a factor is almost the same, independent of the type of strength property.

9.2.1   WOOD STRUCTURE

Strength is affected by wood species, the age of the trees used, and how this affects the juvenile and adult wood proportions, growth conditions such as climate and soil, and wood characteristics such as reaction wood and fibre orientation in the sawn timber, which is caused by the spiral grain in the tree and sawing pattern.

9.2.1.1   Fibre and Growth-ring Orientation

Wood is an orthotropic material, *i.e.* the strength properties are direction-dependent for sawn timber and wood-based materials. For solid wood, the strength (σ) acts in three principal directions, *i.e.* in the longitudinal (L), radial (R), and tangential (T) directions, and the following relation is valid:

$$\sigma_L > \sigma_R > \sigma_T \tag{9.2}$$

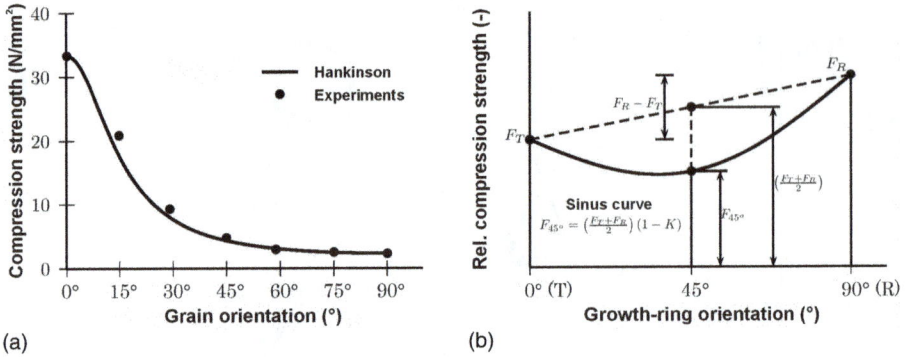

FIGURE 9.2 Influence of deviation between compressive load and (a) fibre/grain orientation in the R/T-L section and (b) growth-ring orientation in the cross-section (R-T section) of solid wood (Bodig and Jayne 1993; Hankinson's formula, see Sections 9.4.3–9.4.5). Grain orientation: $0°$ – load parallel to the fibre direction (L); $90°$ – load perpendicular to the fibre direction. Growth-ring orientation in the cross-section: $0°$ – load parallel to the tangential (T) direction; $90°$ – load parallel to the radial (R) direction.

(Courtesy of Carl Hanser Verlag.)

Therefore, the grain and growth-ring orientations decisively influence strength and elasto-mechanical and rheological properties. In solid wood, high loads can be absorbed mainly parallel to the fibres. As the fibre direction deviates from the direction of the load, the strength of the wood decreases. Even slight deviations from the fibre direction cause a significant decrease in wood strength (Figure 9.2a).

Radial strength is higher than tangential strength; the strength is minimal at about 45 degrees of growth-ring orientation in the cross-section of a piece of sawn timber (Figure 9.2b).

The strength properties of wood-based panels are characterised by their orientation relative to the panel plane: either parallel to the plane (*i.e.*, in-plane properties, which can be distinguished as parallel and perpendicular to the production direction, respectively) or perpendicular to the plane, (*i.e.*, in the thickness direction). The fibre orientation of the layers, particles, and fibres on which the panel material is built affects its strength properties.

Generally, uniaxial stress is still used in practice. However, the characteristic values of wood and wood-based materials in all principal directions (L, R, and T) are required for finite-element (FE) calculations of elasticity and strength.

9.2.1.2   Density

Of all the factors to be considered, the density of the material may provide the most reliable guide to the strength of sawn timber and wood-based materials. Density allows a rough estimate of the strength, as seen in Figure 9.3. When the density increases, the strength increases linearly as the proportion of cell-wall material that can transfer the load increases. The density correlation with strength applies both within one species of wood and between the species. Wood's characteristics also influence its properties (Richter 2014).

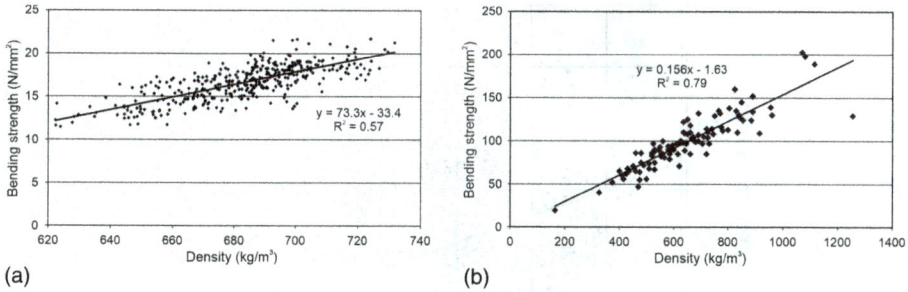

(a)   (b)

FIGURE 9.3   Bending strength as a function of the bulk density: (a) for a particleboard (measurements: ETH Zurich, Wood Physics) and (b) for solid wood of 103 species (data from Sell 1997 in Niemz and Sonderegger 2003).

(Courtesy of Carl Hanser Verlag.)

Wood strength increases with increasing latewood content, as density positively correlates with latewood content. Growth-ring width is a less reliable parameter for strength for most species. Growth factors such as soil, climate, and height above sea level are superimposed on the growth-ring width.

9.2.1.3   Knots

In hardwoods, the wood in knots has approximately 5% higher density than the wood that surrounds the knot. In softwoods, the difference can be up to 150% (Knigge and Schulz 1966). Around the knots within the stem, the fibre orientation varies considerably compared to the fibre orientation in the knot-free regions of the stem, which is nearly parallel to the longitudinal stem. The fibre deviation around knots causes stress concentrations under loading, and the timber is usually damaged or fails in these regions. As the knot-area ratio (KAR) value increases, the tensile, compression and bending strength, and the modulus of elasticity of the wood decrease (Figure 9.4).

In engineered wood products (EWPs) of sawn timber and veneer, the strength-reducing effect of knots is reduced by the so-called lamination effect, *i.e.* layers of timber or veneer joined together so the knotty and other defect regions are distributed in a laminated structure and reinforced by adjacent, mostly defect-free layers. The lamination effect means that, for example, glued-laminated timber has higher characteristic strength values and less variation of the values than sawn timber of the same dimension. For particle- and fibre-based EWPs, the composition of the EWP, *i.e.* type, size, and geometry of the particle itself, and the bonding between particles, determines the strength properties. The density, density profile in the thickness direction of the panel, and solid content of the adhesive eventually used are also crucial for the strength properties.

9.2.1.4   Coefficient of Variation

The coefficient of variation (CV) is the ratio of the standard deviation to the mean. The higher the coefficient of variation, the greater the level of dispersion around the mean. The CV is helpful because the standard deviation of data must always be understood in the context of the mean of the data. In contrast, the actual value of the

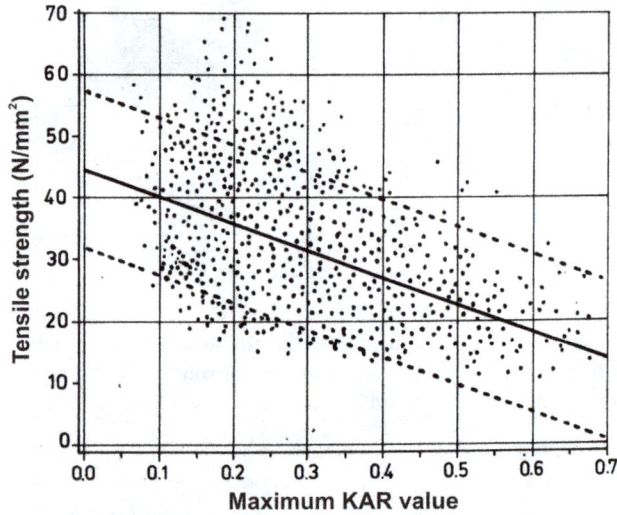

FIGURE 9.4 Influence of the knot-area ratio (KAR value) on the tensile strength of a softwood (data from Görlacher 1990

(Courtesy of Carl Hanser Verlag.)

TABLE 9.2
Reference Values for the Coefficient of Variation (CV) of Some Properties of Solid Wood and Wood-based Materials Loaded in Tension and Bending (DIN 2003)

Tensile Strength	CV	Bending Strength	CV	Density	CV
	(%)		(%)		(%)
Wood, small clear specimens	20	Wood, small clear specimens	6–21	Wood, small clear specimens	5–14
Sawn timber, visually graded	40	Sawn timber, visually graded	14.2	Sawn timber, visually graded	9.7
Plywood	18	Plywood	9.3	Plywood	6.0
Particleboard (OSB)	12	Particleboard (OSB)	17.3	Particleboard (OSB)	9.0
MDF	8	MDF	8–10	MDF	2–6

CV is independent of the unit in which the measurement has been taken, so it is a dimensionless number.

The properties of wood-based materials vary much less than those of solid wood. This is due to homogenisation, which occurs in manufacturing wood-based materials (Table 9.2). Also, the variation depends on the load type. In static calculations, the property variations of wood and wood-based materials are considered by safety factors and characteristic values (5% fractiles); see Tables 9.1 and 9.2.

9.2.2 MOISTURE CONTENT

When the moisture content of wood increases from 0% up to the fibre-saturation point (FSP), the modulus of elasticity and strength decrease. Above the FSP, the strengths are constant because the added water does not penetrate the cell wall and affects its properties (Figure 9.5). The fracture energy increases as the wood moisture increases.

Tensile and shear strength initially increase in the moisture range between approximately 5% and 15% and then decrease until they reach the FSP. For wood-based materials, for example, particleboard, tensile strength, and modulus of elasticity also decrease with increasing moisture content (Figure 9.6).

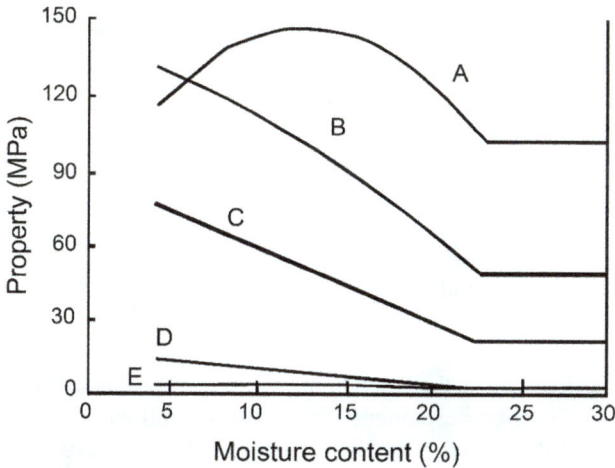

FIGURE 9.5   Influence of moisture content on wood strength (Ross 2010). A – tension parallel to the grain, B – bending, C – compression parallel to the grain, D – compression perpendicular to the grain, and E – tension perpendicular to the grain.

(Courtesy of Carl Hanser Verlag).

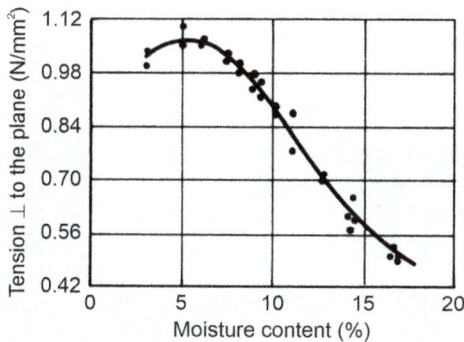

FIGURE 9.6   Influence of moisture content on tensile strength perpendicular to the plane of particleboards (data from Halligan and Schniewind 1974.

(Courtesy of Carl Hanser Verlag.)

Hoffmeyer (in Neuhaus 2011) indicates, for example, the following change in the strength of solid wood per each percentage-point change of moisture content in the moisture range between 8% and 20%:

- 6.0% in compression strength in the fibre direction,
- 5.0% in compression strength perpendicular to the fibre direction,
- 4.0% in bending,
- 2.5% in tension parallel to the fibre direction,
- 2.0% in tension perpendicular to the fibre direction, and
- 2.5% in shear.

Above 20% moisture content, the relationship is no longer strictly linear.

An increase in temperature generally causes a decrease in strength and the modulus of elasticity for wood and wood-based materials (Figures 9.7 and 9.8). For softwood sawn timber at a moisture content of 10–15%, Glos (in Neuhaus 2011) gives the following values of strength reduction per 10 °C temperature increase, starting at 20 °C:

- 5% in bending,
- 5% in compression, and
- 1% in tension.

In addition, the interaction between temperature and moisture content must be considered. As the moisture content increases, the influence of the temperature increases. On the other hand, very low temperatures below zero lead to considerable embrittlement of the wood. This effect is used, for example, in wood chipping to reduce the cutting force and improve the chip geometry. In practice, normal temperature fluctuations, especially indoors, are not considered for static strength calculations.

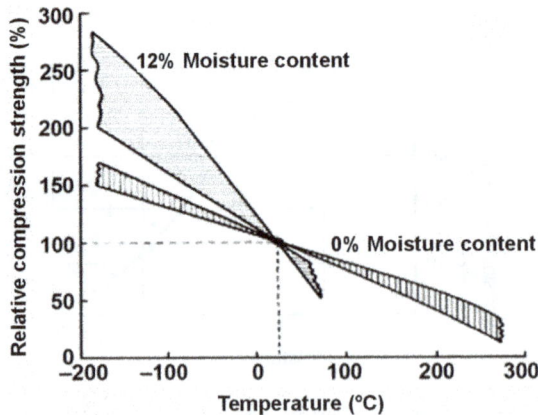

FIGURE 9.7 Influence of temperature on compression strength of wood (data from Ross 2010, courtesy of Carl Hanser Verlag).

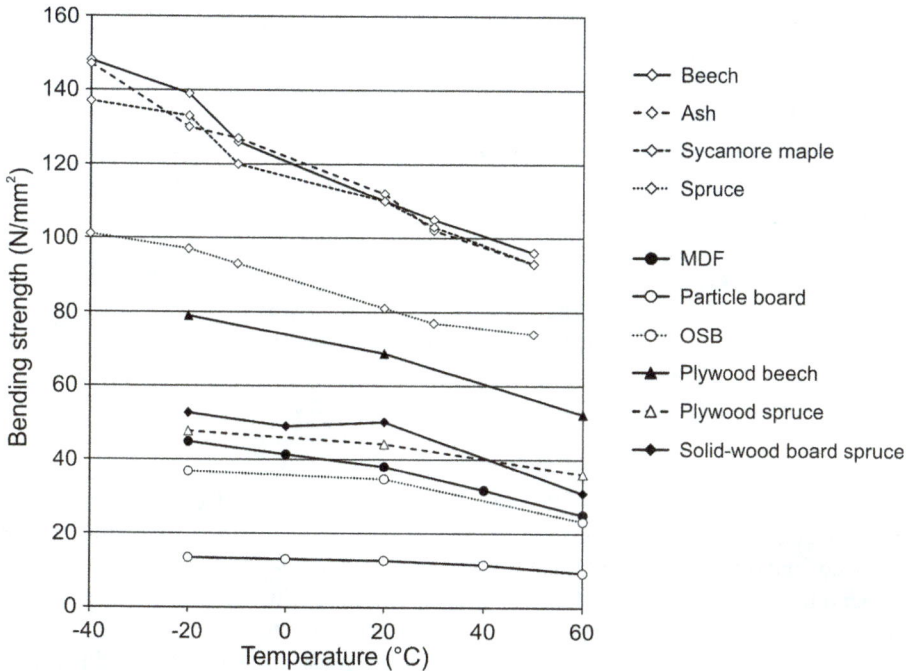

FIGURE 9.8 Bending strength of wood and wood-based materials as a function of temperature (data from Sonderegger and Niemz 2006, Niemz *et al.* 2014.

(Courtesy of Carl Hanser Verlag.)

9.2.3 AGEING

Under dry interior climatic conditions, wood's properties do not change or hardly change over the years (Anon. 1990, Sonderegger 2015). For load-bearing EWPs such as glued-laminated timber, moisture resistance and the durability of the adhesive bond-line are essential. In the dry indoor climate, inappropriate bonds may lead to severe cracking and delamination, especially if the relative humidity varies significantly. This problem is more pronounced for hardwoods than softwoods (Hassani *et al.* 2015).

9.2.4 THE INFLUENCE ON BIOLOGICAL DETERIORATION AND HISTORY OF MECHANICAL AND CLIMATIC STRESSES

The properties of wood and wood-based materials are significantly influenced by fungal or insect infestation and previous mechanical and climatic stresses. Chapter 10 extensively describes how to handle biological deterioration, which is briefly summarised here.

9.2.4.1 Fungus and Insect Damage

The bending and compressive strength of Norway spruce wood attacked by horn-tail (wood wasp) or black spruce beetle decreases independently of the number of boreholes. The compressive strength is reduced by about 10% and the bending strength by up to 30%. Blue-stain and red-striped wood do not affect the strength properties, as these fungi do not destroy the wood material more than colouring it. Wood-destroying fungi such as brown rot, white rot, and soft rot cause a significant loss of strength, whereby the fracture pattern changes:

- Brown rot degrades polysaccharides, decreasing the wood's density and strength. The fracture surfaces are cubical (short-fibred fracture).
- White rot degrades both polysaccharides and lignin. The fracture surfaces are cubical, and both strength and density decrease, and
- Soft rot degrades polysaccharides, decreasing the wood's strength, but the mass (density) loss is low.
- For brown and white rot, the reduction in strength correlates with mass loss.

9.2.4.2 Steaming and Thermal Treatment

Steaming sawn timber softens the wood, making it easier to form curved shapes. During steaming, the modulus of elasticity, proportional limit, and strength are reduced, and ductility, mainly plastic stretching, is strongly increased, making it easy to bend the straight timber and fix it in a curved shape. After cooling and drying, the properties mainly return to their former levels before steaming.

However, high steaming temperatures can cause mass loss whereby a reduction of modulus of elasticity, impact bending strength, and compressive strength may occur after wood redrying (Kollmann 1955, Anon. 1990).

Thermal modification of timber partially results in a significant reduction of hardness and strength, particularly impact bending strength (Chapter 2).

9.2.5 TESTING METHOD

The testing method significantly influences mechanical properties, namely the specimen geometry, loading duration, speed, and type of load. Therefore, it is essential to follow standards when determining mechanical properties.

9.2.5.1 Loading Rate and Load Duration

The loading rate (speed) significantly affects the properties. Very high loading speeds lead to higher values, and very low loading speeds lead to lower values. The results also differ depending on the total time for the test (load duration). Relative to the short-term standard test, a long-term creep test will get lower values for mechanical properties (Figure 9.9).

9.2.5.2 Type of Load

Strength properties are significantly influenced by the type of load. Thus, tensile strength in the fibre direction in solid wood is about twice as high as compression strength, and bending strength lies between tensile and compression strength. For particle-based panels, compression strength is equal to or higher than tensile strength

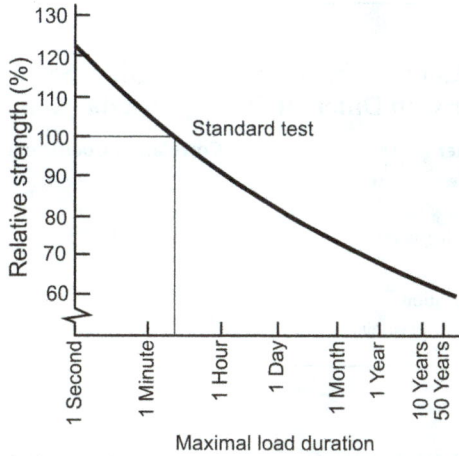

FIGURE 9.9 Influence of load duration on strength (data from Bodig and Jayne 1993. **(Courtesy of Carl Hanser Verlag.)**

in the in-plane direction, and bending strength is higher than tensile and compression strength due to plastic deformation during loading.

In the case of bending, the type of load (3-point or 4-point bending) also influences the test result. For example, in the case of a bending beam with 3-point loading, the ratio of span to thickness of the specimens has a significant effect on the modulus of elasticity because the shear loss is neglected.

9.2.5.3 Sample Geometry: Solid Wood

Material parameters in wood physics relate almost exclusively to small clear specimens. However, the strength properties are generally influenced to a high degree by features such as fibre (grain) orientation, knots, and reaction wood. In addition, the properties are highly dependent on growing conditions and vary between trees and within a single tree. Therefore, the strength properties of sawn timber are lower than those of small, defect-free specimens. Round timber (logs) has about 10% higher strength than sawn timber. That depends on the fact that the fibres are cut during the sawing process, and the choice of sawing pattern causes fibre deviation (slope of grain) in the sawn timber, see Hankinson formula (Equations 9.4, 9.6, and 9.8).

Sawn timber and EWPs such as glued-laminated timber are strength graded, and characteristic strength values are determined and should be used when dimensioning load-bearing constructions. The grading follows the EN 408 standard (CEN 2012). Various test machines are used for industrial wood grading based on the deformation measurement or from the measurement of natural frequency or ultrasound velocity, which determine the modulus of elasticity (*e.g.* Chapters 19 and 20 in Niemz *et al.* 2023). Table 9.3 shows an overview of the correlation between bending strength and sorting according to various criteria of sawn timber.

TABLE 9.3

Correlation of the Bending Strength of Sawn Timber with Different Sorting Criteria (Glos 1982)

Parameter	Correlation Coefficient
Density	0.5
Growth-ring width	0.4
Knots	0.5
Fibre deviation	0.2
Modulus of elasticity	0.7–0.8

Extensive work has been carried out on wood grading softwoods by, for example, Glos and Schulz (1986), Burger and Glos (1996), Steiger (1996), and Fink (2014). Hübner (2013) made extensive investigations on grading hardwoods (common beech and ash).

Various models have been developed for calculating and dimensioning laminated wooden elements, for example, by Ehlbeck (1967), Blaβ (1987), Colling (1990), Görlacher (1990), and Fink (2014). Using finite-element modelling, the strength of glued-laminated timber can be calculated from the properties of the lamellae (density, dimension, knot-area ratio, and modulus of elasticity) and by knowing the properties of the finger jointing if such is used.

9.2.5.4 Sample Geometry: Wood-based Panel Materials

The properties of wood-based materials also depend on the size of the test specimen. Single large particles (*e.g.* OSB) significantly impact strength when testing small specimens. McNatt *et al.* (1990) have shown that Young's modulus of plywood and OSB is higher, and the bending strength is lower than that of small specimens from these materials.

A proposal for testing medium-sized components of wood-based panels is available in the EN 789 standard (CEN 2004b). Böhme (1999) determined the following changes in the properties of medium-sized specimens (the size in the range of one metre in bending and tension) compared to small specimens (400 mm in tension):

- 10% reduction of bending strength,
- 11–12% increase of the modulus of elasticity in bending,
- 1% reduction of tensile strength,
- 18% increase in compression strength,
- 24% reduction of shear strength parallel to the panel plane, and
- 4% reduction of shear strength perpendicular to the panel plane.

Tables 9.4 and 9.5 show the characteristic strength values of solid structural timber and wood-based materials.

TABLE 9.4
Characteristic Strength Values of Structural Sawn Timber According to the EN 338 Standard (CEN 2016a).

Strength Class:	Softwoods			Hardwoods		
	C18	C24	C30	D30	D40	D50
	(N/mm²)			(N/mm²)		
Bending ‖	18	24	30	30	40	50
Tension ‖	10	14.5	19	18	24	30
Tension ⊥	0.4	0.4	0.4	0.6	0.6	0.6
Compression ‖	18	21	24	24	27	30
Compression ⊥	2.2	2.5	2.7	5.3	5.5	6.2
Shear strength (torsion)	3.4	4	4	3.9	4.2	4.5

‖ = in the fibre direction and ⊥ = perpendicular to the fibre direction.

TABLE 9.5
Characteristic Strength Values of Wood-based Materials According to EN 12369 (CEN 2001, 2011a)

Material	Density (kg/m³)	Bending f_m (f_p) (N/mm²)	Tension f_t (N/mm²)	Compression f_c (N/mm²)	Shear f_v (N/mm²)	f_r (N/mm²)
CLT ‖	410	12–35 (10–25)	6–16	10–16	2.5–4.0	1.2–1.6
CLT⊥	410	5–9 (12)	6	10–16	2–5	1.4
Plywood ‖,⊥	350–750	3–80	1.2–40	1.2–40	1.8–7.5	0.4–1.2
OSB/2‖, OSB/3‖	550	14.8–18.0	9.0–9.9	14.8–15.9	6.8	1.0
OSB/2⊥, OSB/3⊥	550	7.4–9.0	6.8–7.2	12.4–12.9	6.8	1.0
OSB/4‖	550	21.0–24.5	10.9–11.9	17.0–18.1	6.9	1.1
OSB/4⊥	550	11.4–13.0	8.0–8.5	13.7–14.3	6.9	1.1
Particle board, type P4	500–650	5.8–14.2	4.4–8.9	6.1–12.0	4.2–6.6	1.0–1.8
Particleboard type P5	500–650	7.5–15.0	5.6–9.4	7.8–12.7	4.4–7.0	1.0–1.9
Particleboard type P6	500–650	10.0–16.5	7.5–10.5	10.4–14.1	5.5–7.8	1.7–1.9
Particleboard type P7	500–650	12.5–18.3	8.0–11.5	13.0–15.5	7.0–8.6	1.8–2.4
Fibreboard, HB.HLA2	800–900	32–37	23–27	24–28	16–19	2.5–3.0

(Continued)

FIGURE 9.5   (Continued)

Material	Density (kg/m³)	Bending f_m (f_p) (N/mm²)	Tension f_t (N/mm²)	Compression f_c (N/mm²)	Shear f_v (N/mm²)	Shear f_r (N/mm²)
Fibreboard, MBH.LA2	600–650	15–17	8–9	8–9	4.5–5.5	0.25–0.30
MDF.LA	500–650	19–21	10–13	10–13	5.0–6.5	
MDF.HLS	500–650	18–22	13–18	13–18	7.0–8.5	

For a description of the various types of wood-based panels, see Section 2.4.

‖ = parallel to the length (production) direction or parallel to the fibre direction of the panel cover layer; ⊥ = perpendicular to the length (production) direction or perpendicular to the fibre direction of the panel cover layer; f_m = bending strength perpendicular to the plane; f_p = bending strength in the plane; f_t, f_c = tensile and compressive strength in the plane; f_v = shearing perpendicular to the plane; f_r = shearing in the plane.

9.3   FRACTURE BEHAVIOUR OF WOOD AND WOOD-BASED MATERIALS

9.3.1   SAWN TIMBER

Extensive work has been carried out on the fracture behaviour of wood and the influence of wood structural elements, such as rays. The first works were published in the 1950s by Kisser and Steininger (1952), followed by studies by, for example, DeBaise *et al.* (1966), Kucera and Bariska (1982), Mindess and Bentur (1986), and Patton-Mallory and Cramer (1987). Smith *et al.* (2003) have summarised the early works in the fracture behaviour of wood.

DeBaise *et al.* (1966) defined three stages of a breakage mechanism in wood under load:

- a crack is initiated,
- crack growth, and
- unstable crack growth and failure.

The failure process depends significantly on the following:

- type of load,
- wood structure, for example, fibre orientation,
- relative humidity and temperature of the ambient air, which affect the moisture content of the wood, and
- properties of the bond-line in adhesively bonded wood.

The ultimate strain at fracture in tension is generally between 0.6% and 1.7% for solid wood and 0.6% and 1.0% for wood-based materials. Anisotropy, moisture content, and temperature significantly influence the strain level at breakage.

9.3.2 Wood-based Panel Materials

9.3.2.1 Cross-wise Laminated Wood

Laminated wood-based materials such as cross-laminated timber (CLT) and plywood with cross-wise oriented layers are characterised by the shear failure of the transverse layers, the so-called rolling shear (Figure 9.10). This is due to the low shear modulus and shear strength in the RT plane, especially for low-density coniferous wood such as Norway spruce and Scots pine.

Curved glued-laminated timber can also fail due to transverse tensile stresses. On the other hand, rolling-shear failure is less pronounced in testing entire panels than in bending beams or small clear specimens (Czaderski *et al.* 2007, Steiger *et al.* 2012).

The proportion of wood failure is an essential criterion for assessing the quality (moisture resistance and strength) of the adhesive bond-line in glued-laminated timber. The bond-line quality of glued-laminated timber is evaluated through a longitudinal tensile shear strength test and a delamination resistance test according to the EN 302-1 (CEN 2023a) and EN 302-2 (CEN 2017) standards, respectively.

Insufficient bonding quality, low moisture-resistant adhesives, and substantial changes in ambient air relative humidity can lead to delamination of adhesive joints even after several years or decades in use (Ammann 2015).

Factors influencing the fracture behaviour of solid wood-based materials are:

- wood properties (modulus of elasticity, strength, growth ring orientation, swelling and shrinkage), especially the growth-ring orientation in the RT plane,
- thickness and orientation of the lamellae,
- moisture-content differences between the lamellae during glueing,
- adhesive quality (level of moisture resistance), and
- large fluctuations in the relative humidity of the ambient air.

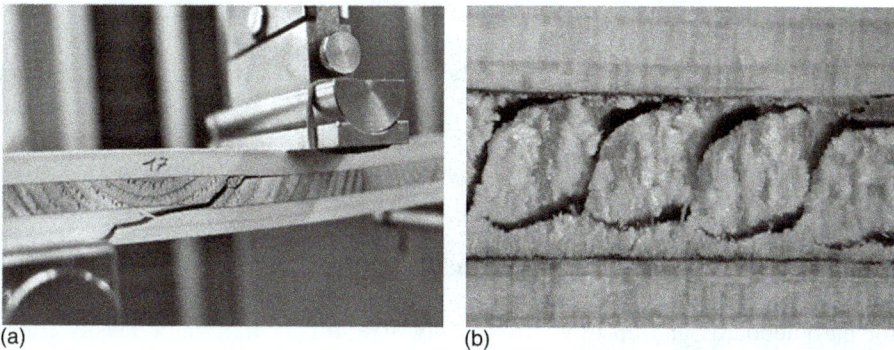

(a)   (b)

FIGURE 9.10 Failure of a three-layer cross-laminated solid-wood panel: (a) a crack in the panel under bending and (b) rolling-shear (the RT plane) damage in the middle layer.

(Courtesy of P. Niemz.)

9.3.2.2   Particle-based Materials

In contrast to solid wood, particle-based materials consist of a network of intersecting and overlapping particles interlinked by adhesive joints. Macroscopic cavities are located between the particles. The proportion of cavities in, for example, particleboard is up to 30% by volume.

The deformation and the macroscopic fracture of particle-based materials are a result of the interaction between:

- elastic and plastic deformation of the particles,
- elastic and plastic deformations of the inter-particle connections (adhesive joints),
- microfractures of particles in adhesive joints and their interfaces, and
- movements between the particles in the panel.

The panel's deformation and fractures are mainly determined by the structural composition of the particle material (morphology of the particles, type and content of adhesive, degree of particle orientation, layer structure) and the moisture content.

Fracturing under load may begin at such low stress levels as 20% of maximum bending strength in the form of local microcracks and particle displacements.

9.4   STRENGTH OF WOOD AND WOOD-BASED MATERIALS

9.4.1   OVERVIEW

Strength is the stress at which the material fails (Figure 9.11). Up to the proportional limit σ_P, there is a linear relationship between stress and strain. The ultimate strain at the failure at tensile load in the fibre direction and perpendicular is about 0.7–1.0%, depending on the moisture content of the wood.

Tensile strength in the fibre direction of solid wood is about twice as high as the compression strength (Figure 9.11a). In the case of wood-based particle materials, compression strength is equal to or higher than tensile strength, as shown in Figure 9.11b (Plath 1971, Schreiber 2007).

Perpendicular to the fibres (Figures 9.11c, d), solid wood collapses in the cell structure under compression, and a maximal strain must be defined to determine the ultimate strength (*e.g.* 2% or 5%). Initially, the less dense earlywood will deform and collapse (densification). When all earlywood reaches the density of the latewood, the latewood will also deform and collapse. The stress increases proportionally with the density. For example, Norway spruce can be densified in the radial direction up to 1200 kg/m^3 and sometimes higher.

In the case of particle-based materials, the particles are strongly compressed perpendicular to the plane (*e.g.* perpendicular to the fibre direction) during hot pressing. Compared to uncompressed solid wood, the densification is about 50% for particleboards and up to approximately 80% for medium/high-density fibreboard (MDF/HDF).

Figure 9.12 defines the yield stress (σ_y) from the stress-strain diagram. The yield strength or yield stress is a material property and is the stress corresponding to the yield point at which the material begins to deform plastically. The yield strength is

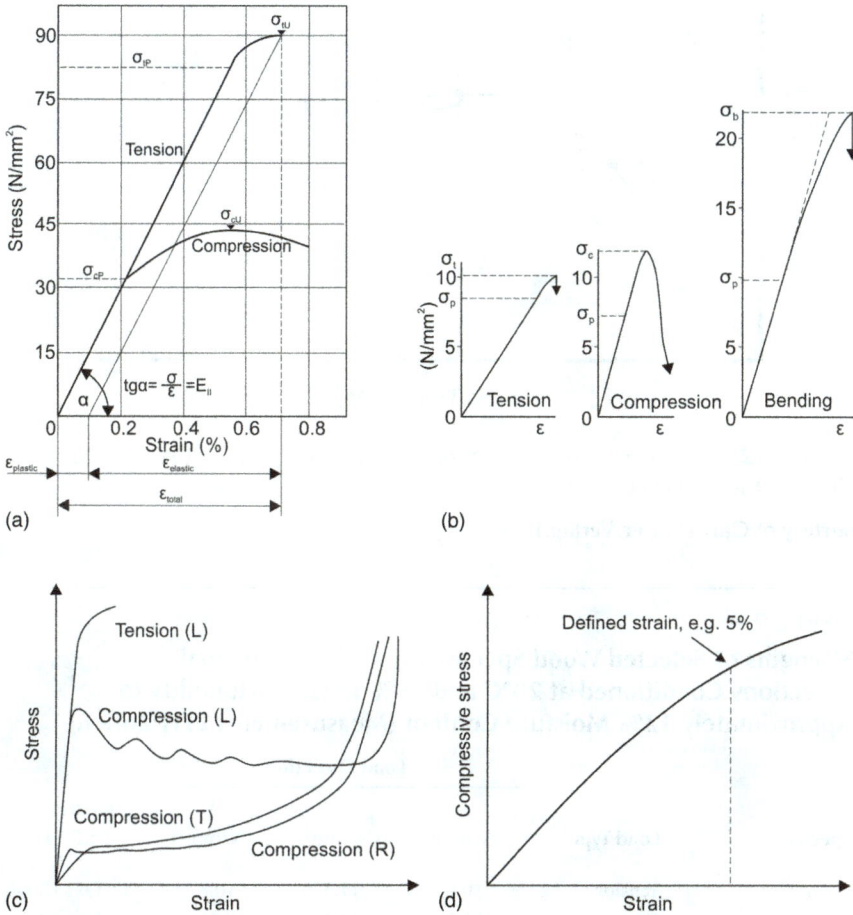

FIGURE 9.11 Stress-strain diagrams for solid wood and wood-based materials: (a) wood in the fibre direction, (b) loading curves parallel to the plane for particleboards, (c) comparison of solid wood parallel and perpendicular to the fibre direction, and (d) definition of maximum strain at compressive stress perpendicular to the fibre direction.

(Courtesy of Carl Hanser Verlag.)

typically defined by the "0.2% offset strain". The yield strength at 0.2% offset is determined by finding the intersection of the stress-strain curve with a line parallel to the initial slope of the curve, which intercepts the abscissa at 0.2% (Figure 9.12). Tables 9.6 and 9.7 show strength values for some wood species and wood-based materials.

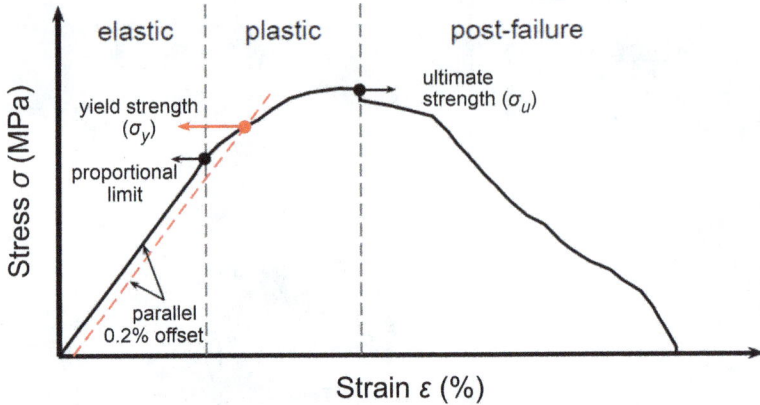

FIGURE 9.12 A schematic stress-strain diagram based on the ASTM-E8/E8M standard
(ASTM 2011 in Hering *et al.* 2012.

(Courtesy of Carl Hanser Verlag.)

TABLE 9.6
Strengths of Selected Wood Species in the Three Principal Directions Conditioned at 20°C and 65% Relative Humidity to Approximately 12% Moisture Content (Measurements ETH Zurich)

Species	Load Type	L (N/mm²)	R (N/mm²)	T (N/mm²)	Relation T : R : L
Cherry	Tension	109.0	17.3	10.8	1 : 1.6 : 10.1
	Compression	53.5	14.4	9.5	1 : 1.5 : 5.6
Common ash	Tension	130.0	12.5	10.1	1 : 1.2 : 12.9
	Compression	43.4	10.5	10.0	1 : 1.0 : 4.3
Common beech	Tension	96.7	14.7	8.9	1 : 1.7 : 10.9
	Compression	45.0	11.0	6.0	1 : 1.8 : 7.5
Common oak	Tension	73.0	6.0	7.8	1 : 0.8 : 9.4
	Compression	47.9	10.6	9.0	1 : 1.2 : 5.3
Maple	Tension	112	16.2	8.9	1 : 1.8 : 12.6
	Compression	61.5	15.4	10.3	1 : 1.5 : 6.0
Norway spruce	Tension	87.2	4.0	3.1	1 : 1.3 : 28.4
	Compression	40.2	4.1	4.2	1 : 1.0 : 9.6
Walnut	Tension	89.1	10.8	8.9	1 : 1.2 : 10.0
	Compression	60.4	13.4	11.9	1 : 1.1 : 5.1

Note: Load directions: L – longitudinal, R – radial, and T – tangential.

TABLE 9.7
Strengths of Particleboard and Medium-density Fibreboard (MDF)

Material	Density (kg/m³)	Load Type	Load Direction x (N/mm²)	y (N/mm²)	z (N/mm²)	Relation x : y : z (-)
Particleboard	660	Tension	6.3	5.7	0.5	14 : 13 : 1
		Compression	10.7	10.6		
MDF	742	Tension	20.6	20.3	0.6	34 : 34 : 1
		Compression	20.3	20.4		

Load directions: in the plane of the board, where x is in the direction of the production flow, and y is perpendicular to the flow of the production line; the z-direction is perpendicular to the plane (thickness direction), *i.e.* internal bond strength (IB).

9.4.2 PLASTIC PROPERTIES

Wood has a ductile failure behaviour perpendicular to the fibre direction under compression load, but is more brittle under tensile load. The plastic deformation behaviour is, in general, very small.

9.4.3 TENSILE STRENGTH

9.4.3.1 Influence of Structural Level

Test of the mechanical properties of wood can be performed on different size scales:

- full-size members (components),
- small defect-free specimens,
- tissue structures,
- cell structures (fibres),
- cell-wall structure, and
- biochemical level.

In civil engineering, full-size specimen properties are used. Small clear specimens are primarily used in wood physics, but more minor scales, down to cell walls and biochemical levels, are occasionally also used.

Wood has a very high strength in the fibre direction due to the covalent bonds within the cellulose molecules. According to Kollmann and Côté (1968), the tensile strength of the cell wall is in the range of 200 MPa to 1300 MPa. The strength of the cell wall for Norway spruce and yew is 1000 MPa and 800 MPa, respectively (Keunecke 2008). For clear small specimens (approx. 2x2 cm in cross-section size), the tensile strength is 80–90 MPa for Norway spruce and 100 MPa for yew, *i.e.* 10

TABLE 9.8
Mechanical Properties (Mean Values) of the Modulus of Elasticity (N/mm²) of Norway Spruce and Yew at Different Structural Levels (Keunecke *et al.* 2008)

Calculation Based on:	Fibres		Tissues		Small Clear Specimens	
	Cell-wall Area	Cross-section	Cell-wall Area	Cross-section	Cell-wall Area	Cross-section
Norway spruce	26,200	n. t.	29,400	9900	28,100	12,100
Yew	13,900	n. t.	15,600	7000	14,300	9700

Note: n. t. – not tested.

TABLE 9.9
Modulus of Elasticity (MOE) and Strength of Wood (Mitchell, Adapted from Zimmermann 2015)

Specimen Type	MOE (N/mm²)	Strength (N/mm²)
Full-size members, board, beam	11,000	25
Small, clear specimens	11,000	90
Mechanically separated single fibre (based on external volume)	40,000	400
Fibril aggregates	70,000	700
Crystalline areas	130,000–250,000	8000–10,000

times lower than for the cell-wall material. The difference in modulus of elasticity between different structural levels is insignificant (Tables 9.8 and 9.9).

In wood-based materials, the transverse tensile strength is an important quality factor for the bond quality. For particleboard and fibreboard, the tensile strength perpendicular to the plane of the board (internal bond strength) is determined according to the EN 319 standard (CEN 1993a). The tensile strength of particleboards used in humid conditions is determined according to the EN 312 standard (CEN 2010). The tensile strength perpendicular to the plane after boiling for two hours in water indicates the weather resistance of the panels.

9.4.3.2 Strength Parallel and Perpendicular to the Fibre Orientation

The tensile strength is the resistance of wood or wood-based materials to breakage under tensile load:

$$\sigma_{tU} = \frac{F}{A} \tag{9.3}$$

where

σ_{tU} is the tensile strength (N/mm²)
F is the ultimate load at fracture (N)
A is the fracture area perpendicular to the load (mm²)

The tensile strength can be determined as follows:

- parallel to the grain (in fibre direction) or parallel to the board plane, or
- perpendicular to the fibre direction (*e.g.* in radial or tangential direction) or perpendicular to the plane (Figure 9.13).

Various standards should be used for tests on small clear specimens, for example, the ISO 13061-6 (ISO 2014) or DIN 52188 (DIN 1979a) standards. For tests in the direction perpendicular to the plane of panels and perpendicular to the fibre direction of solid wood, the specimen is usually designed as shown in Figure 9.13a, b. For tensile strength tests parallel to the fibre direction, so-called dog-bone-shaped specimens are often used, as shown in Figure 9.13c.

9.4.3.3 Tensile Strength: Influencing Factors and Material Parameters

Important factors influencing the tensile strength are:

- density, *i.e.* as the density increases, the tensile strength increases,
- fibre orientation, *i.e.* as the grain angle increases, the tensile strength decreases,
- knots and cracks, *i.e.* knots and cracks cause a substantial decrease in tensile strength; wood with knots only has a tensile strength of 15% to 20% of the tensile strength of wood without knots,

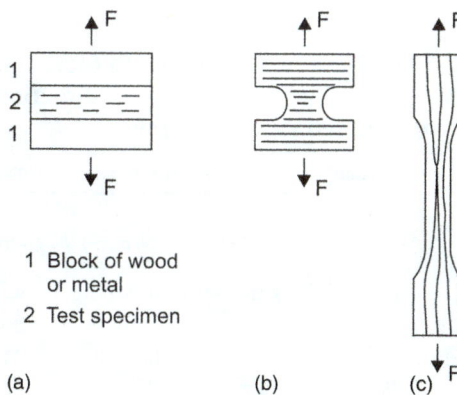

1 Block of wood
  or metal

2 Test specimen

(a)   (b)   (c)

FIGURE 9.13 Specimens for tensile test: (a) particle- and fibre-based panels (*e.g.* particleboard, MDF) perpendicular to the fibre plane, (b) solid wood perpendicular to the fibre direction, and c) solid wood in the fibre direction (based on Ozyhar 2013.

(Courtesy of Carl Hanser Verlag.)

- moisture content, *i.e.* with increasing moisture content, the tensile strength initially increases from the dry state (0%) to approximately 5% and then falls until the fibre saturation point is reached; each percentage point increase in moisture content reduces the tensile strength by about three percentage points, and
- temperature, *i.e.* with increasing temperature, the tensile strength decreases.

The tensile strength perpendicular to the fibre direction is only 3% to 4% of the longitudinal tensile strength. The tensile strength in wood with deviated fibre orientation can be calculated by the Hankinson equation (Hankinson 1921):

$$\sigma_{tU\varphi} = \frac{\sigma_{tU\parallel} \cdot \sigma_{tU\perp}}{\sigma_{tU\parallel} \cdot \sin^n \varphi + \sigma_{tU\perp} \cdot \cos^n \varphi} \tag{9.4}$$

where

φ is the fibre orientation (grain angle)
n is a constant between 1.5 and 2.0
$\sigma_{tU\parallel}$ is the tensile strength parallel to the fibre
$\sigma_{tU\perp}$ is the tensile strength perpendicular to the fibre

The growth-ring orientation in the cross-section of the wood also influences the strength. The tensile strength is more significant in the radial direction than in the tangential (generally, radial is about twice as large as the tangential strength), and a minimum is located at about 45 degrees growth-ring orientation (Figure 9.2b).

The particle geometry, density, and solid resin content of particleboard are crucial for the tensile strength perpendicular to the board plane. Table 9.10 gives values of the tensile strength of wood and wood-based materials.

TABLE 9.10
Strength Parameters of Wood and Wood-based Materials. Parallel (//) and Perpendicular (⊥) to the Fibre Direction or the Plane of a Wood-based Panel Material

Material	Bending // (N/mm²)	Tensile Strength // (N/mm²)	Tensile Strength ⊥ (N/mm²)	Compression Strength // (N/mm²)	Compression Strength ⊥ (N/mm²)
Particleboard	15–25	8–10	0.35–0.40	8–16	
Plywood	30–60	30–60		20–40	
High-density fibreboard	45–50	20–24	0.8–1.0	23–26	
Norway spruce	78	90	2.7	43	5.8
Scots pine	87	105	3.0	55	7.7
Black poplar	60	67	2.3	34	
European oak	94	90	4.0	60	11.0
Common ash	120	165	7.0	52	11.0
Bamboo	84–270			20–40	

9.4.4 COMPRESSION STRENGTH

Compression strength is the resistance of wood or wood-based materials against breakage under compressive load parallel or perpendicular to the fibre direction or plane of a wood-based panel material. The results are calculated from the maximum breaking load (ultimate stress) and the area perpendicular to the load:

$$\sigma_{cU} = \frac{F}{A} \tag{9.5}$$

where

σ_{cU} is the compression strength (N/mm²)
F is the ultimate stress or stress at defined strain when loaded perpendicular to the fibre or plane of a panel material (N)
A is the cross-sectional area (mm²)

9.4.4.1 Testing Methods

The compression strength of wood is tested, for example, according to the DIN 52185 standard (DIN 1976) parallel to the grain and according to the DIN 52192 standard (DIN 1979c) perpendicular to the grain. In DIN 52192, tests with local loads are used, for example, to evaluate the compression strength of railway sleepers (Larsen 2008).

Compression loads parallel to the fibre orientation or to the plane of a wood-based panel material result in a clear failure of the structure of the test specimen. The fracture pattern is strongly influenced by the structure of the specimen, especially in wood-based materials.

In the case of compressive stress perpendicular to the fibre orientation or the plane of a wood-based panel material, the load can be applied over the entire section of the specimen (Figure 9.14) without a definitive failure. The structure of the specimen is only compressed (densified). Therefore, the yield point is used as a measure of compressive strength, which marks in the stress-strain diagram the point at which the deformation of the test specimen significantly increases, for example, 1% according to EN 408 (CEN 2012), 2% or 5% compression. Perpendicular to the grain direction, plastic deformation occurs under compressive loading (cf. Figures 9.11 and 9.12).

9.4.4.2 Compression Strength: Influencing Factors and Material Parameters

Key factors influencing the compression strength are:

- the density, i.e. as the density increases, the compression strength increases,
- the wood species, i.e. the species indirectly influences the compression strength via the density and structural parameters such as fibre length, lignin content, and latewood content,
- the fibre orientation, i.e. the compression strength decreases as the grain angle increases,
- deviations in the microstructure such as uneven structure, knots, cracks, etc.,
- the moisture content, i.e. as the moisture content increases, the compression strength decreases, and
- the temperature, i.e. as the temperature increases, the compression strength decreases.

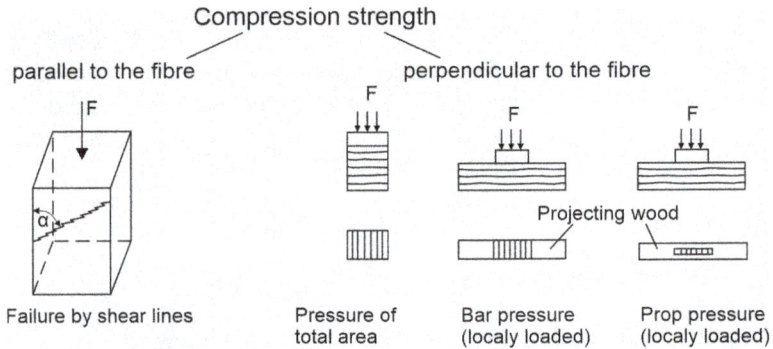

FIGURE 9.14 Principles of testing the compression strength of wood and wood-based materials. In failure under loading parallel to the fibre, so-called shear lines develop. On the longitudinal radial (LR) plane, the compression crease (shear line) runs horizontally. In contrast, on the longitudinal tangential (LT) plane, the crease is inclined (α) about 65° to the longitudinal axis of the wood.

(Courtesy of Carl Hanser Verlag.)

The compression strength in wood with deviation in fibre orientation can be calculated by the Hankinson equation, *cf.* Equations 9.4 and 9.8 (Hankinson 1921):

$$\sigma_{cU\varphi} = \frac{\sigma_{cU\parallel} \cdot \sigma_{cU\perp}}{\sigma_{cU\parallel} \cdot \sin^n \varphi + \sigma_{cU\perp} \cdot \cos^n \varphi} \qquad (9.6)$$

where

φ is the fibre orientation (grain angle)
n is a constant between 1.5 and 2.0
$\sigma_{cU\parallel}$ is the compression strength in grain direction
$\sigma_{cU\perp}$ is the compression strength perpendicular to the grain

Table 9.10 shows the compression strength of solid wood and wood-based materials. For solid wood, the tensile strength is about twice as high as the compressive strength; for particle-based panel materials, the compression strength is equal to or greater than the tensile strength.

9.4.5 BENDING STRENGTH

The bending strength, also named flexural strength or modulus of rupture, is the resistance of wood and wood-based materials against failure under bending load. It is a frequently used material characteristic for wood and wood-based materials, which is defined as:

$$\sigma_{bU} = \frac{M_b}{W_b} \qquad (9.7)$$

where

σ_{bU} is the bending strength (N/mm²)
M_b is the bending moment (N mm)
W_b is the section modulus (mm³)

9.4.5.1 Testing Methods

Bending strength is determined for solid wood according to the DIN 52186 standard (DIN 1978) and particleboard according to the EN 310 standard (CEN 1993b). Figure 9.15 shows schematically the test arrangement for three- and four-point bending tests.

The stress distribution over the specimen cross-section in solid wood is shown in Figure 9.16. A linear stress distribution occurs only if the proportional limit is not exceeded. At higher loads, there is a non-uniform stress distribution across the specimen's cross-section. This is due to the differences in stress-strain behaviour and the significant difference in tensile and compressive strength.

In the bent beam's compressed region (zone), the stress increases up to the ultimate compression strength. In the tension region, the stress increases further and approaches the ultimate tensile strength. Therefore, the neutral axis is shifted towards the tensile region as the load increases. If the wood is without features such as knots, the bending strength will be between the tensile and compression

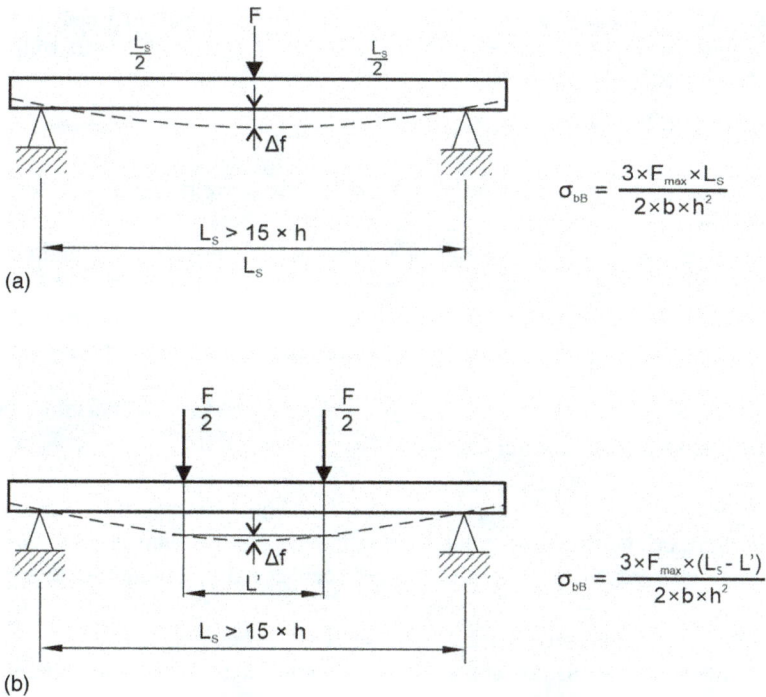

$$\sigma_{bB} = \frac{3 \times F_{max} \times L_s}{2 \times b \times h^2}$$

$L_s > 15 \times h$

L_s

(a)

$$\sigma_{bB} = \frac{3 \times F_{max} \times (L_s - L')}{2 \times b \times h^2}$$

L'

$L_s > 15 \times h$

(b)

FIGURE 9.15 Different setups for bending strength testing: (a) three-point bending and (b) four-point bending, where h is the height, and b is the width of the cross-section of the bending rod.

(Courtesy of Carl Hanser Verlag.)

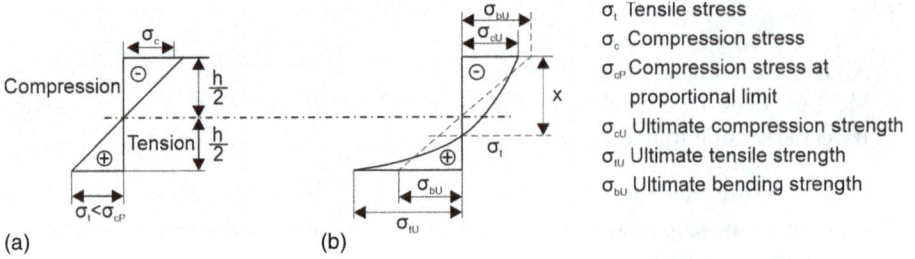

FIGURE 9.16 Stress distribution over the cross-section when bending wood as a function of load: (a) load below the proportional limit and (b) load well above the proportional limit.

(Courtesy of Carl Hanser Verlag.)

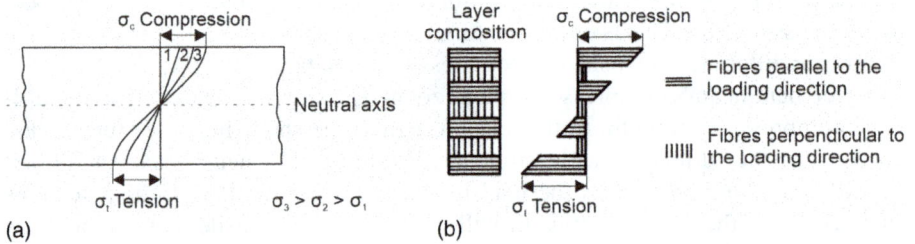

FIGURE 9.17 Stress distribution across the cross-section of panel-shaped wood-based material when bent: (a) particleboard and MDF (1–3 the bending load increases with increasing numbers), and (b) plywood and CLT.

(Courtesy of Carl Hanser Verlag.)

strength of the same type of specimen. Knots often cause premature failure in the tension zone.

The neutral axis will not change the location for particleboard and medium-density fibreboard (MDF) as the load changes, as shown in Figure 9.17 (Niemz *et al.* 2007). In this type of panel material, the tensile and compression strengths have similar values, or sometimes, the compression strength can be slightly higher.

9.4.5.2 Bending Strength: Influencing Factors and Material Parameters
Essential factors of the bending strength of wood are:

- the density, *i.e.* as the density increases, the bending strength increases,
- the fibre orientation, *i.e.* an increase in the grain angle causes a decrease in bending strength (an angle of 15 degrees causes a drop in strength to around 60%),
- knots and cracks cause a substantial decrease in bending strength,
- an increase in moisture content causes a significant reduction in bending strength; when the moisture content increases by 1%, bending strength is reduced by approximately 4%, and
- with increasing temperature, the bending strength decreases.

The bending strength in wood with deviation in fibre orientation can be calculated by the Hankinson equation, *cf.* Equations 9.4 and 9.6 (Hankinson 1921):

$$\sigma_{bU\varphi} = \frac{\sigma_{bU\|} \cdot \sigma_{bU\perp}}{\sigma_{bU\|} \cdot \sin^n \varphi + \sigma_{bU\perp} \cdot \cos^n \varphi} \tag{9.8}$$

where

φ is the fibre orientation (grain angle)
n is a constant between 1.5 and 2.0
$\sigma_{bU\|}$ is the bending strength in the grain direction ($\varphi = 0°$)
$\sigma_{bU\perp}$ is the bending strength perpendicular to the grain ($\varphi = 90°$)

The bending strength of a particleboard is determined by its particle geometry, density, and density profile. Like solid wood, changes in the climate affect bending strength. Table 9.10 gives reference values for the bending strength of wood and wood-based materials.

9.4.6   SHEAR STRENGTH

9.4.6.1   Parameters

Shear strength is the resistance that a body opposes to the displacement of two adjacent surfaces:

$$\tau = \frac{F}{A} \tag{9.9}$$

where

τ is the shear strength (N/mm²)
F is the ultimate load (N)
A is the shear area (mm²)

Shear strength parallel to the grain is related to torsional properties. Shear is complicated and superposed mainly by bending and compression (Kollmann and Côté 1968). Keylwerth (1945) specifies 18 possible load cases. In the cross-section (plane of shearing), shear is complex to test due to substantial wood densification perpendicular to the fibres (Niemz *et al.* 2023).

9.4.6.2   Testing Methods

Shear strength is determined for solid wood according to the DIN 52187 standard (DIN 1979b) and for particleboard, oriented strand board (OSB), and medium-density fibreboard (MDF) according to the DIN 52367 standard (DIN 2017). Additional information on testing can be found in Schulte (1997) and Dunky and Niemz (2002).

Based on the shape of the test specimen, more or less intense secondary stresses (bending, tensile, compression) occur apart from shear stresses, which can make the measurement result unclear. Shear stresses also occur in adhesive bond-lines and welded wood.

Factors influencing shear strength are:

- density, *i.e.* as the density increases, the shear strength increases,
- moisture content: up to a moisture content of about 5%, the shear strength of the wood increases and then falls until it reaches the fibre saturation point,
- the direction of load in relation to the principal anatomical orientations, and
- the influence of the grain angle applies to the Hankinson equation (see Equation 9.8) with n=2.5–3 (for experimental values, see Vorreiter 1949).

In practical applications, the shear strength parallel (*//*) with the fibre orientation is more critical than perpendicular ($\perp$) to the fibre (Figure 9.18). Vorreiter (1949) gives the ratio of $\tau\text{\textbardbl}{:}\tau\perp$ to 1.2–1.6. A distinction must be made between tangential and radial directions for shear stress parallel to the grain. The shear strength in the tangential direction is, in most cases, larger than that in the radial direction due to the influence of the rays.

Shear properties of particleboards are significantly influenced by particle geometry, density, and adhesive content (Niemz and Bauer 1991). The shear strength of particleboards perpendicular to the plane of the panel is approximately 10−18 MPa, parallel to the plane 3−5 MPa (Niemz and Sonderegger 2021). Extensive

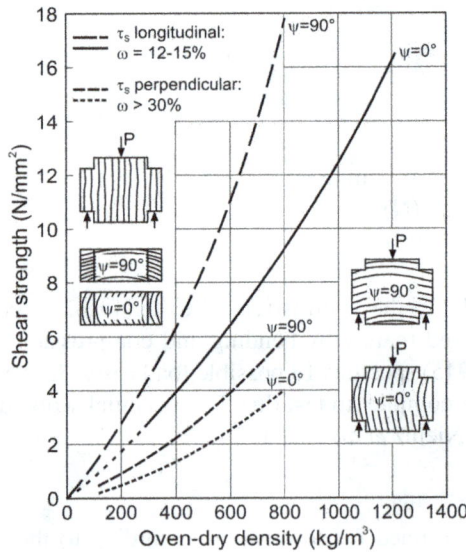

FIGURE 9.18 Influence of density and growth ring orientation on the shear strength of solid wood longitudinal and perpendicular to the grain (Vorreiter 1949). τ_s – shear strength, ω – moisture content, and growth-ring orientation on the cross-section of the specimen.

(Courtesy of Carl Hanser Verlag.)

TABLE 9.11

Shear Strength, Torsional Strength, and Cleavage Strength of Wood (Knigge and Schulz 1966, Niemz and Sonderegger 2021)

Species	Shear Strength				Torsional Strength	Cleavage Strength	
	$\tau//$ (LT) (N/mm²)	$\tau//$ (LR) (N/mm²)	$\tau\perp$ (RL) (N/mm²)	$\tau\perp$ (TL) (N/mm²)	$\tau_t//$ (N/mm²)	$\sigma_{s\,(T)}$ (N/mm²)	$\sigma_{s\,(R)}$ (N/mm²)
Black polar	5.0	5.0				0.74	0.51
Common beech	15.7–19.7	12.7–14.8	5.7–7.6	5.5–6.6	15	0.35	0.45
Common oak	13.5	12.4	4.9	4.1	20	0.53	0.48
Maple	17.9	14.3	7.3	6.6	26	1.60	1.00
Norway spruce	9.1–9.3	9.5–10.5	1.8	2.0	9	0.34	0.25
Scots pine	10.0	10.0			16	0.37	0.36

Note: Parallel (//) and perpendicular (⊥) to the fibre direction or the plane of a wood-based panel material. L = longitudinal (in fibre direction), R = radial, T = tangential, and the shear planes are indicated in parenthesis where the first index is the loading direction.

investigations on particleboards have been made by for example, Kruse (1993). Shear strength values of various wood species are shown in Table 9.11.

9.4.7 TORSION STRENGTH

The torsional strength is the resistance a rod (beam) opposes to breakage by twisting. Figure 9.19 shows schematically the test carried out on a cantilever rod. The rod can be stressed to torsion parallel or perpendicular to the fibre direction. The stress parallel to the fibre leads to a splitting of the rod in the longitudinal direction.

$$\tau_t = \frac{M_t}{W_t} \qquad (9.10)$$

where

τ_t is the torsional strength (Pa)
M_t is the torsional moment (N m)
W_t is the torsion resistance moment (m³)

Extensive work was carried out for example, by Höring (in Kollmann and Côté (1968)) and Chen (2002, 2006). If the torsional stress is within the linear-elastic range, the torsion modulus can be determined during the test (Neuhaus 1981).

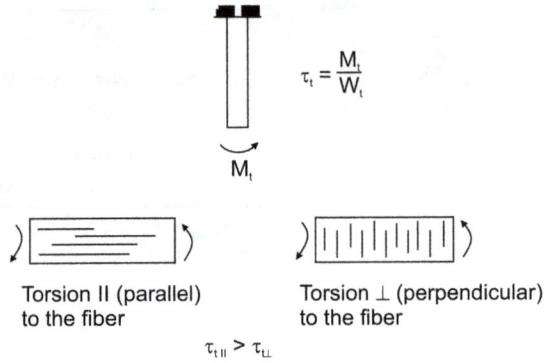

$$\tau_t = \frac{M_t}{W_t}$$

M_t

Torsion II (parallel)   Torsion ⊥ (perpendicular)
to the fiber   to the fiber

$$\tau_{t\,\text{II}} > \tau_{t\perp}$$

FIGURE 9.19   Torsional-strength testing of wood.

(Courtesy of Carl Hanser Verlag.)

The torsion modulus is:

$$G = \frac{M_t \cdot l}{\varphi \cdot I_t} \qquad (9.11)$$

where

G is the torsion modulus/torsional shear modulus (Pa)
l is the length of the specimen (m)
φ is the angle of twist/torsion (-)
I_t is the polar moment of inertia (m^4)

The main factors influencing torsional strength are:

- density: as the density of the wood increases, the torsional modulus increases,
- species: hardwoods have a higher torsional strength than softwoods,
- moisture content: as the moisture content increases, the torsional strength decreases,
- fibre orientation, and
- the principal direction in the specimen to the torsional moment.

Numerical values of the torsional strength of wood are given in Table 9.11.

9.4.8   CLEAVAGE STRENGTH (SPLITTING RESISTANCE)

9.4.8.1   Parameters and Testing Methods

Cleavage strength is the resistance of wood against splitting into two parts by wedge-shaped tools or two forces acting perpendicular to the longitudinal axis. Wood with high cleavage strength is, therefore, difficult to split. Figure 9.20 schematically shows the test arrangement, and Table 9.11 shows some cleavage values.

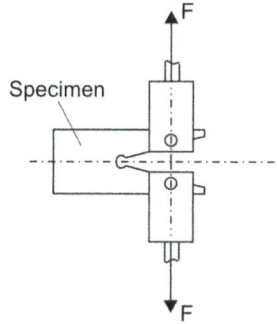

FIGURE 9.20 Specimen for the cleavage test.

(Courtesy of Carl Hanser Verlag.)

Cleavage strength is determined according to:

$$\sigma_{sp} = \frac{F}{b} \tag{9.12}$$

or

$$\sigma'_{sp} = \frac{F}{l \cdot b} \tag{9.13}$$

where

σ_{sp} is the cleavage strength (N/mm)
σ'_{sp} is the cleavage strength (N/mm²)
F is the ultimate load (N)
l is the sample length (mm)
b is the sample width (mm)

9.4.8.2 Influencing Factors and Material Parameters

The cleavage strength is significantly influenced by the angle at which the splitting plane runs to the fibre and growth-ring orientations. The lowest cleavage strength is in the radial direction, where the cleavage plane runs in the direction of the wood rays. Wood with broad rays usually has a lower cleavage strength than wood with small rays. As the density increases, the cleavage strength in hardwoods increases more than in softwoods. The cleavage strength increases to a moisture content of approximately 15%, decreasing at further increased moisture content. Features such as fibre deviation and knots can significantly affect the cleavage strength.

9.4.9 Nail and Screw Withdrawal Resistance

9.4.9.1 Parameters and Testing

The withdrawal resistance of nails and screws is an essential feature for the use of wood and wood-based materials.

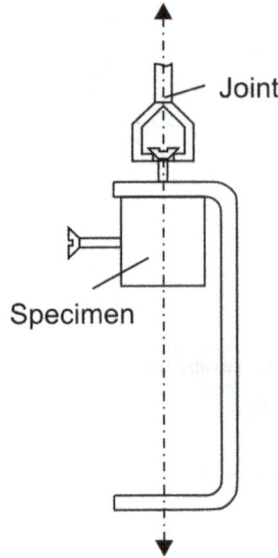

FIGURE 9.21   Schematic representation of the test of nail and screw withdrawal resistance.

(Courtesy of Carl Hanser Verlag.)

Analogously, the nail pull-out resistance is the force required to extract a nail based on the cross-sectional area of the nail (N/mm²) or the nail-in depth (N/mm).

The screw withdrawal resistance is the ultimate load required to pull out a screw about the screw-in depth (N/mm). The screw withdrawal resistance value depends on the test method. The EN 1382 standard (CEN 2016b) is used for solid wood, and the EN 320 standard (CEN 2011b) for particleboard and fibreboard.

Figure 9.21 shows schematically the test arrangement of nail and screw withdrawal resistance.

9.4.9.2   Influencing Factors and Material Parameters

The nail and screw withdrawal resistance of wood depends on the density and the cutting direction. For particle-based panel materials, the density profile in the thickness direction of the panel, the type of particles (geometry and size), the adhesive content, and the transverse tensile strength influence the withdrawal strength (Niemz and Bauer 1990). The impact or screwing direction (parallel or perpendicular to the plane or fibre direction) is also a significant influence on the nail or screw pull-out resistance. In the case of wood, the structure is a decisive influencing factor, for example, the nail or screw withdrawal resistance increases with increasing density, but also, the nail and screw shape affect the measurement result. Nail and screw withdrawal resistance values of wood and wood-based materials are given in Table 9.12.

TABLE 9.12
Nail and Screw Withdrawal Resistance of Wood and Wood-based materials. Parallel (//) and Perpendicular (⊥) to the Fibre Direction or the Plane of a Wood-based Panel Material

Material	Density (kg/m³)	Nail Withdrawal Resistance[a]		Screw Withdrawal Resistance[a]	
		// (N/mm²)	⊥ (N/mm²)	// (N/mm)	⊥ (N/mm)
Softwood	500	1.5–2.2	2.5–4.0	60–80	80–110
Hardwood	650–750			100–150	170–200
Particleboard	600	0.8–1.5	1.2–2.0		
Particleboard[1]	650–700			35–60	50–80
MDF[1]	730–780			40–70	55–85

[a] Tested using a wood screw, form B (particleboard screw), 4 x 40 mm; pre-drilled: 0.8 x screw diameter (Anon. 1990).

9.4.10 HARDNESS AND ABRASION RESISTANCE

9.4.10.1 Hardness

Hardness is the resistance that wood or wood-based materials oppose in penetrating a harder body. According to the speed of the applied load for the test, a distinction can be made between:

- static test methods (slow build-up of strength) and
- dynamic test methods (sudden impact).

9.4.10.2 Static Hardness Test

According to the EN 1534 standard (CEN 2020), the Brinell test is the most used static test method. A polished steel ball with a 2.5, 5, or 10 mm diameter is pressed into the specimen with a defined force within a specific time frame. The diameter of the indented spherical mark in the wood is measured. From the diameter of the sphere D, the force F and the diameter of the spherical cup d, the Brinell hardness H_B is then calculated as follows:

$$H_B = \frac{2 \cdot F}{\pi \cdot D \cdot \left(D - \sqrt{D^2 - d^2} \right)} \tag{9.14}$$

where

H_B is the the Brinell hardness (N/mm²)
F is the applied force in the range of 100–1000 N
π is the the constant pi (-)
D is the diameter of the steel ball (mm)
d is the diameter of the indentation (mm)

Often, the hardness is calculated from the penetration depth h instead of the indentation diameter, which can be measured as the movement of the apparatus that controls the application of force:

$$H_B = \frac{F}{D \cdot \pi \cdot h} \qquad (9.15)$$

A further method for determining the hardness of wood and wood-based materials is Janka hardness. In this hardness test, a polished steel ball with a diameter of 11.284 mm is pressed into the wood until the entire diameter is penetrated. The required penetration force directly indicates hardness. This method is often used in the Americas.

9.4.10.3 Influencing Factors and Material Parameters

Parameters that influence the hardness of wood are:

- density, *i.e.* as the density increases, the hardness will increase (Figure 9.22),
- principal section of the wood, *i.e.* the hardness of the cross-section is approximately 2.5 times higher compared to the transverse sections,
- moisture content, *i.e.* the hardness increases with decreasing moisture content, and
- the level of the test load, *i.e.* the wood is densified during the test and the hardness increases with increased densification.

Table 9.13 gives information on the moisture dependence of the Brinell hardness on selected wood species and wood-based materials. Schwab (1990) has presented an overview of the hardness of various wood species used in flooring. According to Koch *et al.* (2018), bamboo with a density between 500 kg/m³ and 900 kg/m³ has a Brinell hardness of 60–80 N/mm² and a Janka hardness of 17–25 kN.

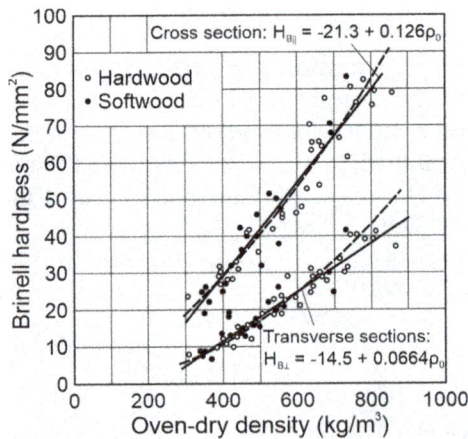

FIGURE 9.22 Brinell hardness as a function of the oven-dry density ρ_0 (data from Kollmann 1951.

(Courtesy of Carl Hanser Verlag.)

TABLE 9.13

Brinell Hardness (H_B) of Selected Wood Species and Wood-based Materials at 20 °C and Relative Humidity of 35%, 65% or 80%, Respectively (Niemz and Sonderegger 2007, Sonderegger and Niemz 2009)

Type of Wood	Density (kg/m³)	$H_{B,35}$ (N/mm²)	$H_{B,65}$ (N/mm²)	$H_{B,80}$ (N/mm²)
European beech	690–750	41	32	26
European larch	620–650	36	31	25
European oak	670–690	35	30	24
Red ironwood	1030–1110	99	73	62
Scots pine	510–570	21	20	17
Plywood of common beech	730–780	27–32	26–30	24–26
Particleboard, t=10–19 mm	630–700	21–25	18–22	15–17
Particleboard, t=25–40 mm	610–620	28–30	26–28	18–19
Oriented strand board (OSB)	620–650	29–35	28–32	19–25
Medium-density fibreboard (MDF), t=10–25 mm	730–800	28–62	26–56	19–39
Low-density fibreboard (LDF)	530	19	17	13

Note: t – thickness of the panel.

9.4.10.4 Abrasion Resistance

9.4.10.4.1 Sandblast Method

In this method, sand of specific particle size is applied to a defined wood-specimen surface using a sandblast blower, and the loss of mass (g/cm² treated surface) or volume (in cm³) of wood is measured. Due to the differences in hardness between earlywood and latewood, there is an uneven removal over the surface (greater removal of the less dense earlywood). Differences also exist between the three principal sections of the wood. According to studies by Schulz (1985), hardwood and softwood show characteristic hardness profiles during sandblasting, which can be used to divide the growth rings in earlywood and latewood regions.

9.4.10.4.2 Sanding-resistance Method

The sanding method allows more practical abrasion resistance testing than sandblasting, and this test method is used for both wood and wood-based materials. Two parameters can be used to measure abrasion:

1)   the abrasion coefficient related to the dimension of the test specimen t_a:

$$t_a = a \cdot \frac{m_1 - m_2}{m_1} \qquad (9.16)$$

and

2)   the mass-related coefficient of abrasion t_m:

$$t_m = \frac{m_1 - m_2}{m_1} \qquad (9.17)$$

where

a is the thickness of the specimen
m_1 is the mass of the specimen before the test
m_2 is the mass of the specimen after the test

The abrasion resistance can be determined by rolling two abrasives covered with abrasive paper on the specimen surface. The abrasion index (in g/100 revolutions) for laminate flooring is tested according to the EN 13329 standard (CEN 2023b) and for high-pressure laminate (HPL) according to the EN 438 T2 standard (CEN 2016c).

The following parameters significantly influence the abrasion resistance of wood:

• density, *i.e.* abrasion resistance increases with increasing density,
• moisture content, and
• cutting direction, *i.e.* the abrasion resistance, is more significant in the radial and tangential directions than in the cross-section.

9.4.11   IMPACT BENDING STRENGTH

Impact bending strength is related to the dynamic wood properties. It describes the behaviour of wood and wood-based materials against an impact load or stress. This measure is used in practice to test the material of tool handles. The impact bending strength is calculated according to the ISO 13061-10 standard (ISO 2017):

$$A_W = 1000 \cdot \frac{Q}{b \cdot h} \qquad (9.18)$$

where

A_w is the impact bending strength (kJ/mm²)
Q is the energy required for fracture of the test piece (J)
b, h are the dimensions of the specimen in the radial and tangential directions (mm)

Impact bending strength is affected by wood species, density, moisture content, fibre orientation, and fungal attack (Figure 9.23).

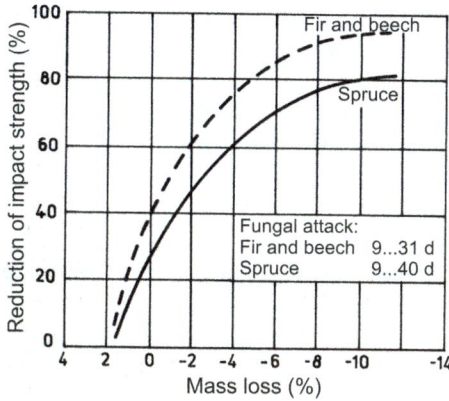

FIGURE 9.23 Impact bending strength as a function of mass loss due to fungal attack in up to 40 days (data from Vorreiter 1949).

(Courtesy of Carl Hanser Verlag.)

9.4.12 FATIGUE

9.4.12.1 Parameters and Testing

Repeated loading and unloading of wood and wooden materials, *i.e.* dynamic or cyclic loading, leads to fatigue. The fatigue strength values are significantly lower than those achieved in static tests, and a failure can occur even if the strength determined in the short-term test has yet to be reached (Figure 9.24). A limit of fatigue strength is controversially discussed in materials science. Often, fatigue strength is defined as related to a certain number of load cycles, for example, one million cycles. The fatigue strength is determined in the Wöhler test (Figure 9.24).

Ross (2010) described that time under intermittent loading has a cumulative effect. In tests where a constant load is periodically placed on a beam and then removed, the

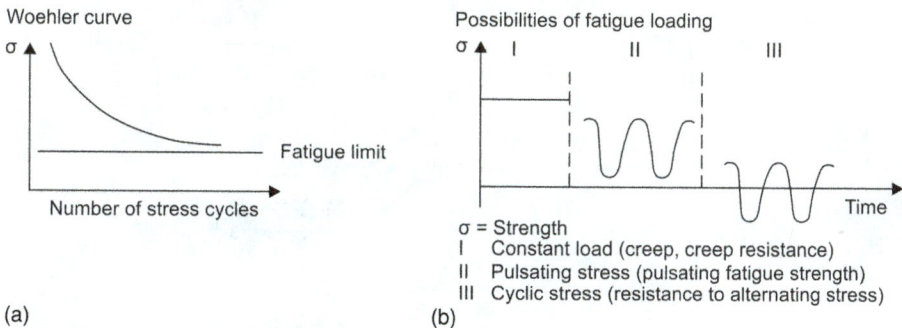

FIGURE 9.24 Test of wood's fatigue: (a) Wöhler curve (also called an S-N curve) and (b) different load cases for fatigue tests.

(Courtesy of Carl Hanser Verlag.)

cumulative time the load was applied to the beam before failure was equal to the time to failure for a similar beam under the same load applied continuously.

Partially, fatigue strength under continuous load is also referred to as fatigue in static loading (duration of the load (DOL) test); see Section 8.6.5 (Smith *et al.* 2003).

Figure 9.24 shows schematically possible load cases with repeated loading and unloading cycles. As a result, a distinction is made between two limit states in terms of strength:

- pulsating load, *i.e.* the stress changes between zero and a value larger than zero and
- alternating load, *i.e.* the stress changes between two equal limits with values above and below 0, for example, tension and compression.

Both load cases lead to fatigue with an increasing number of load cycles and, as a result, material failure. Also significant is the height of the load compared to the strength in the static short-term test.

In practice, fatigue stresses occur in structures exposed to constantly changing wind loads, typically bridges, electricity pylons, towers, high-rise buildings, aeroplanes, and windmill blades.

Fatigue strength: Material parameters and influencing factors

The fatigue bending strength in solid wood is 25% to 40% of the bending strength in the static short-term test, and the corresponding range for wood-based materials is 20% to 30%. There is very little data on wood fatigue, and old statements from the Second World War, when timber was used in aeroplanes, are often used. Mohr (2001) provides a review of the state of knowledge, describing various wood species and their application in construction practice.

According to Reichel (2015), the number of load cycles until the fracture depends on frequency. During the test, the specimens are exposed to different stress levels

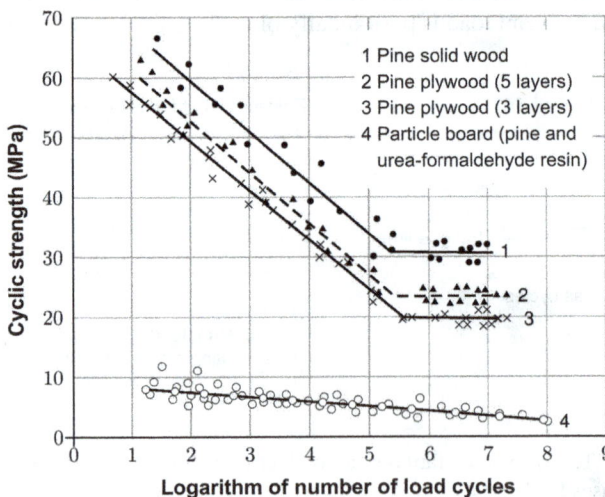

FIGURE 9.25 Fatigue of Scots pine wood and wood-based materials (data from Gillwald 1966).

(Courtesy of Carl Hanser Verlag.)

TABLE 9.14
Reduction Coefficients for Dynamic Fatigue Strength According to the SIA 265/1 Standard (SIA 2003)

Loading Mode	$k_{fat}\infty$
Compression	1.0
Tension	0.5
Bending	0.5
Cyclic tension-compression	0.5
Shear	0.3

until failure, the number of load cycles to failure is determined, and the strength is plotted against the number of load cycles or their logarithms (Figure 9.25). Such loads are also typical for tensile shear specimens to study the adhesive-bond quality according to the EN 302-1 standard (CEN 2023a).

The Swiss standard SIA 265/1 (SIA 2003) proposes reduction coefficients for fatigue strength in timber construction (Table 9.14). Bodig and Jayne (1993) present the fatigue bending strength of several species. The endurance ratio, *i.e.* the fatigue bending strength to static bending strength, is between 0.22 and 0.38. This means a reduction to about one-quarter to one-third of the ultimate static stress.

REFERENCES

Ammann S.D. (2015). *Mechanical Performance of Glue Joints in Structural Hardwood Elements*. Doctoral Thesis, ETH Zürich, Zurich, Switzerland.

Anon. (1990). *Lexikon der Holztechnik*. (4th Ed.), Fachbuchverlag, Leipzig, Germany.

Blaß H.J. (1987). *Tragfähigkeit von Druckstäben aus Brettschichtholz unter Berücksichtigung streuender Einflussgrössen*. Doctoral Thesis, University of Karlsruhe, Karlsruhe, Germany.

Bodig J. & Jayne B.A. (1993). *Mechanics of Wood and Wood Composites*. (2nd Ed.), Krieger Publishing Company, Malabar (FL), USA, (21+712 pp.).

Burger N. & Glos P. (1996). Einfluss der Holzabmessungen auf die Zugfestigkeit von Bauschnittholz. *Holz als Roh- und Werkstoff*, 54(5), 333–340.

Böhme C. (1999). Einfluss der Prüfkörperabmessungen bei Spanplatten. *WKI-Kurzbericht, 21, 22, 23*, Braunschweig, Germany.

CEN (1993a). EN 319: Particleboards and fibreboards - Determination of tensile strength perpendicular to the plane of the board. *The European Committee for Standardization (CEN)*, Brussels, Belgium.

CEN (1993b). EN 310: Wood-based panels - Determination of modulus of elasticity in bending and of bending strength. *The European Committee for Standardization (CEN)*, Brussels, Belgium.

CEN (2001). EN 12369-1: Wood-based panels - Characteristic values for structural design - Part 1: OSB, particleboard and fibreboard. *The European Committee for Standardization (CEN)*, Brussels, Belgium.

CEN (2004a). EN 1995-1-1: 2004+A 1: Eurocode 5: Design of timber structures - Part 1-1: General Common rules and rules for buildings. *The European Committee for Standardization (CEN)*, Brussels, Belgium.

CEN (2004b). *EN 789: Timber structures - Test methods - Determination of mechanical properties of wood based panels.* The European Committee for Standardization (CEN), Brussels, Belgium.

CEN (2010). EN 312: Particleboards. *Specifications. The European Committee for Standardization (CEN)*, Brussels, Belgium.

CEN (2011a). EN 12369-2: Wood-based panels. Characteristic values for structural design – Plywood. *The European Committee for Standardization (CEN)*, Brussels, Belgium.

CEN (2011b). EN 320: Particleboards and fibreboards - Determination of resistance to axial withdrawal of screws. *The European Committee for Standardization (CEN)*, Brussels, Belgium.

CEN (2012). EN 408: Timber structures - Structural timber and glued laminated timber - Determination of some physical and mechanical properties (includes Amendment A1:2012). *The European Committee for Standardization (CEN)*, Brussels, Belgium.

CEN (2016a). EN 338: Structural timber - Strength classes. *The European Committee for Standardization (CEN)*, Brussels, Belgium.

CEN (2016b). EN 1382: Timber Structures. Test methods. Withdrawal capacity of timber fasteners. *The European Committee for Standardization (CEN)*, Brussels, Belgium.

CEN (2016c). EN 438: High-pressure decorative laminates (HPL) - Sheets based on thermosetting resins (usually called laminates) - Part 1: Introduction and general information. *The European Committee for Standardization (CEN)*, Brussels, Belgium.

CEN (2017). EN 302-2: Adhesives for load-bearing timber structures - Test methods - Part 2: Determination of resistance to delamination. *The European Committee for Standardization (CEN)*, Brussels, Belgium.

CEN (2020). EN 1534: Wood flooring and parquet - Determination of resistance to indentation — Test method. *The European Committee for Standardization (CEN)*, Brussels, Belgium.

CEN (2023a). EN 302-1: Adhesives for load-bearing timber structures - Test methods - Part 1: Determination of longitudinal tensile shear strength. *The European Committee for Standardization (CEN)*, Brussels, Belgium.

CEN (2023b). EN 13329: Laminate floor coverings - Specifications, requirements and test methods. *The European Committee for Standardization (CEN)*, Brussels, Belgium.

Chen Z. (2002). *Torsional Fatigue of Wood.* Doctoral Thesis, South Bank University, London, UK.

Chen Z., Gabbitas B. & Hunt D. (2006). The fracture of wood under torsional loading: *Journal of Material Science*, 41(21), 7247–7259.

Colling F. (1990). *Tragfähigkeit von Biegeträgern aus Brettschichtholz in Abhängigkeit von den festigkeitsrelevanten Einflussgrössen.* Doctoral Thesis, University of Karlsruhe, Karlsruhe, Germany.

Czaderski C., Steiger R., Howald M., Olia S., Gülzow A. & Niemz P. (2007). Versuche und Berechnungen an allseitig gelagerten 3-schichtigen Massivholzplatten. *Holz als Roh- und Werkstoff*, 65(5), 383–402.

DeBaise G.R., Porter A.W. & Pentoney R.E. (1966). Morphology and mechanics of wood fracture. *Materials Research and Standards*, 6(10), 493–499.

Denzler, J.K. (2007). *Modellierung des Größeneffektes bei biegebeanspruchtem Fichtenschnittholz.* Doctoral Thesis, Technical University of Munich (TUM), Munich, Germany.

DIN (1976). DIN 52185: Testing of wood; compression test parallel to grain. *Deutsches Institut für Normung (DIN)*, Berlin, Germany.

DIN (1978). DIN 52186: Testing of wood; bending test. *Deutsches Institut für Normung (DIN)*, Berlin, Germany.

DIN (1979a). DIN 52188: Testing of wood; determination of ultimate tensile stress parallel to grain. *Deutsches Institut für Normung (DIN)*, Berlin, Germany.

DIN (1979b). DIN 52187: Testing of wood; determination of ultimate shearing stress parallel to grain. *Deutsches Institut für Normung (DIN)*, Berlin, Germany.

DIN (1979c). DIN 52192: Testing of wood; compression test perpendicular to grain. *Deutsches Institut für Normung (DIN)*, Berlin, Germany.

DIN (2003). DIN 68364: Properties of wood species - Density, modulus of elasticity and strength. *Deutsches Institut für Normung (DIN)*, Berlin, Germany.

DIN (2017). DIN 52367: Wood-based panels - Determination of shear strength parallel to the surface. *Deutsches Institut für Normung (DIN)*, Berlin, Germany.

Dunky M. & Niemz P. (2002). *Holzwerkstoffe und Leime: Technologie und Einflussfaktoren.* Springer, Berlin, Germany.

Ehlbeck J. (1967). *Durchbiegung und Spannungen von Biegeträgern aus Holz unter Berücksichtigung der Schubverformung.* Doctoral Thesis, Universität Karlsruhe, Karlsruhe, Germany.

Fink G. (2014). *Influence of Varying Material Properties on the Load-Bearing Capacity of Glued Laminated Timber.* Doctoral Thesis, ETH Zürich, Zurich, Switzerland.

Gillwald W. (1966). Untersuchungen über die Dauerfestigkeit von mehrschichtigen Spanplatten. *Holz als Roh- und Werkstoff*, 24(10), 445–449.

Glos P. (1982). Die maschinelle Festigkeitssortierung von Schnittholz. *Holz-Zentralblatt*, 108, 153–155.

Glos P. & Schulz H. (1986). Qualität und Festigkeit von Bauschnittholz aus Waldschadensgebieten. *Holz als Roh- und Werkstoff*, 44(8), 293–298.

Görlacher R. (1990). *Klassifizierung von Brettschichtholzlamellen durch Messung von Longitudinalschwingungen.* Doctoral Thesis, University of Karlsruhe, Karlsruhe, Germany.

Halligan A.F. & Schniewind A.P. (1974). Prediction of particleboard mechanical properties at various moisture contents. *Wood Science and Technology*, 8(1), 68–78.

Hankinson R.L. (1921). Investigation of crushing strength of spruce at varying angles of grain. *Air Service Information Circular*, 3(259), 3–15.

Hassani M.M., Wittel F. K., Hering S. & Herrmann H.J. (2015). Rheological model for wood. *Computer Methods in Applied Mechanics and Engineering*, 283(1), 1032–1060.

Hering S., Saft S., Resch E., Niemz P. & Kaliske M. (2012). Characterisation of moisture-dependent plasticity of beech wood and its application to a multi-surface plasticity model. *Holzforschung*, 66(3), 373–380.

Hübner U. (2013). *Mechanische Kenngrößen von Buchen-, Eschen- und Robinienholz für lastabtragende Bauteile.* Doctoral Thesis, Technical University of Graz, Graz, Austria.

ISO (2014). *ISO 13061-6: Physical and mechanical properties of wood – Test methods for small clear wood specimens. Part 6: Determination of ultimate tensile stress parallel to grain.* International Organization for Standardization (ISO), Geneva, Switzerland.

ISO (2017). *ISO 13061-10: Physical and mechanical properties of wood - Test methods for small clear wood specimens - Part 10: Determination of impact bending strength.* International Organization for Standardization (ISO), Geneva, Switzerland.

Keunecke D. (2008). *Elasto-Mechanical Characterisation of Yew and Spruce Wood with Regard to Structure-Property Relationships.* Doctoral Thesis, ETH Zürich, Zurich, Switzerland.

Keunecke D., Hering S. & Niemz P. (2008). Three-dimensional elastic behaviour of common yew and Norway spruce. *Wood Science and Technology*, 42(8), 633–647.

Keylwerth, R. (1945). Zur Scherfestigkeit des Holzes. Unveröffentlichte Messungen der Reichsanstalt für Holzforschung Eberswalde, No. 1/1945, Eberswalde, Germany.

Kisser J. & Steininger A. (1952). Makroskopische und mikroskopische Strukturänderungen bei der Biegebeanspruchung von Holz. *Holz als Roh- und Werkstoff*, 10(11), 415–421.

Knigge W. & Schulz H. (1966). *Grundriss der Forstbenutzung.* Parey, Hamburg, Germany, (584 pp.).

Koch G., Oelker M. & Richter H. (2018). MacroHOLZdata: Illustrations, identification, and information retrieval. Available online: https://www.delta-intkey.com. [Retrieved September 20, 2024].

Kollmann F. (1951). *Technologie des Holzes und der Holzwerkstoffe: Band 1: Anatomie und Pathologie, Chemie, Physik, Elastizität und Festigkeit.* Springer Verlag, Berlin, Germany, (19+1051 pp.).

Kollmann F. (1955). *Technologie des Holzes und der Holzwerkstoffe: Band 2: Holzschutz, Oberflächenbehandlung, Trocknung und Dämpfen, Veredelung, Holzwerkstoffe, spanabhebende und spanlose Holzbearbeitung, Holzverbindungen.* Springer Verlag, Berlin, Germany, (36+1201 pp.).

Kollmann F. & Côté Jr. W.A. (1968). *Principles of Wood Science and Technology. Part 1: Solid Wood.* Springer, Berlin, Heidelberg, Germany, (12+592 pp.).

Kruse K. (1993). *Untersuchungen verschiedener Einflussgrössen auf die zerstörungsfreie Werkstoffprüfung von Holzwerkstoffen mit Ultraschall.* Doctoral Thesis, University of Hamburg, Hamburg, Germany.

Kucera L.J. & Bariska M. (1982). On the fracture morphology in wood. Part 1: A SEM-study of deformations in wood of spruce and aspen upon ultimate axial compression load. *Wood Science and Technology,* 16(4), 241–259.

Larsen H.J., Leijten A.J.M. & van der Put, T.A.C.M. (2008). The design rules in Eurocode 5 for compression perpendicular to the grain - Continuous and semi continuous supported beams. In: *Proceeding of the CIB-W18, 41st meeting, 24-28 August, St. Andrews,* Canada, (pp. 1–12).

McNatt J.D., Wellwood R.W. & Bach L. (1990). Relationships between small-specimen and large panel bending tests on structural wood-based panels. *Forest Products Journal,* 40(9), 10–16.

Mindess S. & Bentur A. (1986). Crack propagation in notched wood specimens with different grain orientations. *Wood Science and Technology,* 20(2), 145–155.

Mohr B. (2001). Zur Interaktion der Einflüsse aus Dauerstandbelastung und Ermüdungsbeanspruchung im Ingenieurholzbau. *Berichte aus dem konstruktiven Ingenieurbau, 10.* Technical University of Munich (TUM), Munich, Germany.

Neuhaus F. (1981). *Elastizitätszahlen von Fichtenholz in Abhängigkeit von der Holzfeuchtigkeit.* Doctoral Thesis, University of Bochum, Bochum, Germany.

Neuhaus H. (2011). *Ingenieurholzbau.* Vieweg +Teubner, Wiesbaden, Germany.

Niemz P. & Bauer S. (1990). Beziehungen zwischen Struktur und Eigenschaften von Spanplatten. Teil 1: Schraubenausziehwiderstand. *Holzforschung und Holzverwertung,* 42(5), 361–364.

Niemz P. & Bauer S. (1991). Beziehungen zwischen Struktur und Eigenschaften von Spanplatten. Teil 2: Schubmodul, Scherfestigkeit, Biegefestigkeit. *Holzforschung und Holzverwertung,* 43(2), 68–70.

Niemz P. & Sonderegger W. (2003). Untersuchungen zur Korrelation ausgewählter Holzeigenschaften untereinander und mit der Rohdichte unter Verwendung von 103 Holzarten. *Schweizerische Zeitschrift für Forstwesen,* 154(12), 489–493.

Niemz, P. & Sonderegger, W. (2021). *Physik des Holzes und der Holzwerkstoffe.* (2nd Ed.), Carl Hanser Verlag GmbH & Company KG, Munich, Germany.

Niemz P. & Sonderegger W. (2007). Feuchte und Härte beim Holz eng korreliert. Vergleichende Untersuchungen der Brinellhärte bei variabler Holzfeuchte. *Holz-Zentralblatt,* 133(47), 1327.

Niemz P., Schreiber J., Naumann J. & Stockmann M. (2007). Experimentelle Ermittlung der Dehnungen im Probenquerschnitt bei Biegebelastung von Holzpartikelwerkstoffen. *Holz als Roh- und Werkstoff,* 65(6), 459–468.

Niemz P, Hug S. & Schnider T. (2014). Einfluss der Temperatur auf ausgewählte mechanische Eigenschaften von Esche, Buche, Ahorn und Fichte. *Forstarchiv,* 85(5), 163–168.

Niemz P., Teischinger A. & Sandberg D. (Eds.) (2023). *Springer Handbook of Wood Science and Technology.* Springer Nature, Cham, Switzerland, (XXV+2069 pp.).

Niemz P., Teischinger A. & Sandberg D. (Eds.) (2025). *Wood Material and Processing Data: The Most Relevant Data, Tables, and Figures.* Springer Nature, Cham, Switzerland, (XII+284 pp.).

Nördlinger H. (1860). *Die Technischen Eigenschaften der Hölzer für Forst- und Baubeamte, Technologen und Gewerbetreibende.* J. G. Cotta'scher Verlag, Stuttgart, Germany,

Ozyhar T. (2013). *Moisture and Time-dependent Orthotropic Mechanical Characterization of Beech Wood.* Doctoral Thesis, ETH Zürich, Zurich, Switzerland.

Patton-Mallory M. & Cramer S.M. (1987). Fracture mechanics: A tool for predicting wood component strength. *Forest Products Journal*, 37(7/8), 39–47.

Plath E. (1971). Beitrag zur Mechanik der Holzspanplatten. *Holz als Roh- und Werkstoff*, 29(10), 377–382.

Richter C. (2014). *Wood Characteristics.* Springer Nature, Cham, Switzerland, (10+222 pp.).

Reichel S. (2015). *Modellierung und Simulation hygro-mechanisch beanspruchter Strukturen aus Holz im Kurz- und Langzeitbereich.* Doctoral Thesis, TU Dresden, Dresden, Germany.

Ross R. (2010). *Wood Handbook. Wood as an Engineering Material.* (Centennial Ed.), US Department of Agriculture, Forest Service, Forest Products Laboratory, Madison (WI), USA., (509 pp.).

Scheffler M. (2000). *Bruchmechanische Untersuchungen zur Trocknungsrißbildung an Laubholz.* Doctoral Thesis, TU Dresden, Dresden, Germany.

Schreiber J., Niemz P. & Mannes D. (2007). Vergleichende Untersuchungen zu ausgewählten Eigenschaften von Holzpartikelwerkstoffen bei unterschiedlicher Belastungsart. *Holztechnologie*, 48(1), 1–10.

Schulz H. (1985). Härteprofile als Hinweis auf verschiedene Festigkeitssysteme im Holz. *Holz als Roh- und Werkstoff*, 43(6), 215–222.

Schulte M. (1997). *Zerstörungsfreie Prüfung elastomechanischer Eigenschaften von Holzwerkstoffplatten durch Auswertung des Eigenschwingverhaltens und Vergleich mit zerstörenden, statischen Prüfmethoden.* Doctoral Thesis, Universität Hamburg, Hamburg, Germany.

Schwab E. (1990). Die Härte von Laubhölzern für die Parkettherstellung. *Holz als Roh- und Werkstoff*, 48(2), 47–51.

Sell J. (1997). *Eigenschaften und Kenngrössen von Holzarten* (4th Ed.), Baufachverlag, Dietikon, Switzerland, (140 p.).

SIA (2003). *SIA 265: Timber Structures.* Swiss Society of Engineers and Architects. Zurich, Switzerland.

Smith I., Landis E. & Gong M. (2003). *Fracture and Fatigue in Wood.* Wiley, Chichester, UK., (248 pp.).

Sonderegger W. & Niemz P. (2006). Der Einfluss der Temperatur auf die Biegefestigkeit und den E-Modul bei verschiedenen Holzwerkstoffen. *Holz als Roh- und Werkstoff*, 64(5), 385–391.

Sonderegger W. & Niemz P. (2009). Untersuchungen zur Bestimmung der Brinellhärte von Holzwerkstoffen bei variabler Holzfeuchte. *Holztechnologie*, 50(1), 11–16.

Sonderegger W., Kránitz K., Bues C.-T. & Niemz P. (2015). Aging effects on physical and mechanical properties of spruce, fir and oak wood. *Journal of Cultural Heritage*, 16(6), 883–889.

Steiger, R. (1996). *Mechanische Eigenschaften von Schweizer Fichten-Bauholz bei Biege-, Zug-, Druck- und kombinierter M/N Beanspruchung: Sortierung von Rund- und Schnittholz mittels Ultraschall.* Doctoral Thesis, ETH Zürich, Zurich, Switzerland.

Steiger R., Gülzow A., Czaderski C., Howald M. & Niemz P. (2012). Comparison of bending stiffness of cross-laminated solid timber derived by modal analysis of full panels and by bending tests of strip-shaped specimens. *European Journal of Wood and Wood Products*, 70(1/3), 141–153.

Vorreiter L. (1949). *Holztechnologisches Handbuch: Band 1.* Fromme, Wien, Austria.

Zimmermann T. (2015). *Fortschritte in der Nanocelluloseforschung.* Tagungsband 3. Holzanatomisches Kolloquium, IHD Dresden, Dresden, Germany.

10 Durability of Wood and Wood-based Materials

C. Brischke and M. Humar

10.1 BIOTIC DEGRADATION AGENTS

10.1.1 FUNGI

Fungi form their own kingdom; they are neither plants nor animals. Fungi are sedentary like plants, but are incapable of performing photosynthesis and are, therefore, carbon heterotrophic. Fungi colonising wood can generally be divided into wood-discolouring fungi and wood-destroying fungi, where the latter can lead to significant damage to the wood material. In contrast, damage by wood-discolouring fungi, such as staining fungi and moulds, is only aesthetic.

There is no generally accepted system for classifying various wood-discolouring and wood-damaging fungi, and the applied classifications used have repeatedly changed. Table 10.1 shows the major fungal groups and identifies essential examples for each. Several compendia are available that contain more detailed information on systematics, morphology, physiology, occurrence, and unique characteristics (*e.g.*, Bravery *et al*. 1987, Rayner and Boddy 1988, Eaton and Hale 1993, Unger *et al*. 2001, Huckfeldt and Schmidt 2006, Schmidt 2006, Zabel and Morrell 2012, Schwarze *et al*. 2013). The following sections briefly describes only the most relevant parameters related to fungal discolouration and wood decay.

10.1.1.1 Wood-destroying Fungi

Wood-destroying fungi are capable of causing massive damage to wooden structures and other wooden components. Breaking down wood substances causes a mass loss and a loss of the elasto-mechanical properties of wood (Wilcox 1978, Winandy and Morrell 1993). The structural integrity of wood is also affected, and wood becomes more brittle (Brischke *et al*. 2006a). There are three types of rot depending on their appearance, the mechanisms of wood decomposition, and partly on the physiological needs of the respective decay fungi. These are brown rot, white rot, and soft rot, which can also occur together or one after the other during succession (Schmidt 2006, Fukasawa and Matsukura 2021, Schrader *et al*. 2024). Sometimes features of more than one rot type are caused by a fungus (Kleist and Schmitt 2001, Huckfeldt and Brischke 2024).

Brown rot causing fungi predominantly degrade hemicelluloses and cellulose but not lignin. Consequently, wood degraded by brown rot fungi has a brown to black-brown colour due to the absence of cellulose and remaining brown lignin. Often, the substrate shows superficial deformation and cubic fractures due to cleavage of cellulose chains. Most brown rot fungi attack preferably softwoods, while hardwoods are more frequently degraded by white and soft rot fungi. White rot fungi

                    DOI: 10.1201/9781003411994-10

TABLE 10.1

Discolouring and Destroying Fungi that Attack Wood and Other Bio-based Building Materials

Destroying Fungi			Staining Fungi		
Brown-rot Fungi	White-rot Fungi	Soft-rot Fungi	Blue-stain Fungi	Other staining Fungi	Surface Moulds
Basidiomycetes	Basidiomycetes (Ascomycetes)	Ascomycetes, Deuteromycetes	Ascomycetes, Deuteromycetes	Ascomycetes, Deuteromycetes	Ascomycetes, Deuteromycetes
Coniophora puteana, Serpula lacrymans, Rhodonia placenta, Gloeophyllum trabeum, G. sepiarium, G. abietinum, Antrodia vaillantii, Dacrymyces stillatus	Trametes versicolor, Donkioporia expansa, Schizophyllum commune, Phanerochaete chrysosporium, Pleurotus ostreatus	Chaetomium globosum, Phialophora spp. Monodyctis spp. Humicola grisea, Petriella setifera, Lecythophora mutabilis	Aureobasidium pullulans, Ceratocystis spp. Ophiostoma spp. Ceratocystiopsis spp. Scleroderma pithyophila	Discula spp. Arthrographis cuboid, Chloro-ciboria aeruginosa	Paecilomyces variotii, Aspergillus niger, Trichoderma spp. Bisporia spp., Penicillium spp.

can degrade all three main constituents of the wooden cell wall, *i.e.*, hemicelluloses, cellulose, and lignin, either simultaneously or successively. Wood degraded by white rot fungi looks bleached and whitish and suffers from fibrous fractures. The different appearance of the three rot types is illustrated in Figures 10.1–10.3.

A third group of rot fungi causes soft rot, which is considered a separate rot type. However, the fungi have the same enzymes as brown rot fungi and can degrade polysaccharides and lignin, but lignin is degraded at a later stage, and decay generally occurs at a slower rate.

FIGURE 10.1   Brown-rot decay.

(Courtesy of C. Brischke and M. Humar.)

FIGURE 10.2   White-rot decay.

(Courtesy of C. Brischke and M. Humar.)

FIGURE 10.3 Soft-rot decay on a wooden post in soil contact.

(Courtesy of C. Brischke and M. Humar.)

The growth of fungi and fungal wood degradation is influenced by numerous parameters such as the availability of nutrients and water, temperature, oxygen, acidity, light, and gravity (Brischke *et al.* 2006a, Schmidt 2006, Marais *et al.* 2022). These must lie within a mostly species-specific spectrum so that fungal growth, wood degradation, or reproduction can occur. If not, fungal wood degradation is inhibited, from which essential wood protection measures can be derived. Therefore, the general rule for protecting wood is to ensure that at least one essential physiological parameter is outside the required range. For instance, keeping moisture away prevents the transport of fungal enzymes that cause the wood to break down.

It is generally accepted that fungal infestation requires a moisture content above cell-wall saturation. However, various fungi can break down wood substance well below cell-wall saturation, at least if there is an external source of moisture nearby. Furthermore, the minimum moisture content for fungal growth is even lower than the threshold for fungal degradation of wood (Huckfeldt and Schmidt 2006, Meyer and Brischke 2015). The temperature requirement for fungal activity depends on moisture conditions and nutrient supply. A distinction must be made between temperatures necessary for growth and decay and lethal temperatures, which are significantly higher depending on the respective experimental conditions and lie between 55 °C and 105 °C (Huckfeldt and Schmidt 2006).

10.1.1.2 Wood-discolouring Fungi

Besides structural changes of building materials, which can significantly affect their functional performance, fungi can affect the quality of building materials by discolouration. Discolouration can occur in living plants (primary stain), harvested raw material (secondary stain), and in-service (tertiary stain) due to different biotic and abiotic causes (Bauch 1984, Koch *et al.* 2002). Wood-discolouring moulds and staining fungi live on nutrients in the living parenchyma cells (Schmidt 2006) and cause little or no damage to the cell wall structure.

Mould fungi cause superficial discolouration but can potentially provoke allergic reactions in inhabitants. Mould can occur indoors and outdoors and is usually related to humidity, ventilation, and condensation issues. It frequently occurs in the building envelope and during wood transport, for example, in closed freight containers over long distances.

Blue stain (sap stain) is a usually blue to black discolouration caused by different Ascomycetes and Deuteromycetes that can be superficial or penetrate deeply into the wood (Ballard *et al.* 1984, Wengert 1997). Besides discolouration, the hyphae of blue stain fungi can penetrate through the pit torus into axial cell elements and consequently increase the permeability of otherwise refractory lignocellulosic materials (Liese and Schmid 1961, Schmidt 2006). Substrates with heterogeneous permeability suffer from a spotted appearance due to local differences in the retention of stains and coatings.

10.1.2 BACTERIA

Compared to decay fungi and wood-destroying insects, bacteria cause less severe damage (Clausen 1996, Embacher *et al.* 2023). However, some types of bacteria have the potential to degrade wood under extreme climatic conditions, for example, at very high moisture contents (Kretschmar *et al.* 2008), low oxygen contents, or high temperatures (Nilsson 2009). Bacteria can also cause discolouration and increase the permeability due to the destruction of pit membranes (Blanchette 2000). They have the potential for detoxification of natural or artificial wood-protecting ingredients (Greaves 1971). Wood-degrading bacteria can be subdivided into erosion bacteria, tunnelling bacteria, and cavity-forming bacteria (Schmidt and Liese 1994, Daniel and Nilsson 1998, Nilsson 2009).

10.1.3 INSECTS

Wood serves as food, shelter, and breeding substrate for insects. The most common wood-destroying insects are termites (Isoptera) and beetle larvae (Coleoptera). Both groups are called wood-eating (xylophagous) insects. This means that they rely on wood or other lignocellulosic materials. Unlike decay fungi, insects are generally less moisture-sensitive but require higher temperatures. The minimum temperatures of relevant wood-boring beetles are around 12–16 °C, but they can cope with wood moisture contents well below cell-wall saturation, *i.e.*, as low as 7–10% (Unger *et al.* 2001). This means insects also threaten wood indoors, while fungi usually rely on liquid water and only attack wood exposed to the weather or otherwise moistened.

Insects transform (metamorphosis) throughout their lives. There are four stages of development (holometabolism) in beetles: egg, larva, pupa, and adult (imago). Regarding wood degradation, the larval stage is the most relevant and takes the longest, *i.e.*, over a decade. Termites, on the other hand, undergo an incomplete transformation (hemimetabolism). They omit the pupal stage, which means that larvae and adults have the same morphology. Wood-destroying beetles represent the greatest insect threat to wood in temperate latitudes. However, this comes from moisture- and warmth-loving termites in tropical and subtropical regions. They can cause enormous damage to wooden components and structures, as exemplarily shown in Figures 10.4 and 10.5.

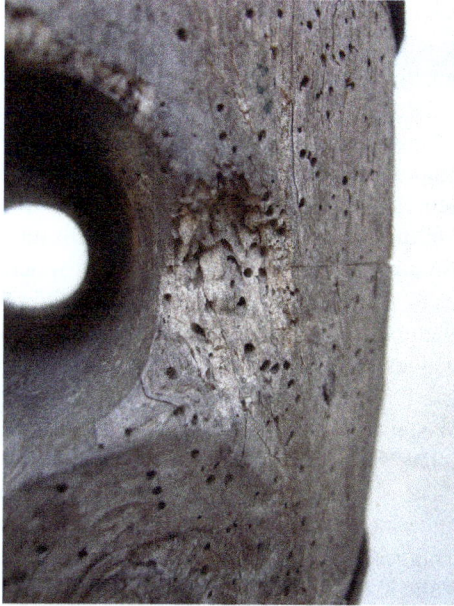

FIGURE 10.4   Bore holes of deathwatch beetle (*Xestobium rufovillosum*) in wood. (Courtesy of C. Brischke and M. Humar.)

FIGURE 10.5   Bore dust beneath wooden boards, indicating severe insect damage. (Courtesy of C. Brischke and M. Humar.)

10.1.4   MARINE BORERS

The most essential wood-destroying marine borers are found among the Bivalvia and Crustaceae. The Teredinidae (shipworms) belonging to the Bivalvia include dangerous pests of objects such as ships, groins, or jetties in salty seawater. The Teredinidae are elongated, whitish animals that depart significantly from the typical bivalve shape. Their distribution depends on the salt content of the water and its temperature. Warm climates favour their life processes, while cold temperatures slow their activity. The most important species belong to the genera *Teredo* and *Bankia*.

Shipworm is the common name of *Teredo navalis* that is distributed in coastal waters of warm and temperate zones. The shipworms are bisexual and produce 15 million eggs annually in three to four batches. After fertilisation in the womb, larvae develop within 14 days and are expelled into the ocean water. The larvae, 0.3 mm long, attach themselves to wood after 1–3 weeks and begin to scrape. Shipworms can penetrate the wood when the shell-like valves have developed into a boring utensil. In contrast to other bivalves, the shipworm uses the bore chips as a food source. The life of a shipworm lasts between one and three years. Shipworms can be found in almost all European coastal areas, more recently also in the Baltic Sea due to increasing salinity (Borges 2014, Borges *et al.* 2014a, 2014b, Appelqvist *et al.* 2015, van Niekerk *et al.* 2022). The shipworm has a worm-like, extended, whitish body and deposits calcareous material on the surface of the borehole. At the head portion, shell-like valves are arranged in a circle. The serrated edges of valves serve as boring instruments and allow the shipworm to bore into the wood rapidly. At the rear end, two snout-like hoses (siphons) protrude into the seawater, and two snout-like hoses (siphons) protrude into the seawater. Adults are about 200–450 mm long.

Wood is gnawed on the surface and studded with circular boreholes of 6–8 mm in diameter. The cross sections of severely damaged wood look like a sieve (Figure 10.6).

FIGURE 10.6   Damage in wooden piles caused by shipworm (*Teredo* spp.) attack.

(Courtesy of C. Brischke and M. Humar.)

FIGURE 10.7   "Hourglass corrosion", damage caused by gribble (*Limnoria* spp.) on wooden piling.

(Courtesy of C. Brischke and M. Humar.)

Shipworms attack both sapwood and heartwood. The excavated burrow is usually lined with a calcareous tube, which allows detection of shipworms by X-ray techniques even if no attack is visible from the outside (EN 275, CEN 1992).

The Limnoriidae (gribble) belonging to the Crustaceae live in shallow coastal waters of cold and temperate zones and can attack submerged wooden structures. They do not penetrate the wood as deep as the Teredinidae and destroy it more slowly. Their tunnels extend parallel to the wood surface, and as the outer layers are lifted off by wave action, the gribble gradually burrows more deeply into the wood. The genus *Limnoria* contains numerous species. One of the most common is *Limnoria lignorum*, often just named gribble. The wood is damaged by round, winding bore tunnels, which usually follow the early wood layers of softwood. Wooden piles being attacked by *Limnoria* in the intertidal zone present the characteristic appearance of the hourglass shape, sometimes named hourglass corrosion (Figure 10.7).

10.2   ABIOTIC DEGRADATION

10.2.1   WEATHERING

Wood exposed to outdoor environments undergoes several weathering processes influenced by biological and non-biological factors. Solar radiation stands out as a significant contributor. When wood is exposed to sunlight, it absorbs solar energy, leading to a swift change in colour. This initial alteration is merely the beginning, as prolonged

FIGURE 10.8   Weathered wooden surface with deep checks and traces of erosion.

(Courtesy of C. Brischke and M. Humar.)

exposure initiates chemical reactions, resulting in more complex modifications and eventual degradation of the wood's surface layer (Ayadi *et al.* 2003). This phenomenon of weathering is driven not only by solar radiation but also by precipitation, which together accelerate the breakdown process (Feist 1990). Within the spectrum of solar radiation, the ultraviolet (UV) portion plays a crucial role despite constituting only 5% of the sun's energy. The impact of UV light on wood is disproportionately large and extensively documented as a primary factor in wood degradation (Hon 2001). The process initiated by UV exposure leaves the wood more vulnerable to biological colonisation. As the wood's surface degrades, it becomes an optimal environment for bacteria and fungi. Staining fungi contribute to the characteristic appearance of weathered wood. The interplay between the lightened colour of photo-degraded wood and the dark hyphae of fungi creates a silvery-grey patina often seen on aged outdoor wood (Hon 2001, Kržišnik *et al.* 2018). This leads to a distinctive weathered appearance (Figure 10.8) that is as much a part of the wood's life cycle as its growth and use.

10.2.2   FORMATION OF CHECKS

Wood absorbs and desorbs water vapour from the surrounding environment as an organic material, directly affecting its density, mechanical properties, and susceptibility to fungal decay. One consequence of wood's sorptive nature is dimensional instability. Due to its anisotropic nature and the interplay of excessive moisture fluctuations, anatomical properties and drying stress, cracks, and checks are formed (Kozlov and Kisternaya 2014). Water and spores accumulate in the cracks during rainfall, forming conditions suitable for fungal decay. In addition to the macrocracks, microcracks could also be formed on the wood surface. Those cracks act like microcapillaries. This is predominately evident in thermally modified wood, where the formation of microcracks results in the hydrophilic surface of weathered wood (Keržič *et al.* 2021).

10.3 BIOLOGICAL DURABILITY

10.3.1 NATURAL DURABILITY

10.3.1.1 Definitions and Test Methods

The biological durability of wood is the "inherent resistance of a wood species or a wood-based material against wood decay organisms […]. This inherent resistance is due to natural components exhibiting different toxicity levels towards biological organisms and/or anatomical particularities or a specific constitution of certain wood-based materials", according to the EN 350 (CEN 2016a) standard. Defined in this way, it becomes evident that durability is a wood property that significantly influences the lifespan of wooden components but cannot be equated with it. In contrast, the service life is defined as the "time after installation during which a building or its parts meets or exceeds the performance requirements" ISO 15686-1 (ISO 2011). Numerous factors influence the service life of wooden components, but biological durability is only one of them. In addition, the exposure conditions of a wooden component affect its service life. Hence, climate, design, work execution level, maintenance, and use conditions affect the service life of wooden components. Wood durability is a material property that can be determined with the help of numerous test methods and against different decay organisms or groups of organisms.

A fundamental distinction is made between laboratory and field test methods, where the latter is preferred since the test scenario is usually more similar to natural exposure conditions. However, such field tests typically take several years, suffer from higher variability, and are less reproducible since the test conditions differ between climatically different locations. Further external parameters can affect the test and its results. In the laboratory, climatic and other external conditions can be adjusted during the test so that they are in the ideal range for the respective decay organisms and thus result in a particularly harsh test, which is not inevitably representative of the outdoor conditions that wood is exposed to. However, test results can be obtained within a few months instead of several years, and the resistance of wood can be tested against single species or pre-defined groups of decay organisms (*e.g.*, soft rot fungi).

The following test methods are frequently used for biological durability testing, and some are part of the instruments of European standardisation, as seen in EN 350 (CEN 2016a). Field tests can be conducted with and without soil and/or seawater contact. Commonly, the durability against decay fungi, including soft rot fungi, is tested in so-called graveyard tests, for example, EN 252 (CEN 2015a), Figure 10.9,

FIGURE 10.9 Graveyard test at the Thünen Institute of Wood Research in Hamburg, Germany: Soil contact field test to determine the durability of wood.

(Courtesy of C. Brischke and M. Humar.)

where stake-shaped specimens are buried to half of their length and decay is assessed regularly, and rated concerning depth and distribution. The rating scheme is between 0 ("sound") and 4 ("failure"). Similarly, decay is assessed with the help of different above-ground test methods such as the horizontal lap-joint method according to EN 12037 (CEN 2022), the L-joint method according to EN 330 (CEN 2015b), or the bundle test method (Brischke *et al.* 2023a, Figure 10.10). Numerous above-ground test methods have been applied and reported in the literature, but few have been standardised so far (Meyer *et al.* 2016, Meyer-Veltrup *et al.* 2017b). The durability of wood against subterranean termites can be tested either in-ground or above. Several methods can be used for the latter, such as the American ground-proximity tests according to AWPA E26 (AWPA 2021). In-ground testing is usually done with the help of graveyard tests such as EN 252 (CEN 2015a). In comparison to the fungal tests, there is an additional requirement. This is the presence of termites, which is limiting such tests to subtropical and tropical regions.

The durability of wood against marine borers can be tested during exposure to seawater according to the EN 275 standard (CEN 1992). For this purpose, wood specimens are exposed on racks in the sea and evaluated yearly for infestation by marine borers such as shipworms and gribble. X-ray scans are taken from each specimen after each exposure interval to detect shipworm attacks, which are almost invisible from the exterior of the specimens (Figure 10.11). Again, a five-step scheme is

FIGURE 10.10 Bundle test at the Thünen Institute of Wood Research in Hamburg, Germany: An above-ground field test to determine wood's durability.

(Courtesy of C. Brischke and M. Humar.)

FIGURE 10.11 Control wood specimen made from Scots pine sapwood after one year of exposure in a seawater test according to EN 275 (CEN 1992) in Hejlsminde, Denmark. Barely affected outer surface (top sample) and bottom sample with severely degraded interior section.

(Courtesy of C. Brischke and M. Humar.)

used to rate the intensity of the attack. The rating refers to the percentage surface area of the specimens infested by marine borers. Laboratory methods with larvae of *Limnoria quadripunctata* are available but are used for screening purposes only (Borges *et al.* 2009).

Solely against the larvae of wood-destroying beetles and wasps, no field test methods have been established yet since these insects are more challenging to control, and infection of wood through them is a complex process that can hardly be reproduced under field conditions (Figure 10.12). So-called semi-field tests were probed by Plarre (2012), who exposed wood samples in an arena where beetle chose different test materials.

In laboratory, the durability against decay fungi is either tested with basidiomycete monocultures or against a set of wood-destroying fungi, for example, by applying a suspension of spores from different soft rot fungi according to the ENV 807 (CEN 2001) standard or by exposure in unsterile soil that is containing natural topsoil or a compost substrate according to CEN/TS 15083-2 (CEN 2005). The latter has the advantage that a greater variety of organisms are present, which may also interact synergistically. These can also include so-called non-target organisms, which do not break down the wood but prepare and thus promote the breakdown by other organisms. Similar to field tests, the results of such "terrestrial micro-coms (TMC)" tests are affected by the natural variability of soil-inhabiting organisms and their

FIGURE 10.12 Ground proximity tests against subterranean termites in Soulac-sur-Mer, France.

(Courtesy of C. Brischke and M. Humar.)

activity. Solely, the microclimatic conditions are defined and usually set to an optimum for fungal decay. Tests with individual fungal species are carried out in incubation jars with a medium containing both water and additional nutrients required for fungal growth and wood decay (Figure 10.13). These can be either malt agar (*e.g.*, agar plate tests according to EN 113-2) (CEN 2021) or sterile soil (soil-block tests according to AWPA E10 (AWPA 2022). Mass loss is the most common measure for durability classification in laboratory tests; sometimes, the loss in modulus of elasticity (MOE) is used alternatively. In contrast, depth and distribution of decay, as indicated through the softening of the wood substance, is the measure of choice in field tests. The latter is used to assign decay ratings to the test specimens. As soon as all test specimens have failed (*i.e.*, reached the highest decay rating), the average lifetime of the specimens can be calculated and used for durability classification. Different measures from selected laboratory and field tests and how they can be assigned to durability classes are summarised in Table 10.2.

The durability of wood and wood-based-materials to attack by *Hylotrupes bajulus*, *Anobium punctatum*, and *Lyctus brunneus* can be tested using procedures based on those in EN 46-1 (CEN 2016b), EN 49-1 (CEN 2016c), and EN 20-1 (CEN 2024a), respectively. These tests are based on exposure of beetle larvae into wood blocks for a pre-defined period. According to EN 350 (CEN 2016a), a wood species or a wood-based material is classified as "not durable" if one or more live insects of the respective test organism are found at the end of the test. If no live insects are found at the end of the respective test, and if the validity criteria for the respective test in the reference species are fulfilled, the test species is classified as "durable". Similarly, the durability of wood against subterranean termites is tested under laboratory conditions according to EN 117 (CEN 2024b), but attack ratings on a scale of 0–4 are given and used for durability classification (Table 10.2).

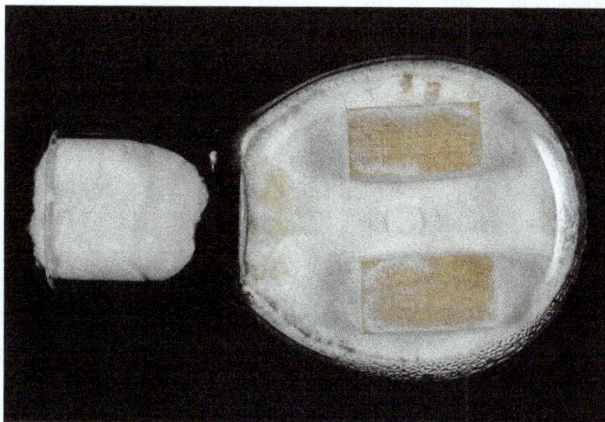

FIGURE 10.13 Laboratory method for determining the biological durability of wood. Specimens incubated in Kolle flasks after inoculation with a white rot fungus (*Trametes versicolor*) on malt agar.

(Courtesy of C. Brischke and M. Humar.)

TABLE 10.2

Different Measures from Laboratory and Field Test Methods and Corresponding Durability Classes (DC) of Wood to Fungal Attack According to EN 350 (CEN 2016a)

DC	Description	Percentage Mass Loss (ML) Based on CEN/ TS 15083-1[a]	x-value[b] Based on CEN/ TS 15083-2 (CEN 2005)	x-value[c] Based on Field Tests According to EN 252 (CEN 2015a)
1	Very durable	ML ≤ 5	x ≤ 0.10	x > 5.0
2	Durable	5 < ML ≤ 10	0.10 < x ≤ 0.20	3.0 < x ≤ 5.0
3	Moderately durable	10 < ML ≤ 15	0.20 < x ≤ 0.45	2.0 < x ≤ 3.0
4	Less durable	15 < ML ≤ 30	0.45 < x ≤ 0.80	1.2 < x ≤ 2.0
5	Not durable	30 < ML	x > 0.80	x ≤ 1.2

[a] CEN/TS 15083-1 has been replaced by EN 113-2 (CEN 2021).

[b] Hardwoods: x = Median mass loss test specimens/median mass loss reference specimens. Softwoods: x-value of test material based on loss of modulus of elasticity (MOE).

[c] x-value = average life of wooden stakes used in-ground expressed relative to the life of the reference stakes

10.3.1.2 Causes of Natural Durability

During heartwood formation, so-called extractive substances are impregnated in the wooden cell walls, increasing their resistance to insects, fungi, and other microorganisms (Taylor *et al.* 2002). In contrast, the extractive content of sapwood and less durable wood species is relatively low. Extractive substances can have a toxic effect or only influence the moisture dynamics of wood (*e.g.*, Stirling and Morris 2006). Both aspects may interact in very durable wood species and thus have a strengthening effect.

It is known that the formation of special anatomical features and the content of extractive substances vary within a stem. As a rule, the juvenile wood has lower extractive contents, while the heartwood closest to the sapwood has very high contents (Hillis 1987, Taylor *et al.* 2002). On the other hand, some woods have a so-called transition wood, which is characterised by less durability than the remaining heartwood. This is known for several tropical wood species (Ali *et al.* 2011, Medeiros Neto *et al.* 2020).

Meyer-Veltrup *et al.* (2017a) clearly distinguished between "inherent protective properties" and the "wetting ability" of wood to predict the outdoor performance of timber in above-ground situations. They also pointed out the need to consider further factors such as the susceptibility to checking, the leachability of ingredients, and the overall ageing of wood.

10.3.1.3 Durability Classes

Biological durability is a relative and, thus, unitless material property and can be based on different measures from different tests. For the sake of simplicity and partly for harmonising the interpretation of test results, durability classes (DCs) were defined, but differently for the different (groups of) decay organisms. While five DCs are distinguished for fungal attacks (Table 10.2), only two and three DCs were defined for attack by termites, beetle larvae, and marine borers, respectively (Table 10.3).

Due to the natural variation of wood ingredients and anatomical differences along the stem axis and the stem radius, natural biological durability is also subject to natural variation. Durability differences can be seen not only between and within species, but also between and within trees (Rudman and Gay 1963, Amusant *et al.* 2004, Brischke and Alfredsen 2023, Brischke et al. 2023b). To consider this natural variation, a sufficient number of replicate specimens should be tested, and a span of DCs rather than one single DC can be assigned to a wood species according to EN 350 (CEN 2016a). In addition, the indicator "v" (= variable) indicates that a species exhibits an unusually high level of variability. The biological durability of selected wood species against different decay organisms is summarised in Table 10.4. Against fungi, none of the European-grown species is considered "very durable" (DC 1); the most durable European-grown species is the black locust, which is assigned to DC 1-2 according to EN 350 (CEN 2016a) as long as its juvenile heartwood is excluded (Brischke *et al.* 2023b, Brischke *et al.* 2024). Only very few wood species are durable against termite and marine borer attacks, among them mainly tropical species, which are highly dense and rich in extractives (Table 10.4), such as Ipé, maçaranduba, and makoré (*Tieghemella heckelii*). Hardwoods are generally not attacked by *Hylotrupes bajulus*, and also most conifers with coloured heartwood are durable against this species, such as Atlas cedar or European larch.

TABLE 10.3

Different Measures from Laboratory and Field Test Methods and Corresponding Durability Classes (DC) of Wood to Insect and Marine Borer Attack According to EN 350 (2016)

DC	Description	Rating based on EN 117 (CEN 2024b)	Number of live insects at the end of test according to EN 20-1 (CEN 2024a), EN 47-2 (CEN 2016d), or EN 49-2 (CEN 2015d)	Results of EN 275 (CEN 1992) tests expressed as x-values
D	Durable	≥ 90% "0 or 1" and max 10% "2"[a]	none	x > 5.0
M	Moderately durable	< 50% "3, 4"	n.a.	3 < x ≤ 5
S	Not durable	≥ 50% "3, 4"	one or more	x ≤ 3

[a] 90% of the test samples rated 0 or 1, and a maximum of 10% of the test samples rated 2 and 0% "3 and 4".

TABLE 10.4

Durability Classes (DC) of Selected Wood Species to Different (Groups of) Wood-destroying Organisms According to EN 350 (CEN 2016a)

Wood Species	Scientific Name	Origin	Durability of Heartwood Against				
			Fungi	Hylotrupes	Anobium	Termites	Marine borers
Silver fir	Abies alba	Europe	4 (4)	S	S	S	S
Atlas cedar	Cedrus atlantica	Africa	1–2	D	D	M	S
European larch	Larix decidua	Europe	3–4 (3–4)	D	D	S	S
Norway spruce	Picea abies	Europe	4 (4–5)	S	S	S	S
Scots pine	Pinus sylvestris	Europe	3–4 (2–5)	D	D	S	S
Douglas fir	Pseudotsuga menziesii	North America	3	D	D	S	S
		Europe	3–4 (3–5)				
Western red cedar	Thuja plicata	North America	2	D	D	S	S
		UK	3 (1)				
Norway maple	Acer pseudoplatanus	Europe	5	D	D	S	S
Silver birch	Betula pendula	Europe	5	D	D	S	S
Sweet chestnut	Castanea sativa	Europe	2 (1)	D	D	M	S
Sipo	Entandrophragma utile	Africa	2–3	D	D	M	M
European beech	Fagus sylvatica	Europe	5 (4–5)	D	S	S	S

Ipé	Handroanthus spp.	South America	1	D	D	D
European walnut	Juglans regia	Europe	3	D	S	n.a.
Bongossi	Lophira alata	Africa	2v (1–2)	D	D	M-D
Maçaranduba	Manilkara spp.	South America	1	D	D	D
English oak	Quercus robur	Europe	2–4 (1–2)	D	M	n.a.
Black locust	Robinia pseudoacacia	North America, Europe	1–2 (1–2)	D	D	S
Dark red meranti	Shorea spp. subgen. Rubroshorea p.p.	Asia	2–4	D	M	S
Teak	Tectona grandis	Asia	1–3 (1)	D	M	M-D
Makoré	Tieghemella heckelii	Africa	1	D	D	D

Note: For fungi, two durability classifications are listed, noted as follows: X (Y). The first one is usually derived from the results of laboratory or field tests simulating in-ground situations. The second one is based on the results of laboratory tests aiming to determine the durability against basidiomycete wood-decay fungi.

In addition to EN 350 (CEN 2016a), several lists on natural durability have been published over the last decades, for example, Scheffer and Morrell (1998), the Australian standard AS 5604 (AS 2005), and an overview of tropical species in the web-tool Tropix of CIRAD (Paradis *et al.* 2011). When interpreting any durability class, attention must be paid to which test methods on which the classification is based. Caution is essential between European and non-European durability attributes because most of them are not based on comparable principles. The European standard EN 460 (CEN 2023) provides guidance on the performance of wood under different use conditions. For this purpose, durability classes are compared to the use classes and minimum requirements for durability are defined. If they are adhered to, it is assumed that no wood preservation is necessary to achieve an acceptable service life for the respective component (Section 10.4.3).

10.3.2   WOOD PRESERVATION

Wood in outdoor conditions is exposed to various degradation factors. Most European wood species do not have durable wood, so we must protect wood to slow down degradation processes. Although the first biocides were used in the antique civilisations, modern wood protection started with the industrialisation of Europe. In the 18th and 19th century, there was a huge demand for wood. In this period, the forestation of Europe was at its minimum (Kaplan *et al.* 2009). The use of durable wood was limited to military purposes; therefore, it was essential to develop reliable wood preservatives for infrastructural purposes such as telecommunication poles and railway ties. In parallel, the development in the steel industry enabled the construction of impregnation chambers that can withstand elevated pressure and temperatures.

10.3.2.1   Wood Preservatives

The first industrial wood preservative was creosote oil, which, until recently, was used to impregnate wood. Creosote is a tar-based wood preservative derived from the distillation of coal tar or wood tar. It has been widely used to protect wood where high reliability is required, such as railroad ties (Figure 10.14), utility poles, and marine pilings. Creosote's effectiveness as a wood preservative is due to its high toxicity to many organisms. However, it is also highly toxic to humans and the environment, leading to restrictions and regulations on its use in many European countries.

In this chapter, we tend to focus primarily on wood preservatives that the European Union approves. A good overview of historic wood preservatives can be found in the past literature (Unger *et al.* 2001, Reinprecht 2016). Copper-based wood preservatives are the most essential active ingredients for the impregnation of wood in exposed applications in Europe, *i.e.*, with and without soil contact due to Use Classes 3 and 4 according to EN 335 (CEN 2013); see also Table 10.5. The industrial use of copper compounds for wood preservation began with patenting the Boucherie process, which involves treating freshly cut logs with an aqueous copper(II) sulphate solution. Wood treated this way was not durable for outdoor use because the copper, which provided protection, was quickly leached out. When such issues related to the leaching of copper compounds from wood were resolved in the early 20th century, their use increased significantly. Copper-based preservatives protect wood from fungi and

FIGURE 10.14   Railway on creosote-treated sleepers.

(Courtesy of C. Brischke and M. Humar.)

algae and act as an anti-fouling agent in marine applications. Annually, more than 100,000 tons of copper compounds are used for wood preservation (Preston 2000), and this amount is increasing. The reasons for this include:

- Copper compounds are effective against fungi, bacteria, and algae at relatively low concentrations and do not affect higher plants. At low concentrations, copper is even essential for their growth and development.
- Copper-based preservatives are relatively inexpensive and safer than other protective compounds.
- The ban or stricter control of some traditional organic wood biocides due to their toxicity or environmental unsuitability (pentachlorophenol, dichlorodiphenyltrichloroethane DDT, Lindane, creosote oil, organotin compounds).
- Rapid development of third-world countries and increased demand for preserved wood (Richardson 1997, Humar 2017).

However, wood treated with copper compounds is increasingly threatened by copper-tolerant strains of fungi (Humar *et al.* 2002). Since the first commercial use of copper compounds for wood impregnation using the Boucherie process in 1838, many copper-containing compounds have been developed. One of the significant compounds was ACZOL, developed in 1907. It is a solution of phenol, copper, zinc, and ammonia. Once ammonia evaporated from the wood, insoluble copper and poorly soluble zinc complexes remained in the wood. This compound was used for wood protection for over 30 years (Richardson 1997).

A breakthrough in developing wood preservatives was Bruning's discovery in 1913, which found that chromium compounds significantly improved the fixation of active components and reduced material corrosion during the treatment of preserved

TABLE 10.5
Use Classes According to EN 335 (CEN 2013)

Use Class	General Service Conditions[a]	Occurring Organisms[b,c]				
		Wood Disfiguring Fungi	Wood-destroying Fungi	Beetles	Termites	Marine Organisms
1	Interior, dry	-	-	U	L	-
2	Interior or under roof, not exposed to weather, possibility of condensation	U	U	U	L	-
3	Exterior, without soil contact, exposed to weather. If class-divided: **3.1** limited moist conditions **3.2** persistently moist conditions	U	U	U	L	-
4	Exterior, in contact with soil or freshwater	U	U	U	L	-
5	Permanently or regularly immersed in salt water	U[d]	U[d]	U[d]	L[d]	U

U = is spread all over Europe and in the area of the European Union L = occurs locally all over Europe and in the area of the European Union

[a] There are borderline and extreme cases of using wood and wood products. These can result in allocating a use class that differs from this standard's definition.

[b] Protection against all listed organisms is not absolutely required because they do not occur under all use conditions and at all geographical locations, they are not economically significant, or they are not able to infest specific wood products due to the particular state of the product.

[c] See Annex C in EN 335 (CEN 2013)

[d] The area above the water surface of specific wooden components can be susceptible to all stated organisms.

wood. This discovery enabled a significant commercial expansion of preservatives. The first commercial preparation based on copper sulphate and sodium dichromate was patented by Gilbert Gunn of the Scottish company Celcure in 1926. It performed well in the north, but when tested in the English colonies, it was found that the protected wood was not resistant to termites and tolerant isolates of wood fungi (Eaton and Hale 1993). These problems were largely resolved when the Indian government researcher Sonti Kamesam discovered that chromium fixates copper and arsenic compounds. Aqueous copper sulphate, sodium dichromate, and arsenic pentoxide solution were named ASCU after its components. The American Wood Protection Association (AWPA) later (1953) named this mixture CCA after its main components. At its peak in 1998, the production of CCA compounds amounted to 100,000 tons. In many European countries, arsenic was later replaced with boron compounds. Chromium-based wood preservatives were used until 2006, when most chromium-based compounds were withdrawn from the market. In newer formulations, chromium compounds were successfully replaced with amines, primarily ethanolamine. Boron and quaternary ammonium compounds were added in the first generations to improve insecticidal properties, later replaced by azoles (Preston 2000, Humar *et al.* 2018).

In recent years, the USA have also seen the rise of copper compounds based on copper nanoparticles (micronised copper). A suspension of copper nanoparticles (microparticles) is used for wood impregnation. Nanoparticles are insoluble in water and thus do not leach out of wood after impregnation. Secondary biocides such as quaternary ammonium compounds and azoles are added to these preparations to enhance their performance. This wood preservative has not yet been approved by the EU. This combination is cheaper than copper-ethanolamine wood preservatives, and wood almost retains its original colour. Unfortunately, this combination is available to impregnate permeable wood species such as pine sapwood and beech (Shukla and Kamdem 2023). Besides water-soluble copper-based wood preservatives, oil-based solutions were developed as well. Oil-based copper systems are gaining importance as a possible substitute for creosote for the impregnation of railway ties. The historical oil-based system is copper naphthenate. Copper naphthenate is a copper-based wood preservative derived from the reaction of copper salts with naphthenic acid, a component of petroleum. It is widely used in North America to protect wood from decay, insects, and fungal growth (Lebow 1996). In European oil-based wood preservatives, copper is primarily dissolved in synthetic or tall oil. These wood preservatives are used alone or as a hydrophobic treatment in a dual-step process.

Besides copper, boron is the second most important inorganic active ingredient for wood protection. Borate compounds are among the oldest active ingredients still used for wood protection. Their use is permitted even after introducing the Biocidal Products Regulation BPR 528/2012 (EU 2012). At its peak, over 90% of water-based wood preservatives contain borate compounds in Germany (Peylo and Willeitner 1999). In addition to good diffusivity, which allows for effective protection of refractory wood species, borate compounds have a broad spectrum of activity against insects and fungi. Furthermore, one of the significant properties of borates is their low toxicity to humans. However, the downside of good diffusivity is that borate compounds leach out of the wood, limiting their use to dry conditions or allowing only occasional moisture increases. Borate compounds act as effective fungicides and insecticides even at low concentrations. According to current knowledge, no wood-decaying fungus is tolerant to boron and cannot degrade wood treated with borate preparations; therefore, boron is an essential additive in copper-based wood preservatives (Caldeira 2010, Lesar *et al.* 2010). However, using borate compounds as wood preservatives is limited due to poor fixation of boron in the wood and its natural solubility in water, resulting in intense leaching of boron from the wood. Therefore, borate compounds are used alone only for Use Classes 1 and 2, with a limited risk of leaching. However, the importance of boron in wood protection will likely decline in the future, as boron has been listed as a substance of high concern (ECHA 2020). However, boron is the only active ingredient suitable for remedially treating degraded mass timber.

Besides inorganic wood preservatives, organic ones are used as well. Except for creosote, organic wood preservatives are used as co-biocides in copper-based solutions (quaternary ammonium compounds, azoles) or for wood protection in less exposed applications (Use Classes 1, 2 and 3.1, see Table 10.5), such as window frames, cladding, and garden furniture. The most important organic active ingredients are: triazoles, carbamates, pyrethroids, quaternary ammonium compounds, and juvenile hormones.

Triazoles are excellent and well-established fungicides used for almost 30 years to protect windows and exterior doors. These biocides have successfully replaced the banned pentachlorophenol. Propiconazole and tebuconazole are the most commonly used fungicides for wood protection. Both active ingredients are stable, and a very positive characteristic is that they do not leach from the wood. They are commercially available in formulations that include IPBC (3-iodo-2-propynyl butyl carbamate) and permethrin. Triazoles are under significant pressure, as there have been attempts to ban them due to environmental concerns. Since the industry has no alternative, their use has been extended until 2028.

Carbamates' most crucial active substance is IPBC, which has been used since 1984 to effectively prevent the development of fungi and mould on protected wood. Today, it is mainly used for protecting claddings, window frames, and garden furniture. It is also added to paints and varnishes for outdoor use. IPBC is one of the most environmentally friendly organic fungicides used for wood protection, and it will likely stay in use in the decades to come (Reinprecht 2016).

The fungicidal activity of quaternary ammonium compounds (QACs) has been known since 1965, but their use has not become widespread due to cheaper and more effective inorganic protective agents. There are more than 20 QACs on the market. QACs are divided into two groups: the first includes primary, secondary, and tertiary amines, and the second includes quaternary ammonium compounds. Due to the ban on using chromium salts for wood protection, QACs have started to be added to water-soluble copper preparations. Quaternary ammonium compounds are now widely used for wood protection because they have a broad spectrum of activity (fungicidal, bactericidal, termiticidal, algicidal), can be combined with inorganic active components, and have low toxicity to mammals. They are used alone only to protect less exposed wood or for protecting wood during transport, as these compounds are highly susceptible to bacterial degradation (Unger *et al.* 2001).

Pyrethroids are synthetic analogues of pyrethrins, which are accumulated in the flower heads of the plant Dalmatian chrysanthemum (*Tanacetum cinerariifolium*). Natural pyrethrins are a mixture of six esters of chrysanthemum and pyrethric acid. Both natural pyrethrins and synthetic pyrethroids are very effective insecticides for a wide range of insects. In addition to wood protection, they are also used in agriculture, horticulture, and veterinary medicine. Natural pyrethrins are less toxic to mammals but, unfortunately, less stable. Their extraction is significantly more expensive than the synthesis of pyrethroids. Cypermethrin is the most used pyrethroid, often combined with azole-based preparations and IPBC. Juvenile hormones are emerging insecticides that are slowly replacing pyrethroids in some applications. Juvenile hormones (JHs) are a group of hormones in insects that play crucial roles in regulating development, reproduction, and other physiological functions. These compounds exploit the critical role that juvenile hormones play in insect development and reproduction to disrupt the life cycle of pest insects. Juvenile hormone analogues are valued for their selective action against insects, minimal environmental persistence, and reduced risk of developing resistance compared to traditional insecticides. This makes them an essential tool in sustainable pest management practices (Cornette *et al.* 2008).

10.3.2.2 Treatment Processes

Several techniques exist for applying wood preservatives. The impregnation process aims to deliver the required amount (retention) of active ingredients to the specified depth (penetration). The technical specifications provided by the final users, with the help of certification agencies like the Nordic Wood Preservation Council (NWPC 2023), consider these parameters.

Usually, superficial treatments are sufficient for sheltered applications without direct weathering. Brushing is the preferred technology for artisanal non-industrial processes and do-it-yourself processes. Brushing is a straightforward procedure that does not require much investment. The penetration of the wood preservatives to brushed timber is typically below 1 mm, while the uptake of preservatives is between 100 g/m^2 and 200 g/m^2. On the other hand, short-term immersion and spraying provide similar results as brushing. However, more infrastructure is required for immersion processes. Short-term immersion is typically used to protect the wood during transport and prevent the growth of moulding and staining fungi. In case of prolonged immersion, penetration up to 3 mm can be achieved in Norway spruce (Humar and Lesar 2009). Spraying for the superficial treatment of wood protects roof timbers during renovation. The main drawback of spraying is associated with environmental hazards since a considerable part of the biocides could be lost during the application. In the last decades, a vacuum chamber has been developed, where spraying is performed. There are no losses of biocides, and penetration and retention are almost in the range of vacuum-pressure processes. The main advantage of this process is that the technical equipment is cheaper, and the volume of biocides required is lower.

Vacuum pressure treatment is the only technology that enables the treaters to meet requirements for fully weathered wood. These treatments enable full sapwood penetration with water-based and solvent-based preservatives. Several processes can be utilised in respective chambers, namely, the Bethell full-cell process, the Rüping empty cell process, the Lowry process, the vacuum process, the Boulton process, and the oscillation process. The Bethell process was developed first and is still used today to treat wood with water-based wood preservatives. The process is also called the full-cell method, in which the purpose is to fill the wood cells with the impregnation agent to receive the highest possible amount of impregnant in the wood. In European beech and Scots pine sapwood, preservative uptakes of up to 600 kg/m^3 can be achieved. The process is based on placing the wood in a sealed cylinder and applying a vacuum to remove air and moisture. A preservative solution is introduced, and high pressure is applied to force the preservative deep into the wood cell lumina. After pressure release, the excess preservative is drained, and a final vacuum removes any surface residue. The process ensures that the preservative penetrates deeply into the wood, providing long-lasting protection. The respective impregnation regime can be used with various types of wood and preservatives, making it suitable for different applications and environments (Eaton and Hale 1993).

Rüping treatment, also known as the empty-cell process, is a wood preservation method designed to protect wood from decay and insect damage using fewer preservatives than the full-cell (Bethell) process. It was designed to treat wood with creosote or similar oil-based wood preservatives. The respective process is used to treat wood with some modification agents. The typical retention of creosote is between 120 kg/m^3 and 140 kg/m^3. This process results in lighter-treated wood that is less saturated with preservatives yet still protected from decay and insects. During Rüping treatment, wood is placed in a treatment cylinder, and air pressure is applied to compress the air within the wood cells. The preservative solution is then introduced while maintaining air pressure, allowing the air in the wood cells to act as a buffer. When the pressure is released, the compressed air in the wood cells expands, forcing out some preservatives and leaving it mainly on the cell walls. A final vacuum may be applied to remove excess preservatives from the surface. One of the drawbacks of the respective process is that more expensive equipment is utilised. Thus, the process is less frequently used than the Bethell process. The Lowrey process is a simplified version of the Rüping treatment that can be utilised in the classical vacuum-pressure cylinder used for Bethell treatment.

Some attempts have been made to develop alternative treatments for refractory wood species, such as Norway spruce. The oscillatory process is a wood preservation method that involves alternating cycles of applying and releasing pressure to improve preservative penetration. This alternating pressure cycle helps the preservative penetrate by exploiting the expansion and contraction of air within the cells (Augustina et al. 2023).

Due to the pressure on the volatile organic solvents in wood processing, an alternative technology was developed that enabled wood penetration with organic active ingredients without organic solvents. Supercritical impregnation of wood is an advanced preservation method utilising supercritical fluids, typically supercritical carbon dioxide (scCO$_2$), to deliver preservatives through the wood cross-section. One of the advantages of this technique is that it can also be applied to glue-laminated timber. This process utilises scCO$_2$, which exhibits liquid-like density and gas-like viscosity and is used to dissolve and transport active ingredients (tebuconazole, propiconazole, silanes, siloxanes…) through the cross-section of wood. The supercritical fluid penetrates the wood efficiently due to its high diffusivity and low surface tension, ensuring thorough and uniform distribution of the active ingredients. Once the desired level of impregnation is achieved, the pressure is gradually released, causing the scCO$_2$ to revert to a gaseous state and leave the wood while the preservative remains. This method is environmentally friendly and effective (Kjellow and Henriksen 2009).

In the past, predominately after WWII, when much infrastructure was destroyed, alternative techniques for wood impregnation were developed. One of these treatments is diffusion treatment. Diffusion treatment of wood involves applying a preservative solution to the surface of wood, allowing the chemicals to move inwards through the wood's moisture content over time. Therefore, wood moisture content needs to be above cell-wall saturation. The preservative diffuses gradually, driven by the concentration gradient, penetrating deeper as it moves along with the water in wood. This process can take several weeks or months, depending on the moisture content and environmental conditions such as temperature and humidity. Nowadays, this process is predominately helpful for the treatment of refractory wood species that are difficult to impregnate with pressure-based methods. Diffusion treatment is often used for in situ wood preservation of mass timber that has been infested by wood decay or insects. Boron-glycol treatments are the most crucial wood preservatives for remedial treatment (Gezer *et al.* 1999).

Another historical treatment that was used for the treatment of wood is the Boucherie process. This wood preservation method is used to treat freshly felled green wood by injecting preservative solutions through the sapwood. The process involves attaching a hose to one end of the log and using hydraulic pressure to force the preservative into the permeable, water-saturated sapwood. As the preservative moves through the wood, it displaces the sap, ensuring even distribution of the active ingredients throughout the sapwood. This method efficiently treats long poles, exploiting the natural capillary action of conductive tissues. Due to environmental concerns, the Boucherie process is rarely used (Kervina-Hamović 1990).

10.3.2.3   Requirements and Regulations

Wood preservation in the European Union (EU) is subject to stringent regulations. Initially, wood protection was dominated mainly by chemicals such as chromated copper arsenate (CCA), pentachlorophenol (PCP), and creosote. For instance, CCA has been widely utilised in Europe since the early 1950s. However, during the 1990s, the use of these prevalent wood preservatives (CCA, creosote, PCP) was restricted or even banned in most EU member states (Willeitner 2001). A significant development was the introduction of the Biocidal Products Directive (BPD 98/08/EC), which reduced the number of permissible activities in the wood preservation market from 81 to 39. This directive was later replaced by the Biocidal Product Regulation BPR 528/2012 (EU 2012). The BPR governs the placement of biocidal products on the market and their use within the European Union.

Its primary principle is to ensure that biocidal products, which include substances or mixtures intended to destroy, deter, render harmless, or exert a controlling effect on harmful organisms, are safe and effective for their intended use. The regulation requires thorough assessment and approval of active substances used in biocidal products before they can be marketed. All products approved by the European Chemical Agency are approved in all EU member countries. Thus, all biocidal products must be evaluated for their efficacy and safety for humans, animals, and the environment, and they must comply with specified labelling and packaging requirements. The BPR aims to harmonise biocidal product regulations across the EU while promoting the use of safer alternatives and minimising risks associated with biocidal use.

Due to the high costs associated with registering new active substances (biocides), approximately 8 million € per new active ingredient (Jones and Brischke 2017), there has been minimal development in biocidal wood protection over the past decade. Thus, only one active ingredient has been introduced after the full implementation of the EU directive. Among the respective active ingredients, 13 meet exclusion or substitution criteria (*e.g.*, fenpropimorph, propiconazole, tebuconazole), six of them are pure insecticides (*e.g.*, permethrin, cypermethrin), seven of them are suitable for niche applications only (*e.g.*, potassium sorbate, dazomet, dichlofluanid), so, the number of active ingredients is expected to decline in the future years. Currently, only copper compounds (five biocides), quaternary ammonium compounds (five biocides), IPBC, and DCOIT are foreseen to remain on the market with minimal limitations. This will present many challenges to the wood preservation industry. Therefore, there is a need to develop new, non-biocidal treatments for wood, such as wood modification.

10.3.2.4 Health and Environmental Issues

Paracelsus stated, "All things are poison, and nothing is without poison; only the dose makes a thing, not a poison!" This quote is also valid for wood preservatives. Nowadays, the current wood preservatives are much safer than the preservatives that were utilised decades ago. All active ingredients must withstand severe chemical and biological testing before approval. Wood preservatives threaten human health and the environment in the production, impregnation, usage, and disposal phases. The design of new products must consider all these aspects when commercialising. In recent years, particular emphasis has been given to the management of waste-treated wood. Wood that was treated with biocides decades ago is entering waste streams today. The idea behind state-of-the-art management principles is to remove wood treated with biocides from waste streams. Therefore, waste-treated wood should be incinerated in approved industrial facilities and should not be reused or recycled to limit possible emissions.

10.3.2.5 Wood Modification and Water Repellents

Although the first ideas of wood modification were developed more than 100 years ago, the major processes were commercialised in the 21st century. Key developments and commercialisation started in Europe and spread throughout the world. Key reasons for wood modification are increased environmental awareness, negative attitude towards biocides, protection of tropical forests, negative image of synthetic polymers, and lack of durable wood species in Europe. Wood modification refers to various treatments and processes applied to wood to enhance its relevant properties (durability, service life, sorption properties, mechanical properties) and performance. Critical techniques for wood modification include thermal modification, chemical modification (acetylation), and impregnation with resins or other substances (furfuryl alcohol, 1,3-dimethylol-4,5-dihydroxyethyleneurea DMDHEU). These modifications extend the usability of wood in different environments and applications, making it a more versatile and sustainable material. The key idea of the modification is to transform relatively cheap, fast-growing wood with low durability into high-performing material. Various poplar species, pine species, and European beech are being utilised. Radiata pine is of great interest, as it is easily available in sufficient quantities and has a defined quality.

Although more than 1000 processes are proposed in the scientific literature, only a few are commercialised. Based on the data provided by Jones *et al.* (2019), almost 0.8 million m^3 of modified wood is produced annually, namely thermally modified wood (600,000 m^3), acetylated wood (100,000 m^3), furfurylated wood (43,000 m^3), and other treatments (33,000 m^3). The importance of modified wood in Europe is increasing. At least one commercial plant is reported in 31 of 44 European states. However, the capacities of modification facilities are still significantly smaller compared to the impregnation plants and other wood processing facilities, such as sawmills.

Thermal modification is the most essential process for wood modification in Europe. Thermal modification aims to treat the wood at elevated temperatures (160–240 °C) in semi-anoxic conditions to alter its chemical and physical properties. The primary goal is to enhance wood's biological durability and dimensional stability without the use of harmful chemicals. During the process, the wood undergoes structural changes, reducing its hygroscopicity (ability to absorb water) and increasing its dimensional stability. Increased modification temperature improves durability and sorption properties while decreasing the mechanical properties, predominately the impact bending. This makes thermally modified wood less prone to warping, swelling, and shrinking. Two key processes are operating in Europe: open processes in saturated steam (*e.g.*, ThermoWood® process) and closed processes in elevated pressure (*e.g.*, VTT 2 process). Thermally modified wood is predominately used for cladding, decking, windows, sauna equipment, and garden furniture (Figure 10.15). Thermal modification results in reduced density and reduced equilibrium moisture content, which positively influences heat conductivity. Using this wood, window producers can meet passive house standards, which makes these windows affordable for the customers. Due to reduced mechanical properties, thermally modified wood cannot be used for load-bearing and construction applications. Nowadays, there are quite some researchers combining thermal modifications with other treatments such as waxes, bicine, tricine, and oils to obtain synergistic effects between various modification treatments.

The second most important modification treatment is acetylation. Acetylation of wood is a chemical modification process that enhances its durability. It involves treating the wood with acetic anhydride, which reacts with the hydroxyl groups in the wood's cell walls. This reaction replaces the hydroxyl groups with acetyl groups, reducing the ability of wood to absorb water and increasing its dimensional stability. Due to the reduced moisture content, acetylated wood becomes more resistant to decay, insects, and fungal attacks. Acetylated wood has been proven to perform better in marine applications as well. Acetylation significantly extends the service life of wood, while this process does not reduce mechanical properties. Acetylated wood was

FIGURE 10.15   Façade made from thermally modified timber.

(Courtesy of C. Brischke and M. Humar.)

used to construct bridges, cladding, decking, jetties, garden furniture, and window frames (Figure 10.16). Acetylated wood has been commercially produced in Europe and the USA. In Europe, it is commercialised under the name Accoya. Residues from the production of modified wood are utilised to produce high-performance medium-density fibreboard (MDF) that is commercialised under the name Tricoya®.

The third critical commercial process is the modification of wood with furfural alcohol. Two furfurylation processes are being implemented in Europe: Kebony and Noble Wood. Furfurylation of timber is an impregnation-based modification process that enhances the wood's durability, mechanical properties, and sorption properties. It is based on impregnating the wood with furfuryl alcohol, a resin derived from agricultural byproducts like corncobs, wood chemical processing, and sugarcane bagasse. During the treatment, the furfuryl alcohol polymerises within the wood's cell structure, permanently altering its properties. This process significantly increases the wood's hardness, dimensional stability, moisture performance, and durability. Furfurylated wood proved practical in-ground applications against termites and marine borers. Material is considered environmentally friendly, using a bio-based resin and not releasing harmful chemicals. The result of the modification is a high-performance wood-based material suitable for demanding outdoor applications, including outdoor furniture, decking, and construction applications.

One drawback of modified wood is its high price, limiting its broader use. Therefore, there is a commercial need to develop processes with a better price-per-formance ratio. One of the alternatives to improve the service life of the less durable

FIGURE 10.16   Façade of the Astrup Fearnley Museum of Modern Art, Oslo, Norway, made from acetylated wood.

(Courtesy of C. Brischke and M. Humar.)

wood species is the application of hydrophobic agents. The performance of wood in outdoor applications is a function of the presence of biologically active agents and water performance. If wood remains dry during rainfall events, the decay cannot proceed. To improve the performance of wood, oils and other methods are being utilised. Sometimes oils are combined with a prior biocidal treatment (dual treatment) known as the Royal process. There are various oils used for the treatment of wood. As natural oils are more expensive than synthetic oils, they are frequently combined with synthetic ones.

10.4   DECAY HAZARD

10.4.1   CLIMATE

Wood moisture and wood temperature largely determine the risk to wood from wood-destroying fungi and insects (Creffield 1991, Brischke *et al.* 2006a). The risk to wood from other wood-destroying organisms, such as shipworms, is determined by different additional factors, such as the salinity of water. The threat to wood is, therefore, largely dependent on the geographical location, as the local climate and other site-specific factors influence the activity of wood-destroying organisms. The moisture and temperature conditions within wooden components and structures are always the result of the interaction between climatic conditions, construction, and conditions of use. In this regard, the climate can be viewed at different levels, *i.e.*, macro, meso, local, and material climate. The latter are the moisture content and wood temperature over time (Brischke *et al.* 2006b). The individual parameters that determine the climate have different effects in the various use classes according to the EN 335 standard (CEN 2013). While wood in soil contact, *i.e.* Use Class 4 according to EN 335 (CEN 2013) is almost always permanently wet – or at least moist enough to allow fungal decay – wood exposed above ground in Use Classe 2 or 3 is subject to alternating wet and dry periods. Consequently, the effect of temperature is more pronounced when the wood is in soil contact. In above-ground applications, the time of wetness, *i.e.*, the cumulated periods of moisture conditions favourable for fungal decay, is the decisive factor (Brischke and Rapp 2008a, Isaksson *et al.* 2013, Van Acker 2017). The interaction of these two factors can be represented in the form of dose-response relationships. It can also be used to model and predict the service life of wooden components (Brischke and Rapp 2008b, Niklewski *et al.* 2021). However, the prerequisite for this is that the requirements of the different groups of organisms are known and sufficiently considered. Numerous authors reported on the moisture and temperature requirements of various decay fungi and wood-destroying insects (*e.g.*, Huckfeldt and Schmidt 2006, Meyer and Brischke 2015), which had later been used for decay modelling and the creation of decay hazard maps (Scheffer 1971, Carll 2009, Kim *et al.* 2011, van Niekerk *et al.* 2022). Pioneer work has been done by Theodore Scheffer, who developed a climate index for estimating decay hazards to wood exposed outdoors above ground (Scheffer 1971). The Scheffer Climate Index (SCI) has later been applied to different regions of the world and inspired others to refine respective hazard maps using dosimeter-based decay models (*e.g.*, Morris and Wang 2011, Helali *et al.* 2021, Brimblecombe and Richards 2023). The SCI and above ground decay models have been compared for Europe by van Niekerk *et al.* (2022), and respective maps are shown in Figure 10.17 (Hosseini *et al.* 2025).

On a local scale, the climate is affected by topography (altitude, orientation of valleys, and wind barriers), the presence of water bodies, or sealed surfaces; on a microscale, the climate can also be influenced by vegetation, shading trees and buildings, orientation, local sources of heat and humidity, and the construction itself. The different climate-related parameters indirectly influence wood moisture and temperature and their progression over time, the material climate (Brischke *et al.* 2006a). Some elements that characterise the landscape can positively or negatively influence the climate-induced risk to wooden components. For instance, trees directly near a building can shelter wooden cladding from wind-driven rain (WDR). However, the same trees may prevent this cladding from re-drying due to shading and hindered ventilation during dry periods. A quantitative assessment of the effect of climate-related influence factors is therefore rather complex and requires comprehensive simulations (Teasdale-St-Hilaire and Derome 2007, Ott et al. 2015, Niklewski *et al.* 2023) or extensive long-term measurements of the material climate under different exposure scenarios (Isaksson and Thelandersson 2013, Bornemann *et al.* 2014, Tariku *et al.* 2015).

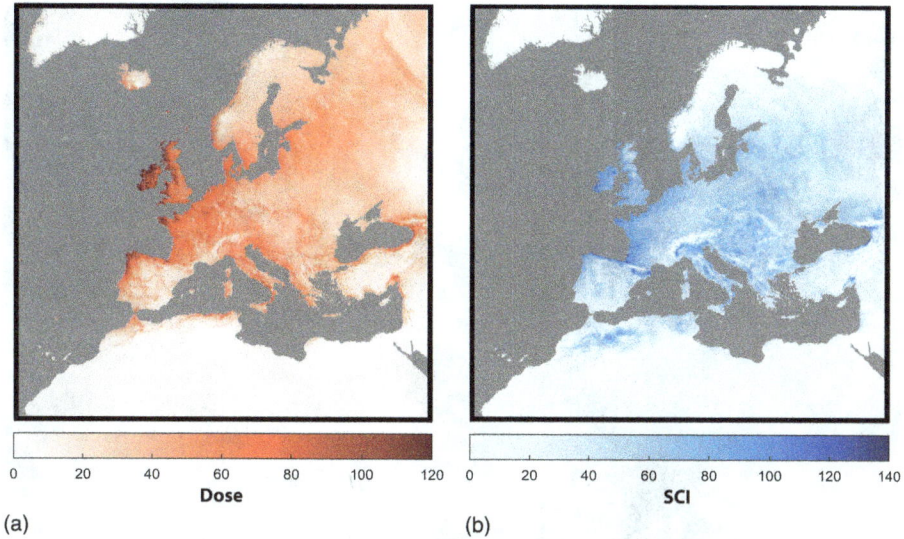

FIGURE 10.17 (a) Mapping the fungal decay hazard in Europe based on contemporary climate data and (b) Scheffer climate index. Cumulative dose in days, *i.e.*, days with favourable conditions for fungal decay. The dataset from CERRA, Copernicus European Regional ReAnalysis (open use Verrelle *et al.* 2022).

10.4.2 DESIGN

In addition to the microclimate to which it is exposed, the material climate in a wooden component is determined by its design and the design of the structure of which it is a part. While rainwater can run off vertical or inclined surfaces (Figure 10.18), it often remains on horizontal surfaces for a long time and can be absorbed by the wood. Water absorption via exposed end grain surfaces is generally higher than via side timbers (Figure 10.19). Even higher is the risk of moisture uptake and accumulation when so-called water traps are formed (Figure 10.20).

Wood can often be protected quickly and effectively by design. In this way, the need for the use of wood preservatives containing biocides can be reduced and sometimes even eliminated. Section 10.5 provides an overview of constructive wood protection measures. Suppose the influence of such measures shall be quantified to predict the service life of a wooden component. In that case, this can be done analogously to the influence of the climate using dose-response relationships. The potential of a particular design detail to trap moisture can be quantified either with the help of time series from material climate monitoring (Isaksson and Thelandersson 2013, Niklewski *et al.* 2018) or simulation software (Fortino *et al.* 2019), where the latter can also include the building envelope (Mundt-Petersen 2015). The relative dose of any design details can be quantified, for example, the details of a terrace decking, where a horizontal free ventilated decking board has been used as a reference, according to Isaksson *et al.* (2016). Accordingly, the effect of other structural elements and design solutions can be quantified with the help of factorisation (*e.g.*, Pousette *et al.* 2017, Niklewski *et al.* 2021). The method of factorisation in combination with

FIGURE 10.18 Wooden shingles on the roof of historic buildings. The rainwater can run off due to inclined application.

(Courtesy of C. Brischke and M. Humar.)

(a)   (b)

FIGURE 10.19 Outdoor use of wood: (a) walkway showing open end-grain surface with high potential of wetting, and (b) additional checks can lead to moisture accumulation.

(Courtesy of C. Brischke and M. Humar.)

dose-response models to describe the relationship between exposure and resistance has now also found its way into software tools for calculating the expected service life of wooden components, such as the CLICK*design* software: https://jklewski. github.io/CLICKdesign/

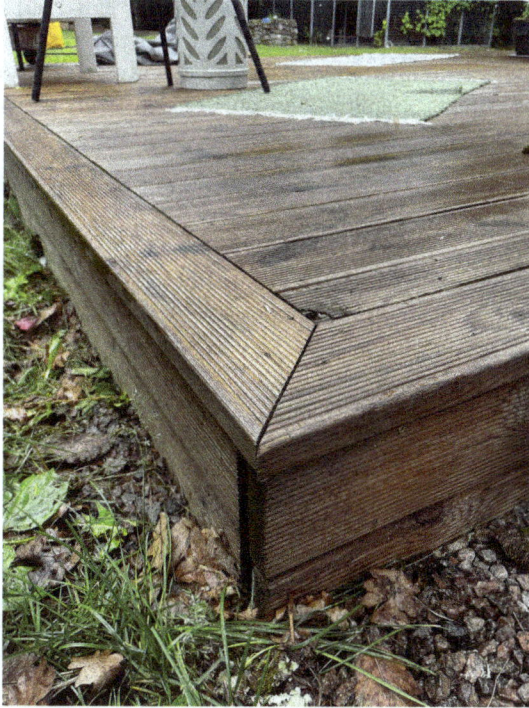

FIGURE 10.20 Terrace decking with different contact faces (side grain – side grain; end-grain – side grain; end grain – end-grain) having a risk of water trapping.

(Courtesy of C. Brischke and M. Humar.)

10.4.3 USE CLASSES

The conditions under which wood can be used differ regarding the organisms that may occur and the risk determined by moisture and temperature. These deviating living conditions allow a principal differentiation of organism groups referring to the exposure conditions of wooden components as reflected by the use class approach described in the EN 335 standard (CEN 2013), where six Use Classes (UCs) are defined according to the respective moisture regime and the potential presence of wood-degrading organisms (Table 10.5). The respective moisture conditions characterise UC 2, UC 3.1, and UC 3.2. In contrast, UC 4 ("ground contact") is considered permanently wet and equalised with "fresh-water contact". Also, the moisture conditions between UC 4 and UC 5 are similar, but in "seawater contact", marine borers such as shipworms and gribble do occur if salinity allows it.

The European UC system can be used to describe the exposure conditions of wooden components in a simplified way. In the second step, durability requirements can be formulated for each UC. For instance, minimum DCs are defined for each UC according to EN 460 (CEN 2023); if these are adhered to, acceptable service life can be expected without the component being preserved. However, the class – class approach of durability guidance is not suitable for concrete service life prediction of wooden components or buildings since it is not quantitative, and several factors such as design detailing and climatic conditions are not or at least only partly considered.

10.5   WOOD PROTECTION BY DESIGN AND MAINTENANCE

10.5.1   BUILDING CODES

Building codes related to wood durability in Europe are generally covered under European standards and national regulations, focusing on the protection, treatment, and appropriate use of wood in construction. These standards address the durability of wood in terms of its exposure to moisture, biological agents (such as fungi, bacteria, insects, termites, and marine borers), and other environmental conditions. Essential standards that address wood durability across Europe are provided in Table (10.6).

TABLE 10.6
Overview of Important Standards and Regulations Associated with Wood Durability and Protection

Standard/code Name	Description
EN 335: Durability of wood and wood-based products	EN 335 (CEN 2013) defines the use classes (UC) for wood and wood-based products based on their risk of decay and insect attack. There are five classes (UC 1 to 5), depending on whether the wood is used indoors, outdoors, or in contact with water or soil.
EN 350: Durability of wood and wood-based products – testing and classification	The EN 350 standard (CEN 2016a) guides methods for determining and classifying the durability of wood and wood-based materials against biological wood-destroying agents. Different durability classes (DC) are defined for decay fungi (DC 1–5), wood-boring beetles (DC S and DC D), termites, and marine borers (DC S, DC M, and DC D). Data from more than 200 wood species are provided.
EN 460: Durability of wood – natural durability of solid wood	The EN 460 standard (CEN 2023) guides the selection of wood and wood-based products for use in situations where they can be degraded by fungi or wood-destroying insects. This guidance includes information on factors that can influence the service life of a wood or wood-based product when considering biological degradation.
EN 599-1: Wood preservatives – performance of preventive wood preservatives	The EN 599-1 standard (CEN 2014) provides specifications for the testing and approval of wood preservatives used to enhance durability.
	It also helps determine the performance of different wood preservative treatments for various applications, particularly for wood exposed to higher moisture and environmental stress.

(Continued)

TABLE 10.6 (Continued)

Standard/code Name	Description
EN 1995-1-1 (Eurocode 5): Design of timber structures	Eurocode 5 (CEN 2010) is the critical European standard for the structural design of timber structures. While it primarily focuses on structural integrity, it includes provisions regarding wood durability.It outlines measures to protect wood from environmental factors such as moisture and biological degradation, mainly through proper detailing and design (*e.g.*, allowing ventilation and avoiding water traps).
National Codes and Regulations In addition to European standards, individual countries may have specific building codes or regulations related to wood durability:	Germany: The DIN 68800-1 standard (DIN 2019) governs wood protection against decay, providing rules for structural design and preservation methods.
	France: The NF DTU 31.2 (NF DTU 2019) standard governs the use of wood in construction, including specific requirements for protection against decay and pests.
	United Kingdom: The British Standard BS 8417 (BS 2024) recommends wood preservation and specifies the desired service life for treated wood.
	Nordic Wood Preservation Council (NTR) in the Scandinavian countries: Due to their harsh climates, these countries often have stricter codes, with detailed rules on wood selection, treatment, and design for durability.
EN 13986: Wood-based panels for use in construction	The EN 13986 standard (CEN 2015c) specifies requirements for wood-based panels, such as plywood and particleboard, to ensure their durability and suitability for construction under various conditions.
Sustainability and Environmental Considerations	EUTR (EU Timber Regulation) ensures that wood used in construction is sourced from legal and sustainable operations. By promoting high-quality, properly harvested timber, EUTR contributes to long-term durability.
CE marking	CE Marking is required for wood products used in construction, ensuring they meet essential safety and performance standards, including durability.

Note: DC – Durability Class and UC – Use Class.

These European standards and building codes emphasise the need for thoughtful design, material selection, and protective treatments to ensure wood's long-term durability in various environmental conditions. Predominantly, EN 460 (CEN 2023) guides the choice of wood and wood-based products for use when they can be degraded by fungi, wood-destroying insects, or marine borers. This guidance includes information on factors that can influence the service life of a wood or wood-based product when considering biological degradation. In many end uses, design, workmanship, and maintenance will also affect the service life of the wood or wood-based product. The EN 460 standard is a step towards evaluating the service life of a wood product, considering 1) the durability characteristics of the glue used in wood-based products and 2) the aesthetic function of wood products (discolouration, surface weathering, mould).

10.5.2 GOOD PRACTICE

Wood durability is frequently very romanticised. Sometimes, the selection of the materials and the execution of the details rely on the experiences from previous centuries. However, designers and architects frequently need to remember that the climate has changed in recent decades, which influences the wood quality, availability, presence of degradation organisms, and decay patterns (van Niekerk *et al.* 2022). In addition, building standards in the past decades were much lower than today. Nowadays, buildings are airtight, with permeable and non-permeable membranes controlling moisture diffusion. In the past, construction with wood was different. There was a lot of air draft, which, on the one hand, reduced building comfort, but on the other hand, enabled wooden construction drying and prevented condensation. So, the experiences from the past can only sometimes be transferred to modern constructions.

However, the basic principles of wood protection by design remained the same: umbrella and boots. The umbrella (roof) protects the building from rainfall, while the boots (foundation, anchor) preserve the wood from absorbing water from the soil. Therefore, protection by design is often synonymous with "moisture protection", aimed at preventing moisture-induced risks such as decay or biotic attack. Two fundamental principles are: 1) keeping water away from the structure (Figure 10.21),

FIGURE 10.21   Noise barrier, sheltered from rain and protected from splash water due to a concrete base.

(Courtesy of C. Brischke and M. Humar.)

and if that is not possible, 2) removing water from the structure as quickly and effectively as possible (Figure 10.22). It is generally agreed that biocidal treatments should only be applied to wood when construction measures are insufficient to provide protection.

In addition to various design solutions, water can be kept away from organic building materials through coatings and covers (physical protection). Coatings have long been influential in moisture protection for timber products, such as window joinery or cladding. However, their effectiveness depends on their durability and the local climate. When coatings become damaged, they can trap water, negating their protective benefits. For example, thick organic coatings are highly effective in the United Kingdom and northern Europe but are significantly less efficient in southern Europe's warmer, sunnier climates.

Most organic building materials are vulnerable to moisture, and wetting is a prerequisite for biotic decay. Therefore, moisture protection is essential for wood and other bio-based building materials. The first and most crucial step in any protection strategy by design is to prevent water from reaching the structure. Moisture can come from rain, hail, or snow, often combined with wind from wind-driven rain (WDR). Additionally, moisture can arise from air humidity, with water vapour condensing indoors, outdoors, and – most critically – within the building envelope, where re-drying is usually inhibited. Water can also rise from the soil or other porous foundations.

(a)   (b)

FIGURE 10.22 Experimental building made of round timber: (a) exterior wall and (b) slotted timber on the underside to prevent cracking and allow water to drain off.

(Courtesy of C. Brischke and M. Humar.)

The general principles of the protection by design are:

1) Raise the structures from the ground. If possible, wood should not be in contact with or close to the ground. To protect the wood from water splash, the suggested distance from the ground is 300 mm. This can be performed with metal anchors or concrete foundations.
2) Protect the end grains of the wood. Wood has an anisotropic nature and is thus much more permeable in axial direction than tangential or radial. If the axial planes are protected, water penetration into the wood element is limited. Protection can be performed through coverage with wooden, metal, or plastic plates.
3) Orientate wooden elements vertically rather than horizontally. Water can run faster from vertically oriented elements than horizontally oriented ones if the axial planes are not exposed to the water.
4) Avoid tight gaps. The gap between the elements should be at least 8 mm to prevent capillary water uptake. Drying of water in the gaps takes longer, so the water is usually absorbed into the wood.
5) Be careful if wood is in contact with metal or concrete. Concrete can be a good source of moisture, so sealants are advised. On the other hand, water is more likely to condense on metal parts and drip into wood.
6) There should be an air layer behind every façade (ensure ventilation).
7) If possible, avoid houses without roof overhangs.
8) Pay attention to the building physics to avoid possible water condensation in the wall or ceiling.
9) Avoid completely horizontal positions on open areas or enable adequate drainage.
10) Thoroughly consider wood's dimensional stability. Wood planks can wrap around or crack, forming water traps. Cracks are frequently the first entry point for water and fungal decay.
11) Be careful about additional construction works on the building. Mounting of the solar panels can damage the membrane on the roof, which could lead to degradation.
12) Regularly inspect the building's water installation. Dripping water can cause considerable decay, and regular inspections should be provided to detect decay in its early stages.
13) Install moisture logging sensors on critical points within the building and link them with the early warning system.
14) Do not use infested old wood as firewood in the fireplace.

REFERENCES

Ali A.C., Júnior E.U., Råberg U. & Terziev N. (2011). Comparative natural durability of five wood species from Mozambique. *International Biodeterioration & Biodegradation*, 65(6), 768–776.

Amusant N., Beauchêne J., Fournier M., Janin G. & Thevenon M.F. (2004). Decay resistance in *Dicorynia guianensis* Amsh.: analysis of inter-tree and intra-tree variability and relations with wood colour. *Annals of Forest Science*, 61(4), 373–380.

Appelqvist C., Havenhand J.N. & Toth G.B. (2015). Distribution and abundance of teredinid recruits along the Swedish coast – are shipworms invading the Baltic Sea? *Journal of the Marine Biological Association of the United Kingdom*, 95(4), 783–790.

AS (2005). *AS 5604: Timber—Natural durability ratings*. Australian Standard. Standards Australia, Sydney, Australia.

Augustina S., Dwianto W., Wahyudi I., Syafii W., Gérardin P. & Marbun S.D. (2023). Wood impregnation in relation to its mechanisms and properties enhancement. *BioResources*, 18(2), 4332–4372.

AWPA (2021). *AWPA E 26: Standard field test for evaluation of wood preservatives to be used for interior applications (UC1 and UC2); ground proximity termite test*. American Wood Protection Association, Clermont, USA.

AWPA (2022). *AWPA E 10: Laboratory method for evaluating the decay resistance of wood-based materials against pure basidiomycete cultures: Soil/block test*. American Wood Protection Association, Clermont, USA.

Ayadi N., Lejeune F., Charrier F., Charrier B. & Merlin A. (2003). Color stability of heat-treated wood during artificial weathering. *Holz als Roh- und Werkstoff*, 61(3), 221–226.

Ballard R.G., Walsh M.A. & Cole W.E. (1984). The penetration and growth of blue-stain fungi in the sapwood of lodgepole pine attacked by mountain pine beetle. *Canadian Journal of Botany*, 62(8), 1724–1729.

Bauch J. (1984). Development and characteristics of discolored wood. *IAWA Bulletin*, 5(2), 91–98.

BS (2024). BS 8417:2024 – TC: Preservation of wood. *Code of practice*. British Standards Institution, London, UK.

Blanchette R.A. (2000). A review of microbial deterioration found in archaeological wood from different environments. *International Biodeterioration and Biodegradation*, 46(3), 189–204.

Borges L.M.S. (2014). Biodegradation of wood exposed in the marine environment: Evaluation of the hazard posed by marine wood-borers in fifteen European sites. *International Biodeterioration and Biodegradation*, 96(12), 97–104.

Borges L.M.S., Cragg S.M. & Busch S. (2009). A laboratory assay for measuring feeding and mortality of the marine wood borer *Limnoria* under forced feeding conditions: A basis for a standard test method. *International Biodeterioration & Biodegradation*, 63(3), 289–296.

Borges L.M.S., Merckelbach L.M. & Cragg S.M. (2014a). Biogeography of wood-boring crustaceans (Isopoda: Limnoriidae) established in European coastal waters. *PloS one*, 9, Article ID: e109593.

Borges L.M.S., Merckelbach L.M., Sampaio Í. & Cragg S.M. (2014b). Diversity, environmental requirements, and biogeography of bivalve wood-borers (*Teredinidae*) in European coastal waters. *Frontiers in Zoology*, 11(1), 1–13.

Bornemann T., Brischke C. & Alfredsen G. (2014). Decay of wooden commodities–moisture risk analysis, service life prediction and performance assessment in the field. *Wood Material Science & Engineering*, 9(3), 144–155.

Bravery A.F., Berry R.W., Carey J.K. & Cooper D.E. (1987). *Recognising Wood Rot and Insect Damage in Buildings*. BRE Report BR453, Building Research Establishment (BRE), Watford, UK.

Brimblecombe P. & Richards J. (2023). Köppen climates and Scheffer index as indicators of timber risk in Europe (1901–2020). *Heritage Science*, 11(1), Article ID: 148.

Brischke C. & Rapp A.O. (2008a). Influence of wood moisture content and wood temperature on fungal decay in the field: observations in different micro-climates. *Wood Science and Technology*, 42(8), 663–677.

Brischke C. & Rapp A.O. (2008b). Dose–response relationships between wood moisture content, wood temperature and fungal decay determined for 23 European field test sites. *Wood Science and Technology*, 42(6), 507–518.

Brischke C. & Alfredsen G. (2023). Biological durability of pine wood. *Wood Material Science & Engineering*, 18(3), 1050–1064.

Brischke C., Bayerbach R. & Rapp A.O. (2006a). Decay-influencing factors: A basis for service life prediction of wood and wood-based products. *Wood Material Science & Engineering*, 1(3/4), 91–107.

Brischke C., Welzbacher C.R. & Rapp A.O. (2006b). Detection of fungal decay by high-energy multiple impact (HEMI) testing. *Holzforschung*, 60(2), 217–222.

Brischke C., Alfredsen G., Emmerich L., Humar M. & Meyer-Veltrup L. (2023a). Durability of wood exposed above ground - Experience with the bundle test method. *Forests*, 14(7), Article ID: 1460.

Brischke C., Haase F., Bächle L. & Bollmus S. (2023b). Statistical analysis of wood durability data and its effect on a standardised classification scheme. *Standards*, 3(2), 210–226.

Brischke C., Stolze H., Koddenberg T., Vek V., Caesar C.M.C., Steffen B., Taylor A.M. & Humar M. (2024). Origin-specific differences in the durability of black locust (*Robinia pseudoacacia*) wood against wood-destroying basidiomycetes. *Wood Science and Technology*, 58(4), 1427–1449.

Caldeira F. (2010). Boron in wood preservation: A review in its physico-chemical aspects. *Silva Lusitana*, 18(2), 179–196.

Carll C. (2009). *Decay Hazard (Scheffer) Index Values Calculated from 1971-2000 Climate Normal Data*. Report FPL-GTR-179, US Department of Agriculture, Forest Service, Forest Products Laboratory, Madison (WI), USA, (17 pp.).

CEN (1992). *EN 275: Wood preservatives; determination of the protective effectiveness against marine borers*. European Committee for Standardization, Brussels, Belgium.

CEN (2001). *ENV 807: Wood preservatives - Determination of the effectiveness against soft rotting micro-fungi and other soil inhabiting micro-organisms*. European Committee for Standardization, Brussels, Belgium.

CEN (2005). *CEN/TS 15083-2: Durability of wood and wood-based products - Determination of the natural durability of solid wood against wood-destroying fungi, test methods - Part 2: Soft rotting micro-fungi*. European Committee for Standardization, Brussels, Belgium.

CEN (2010). *EN 1995-1-1: Eurocode 5: Design of timber structures - Part 1-1: General - Common rules and rules for buildings*. European Committee for Standardization, Brussels, Belgium.

CEN (2013). EN 335 (2013). *Durability of wood and wood-based products - Use classes: definitions, application to solid wood and wood-based products*. European Committee for Standardization, Brussels, Belgium.

CEN (2014). *EN 599-1: Durability of wood and wood-based products - Efficacy of preventive wood preservatives as determined by biological tests - Part 1: Specification according to use class*. European Committee for Standardization, Brussels, Belgium.

CEN (2015a). *EN 252: Field test method for determining the relative protective effectiveness of a wood preservative in ground contact*. European Committee for Standardization, Brussels.

CEN (2015b). *EN 330: Wood preservatives - Determination of the relative protective effectiveness of a wood preservative for use under a coating and exposed out-of-ground contact - Field test: L-joint method*. European Committee for Standardization, Brussels, Belgium.

CEN (2015c). *EN 13986: Wood-based panels for use in construction - Characteristics, evaluation of conformity and marking*. European Committee for Standardization, Brussels, Belgium.

CEN (2015d). *EN 49-2: Wood preservatives - Determination of the protective effectiveness against Anobium punctatum (De Geer) by egg-laying and larval survival - Part 2: Application by impregnation (Laboratory method)*. European Committee for Standardization, Brussels, Belgium.

CEN (2016a). *EN 350: Durability of wood and wood-based products - Testing and classification of the durability to biological agents of wood and wood-based materials.* European Committee for Standardization, Brussels, Belgium.

CEN (2016b). *EN 46-1: Wood preservatives - Determination of the preventive action against recently hatched larvae of Hylotrupes bajulus (Linnaeus) - Part 1: Application by surface treatment (laboratory method).* European Committee for Standardization, Brussels, Belgium.

CEN (2016c). *EN 49-1: Wood preservatives - Determination of the protective effectiveness against Anobium punctatum (De Geer) by egg-laying and larval survival - Part 1: Application by surface treatment (Laboratory method).* European Committee for Standardization, Brussels, Belgium.

CEN (2016d). *EN 47-2: Wood preservatives - Determination of the toxic values against larvae of Hylotrupes bajulus (Linnaeus) - (Laboratory method).* European Committee for Standardization, Brussels, Belgium.

CEN (2021). *EN 113-2: Durability of wood and wood-based products - Test method against wood destroying basidiomycetes - Part 2: Assessment of inherent or enhanced durability.* European Committee for Standardization, Brussels, Belgium.

CEN (2022). *EN 12037: Wood preservatives - Field test method for determining the relative protective effectiveness of a wood preservative exposed out of ground contact - Horizontal lap-joint method.* European Committee for Standardization, Brussels, Belgium.

CEN (2023). *EN 460: Durability of wood and wood-based products - Guidance on performance.* European Committee for Standardization, Brussels, Belgium.

CEN (2024a). *EN 20-1: Wood preservatives - Determination of the protective effectiveness against Lyctus brunneus (Stephens) - Part 1: Application by surface treatment (laboratory method).* European Committee for Standardization, Brussels, Belgium.

CEN (2024b). *EN 117: Wood preservatives - Determination of toxic values against Reticulitermes species (European termites) (Laboratory method).* European Committee for Standardization, Brussels, Belgium.

Clausen C.A. (1996). Bacterial associations with decaying wood: a review. *International Biodeterioration & Biodegradation*, 37(1/2), 101–107.

Cornette R., Gotoh H., Koshikawa S. & Miura T. (2008). Juvenile hormone titers and caste differentiation in the damp-wood termite *Hodotermopsis sjostedti* (Isoptera, Termopsidae). *Journal of Insect Physiology*, 54(6), 922–930.

Creffield J.W. (1991). *Wood Destroying Insects: Wood Borers and Termites.* CSIRO, Melbourne, Australia, (44 pp.).

Daniel G. & Nilsson T. (1998). Developments in the Study of Soft Rot and Bacterial Decay. In: Bruce A. & Palfreyman J.W. (Eds.) *Forest Products Biotechnology.* Taylor and Francis, London, UK, (pp. 37–62 pp.).

DIN (2019). DIN 68800-1: Wood preservation - Part 1: General. *Deutschen Institut für Normung (DIN)*, Berlin, Germany.

Eaton R.A. & Hale M.D. (1993). *Wood: Decay, Pests and Protection.* Chapman and Hall Ltd, London, UK, (546 pp.).

ECHA (2020). Candidate List of Substances of Very High Concern for Authorisation. *European Chemicals Agency (ECHA)*, Helsinki, Finland.

Embacher J., Zeilinger S., Kirchmair M., Rodriguez-R, L.M. & Neuhauser S. (2023). Wood decay fungi and their bacterial interaction partners in the built environment–A systematic review on fungal bacteria interactions in dead wood and timber. *Fungal Biology Reviews*, 45, Article ID: 100305.

EU (2012). Regulation (EU) No. 528/2012 of the European Parliament and the Council of 22 May 2012 concerning the making available on the market and use of biocidal products. *Official Journal of the European Communities*, L 269(528), 1–15.

Feist W.C. (1990). Outdoor Wood Weathering and Protection, In: Rowell R.M. & Barbour R.J. (Eds.), *Archaeological Wood Properties, Chemistry, and Preservation.* American Chemical Society, Washington DC, USA, (472 pp.).

Liese, W., & Schmid, R. (1961). Light and electron microscopical studies on the growth of blue stain fungi in pine und spruce. *Holz als Roh- und Werkstoff*, 19, 329–337.

Fortino S., Hradil P., Genoese A., Genoese A. & Pousette A. (2019). Numerical hygro-thermal analysis of coated wooden bridge members exposed to Northern European climates. *Construction and Building Materials*, 208, 492–505.

Fukasawa Y. & Matsukura K. (2021). Decay stages of wood and associated fungal communities characterise diversity–decomposition relationships. *Scientific Reports*, 11(1), Article ID: 8972.

Gezer E.D., Michael J.H. & Morrell J.J. (1999). Effects of glycol on leachability and efficacy of boron wood preservatives. *Wood and Fiber Science*, 31(2), 136–142.

Greaves H. (1971). The bacterial factor in wood decay. *Wood Science and Technology*, 5(1), 6–16.

Helali J., Momenzadeh H., Saeidi V., Brischke C., Ebrahimi G. & Lotfi M. (2021). Decadal variations of wood decay hazard and El Nino Southern oscillation phases in Iran. *Frontiers in Forests and Global Change*, 4, Article ID: 693833.

Hillis W.E. (1987). *Heartwood and Tree Exudates*. Springer, New York, USA, (268 pp.).

Hon D.N.-S. (2001). Weathering and photochemistry of wood. In: Hon, D.N.-S. & Shiraishi N. (Eds.), *Wood and Cellulosic Chemistry*. (2nd Ed.), Marcel Dekker, New York, USA, (pp. 513–546).

Hosseini H., Iannacone L., Brischke C. & Niklewski J. (2025). Quantifying spatiotemporal variation of wood decay risk using a data-driven moisture content model and multi-scale weather datasets across Europe and Scandinavia. pre-print manuscript available at: https://papers.ssrn.com/sol3/papers.cfm?abstract_id=5021825

Huckfeldt T. & Schmidt O. (2006). *Hausfäule- und Bauholzpilze – Diagnose und Sanierung*. Rudolf Müller, Cologne, Germany, (610 pp.).

Huckfeldt T. & Brischke C. (2024). Fäuleschäden an Holzspielplätzen und ihre Vermeidung - Theorie und Praxisbeispiele. In: *Europäisches Institut für Postgraduale Bildung an der TU Dresden Tagungsband des EIPOS-Sachverständigentages Holzschutz 2024: Beiträge aus Praxis, Forschung und Weiterbildung*. Fraunhofer IRB Verlag, Stuttgart, Germany.

Humar M. (2017). Protection of the bio-based material. In: Dennis J. & Brischke C. (Eds.), *Performance of Bio-based Building Materials*. Woodhead Publishing, Sawston, Cambridge, UK, (pp. 187–247).

Humar M. & Lesar B. (2009). Influence of dipping time on uptake of preservative solution, adsorption, penetration and fixation of copper-ethanolamine based wood preservatives. *European Journal of Wood and Wood Products*, 67(3), 265–270.

Humar M., Petrič M., Pohleven F., Šentjurc M. & Kalan P. (2002). Changes in EPR spectra of wood impregnated with copper-based preservatives during exposure to several wood-rotting fungi. *Holzforschung*, 56(3), 229–238.

Humar M., Lesar B., Thaler N., Kržišnik D., Kregar N. & Drnovšek S. (2018). Quality of copper impregnated wood in Slovenian hardware stores. *Drvna Industrija*, 69(2), 121–126.

Isaksson T. & Thelandersson S. (2013). Experimental investigation on the effect of detail design on wood moisture content in outdoor above ground applications. *Building and Environment*, 59, 239–249.

Isaksson T., Brischke C. & Thelandersson S. (2013). Development of decay performance models for outdoor timber structures. *Materials and Structures*, 46, 1209–1225.

Isaksson T., Thelandersson S., Jermer J. & Brischke C. (2016). *Service Life of Wood in Outdoor Above Ground Applications: Engineering Design Guideline – Background Document*. Report TVBK-3067. Lund University, Structural Engineering, Lund, Sweden, (35 pp.).

ISO (2011). ISO 15686-1: Buildings and constructed assets. Service life planning. Part 1: General principles and framework. *International Standardisation Organization (ISO)*, Geneva, Switzerland.

Jones D. & Brischke C. (2017). Performance of Bio-based Building Materials. In *Performance of Bio-based Building Materials*. Woodhead Publishing, (650 pp.).

Jones D., Sandberg D., Goli G. & Todaro L. (2019). *Wood Modification in Europe: A State-of-the-Art About Processes, Products, Applications*. Firenze University Press, Florence, Italy, (113 pp.).

Kaplan J.O., Krumhardt K.M. & Zimmermann N. (2009). The prehistoric and preindustrial deforestation of Europe. *Quaternary Science Reviews*, 28(27/28), 3016–3034.

Kervina-Hamović L. (1990). Zascita Lesa. *Biotehniska fakulteta, VTOZD za lesarstvo*, University of Ljubljana, Ljubljana, Slovenia, (126 pp.).

Kim T.G., Ra J.B., Kang S.M. & Wang J. (2011). Determination of decay hazard index (Scheffer index) in Korea for exterior above-ground wood. *Journal of the Korean Wood Science and Technology*, 39, 531–537.

Kjellow A.W. & Henriksen O. (2009). Supercritical wood impregnation. *Journal of Supercritical Fluids*, 50(3), 297–304.

Koch G., Bauch J., Puls J. & Welling J. (2002). Ursachen und wirtschaftliche Bedeutung von Holzverfärbungen. *Allgemeine Forstzeitschrift*, 57, 315–318.

Kozlov V. & Kisternaya M. (2014). Sorption properties of historic and recent pine wood. *International Biodeterioration and Biodegradation*, 86(B), 153–157.

Kretschmar E.I., Gelbrich J., Militz H. & Lamersdorf N. (2008). Studying bacterial wood decay under low oxygen conditions - results of microcosm experiments. *International Biodeterioration and Biodegradation*, 61(1), 69–84.

Keržič E., Lesar B. & Humar M. (2021). Influence of weathering on surface roughness of thermally modified wood. *BioResources*, 16(3), 4675–4692.

Kleist G. & Schmitt U. (2001). Characterisation of a soft rot-like decay pattern caused by *Coniophora puteana* (Schum.) Karst. in sapelli wood (*Entandrophragma cylindricum* Sprague). *Holzforschung* 55(6), 573–578.

Kržišnik D., Lesar B., Thaler N. & Humar M. (2018). Influence of natural and artificial weathering on the colour change of different wood and wood-based materials. *Forests*, 9(8), 1–22.

Lebow S. (1996). *Leaching of Wood Preservative Components and Their Mobility in the Environment Summary of Pertinent Literature*. General Technical Report FPL-GTR-93, US Department of Agriculture, Forest Service, Forest Products Laboratory, Madison (WI), USA, (36 pp.).

Lesar, B., Humar, M. & Kralj, P. (2010). Influence of montan wax emulsions on leaching dynamics of boric acid from impregnated wood. *Wood Research*, 55(1), 93–100.

Marais B.N., Brischke C. & Militz H. (2022). Wood durability in terrestrial and aquatic environments – A review of biotic and abiotic influence factors. *Wood Material Science & Engineering*, 17(2), 82–105.

Medeiros Neto P.N.D., Paes J.B., Oliveira J.T.D.S., da Silva J.G.M., Coelho J.C.F. & Ribeiro L.D.S. (2020). Durability of eucalypt wood in soil bed and field decay tests. *Maderas. Ciencia y tecnología*, 22(4), 447–456.

Meyer L. & Brischke C. (2015). Fungal decay at different moisture levels of selected European-grown wood species. *International Biodeterioration and Biodegradation*, 103(7), 23–29.

Meyer L., Brischke C. & Preston A. (2016). Testing the durability of timber above ground: A review on methodology. *Wood Material Science & Engineering*, 11(5), 283–304.

Meyer-Veltrup L., Brischke C., Alfredsen G., Humar M., Flæte P.O., Isaksson T., Larsson Brelid P., Westin M. & Jermer J. (2017a). The combined effect of wetting ability and durability on outdoor performance of wood: development and verification of a new prediction approach. *Wood Science and Technology*, 51(3), 615–637.

Meyer-Veltrup L., Brischke C. & Källander B. (2017b). Testing the durability of timber above ground: evaluation of different test methods. *European Journal of Wood and Wood Products*, 75(3), 291–304.

Morris P.I. & Wang J. (2011). Scheffer index as preferred method to define decay risk zones for above ground wood in building codes. *International Wood Products Journal*, 2(2), 67–70.

Mundt Petersen S. (2015). *Moisture Safety in Wood Frame Buildings-Blind evaluation of the hygrothermal calculation tool WUFI using field measurements and determination of factors affecting the moisture safety.* Report TVBH-1021, Doctoral Thesis, Building Physics, Lund University, Lund, Sweden.

NF DTU (2019). *NF DTU 31.2: Building works - Timber frame houses and buildings construction - Part 1-1: Contract bill of technical model clauses - Part 1-2: General criteria for selection of materials - Part 2: Contract bill of special administrative model clauses.* AFNOR Standardization, Saint-Denis, France.

Niklewski J., Isaksson T., Frühwald Hansson E. & Thelandersson S. (2018). Moisture conditions of rain-exposed glue-laminated timber members: the effect of different detailing. *Wood Material Science & Engineering*, 13(3), 129–140.

Niklewski J., van Niekerk P.B., Brischke C. & Frühwald Hansson E. (2021). Evaluation of moisture and decay models for a new design framework for decay prediction of wood. *Forests*, 12(6), Article ID: 721.

Niklewski J., Sandak J., Van Niekerk P., Brischke C., Acquah R. & Sandak A. (2023). Simplified environmental analysis of the long-term performance of wood cladding and decking. In: *Proceedings of the World Conference on Timber Engineering (WCTE 2023)*, 19-22 June, Oslo, Norway, (pp. 548–557).

Nilsson T. (2009). Biological Wood Degradation. In: Ek M., Gellerstedt G. & Henriksson G. (Eds.). *Volume 1. Wood Chemistry and Wood Biotechnology*. De Gruyter, Berlin, New York, USA, (pp. 219–244).

NWPC (2023). *Wood Preservatives and Other Wood Protection Systems Approved by the Nordic Wood Preservation Council.* Report No. 102, Nordic Wood Preservation Council (NWPC), Stockholm, Sweden, (2 pp.).

Ott S., Tietze A. & Winter S. (2015). Wind driven rain and moisture safety of tall timber houses–evaluation of simulation methods. *Wood Material Science & Engineering*, 10(3), 300–311.

Paradis S., Guibal D., Vernay M., Beauchêne J., Brancheriau L., Châlon I., Daigremont C., Détienne P., Fouquet D., Langbour P., Lotte, S., Méjean C., Thévenon M.F., Thibaut A. & Gérard J. (2011). Tropix 7.0: Caractéristiques technologiques de 245 essences tropicales et tempérées. *[Tropix 7.0: Technological characteristics of 245 tropical and temperate species.]* French Agricultural Research Centre for International Development *(CIRAD)*, Montpellier, France.

Peylo A. & Willeitner H. (1999). Five Years Leaching of Boron. *Report No. IRG/WP 99–30195, International Research Group on Wood Protection (IRG)*, Stockholm, Sweden.

Plarre R. (2012). Reaktion von Hausbockkäfern auf Bauholz unterschiedlicher, Qualitäten im Labor und Freiland. In: *Proceedings Deutsche Holzschutztagung, 28-29 September 2012*, Göttingen, Germany, (pp. 36–45).

Pousette A., Malo K.A., Thelandersson S., Fortino S., Salokangas L. & Wacker J. (2017). *Durable Timber Bridges - Final Report and Guidelines.* SP Rapport 2017:25, RISE Research Institutes of Sweden, Stockholm, Sweden, (177 pp., pp. 17–48).

Preston A.F. (2000). Wood preservation. *Forest Products Journal*, 50(9), 1–12.

Richardson H.W. (1997). Copper Fungicides/Bactericides. In: Richardson H.W. (Ed.), *Handbook of Copper Compounds and Applications: Marcel Dekker*, New York, USA, (432 pp.).

Reinprecht L. (2016). Wood Deterioration, Protection and Maintenance. In: *Wood Deterioration, Protection and Maintenance*. John Wiley & Sons, Hoboken (NJ), USA, (357 pp.).

Rayner A.D. & Boddy L. (1988). Fungal Decomposition of Wood. *Its Biology and Ecology*. John Wiley & Sons, Hoboken (NJ), USA, (587 pp.).

Rudman P. & Gay F.J. (1963). The causes of natural durability in timber. *Holzforschung*, 17(1), 20–25.

Shukla S.R. & Kamdem D.P. (2023). Effect of micronised copper treatments on retention, strength properties, copper leaching and decay resistance of plantation grown *Melia dubia* Cav. wood. *European Journal of Wood and Wood Products*, 81(2), 513–528.

Scheffer T.C. (1971). A climate index for estimating potential for decay in wood structures above ground. *Forest Products Journal*, 21(1), 25–31.

Scheffer T.C. & Morrell J.J. (1998). Natural Durability of Wood: A Worldwide Checklist of Species. *Research Contribution No. 22. College of Forestry, Forest Research Laboratory*, Oregon State University, Corvallis, USA, (62 pp.).

Schmidt O. (2006). *Wood and Tree Fungi – Biology, Damage, Protection, and Use*. Springer, Berlin, Germany, (336 pp.).

Schmidt O. & Liese W. (1994). Occurrence and significance of bacteria in wood. *Holzforschung*, 48(4), 271–277.

Schrader L., Trautner J. & Tebbe C.C. (2024). Identifying environmental factors affecting the microbial community composition on outdoor structural timber. *Applied Microbiology and Biotechnology*, 108(1), Article ID: 254.

Schwarze F.W., Engels J. & Mattheck C. (2013). *Fungal Strategies of Wood Decay in Trees*. Springer, Berlin, Heidelberg, Germany, (185 pp.).

Stirling R. & Morris P.I. (2006). The Influence of Extractives on Western Redcedar's Equilibrium Moisture Content. *Report No. IRG/WP 06-4033, 1International Research Group on Wood Protection (IRG)*, Stockholm, Sweden.

Tariku F., Simpson Y. & Iffa E. (2015). Experimental investigation of the wetting and drying potentials of wood frame walls subjected to vapor diffusion and wind-driven rain loads. *Building and Environment*, 92(10), 368–379.

Taylor A.M., Gartner B.L. & Morrell J.J. (2002). Heartwood formation and natural durability - A review. *Wood and Fiber Science*, 34(4), 587–611.

Teasdale-St-Hilaire A. & Derome D. (2007). Comparison of experimental and numerical results of wood-frame wall assemblies wetted by simulated wind-driven rain infiltration. *Energy and Buildings*, 39(11), 1131–1139.

Unger A., Schniewind A.P. & Unger W. (2001). *Conservation of Wood Artefacts. A Handbook*. Springer, Berlin, Germany, (578 pp.).

Van Acker J. (2017). Moisture dynamics defining service life performance of wood products. *Pro Ligno*, 13(4), 8–26.

van Niekerk P.B., Marais B.N., Brischke C., Borges L.M., Kutnik M., Niklewski J., Ansard D., Humar M., Cragg S.M. & Militz H. (2022). Mapping the biotic degradation hazard of wood in Europe–biophysical background, engineering applications, and climate change-induced prospects. *Holzforschung*, 76(2), 188–210.

Verrelle A., Glinton M., Bazile E., Le Moigne P., Randriamampianina M., Ridal R., Berggren L., Undén P., Schimanke S., Mladek R. & Soci C. (2022). CERRA-Land sub-daily regional reanalysis data for Europe from 1984 to present. *Copernicus Climate Change Service (C3S) Climate Data Store (CDS)*. https://doi.org/10.24381/cds.a7f3cd0b, [Retrieved September 15, 2024].

Willeitner H. (2001). Current national approaches to defining retentions in use. In: *COST E22 Environmental Optimization of Wood Protection. Optimising Treatment Levels and Managing Environmental Risks*, 8 – 9 November, Reinbek, Germany, (6 pp.).

Wengert E.M. (1997). Causes and cures for stains in dried lumber: Sticker stain, chemical stain, iron stain and blue stain. Prevention of discolorations in hardwood and softwood logs and lumber. *Forest Products Society*, Madison (WI), USA, (pp. 12–16).

Wilcox W.W. (1978). Review of literature on the effects of early stages of decay on wood strength. *Wood and Fiber Science*, 9(4), 252–257.

Winandy J.E. & Morrell J.J. (1993). Relationship between incipient decay, strength, and chemical composition of Douglas-fir heartwood. *Wood and Fiber Science*, 25(3), 278–288.

Zabel R.A. & Morrell J.J. (2012). *Wood Microbiology: Decay and its Prevention*. Academic Press, San Diego (CA), USA, (476 pp.).

11 Surface Treatment

G. Grüll, B. Forsthuber, and M. Truskaller

11.1 SELECTION OF WOOD COATINGS

Wood is a natural material with appealing colour and structure, so people desire to maintain its natural appearance even under heavy stresses and direct weathering. Wood and wooden objects must be protected against these influences by surface treatments or coatings. Because of wood's anisotropic nature, surface coatings must have specific properties. The requirements for coatings on wood and wood products for outdoor use differ considerably from those for indoor use.

The surface is the calling card of every object. People meet the surface through their senses, especially sight, touch, and smell, thus perceiving the material as pleasant or unpleasant. Various design options and materials with different properties are available for the surface treatment of wood in interior applications. With the visual and sensory impression, the protective functions of wood coatings in interior applications must also be considered. In exterior applications, the protection of wood against degradation and the technical functions of coatings have priority; visual appearance is significant, too. Sensory properties of exterior surfaces have some relevance only in applications such as decking or exterior furniture, where humans touch or interact with the materials.

This chapter guides the selection of surface treatments for wood in building applications and explains the technical and visual properties of coatings and treated wood surfaces. The furniture area is not covered in this chapter because furniture is not defined as a building component (European Commission 2011). It must be considered that the surfaces of different objects in a room are directly compared, mainly when objects are situated next to each other. This may lead to a direct comparison of furniture with building components such as walls, windows, doors, or floors in optical and sensual impressions. In this respect, the surface quality of objects may differ considerably since substrate quality, machining techniques, coating materials, and application techniques are not the same depending on the purpose and size of the objects. Therefore, some differences must be tolerated.

11.1.1 SELECTION FOR OUTDOOR APPLICATIONS

Wood coatings for exterior applications can be selected according to the EN 927-1 standard (CEN 2013) regarding appearance, *i.e.* build, thickness, opacity, and gloss of the coating, and three distinct end-use categories (dimensionally stable, dimensionally semi-stable, dimensionally non-stable).

The structure refers to the dry film thickness of a coating and is either low (5–20 µm), medium (20–60 µm), or high (>60 µm).

DOI: 10.1201/9781003411994-11

The opacity is determined by the type and concentration of pigments in a coating formulation (Goldschmidt and Streitberger 2014). It can be divided into opaque coatings (the wooden substrate is not visible), semi-transparent (the pattern of the wooden substrate is visible but the coating has a distinct colour), and transparent (almost no pigmentation is used or pigments are so small that the wood colour is unaltered). Finally, the classification of appearance can also be made concerning gloss. It ranges from matt (reflectance value up to 10), via satin (10–35), semi-gloss (35–60), gloss (60–80), up to high gloss (>80). Figure 11.1 shows a variety of possible façade construction and treatment methods on an experimental house. The surfaces are either untreated, treated with semi-transparent, or opaque coatings.

The coating appearance (structure, opacity, and gloss) is an aesthetic factor influencing maintenance intervals (*cf.* Section 11.5.2). In addition to appearance, the right choice of coatings depends mainly on the end-use category. The following sections describe these categories in more detail.

FIGURE 11.1 Examples of different façade constructions and surface treatments on an experimental house (Holzforschung Austria); boards are mounted either horizontally or vertically; surfaces are either untreated or coated with semi-transparent or opaque coatings.

(Courtesy of G. Grüll, B. Forsthuber, M. Truskaller.)

11.1.1.1   Dimensional Stability Categories

Coatings can reduce the speed of moisture changes in wood and increase dimensional stability because of reduced swelling and shrinking (de Meijer and Militz 2001, 2005). Therefore, the dimensional stability categories in EN 927-1 (CEN 2013) are significant guidance for selecting the right coating system for a specific application (Figure 11.2).

Products without dimensional stability requirements (dimensionally non-stable) can shrink and swell almost unhindered. These products include overlapping cladding, fencing, garden sheds, open cladding, and ventilated rain screens. As every piece of sawn timber overlaps its neighbours, a significant swelling is possible without opening a gap between the timber pieces.

Typical examples of semi-stable end-use categories are façades with, for example, tongue-and-groove joints. The façade boards are allowed to move to a certain extent without opening the joints. Other examples of this category are sound-absorbing barriers and garden furniture.

Wood products such as windows and exterior doors are typical examples of the end-use category "dimensionally stable." As only minimal movement is permitted, coatings are required to strongly reduce the water uptake and thus the wood's dimensional changes.

Depending on the application level, different coating systems are recommended (Figure 11.2). Dimensionally stable wood products usually need coatings with high dry-film thicknesses (above 60 µm), as only those ensure the required moisture protection. Low-build coatings (dry film thickness < 20 µm) are only recommended for non-stable construction elements. Medium-build coatings (between 20 and 60 µm) are mainly recommended for products in the semi-stable end-use category. However, medium-build coatings are also suitable for the non-stable and dimensionally stable category, making them quite universal. High-build coatings can also be used for dimensional semi-stable products. However, care must be taken, as water could get trapped in the construction or the wood itself due to the high water-vapour resistance of thick coatings. This is more likely on façades than wood windows, as single coating defects may occur that allow moisture ingress due to mechanical friction in the joints or other mechanical damages due to the large surface area.

11.1.1.2   Coating Systems

Coating systems for exterior applications consist of several different coating products from the same manufacturer. These can be divided into primers, undercoats, and topcoats. Not all products are mandatory. Sometimes, coating systems use a combination of primer and topcoat or just the topcoat. The following section gives a brief overview of the different coating types.

11.1.1.3   Wood Primers

The primary function of wood primers is to provide adequate bonding between the wood substrate and the subsequent finishing coats. Wood primers can also contain fungicides (preservative primers), for example, against blue stain fungi (Sharpe and Dickinson 1992). However, as soon as a coating contains fungicides and has

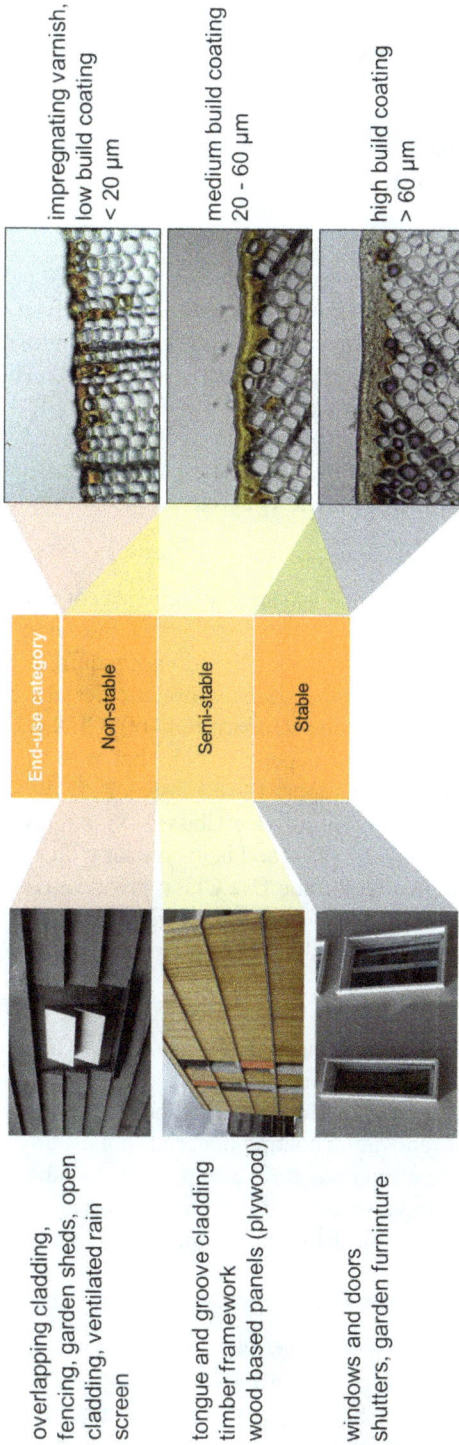

FIGURE 11.2 Selection of exterior wood coatings according to end-use categories and applications.

(Courtesy of G. Grüll, B. Forsthuber, M. Truskaller.)

a declared efficacy against selected microorganisms, the manufacturer must test and label these products appropriately according to the Biocidal Products Directive (BPD) from the European Commission (2012).

11.1.1.4 Undercoats

Undercoats are optional. Their primary purpose is to increase the dry film thickness of the coating. That is why they also contain high pigmentations and are often very brittle.

11.1.1.5 Topcoats

Topcoats or finishes are the final coating applied to wood. While high gloss topcoats often also have an undercoat, lower gloss topcoats are usually formulated to be used directly on wood without or with an appropriate primer. These topcoats benefit significantly from a high elasticity and high weathering resistance. These products can also contain fungicides against, for example, blue stain and mould growth. However, as soon as these fungicides exceed a specific concentration and have a declared efficacy against selected microorganisms, the coating materials are again considered biocidal products and need appropriate tests and labels according to the BPD.

11.1.1.6 Wood Preservation Needs

Depending on the Use Class and the risks and consequences of wood decay, different wood preservatives are recommended when constructive measures alone are insufficient to protect the wood. While interior coatings do not require preservatives, wood in exterior applications should have fungicides or insecticides, depending on the Use Class and application. According to the standard EN 460 (CEN 2023), different Use Classes for wood are defined. These depend on whether wood is only used in the interior (Use Classes 0 and 1) or outdoors (Use Classes 2–5). Whether the wood is sheltered (Use Class 2) or fully exposed (Use Class 3–5) is relevant in the exterior use. Furthermore, it is appropriate if the wood is in contact with soil (Use Class 4) or seawater (Use Class 5). Depending on the Use Class, preservatives against different target organisms (*e.g.* blue stain, wood destroying fungi or insects) are required/recommended. More details are found in the standard or the literature (*cf.* Chapter 10).

11.1.1.7 Application Methods: Industrial and Manual

Wood coatings for exterior use are applied industrially (*e.g.* by the cladding manufacturer) or manually (Bulian and Graystone 2009). In any case, treating the parts when they are cut to length and pre-drilled for fasteners, but before they are mounted, is beneficial. Maintenance or renovation coating materials are usually applied by brushing for on-site treatments. Spraying on-site is also possible but should be avoided due to the high overspray and contamination of the surrounding environment. The coating system must be selected to suit the intended application method.

11.1.2 Selection for Indoor Applications

The position of the surfaces in interior use determines the intensity of physical stresses and the interaction of humans with the objects (Berger *et al.* 2006, Teischinger *et al.* 2012). Timber-building components are used for ceilings, ceiling coverings, columns, walls, wall coverings, windows, doors, and floors. In the order of these uses, stresses and intensity of human interaction increase (Figure 11.3).

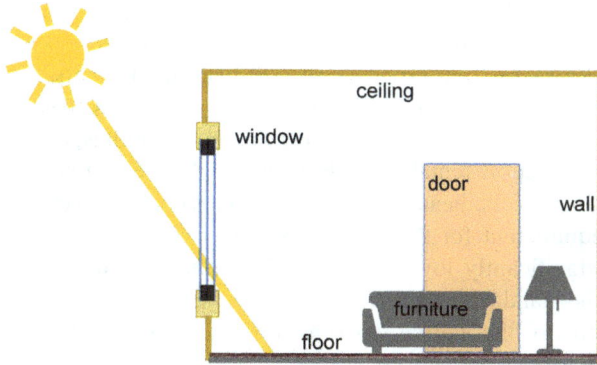

FIGURE 11.3 Timber products in interior use.

(Courtesy of G. Grüll, B. Forsthuber, M. Truskaller.)

The specific needs and requirements for wood surface treatment for these uses are described briefly in the following chapter. Here, load-bearing timber elements, such as cross-laminated timber, glued-laminated timber, and solid structural timber, are treated separately because they constitute the main structure of a building, and surface quality is secondary.

11.1.2.1 Ceilings and Ceiling Coverings

On visible surfaces of ceilings, there is only very little stress on the surfaces. There is no exposure to liquid water or other liquids and no mechanical influences. In some cases, ceilings in bathrooms or kitchens can be exposed to high levels of moisture due to splashing water or water vapour, which can lead to minor discolouration in the case of untreated wood. Accordingly, there are no requirements for the durability or protection of wooden surfaces for ceiling soffits. Wood coatings on ceiling coverings can be used for decorative purposes, and there are no technical requirements.

11.1.2.2 Walls and Wall Coverings

Compared to ceilings, wall surfaces are exposed to higher stresses, but still, these are much less than those on floors. However, especially in the lower areas of interior walls, adverse effects of moisture and pollution of the surfaces must be expected. Touching the walls and occasionally soiling, for example, by spilling tea or coffee, can be regarded as everyday stresses. In the connection region of the floor and wall, a more substantial effect of moisture is expected during the washing of the floors. The plinth protects the wall surface from direct contact with the wash-up water and avoids the mechanical impact of chair backrests by keeping them at a distance. In bathrooms, splashing water must be expected, which leads to higher moisture levels and requires high moisture resistance on the surfaces. In the case of untreated wood, this leads to discolouration in the form of water stains and discolouration within a short period of use. Therefore, applying suitable coatings to wall coverings is advisable to keep the surfaces attractive and permanently beautiful. The requirements for the quality of coatings on wood in the wall area are not regulated by standards.

11.1.2.3   Load-bearing Constructions

In various constructions, cross-laminated timber, glued-laminated timber, and sawn timber are used for structural purposes as load-bearing elements, typically for ceilings, beams, walls, and columns. These products enable us to make buildings of renewable raw materials like wood. In many modern buildings, the surfaces of these parts of the construction remain visible to highlight the engineering performance and, at the same time, the beauty of the wood material. However, surface quality is a secondary requirement for these wood products. Therefore, wood surface quality is usually significantly lower than wood products for decorative applications. According to the standards, the material may include knots, cracks, resin pockets, and other features of the natural wood as long as the strength and stiffness of the construction are given.

In the production of timber construction elements, the timber surfaces are planed by machines. Certain features in the timber may be repaired by inserting new parts of wood, such as dowels from cross-sections of knots, longitudinal wood dowels, or strips. The inserted parts are then levelled. Finger-jointing is a typical longitudinal connection of wooden parts in engineered wood products (EWPs). In strength-grade finger-jointed timber, there must be a clearance at the tips of the finger joints, which remains visible as holes on the surface and cannot be filled entirely by coatings. In the case of cross-laminated timber, a calibration sanding is carried out, which allows abrasive grooves to remain visible at right angles to the direction of the grain due to production. Differences in colour, gloss, and roughness due to the structure of the wood are permitted. In different room climates, cracks can occur in the first period of use due to shrinkage. Coatings cannot avoid cracks and must not be considered a material or producer's fault. These are examples of surface-quality restrictions of load-bearing timber construction elements that the user must accept.

11.1.2.4   Windows and Exterior Doors

As windows and exterior doors separate a building's rooms from the exterior climate, they are considered exterior building components, even though one side is visible inside the room (Sell 1985). Wood coatings for this field of application are intensely standardised in the category of "dimensionally stable wood elements," as described above.

11.1.2.5   Interior Doors

Interior doors' surfaces are exposed to the same stresses as walls. Additional abrasive stress may occur around door handles and locks when a door is used and locked frequently. Wood coatings protect against specific mechanical influences under normal use conditions.

11.1.2.6   Floors

Wooden floors are exposed to very high stresses on the surfaces. Depending on the location, there are further differences in the intensity of use. Floors in dry living rooms are usually exposed to moderate stress. Except for the anteroom, most areas in a living space are hardly walked on with street shoes, are cleaned regularly, and have only short periods of moisture exposure. Normal use must be considered for floors usually used with street shoes but in low traffic, such as in anterooms or kindergartens. For floors with high traffic use with street shoes (*e.g.* school rooms, restaurants, offices, sales premises, and other publicly accessible premises), heavy stresses apply. By using street shoes, the surfaces are usually exposed to greater moisture loads, especially in winter. In addition, contamination of the floors or the soles of shoes with sand or stones can lead to powerful mechanical effects. This can result in scratches, abrasion marks, and worn floor areas. Therefore, selecting suitable materials and coatings is essential, especially in heavily stressed areas. Some standards specify minimum requirements for surfaces and coating materials for wooden floors for these stress classes according to the EN 14354 standard (CEN 2017) and the ÖNORM C 2354 (ASI 2009). Wooden floors with non-film-forming surface treatments can also be used in intensively used areas if suitable materials, application methods, and maintenance concepts are used.

In office workplaces, the load of castors on swivel chairs can cause heavy stress on wooden floors. For this reason, castors with soft running surfaces (type W according to the EN 12529 standard (CEN 2007)) must be used for swivel chairs, which are usually characterised by a different colour than the castor body. Furthermore, floors made of softwood species are not recommended in office workplaces because of their low hardness. Floors made of hardwoods, such as oak, maple, ash, or birch species, are suitable for such applications.

11.2 TYPES OF SURFACE TREATMENTS

11.2.1 FILM THICKNESS AND FILM FORMATION

A distinction is made between film-forming coatings, which form a continuous, clearly perceptible and measurable coating film on wood surfaces, and non-film-forming coatings, which do not create a continuous coating film (Grüll and Hansemann 2023). A continuous coating film can be achieved on planed wood surfaces at dry film thicknesses above approximately 20 µm. Therefore, coatings with less than 20 µm dry film thickness are categorised as non-film-forming. This threshold marks a shift in the properties of coated wood surfaces because non-film-forming coatings cannot flake from the surface during ageing, and they are also easy to maintain and renovate.

Film-forming coatings provide better surface resistance against liquids and better substrate protection against moisture, light, dirt, and erosion, achieving higher durability. They are made by using paints, varnishes, or lacquers. In contrast, non-film-forming coatings use impregnations (waterborne or solvent-based), oils, or waxes (Janesch *et al.* 2020). However, there are also film-forming oils, which are often referred to as hard oils. Table 11.1 gives a general overview of the practical advantages of non-film-forming and film-forming coatings on wood.

For wood species with coarse vessels, such as oak or ash, a distinction is also made between open-pore surface treatments, in which the vessels cut on the wood surface are not entirely filled, and closed-pore surface treatments (Figure 11.4). In the latter, a perfectly smooth surface is produced with a film-forming coating, including fillers, and the coating film covers the pore structure (de Windt *et al.* 2014).

TABLE 11.1

Advantages of Non-film-forming and Film-forming Coatings on Wood

Non-film-forming Coating (*e.g.* Oil and Wax)	Film-forming Coating (*e.g.* Lacquer)
Natural appearance	Adjustable gloss
Partial repair	High resistance
Diffusion open	Easy cleaning
Good wet look with oils	Flexible colour selection
Minor colour change during ageing	Good light protection is possible

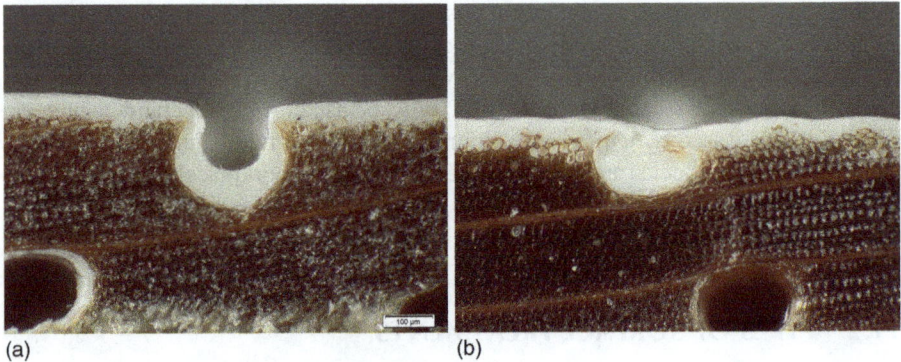

(a)   (b)

FIGURE 11.4 A white-coloured coating on meranti wood: (a) open pore coating and (b) closed pore coating.

(Courtesy of G. Grüll, B. Forsthuber, M. Truskaller.)

In the case of film-forming coatings, the gloss level is adjusted by the proportion of matting agents in the coating material, and it is possible to choose between high-gloss, semi-gloss, and matt surfaces.

Non-film-forming surface treatments involve materials that penetrate and impregnate the wood substrate. Oils can do this very well, enhancing the wood structure's contrast and creating a more intense colour impression ("wet look"). Waxes are often used in combination with oils to create a water-repellent effect and durability with a thin layer on the surface of the wood. The non-film-forming surface treatments of wood usually result in a lower resistance, for example, to water or other liquids. However, they offer the advantage that partial repairs on surfaces are more straightforward than with film-forming surfaces.

The film formation of coating materials on wooden surfaces impacts water-vapour permeability. Film-forming coatings form a higher resistance to vapour diffusion than non-film-forming coatings. In many non-film-forming systems, no water vapour diffusion resistance is measurable. A high permeability of wood surfaces in indoor use can be advantageous for stabilising the indoor climate. The hygroscopic material wood can absorb moisture from the surrounding air and rerelease it in dry indoor air conditions. On the other hand, diffusion resistance can reduce emissions from building materials.

Coating formulation, viscosity, binder content, and the coating material's application rates are the major parameters influencing film formation on wood.

The EN 927-1 standard (CEN 2013) defines film thickness categories for exterior wood coatings as minimal, low, medium, high, and very high build. This may be simplified into four essential categories: impregnations, low, medium, and high-built coatings.

11.2.2 TRANSPARENCY

The pigmentation of coating materials determines the hiding power of the coating (Figure 11.5). In the case of opaque coatings, colour differences of the wood substrate are no longer recognisable. In the case of light opaque colours, predominantly white, cracks in the wood and occasional discolouration of knots or a cut wood fibre can be visible. Varnishes are semi-transparent, which leads to colouring in connection with the substrate through transparent pigments. Differences in the colour of the wood are not fully covered. With white transparent coatings, good colour stability is achieved. Clear coatings do not contain any colouring pigments. To improve colour stability, they are usually equipped with transparent light stabilisers.

11.2.3 COMPOSITION OF COATING MATERIAL

Most coating materials are liquids composed of a series of ingredients according to a defined recipe by the coating manufacturer. These coating formulations consist of three main components: polymers, solvents, pigments/fillers, and a series of additives in small quantities. Polymers and non-volatile additives together form the binder of the coating material, as shown in Figure 11.6. Goldschmidt and Streitberger (2014) provide more details on coating raw materials.

11.2.3.1 Binder Polymers

The backbone of a coating is the polymeric binder, which significantly affects its quality and durability. These binders can be divided into natural binders, modified natural binders, and synthetic binders. Synthetic binders are sometimes highly engineered

(a)   (b)   (c)

FIGURE 11.5 Hiding power and transparency of wood coatings: (a) opaque coating, (b) semi-transparent light brown coating, and (c) clear coating.

(Courtesy of G. Grüll, B. Forsthuber, M. Truskaller.)

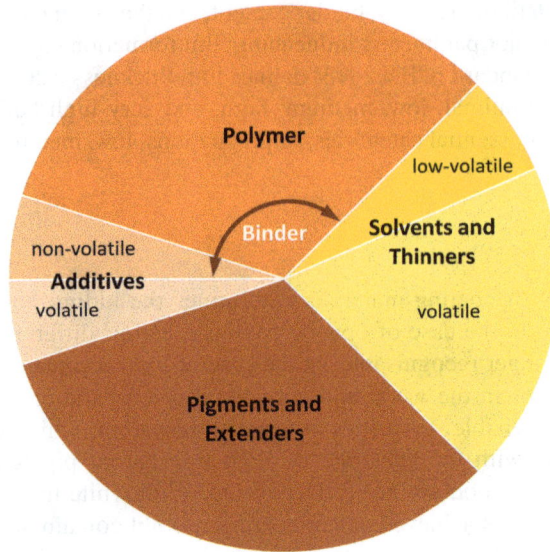

FIGURE 11.6   Components of a coating material composition.

(Adapted from Goldschmidt and Streitberger 2014.)

polymers with outstanding properties. While these binders were produced mainly from petrol-based monomers in the past, substantial efforts are currently being made to obtain the same or similar polymers from bio-based raw material sources. This is realised either by improving traditional ways of surface treatments with oils and waxes (Lesar *et al.* 2011, Humar and Lesar 2013, Teacă *et al.* 2019, Arminger 2021) or by producing synthetic resins from non-fossil but renewable feedstock (Kurimoto *et al.* 2001, Kumar *et al.* 2017, Jung and Baumstark 2018). Table 11.2 lists the most important binder polymer groups for wood coatings in interior and exterior use.

11.2.3.2   Solvents and Thinners

Solvents keep the binder polymer in a liquid state with the desired viscosity needed for a good and easy application to the substrate. Thinners are used additionally to adjust viscosity. Depending on the type and amount of solvents and thinners, coating materials can be solvent-borne, water-borne, or solvent-free.

Solvent-borne systems are classical lacquers or paints with mixtures of petrol fractions of different boiling points to adjust drying speed. The liquid coating materials have a significant smell. Evaporating solvents may be explosive and harmful to the environment and health during application and drying. After drying, a small proportion of the solvents are emitted slowly and may influence interior air quality.

Water-borne systems: Water is not a solvent for the substances used in a coating formulation, but it can be used as a liquid phase and thinner in a coating

---

TABLE 11.2

Most Important Binder Polymer Groups for Wood Coatings

	Interior Coatings	Exterior Coatings
Natural binders		
Drying oils	x	x
Natural resins	x	
Modified natural binders		
Modified oils	x	x
Nitrocellulose	x	
Synthetic binders		
Polyesters	x	
Alkyd resins (oil-modified polyesters)		x
Urea-formaldehyde resins	x	
Polyurethane resins	x	x
Polyacrylate resins	x	x

---

system to reduce organic solvents as much as possible. A small proportion of co-solvents, less than 10% of the total formulation, is still necessary for water-borne coating formulations. Water-borne coating materials are dispersions of binder particles in water; the significant technologies are suspensions of acrylic binder particles (solid in liquid) or emulsions of alkyd binders (liquid in liquid).

Solvent-free systems: The following two coating technologies can be regarded as solvent-free. Acrylic coatings for interior use can be realised as 100% UV systems. The liquid phase consists of oligomers of the acrylic binder and a photo-initiator as an additive. Curing of the coating material is initiated after application to the substrate with UV lamps in an industrial production process (Glöckner *et al.* 2008). The second technology is powder coatings, where dry particles are applied to the substrate by powder spraying, and curing is realised under high temperatures. For wood-based materials, powder coating is limited to medium-density fibreboard (MDF) for interior use.

11.2.3.3 Pigments and Fillers

Pigments and fillers consist mainly of minerals in defined particle sizes mixed into liquid coating formulations. Pigments define the colour and light transmission of the coating film, whereas fillers only have the function of gaining volume and increasing film thickness. The type and content of pigments determine a coating material's transparency and hiding power. Specifically for wood applications, transparent pigments (with low particle size) are used to obtain semi-transparent coatings even at high film thicknesses (high build). When applied to a wood substrate, the selection of pigments determines whether the coating materials result in clear coatings, semi-transparent varnishes, or opaque paints.

11.2.3.4 Functional Additives

Every coating formulation contains a series of additives with a content of a few per cent. These are very important for drying and curing, improving quality, and adding functions to the coating film. Examples of the most common types of additives are listed in Table 11.3.

11.2.4 STAINING AND COLOUR ENHANCEMENT (INTERIOR USE)

A colour design can be done with pigmented varnishes. With opaque pigmentation, the colour differences of the wood substrate are entirely covered, and with transparent pigmentation in varnishes, colour differences of the wood substrate are still recognisable (*e.g.* white varnish). On the other hand, the colour of the wood surfaces is often changed by staining. Different stain systems are used depending on the type of wood and the desired staining pattern (Prieto and Kiene 2007). The natural colour differences can be increased or reduced depending on the stain. Especially in the case of softwoods, it is essential to pay attention to a positive staining pattern (Figure 11.7). This means that the earlywood of the growth rings should remain lighter than the latewood after staining while the contrast is increased. Dyes are deposited on the surface of the wood, and these accumulate more in areas of wood with low density and large pores than in denser regions of the wood. As a result, the earlywood of the growth ring is darkened, and the latewood remains lighter, which leads to a so-called negative staining pattern. In addition, there are specific processes in which only the pores of coarse-vessel wood species, such as oak, are filled with a pigment called pore stain. Stained surfaces are always coated with a clear coat to protect the colouring and maintain a surface resistant to stress.

TABLE 11.3
Main Types of Additives in Wood Coating Formulations

Additive	Substance	Function
Dryer	Cobalt (Co) or manganese (Mn) salts	Catalyses the polymerisation of drying oils
Surfactant	Tensides	Dispersion, wetting in waterborne systems
Initiator	Source of free radicals	Starts curing in UV coatings
Matting agent	Minerals or polymer particles	Gloss adjustment
Radical scavenger	Hindered amine light stabilisers (HALS)	Scavenging of radicals formed by UV radiation in polymers in coating and substrate
UV absorber	Hydroxybenzotriazoles, hydroxybenzotriazines	Light protection of polymers in coating and substrate
Biocide	Biocides	Conservation against microorganisms

Note: UV – Ultraviolet radiation.

(a)   (b)   (c)

FIGURE 11.7   Natural colour differences due to stain: (a) positive stain, (b) negative stain, and (c) pore stain.

(Courtesy of G. Grüll, B. Forsthuber, M. Truskaller.)

11.3   PURPOSE OF COATINGS

Coatings on wooden surfaces have both decorative and protective functions. The technical purpose of coatings is to increase and protect the wood surface against mechanical and chemical degradation and light. At the same time, they are used for colour design by applying pigmented coating materials, and coatings can also control the gloss of a surface. Many types of coatings combine both technical (Figure 11.8) and decorative functions.

Original wooden surfaces, whether rough-sawn, planed, brushed, or sanded, usually have a matt appearance and are very susceptible to staining or discolouration when exposed to liquids, including water, and change colour when exposed to light. Bright wood species (*e.g.* Norway spruce, fir) can darken very quickly, while dark wood species (*e.g.* walnut) can fade under the influence of light (Passauer *et al.* 2015). Wood is a hygroscopic material; in contact with the surface, almost all liquids are quickly absorbed capillary into the cell-wall cavities, making cleaning of uncoated wood impossible. The absorption and release of moisture lead to dimensional changes in wood (swelling when moisture is absorbed and shrinkage when moisture is released). This is undesirable in, for example, windows and doors that need to be dimensionally stable for their function and can lead to problems. Depending on the type of wood, low-density softwood rather than high-density hardwood, the surface is only partially resistant to mechanical damage. Therefore, several properties of untreated wood surfaces are undesirable and detrimental where coatings with their assigned protective functions are used. Summarised to the essentials, coatings create a lasting visual appearance, protect the substrate, and fulfil functional tasks. The influences on and requirements of the surfaces for interior and exterior applications differ. The functions of the coatings must be adapted to the application. The influencing factors and the most critical functions of the coatings are described below according to the application. Studies have shown that natural, sanded wood surfaces are perceived as very pleasant (Truskaller 2017). Particularly for interior applications, efforts are being made to develop coatings that also have a pleasant haptic effect and offer the protective functions mentioned below.

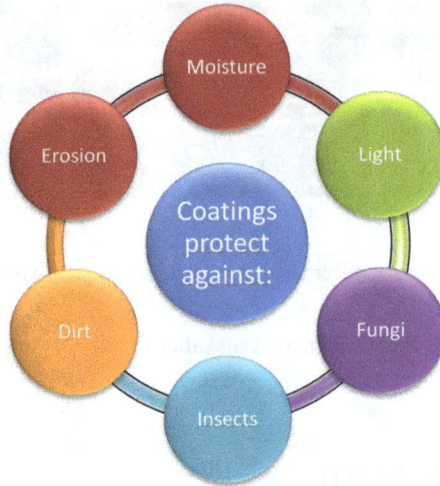

FIGURE 11.8   Protective functions of wood coatings.

(Courtesy of G. Grüll, B. Forsthuber, M. Truskaller.)

11.3.1   Outdoor Use

Wood is a natural organic material that is biologically degraded and decomposed outdoors by organisms and environmental influences over time. It is then reintroduced into the natural cycle through its basic elements (Evans *et al*. 2005). In exterior use, light, humidity, and rain strongly affect unprotected surfaces. There are numerous biotic (including wood-discolouring blue stain and mould fungi, wood-destroying fungi such as brown rot and white rot and insects) and abiotic damaging influences (including high temperatures, weather influences such as UV radiation and water, chemical and mechanical influences) that destroy the wood material. In practice, the intensity of these damaging influences depends on the geographical location and altitude of the building, the orientation of the timber components according to the direction of the compass, the inclination of the surface to the vertical and the structural protection of the surface from direct weathering by canopies and protruding components.

If untreated and uncoated wooden surfaces are exposed to the weather without protection, changes to colour and the surface structure occur relatively quickly. The colour changes are mainly caused by photodegradation, colonisation with microorganisms, and dirt contamination. Initially, an irregular, patchy brown colouring occurs. After prolonged weathering, the wood turns grey. Irregularly weathered components that are partially sheltered (*e.g.* by balconies) also show very irregular discolouration and, consequently, a very patchy external appearance (Figure 11.9).

In the case of uncoated wood surfaces, internal stresses in the wood and the resulting longitudinal end-grain cracks after a short period of weathering due to very rapid

FIGURE 11.9 Irregularly weathered façade due to differences in light exposure, humidity, and rain reaching the surfaces. The façade is partially sheltered by various overhangs such as balconies and sheet metals for fire protection.

(Courtesy of G. Grüll, B. Forsthuber, M. Truskaller.)

moisture absorption and release through the cross-section of the sawn timber. UV radiation and short-wave visible light photolytically cleave the water-soluble lignin as weathering continues. Due to the degradation of the lignin on the weathered wood surface, the cellulose fibre layers of the cell walls lose their cohesion and are eroded by weather influences such as wind, rain and snow, insect damage, and other mechanical influences, particularly in the earlywood regions (Böttcher 2004).

Protecting the wood from damaging influences is necessary to maintain the functionality of timber components over a more extended period, as described below.

11.3.1.1 Moisture Protection

Coating films or the hydrophobic properties of coating materials can protect against increased moisture absorption. Cross-section regions can absorb liquids quickly, but moisture protection can also reduce the formation of surface cracks (de Meijer and Militz 2001). The permeability of coatings is a significant factor in selecting materials to fulfil the dimensional stability categories according to EN 927-1 (CEN 2013).

Moisture protection is essential, as water is usually the precondition for destroying wood (Rapp *et al.* 2000, Viitanen *et al.* 2010, Bornemann *et al.* 2014). Increased wood moisture is a prerequisite for fungal infestation and also for the leaching of wood components such as photolytically split lignin and wood constituents (Rothkamm *et al.* 2003), and also has an unfavourable effect on the adhesive strength of a coating on the substrate (Ahola *et al.* 1999).

The moisture protection of a coating system is mainly influenced by its dry-film thickness and its formulation (Wassipaul and Janotta 1972, Derbyshire and Robson 1999, de Meijer and Militz 2001, Ekstedt 2002, Svane 2004, Gibbons *et al.* 2020). The liquid-water permeability shall be low, and the water-vapour permeability shall be in a range to allow the drying of trapped water (Miklečić and Jirouš-Rajković 2021). In Figure 11.10, the moisture content of Scots pine sapwood panels during field exposure, measured underneath the coating, using coatings with different dry-film thicknesses (EL_1xICP is the lowest, EL_3xICP the highest) is shown compared to uncoated wood (EL_U). The ICP was a standardised coating material that can be characterised as a solvent-borne alkyd stain with red pigmentation. This was applied 1x, 2x, and 3x by brushing. The coatings 2x ICP and 3x ICP reduced the moisture content and its fluctuations in the wooden panels. At the same time, the minor effect of 1x ICP was lost after five months of natural weathering due to the degradation of the coating.

11.3.1.2 Light Protection

The photo-oxidation of the wood components by the short-wave proportions of sunlight is the beginning of the weathering process of wood surfaces. Initially, it causes colour changes (yellowing, browning). This is followed by a degradation of the wood substance in areas close to the surface, which is an essential process in the greying and erosion of the wood (Grüll 2010). Effective light protection of exterior wood surfaces can only be achieved if the coating completely filters out the UV component of the sunlight. In addition, protection against short-wavelength visible light up to approximately 500 nm is required, as this can induce photo-oxidation processes on the wood surface (Kataoka *et al.* 2007).

Many colourful shades can be achieved with opaque pigmented paints. These coatings are also highly durable because they protect the wood well from weather effects, especially harmful UV light. With these systems, the weather resistance of coated wood is comparable to that of automotive paints, which is sometimes used as

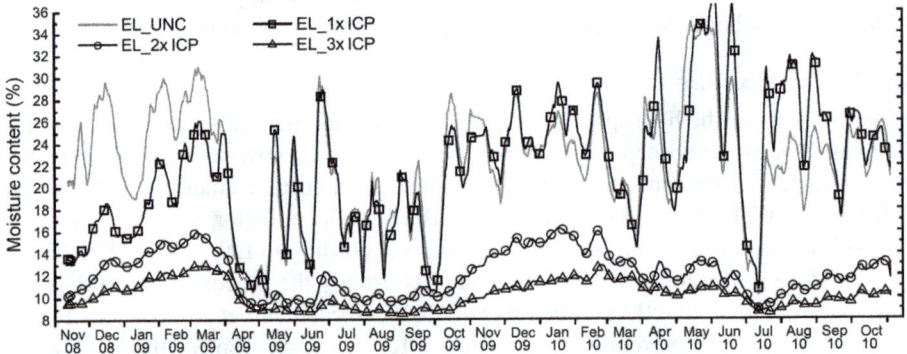

FIGURE 11.10 Moisture content of uncoated and coated Scots pine sapwood panels during exterior exposure (weekly mean values). Coating system with high (EL_1xICP), medium (EL_2xICP), and low (EL_3xICP) water permeability.

(Data from Grüll et al. 2013, courtesy of G. Grüll, B. Forsthuber, M. Truskaller.)

FIGURE 11.11   Rough-sawn timber with opaque paint.
(Courtesy of G. Grüll, B. Forsthuber, M. Truskaller.)

a benchmark for durability. With smoothly planed wood surfaces, the wood structure is covered and no longer visible from a distance. However, a sensible alternative is to coat rough-sawn timber with opaque paints, which leaves the wood structure recognisable and achieves a high level of durability (Figure 11.11).

Rough-sawn timber absorbs a large quantity of coating material, which is well mechanically anchored to the wood surface, and due to the very irregular coating film, large-scale flaking is not expected. Due to the unique structure of the wood, transparent coatings are desired, for which varnishes with transparent iron oxide pigments in various brown, red, and yellow colours are available. Only these transparent pigments absorb enough of the short-wave UV range of light to adequately protect the wood surfaces against degradation processes without further light protection additives. Individual colourless coating systems are available on the market for wood in exterior use, whereby only tested products (two years of outdoor weathering in accordance with the EN 927-3 standard (CEN 2020)) should be used in this product category. They contain transparent light stabilisers that absorb UV light or render aggressive degradation products (radicals) harmless (Forsthuber and Grüll 2010). The weather resistance of coloured pigmented glazes and opaque paints is still unmatched by most transparent systems (Grüll 2010).

11.3.1.3   Protection against Microorganisms

Unprotected wood is attacked by microorganisms such as insects and fungi, mainly when used outdoors. The less durable wood species and the sapwood part of the tree are particularly at risk due to their naturally low durability. Specifications for protecting wood are contained in various wood-protection standards and guidelines. They are divided into preventive protective measures, such as structural wood protection, constructive wood protection, chemical wood protection, and combating wood protection measures.

The structural and constructive wood protection aims to keep water and moisture away from the construction and prevent water traps. Chemical wood preservatives introduce active chemicals into the wood.

The active chemicals are mainly applied to the wooden components using impregnation and primers. Primers are usually part of a coating system completed with a topcoat. In the case of superficial application methods such as painting, rolling, flow coating, or dipping, the impregnation liquid can penetrate a maximum of 2 mm beneath the face and edge surfaces of the timber, but considerably deeper when penetration through the cross-section surface. In addition to impregnations and primers, active ingredients are also found in film-forming coatings, as so-called film preservation. They are intended to serve as a barrier against the superficial growth of mainly blue stain and mould fungi and to maintain the moisture protection function of the coating for longer (Sharpe and Dickinson 1992).

In addition to the protection provided by the active ingredients in the coating materials, film-forming coatings offer a mechanical barrier against the penetration of fungi and insects into the wooden substrate from a specific film thickness. The slightest injuries, coating cracks, and layer-thickness reductions can be entry points for moisture and especially for fungi.

According to the current state of the art, the moisture protection function and the mechanical protection of the coating are not enough protective functions against microorganisms in applications with a high risk of damage. However, the moisture permeability of wood coatings plays a vital role in the moisture balance of timber construction components to lower the moisture-content levels below 20% and also lower the moisture fluctuations (de Meijer and Militz 2005).

11.3.1.4 Physical Protection and Cleanability

Depending on the type of wood, uncoated wood surfaces are only slightly resistant to mechanical influences, although high-density hardwood is often more resistant than low-density softwood. In the case of softwood, there are differences between the less dense earlywood and the considerably denser latewood. Wind and driving rain erode the wood surfaces without a coating, removing the earlywood. Coatings can be a physical barrier that prevents erosion of wood under weathering.

Mechanical influences can be powerful, especially in building components where users frequently interact, such as decking, balcony flower troughs, balcony handrails, etc. Exterior coatings must resist some mechanical influences, and the main goal is to maintain the coating's protection properties and, thus, its visual appearance.

Coating systems must manage the balancing act between high mechanical resistance, often accompanied by an increase in hardness, and the elasticity of the coating required in outdoor applications (Schwalm *et al.* 1997, Baumstark and Tiarks 2002). Exterior surfaces are also exposed to influences such as hail and ice, and hail events are becoming increasingly intense. Here, a high elasticity of coating films is required to prevent cracking.

As mentioned, uncoated wooden surfaces are very prone to soiling. Colouring substances absorbed into deeper layers cannot simply be cleaned from the surfaces. Coatings make the surfaces cleanable when a closed coating film is on the wood surface.

11.3.1.5   Colour Stability and Visual Aspects (Including Grey Varnishes)

The possibility of visual design of the surface and the long-term preservation of the appearance is extremely limited without using coatings. Although it is possible to design the appearance in new condition to a certain extent by selecting wood species and sawn timber according to the quality of the wood (knot-free, knotty), the growth-ring orientation (quarter sawn, half-quarter sawn, flat sawn), the wood pre-treatment (rough-sawn, brushed, planed, sanded), and the wood species, the range of design options, compared to coated surfaces, is small. Above all, it is impossible to maintain the original appearance in the outdoor application without using coatings, as uncoated surfaces turn grey in the long term. With all wood species with different original colour greying due to weathering results in very similar grey shades within a few months or years, and if this greying is not accepted, design aspects may be lost (Sell and Leukens 1971, Rodríguez-Grau *et al.* 2021).

By using special, grey-coloured varnishes, it is possible to create the visual appearance of a greyed surface right from the start (Figure 11.12). In the case of heavily weathered surfaces, the grey-coloured varnish should disappear because of the weathering process, and the wood emerging from underneath should naturally turn grey. In the case of the regions that are protected from weathering, the grey appearance is retained by the greying varnish, which is still intact. Accelerated greying of wood surfaces can also be achieved with iron (II) sulphate ($FeSO_4$) solutions in traditional methods used in Scandinavia (Hundhausen *et al.* 2020).

FIGURE 11.12   Special grey-coloured varnishes on a timber façade.

(Holzforschung Austria, courtesy of G. Grüll, B. Forsthuber, M. Truskaller.)

In the case of coating systems, a fundamental distinction must be made between transparent and semi-transparent (varnish) systems, in which the wood substrate remains recognisable and opaque coating systems (opaque paint). Varnishes are traditionally found in Alpine countries such as Austria, while opaque coatings are particularly common in Scandinavia and the Mediterranean region.

Users should be aware that the choice of colour significantly affects the maintenance interval of the coating, which is explained in more detail in Section 11.5. The choice of colour should also be made in view of the advancing summer warming of buildings. Bright colours reflect radiation more effectively, and surfaces heat up less. Depending on the structure of a timber façade, this can reduce the cooling requirement for buildings in summer (Nusser *et al.* 2024).

11.3.2   INDOOR USE

In interior applications, coatings' technical functions are only needed in areas where stresses occur. This is usually on floors, doors, and lower areas of walls, where heavy stresses may be present, whereas ceilings are not exposed to degrading factors except some diffuse light irradiance. The purpose of coatings on building components in interior use is to enable resistance of the surfaces to prevent spots and discolouration that cannot be removed by cleaning.

11.3.2.1   Resistance against Water and Other Liquids

Coatings prevent dirt absorption and staining on wooden surfaces, making it easier to remove contaminants. This can be achieved by the physical barrier of a film-forming coating, where the resistance of the binder polymer is decisive for the performance of the coated surface. With non-film-forming coatings, the essential quality criterion is high water resistance.

11.3.2.2   Cleaning and Hygiene

Good cleaning properties are an essential factor for obtaining hygienic surfaces. Cleaning should be done only with water and cleaning agents, which the coating manufacturer recommends for the specific product. Strong solvents must be avoided, and oil-treated surfaces may be damaged when high amounts of surfactants are used for cleaning. Disinfectants must be selected carefully before they can be used on coated wood surfaces. Ideally, the coating should be tested for resistance against the disinfectant to be used.

11.3.2.3   Light Protection

All types of wood change colour when exposed to light, with light wood species usually yellowing and dark wood species fading to brighter hues. Most indoor surfaces are exposed to diffuse daylight, while surfaces of floors, walls, and doors close to windows may be irradiated by direct sunlight filtered by the windowpane and shadowing. Artificial light usually does not have a negative effect on wooden surfaces. By choosing surface coatings with pigmentation or light protection finishes, the light-fastness of the surfaces can be improved. The effects of the colour changes become visible when, after a long time, pictures are taken from the wall, furniture is moved, or carpets are removed.

11.3.2.4   Colour and Gloss

With pigments in coating materials, the colour and hiding power can be designed very diversely. Matting agents can adjust the gloss of film-forming coatings. However, high-gloss surfaces cannot be achieved on timber construction elements because deviations in the wood structure (*e.g.* knots, fibre tendencies) are frequent and can lead to differences in gloss.

11.4   COATING APPLICATION

The application of coatings on newly produced wooden components is usually carried out before installation, either by industrial application in a factory or in the workshop of a carpenter or interior decorator. However, it is also possible to apply coatings on site when installed. In the case of traditional wood floors, which are laid as non-treated elements, this is a common operation after filling and sanding the surfaces. Even in the case of industrially oiled surfaces, an oil application is still advisable when installed, for which the manufacturer's recommendations must be observed. Renovations also require an on-site application.

11.4.1   Pre-treatment of Wood

The machining techniques and sharpness of the tools used determine a wood surface's actual structure and quality. A combination of planing and sanding is a standard procedure to prepare wood surfaces for coating application, where the sanding step produces a homogeneous and defined roughness of the surfaces and contributes to better coating adhesion. Blunt machining tools have a fundamentally negative effect on surface integrity and coating performance. Therefore, a timely change of tools is crucial to ensure constant product quality (Böttcher 2004).

11.4.1.1   Machining and Wood Quality

When sawn surfaces are used as a substrate for coatings, such as claddings, the dried sawn timber is split by bandsaws, and the bandsawn surfaces are used as the visible surface of the façade. These machines produce fine-sawn surfaces with a linear cutting pattern, and the surfaces can absorb a high amount of coating that is very well bonded to the wood. The coating thickness varies strongly on these surfaces, and they do not tend to flake in larger areas; instead, the film degrades into small pieces more uniformly.

Planing is usually done in machines with cylindric or helical tools, where feeding speed, tool geometry, sharpness, and rotation speed of the tools significantly influence surface quality (Hernández and Cool 2008a,b).

11.4.1.2   Sanding

Sanding with defined sanding grid sizes is a standard procedure for preparing plane and profiled wood surfaces before coating application. For the highest quality requirements, a series of sanding steps with decreasing grain size is applied where the machines work in, against, and across the grain direction of the wood, with the last step in the grain direction.

11.4.1.3 Structuring

Structural planning techniques are sometimes used to produce ribbed surfaces for decking, which reduces checking (Evans *et al.* 2010). Machining sawn timber surfaces with brushes leads to erosion according to the density variations in the wood structure. Low-density earlywood is eroded deeper, and high-density latewood is less eroded. This leads to a relief of the growth rings where latewood forms peaks and earlywood constitutes valleys of the structure. Hence, surfaces with higher mechanical resistance and interesting tactile properties are achieved (Grüll *et al.* 2020).

11.4.1.4 Cleaning

After machining, loose particles such as sanding dust or loose fibres must be removed from the surface. A clean substrate is a prerequisite for good coating quality.

11.4.2 MANUAL APPLICATION

When small quantities of timber components or various geometries are produced, manual methods for applying the coating materials are preferred because they are flexible and accessible. Professional craftsmen use manual application methods, such as painters, carpenters, joiners, or floorers, but also in do-it-yourself (DIY) work by lay people. The supplier´s technical and safety data sheets for each coating product must be followed. They will provide guidance on the correct and safe use of these products, including the proper method and parameters for application.

Depending on the surface's size and location, the coating product used, and the substrate, technical qualifications are required to apply the coatings evenly and avoid build-up marks at transitions. In the commercial production of wooden components, coating materials are processed and dried in the craftsman's workshop, where suitable conditions and clean rooms are usually available. When on-site application is necessary, climatic conditions may vary, and clean room conditions must also be maintained as well as possible. Floor sealants applied on-site usually also provide good protection for the joints of parquet floors, as they are sealed by fillers and covered with coating material. Nevertheless, the joints of any wooden floor will open and close over its lifetime, which is caused by the constant swelling and shrinkage of the wood depending on the indoor climate over the seasons.

11.4.2.1 Brush

A brush is often used for smaller parts like window frames and profiled and complex surfaces, where a roller is unusable. However, it can also be used on larger surfaces like cladding, decking, and ceilings. It is a material-saving type of application but time-consuming. Depending on the worker's skills and the coating material, brush-strokes may remain visible to some extent. There are many different types and qualities of brushes, and the right choice lies within the skills of the professional or DIY worker.

11.4.2.2 Roller

The roller application is material-saving, like the brush application, but it is less time-consuming in larger areas. Most products suitable for brush application are also appropriate for roller application. Roller application is typical for more extensive and even surfaces like floors, walls, ceilings, decking, cladding, and doors. It is often combined with brushes used for detail work on edges and corners.

11.4.2.3   Spray

With spray application, it is possible to achieve a very high quality of the finish. Spraying is generally less efficient and more material-consuming because of the inevitable overspray. It is usually a fast method when using the right equipment (nozzle size, filters, etc.). The cleaning of the equipment is, in a way, time-consuming. One advantage is the smooth surface and the absence of brush- and roller-strokes. Especially on the construction site, all parts not intended to be coated and close to the processed surface must be covered. Spraying requires the worker to have the skills to produce excellent smooth surfaces. It is suitable for any surface on the floor, wall, ceiling and single parts.

11.4.2.4   Pad Machine

Pad machines are commonly used on flooring to distribute and polish oils and waxes onto the surface. They are also used for cleaning, polishing, and maintaining floors, where different grades of pads are available for each working step. These machines enable the processing of large areas in a uniform quality.

11.4.2.5   Oil Application

Oil can be applied with a brush, roller, or rag. Usually, the excess not impregnated into the wood must be wiped away, which can be done with a pad machine on the flooring. Oily rags must be handled carefully and strictly according to the supplier´s safety instructions because of the risk of spontaneous ignition.

11.4.3   INDUSTRIAL APPLICATION

In industrial applications, surface treatment is carried out under optimal conditions. As a result, the coating properties are tailored to the products' intended application area. In addition, environmentally friendly coating materials are applied to the products using energy-efficient processes for application and curing. This applies particularly to the coating of finished parquet, where high-quality coatings are produced in the production plant.

11.4.3.1   Flow Coating

The flow coating technique is used to apply primers and intermediate coats on parts with flexible geometry. A small volume of coating material is pumped in a circular motion and spread over the pieces to be coated. This method has many advantages over the dipping method used for primer application in industrial manufacturing.

11.4.3.2   Spray

Industrial spraying uses a high degree of automatisation, including sensors, to capture part geometry and enable flexibility in the process. A primary task is avoiding overspray and loss of coating material, a general downside of spray application. Electrostatic spraying and recycling of overspray material can reduce this.

11.4.3.3 Vacuum Coating

Vacuum machines for coating application allow for a very efficient and fast process to treat parts with linear geometry, such as boards, battens, and panel edges.

11.4.3.4 Roller Coating

Industrial roller coating allows for using UV curing coatings with low solvent content or solvent-free coating materials, and the application process reaches very high transfer efficiency.

11.4.3.5 Radiation Curing

UV light, infrared light, or electron beams are possible energy sources for forced drying and curing of specific industrial coating systems. This improves efficiency and speed and the final quality of coatings on wood treated in industrial processes.

11.5 MAINTENANCE OF EXTERIOR WOOD COATINGS

Coatings can considerably increase the aesthetic and, in many instances, the technical service life of wooden building products in exterior applications. However, regular maintenance is needed if coatings are used to protect the wood surface. Applying a maintenance coating at the right time can increase the service life of a coating and, thus, the coated building elements. This is shown exemplarily in Figure 11.13, where a maintenance coating was applied only on one side of a wooden handrail corner after a hail event. It can be seen that the half without proper maintenance was severely degraded after 24 months and would have to be renovated. The adequately maintained half, however, is still fully functional without any signs of cracking, flaking, blue stain, or other visible coating defects.

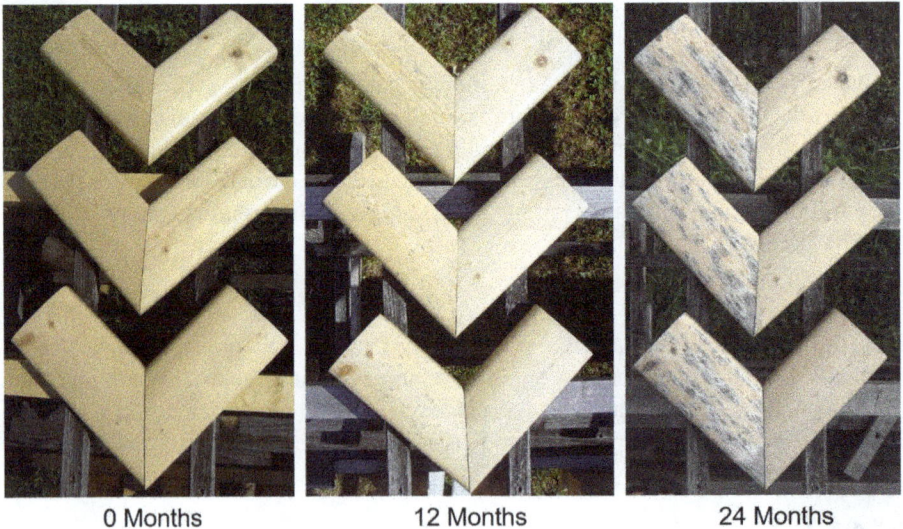

0 Months   12 Months   24 Months

FIGURE 11.13 Clear-coated wooden handrail corners, where only the right side was maintained adequately after a hail event.

(Courtesy of G. Grüll, B. Forsthuber, M. Truskaller.)

A coating system on wood can be maintained and renovated several times, contributing to an extended service life of wood components (Graystone 1985, Grüll 2003, Grüll and Tscherne 2020). While applying a maintenance coating is relatively easy to do (just applying one layer of the same or a compatible coating to the old one), renovation is very costly and includes the removal of the whole coating system, including the degraded parts of the wood. It is, therefore, evident that regular maintenance is preferred over renovation.

11.5.1 TERMINOLOGY FOR INSPECTION, MAINTENANCE, AND RENOVATION

Coated surfaces of wooden components in exterior use require regular maintenance measures. This includes determining and assessing the current condition (inspection or control) and measures to maintain and restore the target condition (care, cleaning, maintenance, and repair).

Inspection or **control** of the surfaces is recommended at least once a year. In addition, the surfaces should be checked for damage after a hail event or other mechanical stresses. These damages can be minimal and just barely visible to the naked eye. The difficulty is determining the right time for a maintenance coat. This is not easy, not only for laymen but also for experts. In most coating systems, there are no apparent signs of this, but visually unrecognisable degradation processes (*e.g.* chemical changes, layer degradation, embrittlement, formation of micro-cracks, degradation of active ingredients, etc.) occur. Depending on the type of coating, especially the kind of binder and the thickness of the layer, different weathering patterns make maintenance coating necessary. In most cases, cracking in the coating film is the decisive sign of losing its functional suitability. Based on the condition found, a decision must be made whether maintenance or renovation is necessary.

Maintenance is a measure to restore the target condition and repair it with relatively little effort. It should be carried out before visible damage to the wooden components and their coating occurs (predictive maintenance). During maintenance work, the original coating is not removed, so this can only be done as long as the original coating is still largely intact and its changes are only minor.

Renovation or **repair** is also a measure to restore the target condition. Compared to maintenance, the required effort is significantly greater (corrective maintenance). The original coating is damaged to such an extent that it must be removed to regain a sound surface of the substrate for a renovation or renewal coating, at least in the area of damage.

Renovation of surface coatings is required for:

*   severe weathering,
*   mechanical injuries,
*   coating cracks with moisture infiltration,
*   greying,
*   flaking,
*   blue stain infestation (wood-discolouring fungi), or
*   rot (wood-destroying fungi).

Wooden parts that show rot must be replaced. It must be checked whether the rot is due to an unfavourable construction. In the given case, the structural conditions must be improved to ensure no prolonged moisture accumulation occurs in the wooden parts in the future.

11.5.2 Planning of Maintenance – Predictive Maintenance

It has already been shown that maintaining a coating can considerably extend its usability. However, renovation is required if the defects are already too severe, as maintenance is no longer sufficient. The questions remain: when is the right time for maintenance, when is it already too late, and when is renovation required? Table 11.4 gives limit states for only visually detectable defects.

TABLE 11.4
Limit States of Coatings for Exterior Applications. The Limit State Determines the Required Maintenance or Renovation Type

Limit State	Film-forming Coatings	Non-film-forming Coatings	State of Coating	Defects
L-E	Esthetical limit, Optical deficiency	Esthetical limit, Optical deficiency	Only optical alterations	Change of gloss Change of colour Growth of algae
L-D1	Maintenance interval	Maintenance interval = renovation interval	Minor defects that do not require the removal of the original coating	Intensive chalking Crack in coating film (without discolouration) Cracks in coating by hail impact (without discolouration) Flaking in single areas (<5 mm², without discolouration around flaked area) Superficial mould, growth/blue stain
L-D2	Renovation interval		Coating degradation	Cracking Blistering Flaking Hail damage Discoloration around cracks Penetrating mould growth/blue stain
L-D3	Decay of wood	Decay of wood	Onset of decay in wood	Brown-rot fungi White-rot fungi Wood-boring insects

Note: Limit states: L-E – no maintenance is required, cleaning or a refreshing coating are possible L-D1 – a maintenance interval with very small cracks without (!) visible discolouration around the cracks, L-D2 – renovation needed, and L-D3 – wooden sections need to be exchanged with new wood.

11.5.2.1 Maintenance Intervals, Inspection

Maintenance intervals for wood coatings can range from 1 year to 15 years, and the actual interval depends on factors such as compass orientation, inclination, coating colour, and dry film thickness. Table 11.5 gives an overview of typical maintenance intervals of wood coatings. Very thin coatings with bright colours (or even transparent coatings) with only small inclination (horizontal exposure) have the shortest maintenance intervals. In contrast, thicker coatings ($\geq 60\mu m$) with 90° inclination (vertical) facing north have expected maintenance intervals of up to 15 years. The influence of the different factors is described in the following sections.

TABLE 11.5

Guidance Values for Typical Maintenance Intervals of Wood Coatings in Exterior Use

Coating (film thickness)	Colour	Sheltering	Exposure Orientation	Maintenance Interval
Impregnation, low build stain (0–30 µm)	Bright	Sheltered	Vertical	3 years
			Horizontal	2 years
		No sheltering	Vertical	1–2 years
			Horizontal	1 year
	Dark	Sheltered	Vertical	3–4 years
			Horizontal	2 years
		No sheltering	Vertical	2 years
			Horizontal	1–2 years
Medium build stain (30–60 µm)	Bright	Sheltered	Vertical	5 years
			Horizontal	3 years
		No sheltering	Vertical	2 years
			Horizontal	1 year
	Dark	Sheltered	Vertical	7 years
			Horizontal	4 years
		No sheltering	Vertical	3 years
			Horizontal	2 years
Opaque paint (>60 µm)	Bright	Sheltered	Vertical	up to 15 years
			Horizontal	7 years
		No sheltering	Vertical	10 years
			Horizontal	5 years
	Dark	Sheltered	Vertical	12 years
			Horizontal	6 years
		No sheltering	Vertical	8 years
			Horizontal	2–4 years

11.5.3   MAINTENANCE PROCEDURES

11.5.3.1   Surface Preparation and Maintenance

Maintenance coatings should be applied to timber-construction elements with a moisture content between 12 and 18% according to the standard ÖNORM B 3430-1 (ASI 2023). Additionally, a specific minimum temperature is required (the minimal film-forming temperature of the coating), and these coating materials should not be used when low temperatures (*e.g.*, overnight) can be expected.

Good adhesion to the old coating is crucial for a maintenance coating. Therefore, it is strongly recommended that the surface is cleaned before the application. For low-build coatings, brushing with a stiff brush in the fibre direction is recommended, while sanding is the preferred method for medium- and high-build coatings. Dust and sanding residues must be removed before applying the maintenance coatings. With semi-transparent coatings, it must be kept in mind that the colour will get darker after painting. Therefore, maintenance coating should be applied to the entire surface rather than only the degraded spots (Figure 11.14).

Adhesion failure between the coating layers can lead to blistering, a typical sign of inadequate surface pre-treatment (see Section 11.4.1).

FIGURE 11.14   Colour change due to maintenance coating (left side of the cladding) and without maintenance coating (right side of the cladding).

(Holzforschung Austria, courtesy of G. Grüll, B. Forsthuber, M. Truskaller.)

11.5.3.2 Renovation

For coating renovation, the old coating must be removed. The remaining wood surface has to be sanded to remove the degraded wood. This is very important, as degraded wood is not a suitable substrate for a coating and will result in premature coating failure.

11.5.4 FACTORS INFLUENCING THE DURABILITY AND MAINTENANCE INTERVALS IN WEATHERING

The factors influencing the durability of coated wood products during exterior weathering are summarised in Figure 11.15 and can be divided between the coating properties (*e.g.* choice of coating binder), wood properties (*e.g.* wood species), and exposure conditions (*e.g.* exposing the coated wood samples 45° against the south).

Regarding the coatings' properties, opaque coatings with white pigmentation and high dry-film thickness provide the most extended maintenance intervals. Opaque coatings offer the best UV-light protection for the wood substrate and the coating itself. Brighter colours reduce surface temperatures and thus reduce the rate of photodegradation of the coatings; therefore, white is the best choice.

In cases where the wood substrate should still be visible, semi-transparent coatings are preferred. Semi-transparent coatings have much shorter maintenance intervals compared to opaque coatings. With semi-transparent coatings, darker colours have longer maintenance intervals than brighter ones, as darker colours provide a higher UV and visible light protection of the wood substrate. Brighter colours are more translucent for UV light; thus, the coating and the wood surface are photodegraded faster. On the other hand, fully transparent coatings are a special case and have the shortest maintenance intervals in exterior use of all coating types. Although

FIGURE 11.15 Factors influencing the durability and maintenance intervals of wood coatings in weathering.

(Courtesy of G. Grüll, B. Forsthuber, M. Truskaller.)

many customers desire them, only the best coatings have the required durability, and frequent maintenance is crucial.

Generally, higher dry-film thickness considerably extends the maintenance intervals of coatings. Higher dry-film thickness leads to higher UV and moisture protection of the wood. While UV protection can never be too good, problems with too high dry-film thickness regarding moisture can arise. While protecting the wood from high water uptake is beneficial, it can pose a problem as soon as water somehow finds its way into the wood. This can happen through cracks originating from sharp edges, joints or some mechanical defects. In these cases, water gets trapped in the wood as drying is hindered, leading to long periods of high moisture content and fungal degradation. This is especially true for coatings with a high *sd value* (equivalent air-layer thickness. The sd value indicates the paint's permeability to water vapour, with a lower value signifying a more breathable paint. Ideally, a breathable paint should have a sd value of 0.5 m or lower. Always check the product specifications or consult the manufacturer to confirm a paint's breathability). It is, therefore, important not to significantly exceed the dry-film thickness for the intended purpose.

Wood quality is another important aspect regarding durability and maintenance intervals. Wood features such as knots or reaction wood can lead to problems in coatings and considerably reduce the maintenance intervals. Although considered well-suited for exterior applications, some wood grades are sometimes challenging. One of these wood species is European larch, which, in some cases, shows coating issues. Also, resin exudations can penetrate the coatings and lead to moisture ingress with subsequent flaking.

Regarding the usage conditions, the compass orientation and inclination are important in the durability and maintenance intervals. Coated wood surfaces facing south usually need much more frequent maintenance than those facing north. Horizontal-oriented surfaces, like terrace flooring, have the shortest maintenance intervals, while vertical faces have considerably longer intervals. Surfaces facing 45° have expected maintenance intervals between the two extremes (Figure 11.16).

In the project "Servowood," some first factors were given about the influence of orientation and inclination (Forsthuber *et al.* 2022). These factors help estimate the relative extension or reduction of the maintenance intervals depending on the dry film thickness, colour, temperature, compass orientation, inclination, and many other factors (Table 11.6).

11.6   USE AND CARE OF INTERIOR WOOD COATINGS

11.6.1   Maintenance

11.6.1.1   Cleaning

Regular cleaning is critical for floors to remove coarse and fine-grained contaminants such as stones, sand, or dust, which can contribute to heavy scratches and abrasion when walked on. Doormats, as coarse and fine dirt traps at building and apartment entrances, are essential to keep the soiling and moisture of floors used with street shoes as low as possible. Dry cleaning by vacuuming or mopping is suitable. During cleaning, water should only act on the surfaces for a short time and in small quantities ("mist-damp"). Especially with oiled surfaces, only suitable cleaning agents may be used because normal cleaning agents can remove the oil or wax from the wood surface.

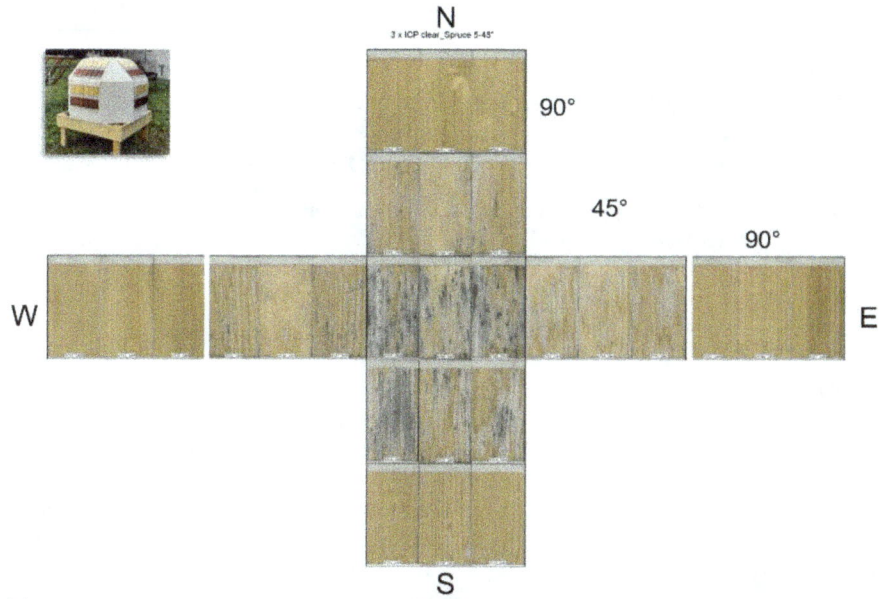

(a)

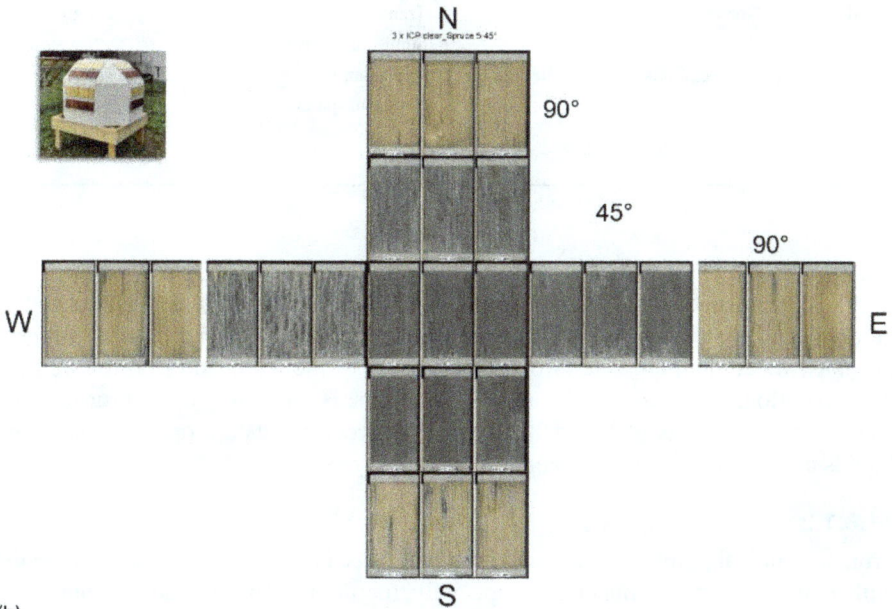

(b)

FIGURE 11.16   Weathering results of the same coating exposed on a multifaceted exposure rack. The coating system is exposed simultaneously on four compass directions (N, S, W, and E) and three inclinations (90° – vertical, 45° and 0° – horizontal). Weathering of a transparent coating system after (a) 24 months and (b) 48 months.

(Courtesy of G. Grüll, B. Forsthuber, M. Truskaller.)

TABLE 11.6

**Characteristic Factor Values from the Project "Servowood" Showing
the Extension (Value >1) or Reduction (Value <1) of Maintenance
Intervals Relative to a Reference (Reference of Factor Value)**

Variable	Characteristic Factor Values	Reference of a Factor Value	Pigmentation	Factor Value	Statistical Significance
Dry-film thickness	3 x coating	2 x coating	Transparent	1.22	Sign.
			Semi-transparent	1.37	Sign.
Orientation	E	S	Transparent	1.22	Sign.
			Semi-transparent	1.00	Not sign.
	N	S	Transparent	1.34	Sign.
			Semi-transparent	1.34	Sign.
	W	S	Transparent	1.01	Not sign.
			Semi-transparent	1.01	Not sign.
Inclination	90° (vertical)	45°	Transparent	1.40	Sign.
			Semi-transparent	1.49	Sign.
	0° (horizontal)	45°	Transparent	0.84	Sign.
			Semi-transparent	0.84	Sign.
Wood species	Norway spruce	Pine	Transparent	1.36	Sign.
			Semi-transparent	0.99	Not sign.
	European oak	Pine	Transparent	0.87	Sign.
			Semi-transparent	1.18	Not sign.

Note: Sign. – significant difference

All unplanned surface contamination, such as spilt liquids, must be removed as quickly as possible. Depending on the film formation of the coating and the tightness of joints, water can penetrate through the surface of the wood and lead to swelling, discolouration, or peeling of film-forming coatings. High-quality coated surfaces are resistant to the effects of liquids. However, the weak points are open joints, where liquids can penetrate the wood and dirt can occur.

11.6.1.2 Care Treatments

From a technical point of view, regular maintenance is necessary for surfaces treated with non-film-forming materials, especially for floors. This includes cleaning to remove dirt and reapplying coating materials, usually oils and waxes, to maintain the surface properties in the long term. The intervals of care depend on the extent of the stress. From a technical point of view, wood surfaces coated with film-forming materials do not require any maintenance.

11.6.2 RENOVATION

With all goods made of solid wood, it is possible to renovate them several times and restore them to a condition that is as good as new. The renovation is usually carried out by sanding off the coating, sanding the wooden substrate to remove any discolouration, and then reapplying the coating. Surfaces with old paints can also be sandblasted and thus lightened, and they are then again suitable substrates for various types of surface treatments. In the case of oiled surfaces, it may be challenging to apply waterborne coating materials during renovations, as the water-repellent properties of oil may affect the subsequent coating's adhesion. Therefore, sticking to the same system when renovating or checking the compatibility is recommended. In addition, old and heavily discoloured wood surfaces can also be treated with opaque paints to create a new colour or a uniform appearance.

11.7 SUSTAINABILITY AND CIRCULAR ECONOMY

The protection of wood surfaces with high-quality coatings is an essential part of the concept of sustainable use of timber in building applications. Storage of CO_2 in timber lasts as long as it is used as material. Combustion or biological decomposition leads to CO_2 emissions in the same quantity stored by photosynthesis when they form inside the tree. Therefore, it is essential to achieve high durability in timber construction because a long service life results in a prolonged storage period (Churkina *et al.* 2020, Mishra *et al.* 2022). At the end of the service life of a product made from newly machined wood, the principles of circular economy and cascading reuse of wood must be emphasised to prolong the storage of greenhouse gases (Reindahl Andersen *et al.* 2018, Dodoo and Muszyński 2021, Ahn *et al.* 2023).

Wood surface treatments and wood coatings, in particular, aim to improve the durability of wood products to prolong their service life. In exterior application, wood coatings counteract the degradation of wood by weathering and microorganisms and retain the integrity of the substrate. It is well documented in historic buildings that exterior wood can remain for decades, and the protective functions of coatings prolong the technical function and the visual appearance of wood constructions (Hill *et al.* 2022, Broda and Hill 2022). The protective layer of wood coatings can be maintained using well-known techniques by local painters. This results in a sustainable concept of care and maintenance for exterior wood, making a very long service life possible.

It must be mentioned that wood coatings can counteract sustainability because synthetic products from fossil feedstock are mainly used at the present state of the art for exterior applications. Coating formulations may contain harmful and toxic substances, and end-of-life scenarios lack options for recycling and reuse. During the last decades, high efforts have been undertaken to develop coating materials that are environmentally friendly and sustainable (Rössler 2022). This includes many aspects. Problems such as safe application processes and use, low emissions of solvents and other substances to the environment, and waste reduction have been addressed in the sector. Because of the relatively high proportion of organic solvents in traditional coating systems, the corresponding environmental regulations (national, European, or international) have been considered in the selection and development of paint systems. Renewable feedstock for raw materials, reduction of and alternatives to substances of concern and fundamental concepts for recycling and reuse have been developed (Kurimoto *et al.* 2001, Kumar *et al.* 2017, Jung and Baumstark 2018). But still, a lot needs to be done. Many of the United Nations' 17 Sustainable Development Goals (SDGs) provide excellent guidance for the future development of these chemical products and formulations, considering the whole life cycle from raw materials sourcing and production, over application and service life, to end-of-life, and back to the cradle. This should be considered for coating materials and coated wood in building applications, where coatings positively contribute to sustainability.

REFERENCES

Ahn N., Bjarvin C., Riggio M., Muszynski L., Schimleck L., Pestana C., Dodoo A. & Puettmann M. (2023). Envisioning mass timber buildings for circularity: Life cycle assessment of a mass timber building with different end-of-life (EoL) and post-EoL options. In: Malo K.A., Nyrud A.Q. & Nore K. (Eds.), *World Conference on Timber Engineering (WCTE 2023), Ås,* Norway, (pp. 3581–3587).

Ahola P., Derbyshire H., Hora G. & de Meijer M. (1999). Water protection of wooden window joinery painted with low organic solvent content paints with known composition. Part 1. Results of inter-laboratory tests. *Holz als Roh- und Werkstoff,* 57(1), 45–50.

Arminger B. (2021). Advanced Organic Coatings for the Hydrophobization of Solid Wood Materials. Doctoral Thesis, *University of Natural Resources and Applied Life Sciences (BOKU),* Vienna, Austria.

Baumstark R. & Tiarks F. (2002). Studies for a new generation of acrylic binders for exterior wood coatings. *Macromolecular Symposia,* 187(1), 177–186.

Berger G., Katz H. & Petutschnigg A.J. (2006). What consumers feel and prefer: Haptic perception of various wood flooring surfaces. *Forest Products Journal,* 56(10), 42–47.

Bornemann T., Brischke C. & Alfredsen G. (2014). Decay of wooden commodities – Moisture risk analysis, service life prediction and performance assessment in the field. *Wood Material Science & Engineering,* 9(3), 144–155.

Broda M. & Hill, C.A.S. (Eds.) (2022). *Historical Wood: Structure, Properties and Conservation.* MDPI, Basel, Switzerland, (317 pp.).

Bulian F. & Graystone J. (2009). *Wood Coatings: Theory and Practice.* Elsevier Science, Amsterdam, Boston, The Netherlands, USA, (320 pp.).

Böttcher P. (2004). *Oberflächenbehandlung von Holz und Holzwerkstoffen.* (1st Ed.), Ulmer, Stuttgart, Germany, (142 pp.).

CEN (2007). EN 12529: Castors for furniture – Castors for swivel chairs – Requirements. *The European Committee for Standardization (CEN).* Brussels, Belgium.

CEN (2013). EN 927-1: Coating materials and coating systems for exterior wood - Part 1: Classification and selection, CEN European Committee for Standardization. *The European Committee for Standardization (CEN).* Brussels, Belgium.

CEN (2017). EN 14354: Wood-based panels - Wood veneer floor coverings, CEN European Committee for Standardization. *The European Committee for Standardization (CEN).* Brussels, Belgium.

CEN (2020). EN 927-3: Paints and varnishes - Coating materials and coating systems for exterior wood - Part 3: Natural weathering test, CEN European Committee for Standardization. *The European Committee for Standardization (CEN)*. Brussels, Belgium.

CEN (2023). EN 460: wood and wood-based products Durability - Guidance on performance, Austrian Standards Institute. *The European Committee for Standardization (CEN)*. Brussels, Belgium.

Churkina G., Organschi A., Reyer C.P.O., Ruff A., Vinke K., Liu Z., Reck B.K., Graedel T.E. & Schellnhuber H.J. (2020). Buildings as a global carbon sink. *Nature Sustainability*, 3(4), 269–276.

de Meijer M. & Militz H. (2001). Moisture transport in coated wood. Part 2: Influence of coating type, film thickness, wood species, temperature and moisture gradient on kinetics of sorption and dimensional change. *Holz als Roh- und Werkstoff*, 58(6), 467–475.

de Meijer M. & Militz H. (2005). Sorption behaviour and dimensional changes of wood-coating composites. *Holzforschung*, 53(5), 553–560.

Derbyshire H. & Robson D.J. (1999). Moisture conditions in coated exterior wood, Part 4: Theoretical basis for observed behaviour. A computer modelling study. *Holz als Roh-und Werkststoff*, 57(2), 105–113.

De Windt I., Van den Bulcke J., Wuijtens I., Coppens H. & Van Acker J. (2014). Outdoor weathering performance parameters of exterior wood coating systems on tropical hardwood substrates. *European Journal of Wood and Wood Products*, 72(2), 261–272.

ASI (2009). *ÖNORM C 2354: Transparent sealing materials for wooden floors and sealings made thereof - Minimum requirements and test methods*. Austria Standards Institute (ASI), Vienna, Austria.

ASI (2023). *ÖNORM B 3430-1: Planning and execution of painting and coating work — Part 1: Coating on wood and wooden materials, metal, plastic, masonry, plaster, concrete and lightweight building boards*. Austria Standards Institute (ASI), Vienna, Austria.

Dodoo A. & Muszyński L. (2021). End-of-life management of cross-laminated timber multi-storey buildings: A case for designing for post-use material recovery and environmental benefits. In: *Proceedings of the World Conference on Timber Engineering (WCTE 2021)*, Santiago de Chile, Chile, (pp. 68–74).

Ekstedt J. (2002). Influence of coating additives on water vapour absorption and desorption in Norway spruce. *Holzforschung*, 56(6), 663–668.

European Commission (2011). CPR construction products regulation. *Official Journal Council Directive*, 89/106/EEC OJ L 88.

European Commission (2012). Regulation (EU) No 528/2012 of the European Parliament and of the Council of 22 May 2012 concerning the making available on the market and use of biocidal products, *Luxembourg*.

Evans P.D., Chowdhury M.J., Mathews B., Schmalzl K., Ayer S., Kiguchi M. & Kataoka Y. (2005). *Weathering and surface protection of wood. In: Handbook of environmental degradation of materials*. William Andrew Pub, Norwich, (NY), USA. (pp. 277–297).

Evans P.D., Cullis I. & Morris P.I. (2010). Checking of profiled southern pine and Pacific silver fir deck boards. *Forest Products Journal*, 60(6), 501–507.

Forsthuber B. & Grüll G. (2010). The effects of HALS in the prevention of photo-degradation of acrylic clear topcoats and wooden surfaces. *Polymer Degradation and Stability*, 95(5), 746–755.

Forsthuber B., Grüll G., Arnold M., Podgorski L., Bulian F., Naden B. & Graystone J. (2022). The influence of exposure orientation and inclination on the service life of wood coatings. In: *Proceeding of the 12th Woodcoatings Congress*, 8-9 November, Amsterdam, The Netherlands.

Gibbons M.J., Nikafshar S., Saravi T., Ohno K., Chandra S. & Nejad M. (2020). Analysis of a wide range of commercial exterior wood coatings. *Coatings*, 10(11), Article ID: 1013.

Glöckner P., Jung T., Struck S. & Studer K. (2008). *Radiation Curing: Coatings and Printing Inks; Technical Basics*, Applications and Trouble Shooting. Vincentz Network GmbH & Co. KG, Hannover, Germany.

Goldschmidt A. & Streitberger H.-J. (2014). *BASF-Handbuch Lackiertechnik.* Vincentz Network GmbH & Co. KG, Hannover, Germany, (863 pp.).

Graystone J. (1985). *The Care and Protection of Wood.* ICI Paints Division, Slough Berkshire, UK.

Grüll G. (2003). Holzfenster – Pflege, Wartung und Renovierung. *Renovation*, (2), 28–30.

Grüll G. (2010). Beschichtungen für Holz im Außenbereich: Zweck und Anwendungsbereiche. In: *Tagungsband Wiener Holzschutztage*, 25–26 November, Holzforschung Austria, Vienna, Austria, (pp. 71–76).

Grüll G., Truskaller M., Podgorski L., Bollmus S., de Windt I. & Suttie E. (2013). Moisture conditions in coated wood panels during 24 months natural weathering at five sites in Europe. *Wood Material Science & Engineering*, 8(2), 95–110.

Grüll G., Truskaller M., Illy A., Fischer A., Schulz P. & Beyer M. (2020). Structured wood surfaces for flooring - Coatings to obtain high resistance and natural look and feel. In: Németh R., Rademacher P., Hansmann C., Bak M. & Báder M. (Eds.), *Proceedings of the 9th Conference on Hardwood Research and Utilization in Europe*, Sopron, Hungary. (Part 1: pp. 90–94).

Grüll G. & Tscherne F. (2020). Wartungsanleitung für Beschichtungen auf Holzoberflächen im Aussenbereich. In: (4th Edn.), Holzforschung Austria, Vienna, Austria.

Grüll G. & Hansemann W. (2023). *Wood, Surface Treatment. Ullmann's Encyclopedia of Industrial Chemistry Online.* Wiley-VCH Verlag GmbH & Co. KGaA.

Hernández R.E. & Cool J. (2008a). Effects of cutting parameters on surface quality of paper birch wood machined across the grain with two planing techniques. *European Journal of Wood and Wood Products*, 66(2), 147–154.

Hernández R.E. & Cool J. (2008b). Evaluation of three surfacing methods on paper birch wood in relation to water- and solvent-borne coating performance. *Wood and Fiber Science*, 40(3), 459–469.

Hill C., Kymäläinen M. & Rautkari L. (2022). Review of the use of solid wood as an external cladding material in the built environment. *Journal of Materials Science*, 57(20), 9031–9076.

Humar M. & Lesar B. (2013). Efficacy of linseed- and tung-oil-treated wood against wood-decay fungi and water uptake. *International Biodeterioration & Biodegradation*, 85(10), 223–227.

Hundhausen U., Mai C., Slabohm M., Gschweidl F. & Schwarzenbrunner R. (2020). The staining effect of iron (II) sulfate on nine different wooden substrates. *Forests*, 11(6), Article ID: 658.

Janesch J., Arminger B., Gindl-Altmutter W. & Hansmann C. (2020). Superhydrophobic coatings on wood made of plant oil and natural wax. *Progress in Organic Coatings*, 148(9), Article ID: 105891.

Jung T. & Baumstark R. (2018). Biomass Balance - a ground-breaking approach for more sustainable wood coatings without compromising on performance. In: *Proceeding of the PRA Wood Coatings Congress*, 23-24 October, Amsterdam, The Netherlands.

Kataoka Y., Kiguchi M., Williams R.S. & Evans P.D. (2007). Violet light causes photodegradation of wood beyond the zone affected by ultraviolet radiation. *Holzforschung*, 61(1), 23–27.

Kumar A., Vlach T., Ryparovà P., Škapin A.S. Kovač, J., Adamopoulos S., Hajek P. & Petrič M. (2017). Influence of liquefied wood polyol on the physical-mechanical and thermal properties of epoxy based polymer. *Polymer Testing*, 64(2), 207–216.

Kurimoto Y., Koizumi A., Doi S., Tamura Y. & Ono H. (2001). Wood species effects on the characteristics of liquefied wood and the properties of polyurethane films prepared from the liquefied wood. *Biomass and Bioenergy*, 21(5), 381–390.

Lesar B., Pavlič M., Petrič M., Škapin A.S. & Humar M. (2011). Wax treatment of wood slows photodegradation. *Polymer Degradation and Stability*, 96(7), 1271–1278.

Miklečić J. & Jirouš-Rajković V. (2021). Effectiveness of finishes in protecting wood from liquid water and water vapor. *Journal of Building Engineering*, 43, Article ID: 102621.

Mishra A., Humpenöder F., Churkina G., Reyer C.P.O., Beier F., Bodirsky B.L., Schellnhuber H.J., Lotze-Campen H. & Popp A. (2022). Land use change and carbon emissions of a transformation to timber cities. *Nature Communications*, 13, Article ID: 4889.

Nusser B., Schumacher M., Lux C. & Tieben J. (2024). Coole Hülle – Holzfassaden bauphysi-kalisch beleuchtet. In: *7th Internationale Fachtagung Bauphysik & Gebäudetechnik (BGT), Forum Holzbau, 24-25 April*, Friedrichshafen, Germany, (pp. 159–171).

Passauer L., Prieto J., Müller M., Rössler M., Schubert J. & Beyer M. (2015). Novel color stabilization concepts for decorative surfaces of native dark wood and thermally modified timber. *Progress in Organic Coatings*, 89, 314–322.

Prieto J. & Kiene J. (2007). *Holzbeschichtung: Chemie und Praxis*. Vincentz Network GmbH & Co. KG, Hannover, Germany, (334 pp.).

Rapp A.O., Peek R.-D. & Sailer M. (2000). Modelling the moisture induced risk of decay for treated and untreated wood above ground. *Holzforschung*, 54(2), 111–118.

Reindahl Andersen M., Beineix O., Romagnoli V., Birnstengel B., de Bruijne A., Holmgren T., Althoff Palm D., Pritchard O., Zotz F. & Weißenbacher J. (2018). Sceening Study on End of Life Treatment of Wood from Doors and Windows: Final Report Ramboll. Available at: dvv.dk/cvm/wp-content/uploads/2020/03/dvv_traevinduer_eol-rapport.pdf

Rodríguez Grau G., Marín Uribe C.R., Pinto González F. & Cortés Rodríguez P. (2021). Native wood revaluation through green gluing: A systematic review. In: *Proceedings of the World Conference on Timber Engineering (WCTE 2021)*, 9-12 August, Santiago de Chile, Chile. (pp. 3282–3289).

Rothkamm M., Hansemann W. & Böttcher P. (2003). *Lackhandbuch Holz*. DRW-Verlag, Leinfelden-Echterdingen, Germany, (280 pp.).

Rössler A. (2022). Sustainable wood coatings – only a dream or realistic? In: *Proceedings of the 12th Woodcoatings Congress*, Vicentz Network, Hannover, Germany.

Schwalm R., Häußling L., Reich W., Beck E., Enenkel P. & Menzel K. (1997). Tuning the mechanical properties of UV coatings towards hard and flexible systems. *Progress in Organic Coatings*, 32(1/4), 191–196.

Sell J. (1985). Physikalische Vorgänge in wetterbeanspruchten Holzbauteilen Fensterrahmen. *European Journal of Wood and Wood Products*, 43(7), 259–267.

Sell J. & Leukens U. (1971). Untersuchungen an bewitterten Holzoberflächen - Zweite Mitteilung: Verwitterungserscheinungen an ungeschützten Hölzern. *European Journal of Wood and Wood Products*, 29(1), 23–31.

Sharpe P.R. & Dickinson D.J. (1992). Blue Stain in Service on Wood Surface Coatings: Part 1. The nutritional Requirements of *Aureobasidium pullulans*. *International Research Group on Wood Protection, Report, IRG/WP/1556-92*, Stockholm, Sweden.

Svane P. (2004). Painted wood stands the test of time. *European Coatings Journal*, 5, 38–42.

Teacă C.-A., Roşu D., Mustaţă F., Rusu,T., Roşu L., Roşca I. & Varganici C.-D. (2019). Natural bio-based products for wood coating and protection against degradation: A Review. *BioResources*, 14(2), 4873–4901.

Teischinger A., Zukal M.L. & Kotradyova V. (2012). Exploring the possibilities of increasing the contact comfort by wooden materials - tactile interaction of man and wood. *Innovation in Woodworking Industry and Engineering Design*, 1(1), 20–24.

Truskaller M. (2017). Haptisches Empfinden von Oberflächen: Holzoberflächen im Vergleich
    zu anderen Materialien. *HFA-Magazin*, (5), 8–9.
Viitanen H., Toratti T., Makkonen L., Peuhkuri R., Ojanen T., Ruokolainen L. & Räisänen
    J. (2010). Towards modelling of decay risk of wooden materials. *Holz als Roh- und
    Werkstoff*, 68(3), 303–313.
Wassipaul F. & Janotta O. (1972). Die Wasserdampfdurchlässigkeit von Anstrichmitteln Teil
    1. *Holzforschung und Holzverwertung*, 24(4), 74–79.

12 Bonding of Wood

P. Niemz

12.1 INTRODUCTION

Bonding is a technology without which modern wood-based materials and multi-storey timber construction would be scarcely feasible, if at all. Most wood-based materials are produced using adhesives, with only a few relying on mechanical connections such as aluminium, steel or wooden dowels, nails, or carpentry-style joints. The intrinsic bonding forces of wood, *i.e.* without any adhesive, are of industrial significance only in the case of fibreboards produced using the wet process (such as panels and moulded parts for automobiles), which are now rarely produced. Research into this field has been ongoing for decades. Hydrogen bonds are employed in fibreboards made using the wet process. Enzyme treatments, particularly on fibres, have been tested at some facilities but have not yet reached industrial standards (Wagenführ 1988, Kharazipour *et al.* 1997).

The development of adhesives began with bio-based raw materials, which were replaced by synthetic adhesives in the 1940s (Dunky and Mittal 2023, Tobisch *et al.* 2023). Thanks to their consistent quality and cost-effectiveness, the widespread use of synthetic adhesives enabled the production of many new products (Zeppenfeld and Grunwald 2005). Challenges with fluctuations in quality and the seasonal availability of agricultural products, such as bio-based raw materials such as tannin and lignin, continue to be significant issues when using bio-based adhesives.

Today, research has shifted back towards bio-based adhesives, reminiscent of the early days of adhesive development with gluten, casein, and tannin adhesives. However, work on many bio-based materials has been ongoing for decades without significant breakthroughs (Dunky and Mittal 2023). In most cases, bio-based adhesives have been blended with synthetic adhesives. An example is adhesives for glued-laminated timber, where a small amount of natural carbon is included in one-component polyurethane (1C-PUR) adhesives. Rapeseed oil may also be added in minor amounts. In particleboard production, tannins are used to a limited extent for specialised products, mainly focusing on a bioproduct (*e.g.* tannins). The quantity available for niche products is often too small, and the quality frequently fluctuates somewhat depending on the season. Lignin, produced in large amounts in paper-pulp production, can be added proportionally to aminoplastics for particleboard manufacturing. Up to 60% lignin can be incorporated into phenolic-based adhesives in plywood production. Further details can be found in Dunky and Mittal (2023), Tobisch *et al.* (2023), and Frihart and Konnerth (2023).

Typical applications for adhesives in timber construction are:

a) load-bearing elements (cross-laminated timber, glued-laminated timber),
b) non-load-bearing elements,
c) veneer and particle-based panel products,

DOI: 10.1201/9781003411994-12

d) joining technology: joining of timber in length, *e.g.* finger and butt joining, and

e) other applications such as:
   a. flooring: bonding of multi-layer flooring, gluing of parquet to the substrate material, such as concrete,
   b. windows and doors: the window glass is often glued, and windows and doors are often glued into the wall openings using polyurethane foam, and
   c. bio-based insulation materials: to manufacture insulation materials with large thicknesses by the wet process; sometimes adhesives are added. Polymeric methyl diphenyl diisocyanate (PMDI) adhesive is used as a bonding agent when the dry process is applied, and larger panel thicknesses are always possible in the wet process.

In the case of glued elements, the type of adhesive and the adhesive joint thickness influence the diffusion behaviour. This applies to plywood cross-laminated timber and other glued products such as window frames (Frühwald 1972, Sonderegger 2011). The diffusion resistance of glued joints is several times higher than that of wood (Frühwald 1972, Sonderegger 2011). The adhesive bond-line thickness also affects the mechanical properties; a thicker bond-line reduces the strength.

It is also essential to check the ageing of bonded elements, especially for load-bearing purposes (halls, bridges, high-rise buildings), as moisture stress (seasonal fluctuations in humidity) and even very low relative humidity can lead to delamination and, in extreme cases, to an impairment of the load-bearing capacity. Examples include the collapse of the hall for winter sports in Bad Reichenhall, Germany, in 2006 due to an unsuitable adhesive and constructional inaccuracies. The adhesive, however, complied with the requirements according to the state of knowledge at the time.

Adhesives are also used on construction sites for assembly (often together with screws) for assemblies and connections. Longitudinal joints by finger jointing, overlapping or butt bonding with two-component polyurethane (2C-PUR) adhesives, several millimetre thick adhesive joints, and a pre-treatment of the surface (*e.g.* the Swiss system TS3) are also often used. Section 12.5 provides an overview of longitudinal joints. Further reading is available in Zeppenfeld and Grunwald (2005), Habenicht (2017), and Dunky and Mittal (2023).

Highly elastic adhesives are used for specific applications, such as parquet flooring or joints that require a certain amount of stretching. These adhesives transfer strains, stress, or even swelling and shrinkage. For example, bonding wooden parquet to concrete with elastic adhesives has been successful for many years.

12.2   ADHESIVES FOR LOAD-BEARING COMPONENTS IN TIMBER CONSTRUCTION

Special, approved adhesives are used for bonding of load-bearing components. Before bonding, the lamellas are often strength-graded by the machine according to the EN 338 standard (CEN 2016a). Structural timber – Strength Classes for glued-laminated timber are defined according to the EN 14080 standard (CEN 2013), *e.g.* GL 24c, GL24h, GL30c, GL30h, GL32c, GL 32h strength classes.

As a result, reliable properties of the components and bonding adapted to the usage class (*i.e.*, the moisture load) are required.

Bonding also makes it possible to produce curved and, in some cases, three-dimensionally shaped beams or plates within a specific framework (Niemz *et al.* 2023). The final processing is often carried out using special CNC machines.

Whenever possible, the bonding is carried out in a timber construction company with the appropriate equipment, such as machines, air-conditioned rooms, and trained specialists. The bonding must, therefore, be adapted to the wood (the type of wood) and the material structure (*e.g.* when bonding hardwood, special softwoods, or combinations of hardwood and softwood). This requires technology, know-how, suitable adhesives, trained (qualified) personnel for bonding, and quality control.

Adhesives must provide durable bonds throughout the structure's lifetime to meet the required Service Class given in EN 1995-1-1 (CEN 2004 National provisions at the place of use may further limit the applicability of an adhesive in a specific Service Class. For the basics of adhesives, requirements concerning these aspects can be found in the standards for the following adhesive families, which apply to glued-laminated timber manufacture:

- phenolic and aminoplastic adhesives (*e.g.* UF, MF, MUF, PRF,) – standard: EN 301 (CEN 2023a),
- moisture-curing one-component polyurethane adhesives (1C-PUR) – standard: EN 15425 CEN (2017a), and
- emulsion polymer isocyanate adhesives (EPI) – standard: EN 16254 (CEN 2023b) (Dunky and Mittal 2023).

The choice of adhesives depends on the Service Class. Depending on the application area, the adhesives have different open, closed, and pressing times. They are used for surfaces and longitudinal bonding (finger-jointing). Adhesive curing occurs at room temperature at MUF/MF on radiofrequency, significantly reducing the curing time. This technology is often used, especially for standard products (mass-produced goods). Radiofrequency is rarely used for curing one-component polyurethane (1C-PUR) (mainly because the high costs of adhesives and radiofrequency techniques are a problem). Bonding for CLT and glued-laminated timber is primarily carried out at room temperature. The temperature in the production hall, especially relative humidity (low RH in the winter), influences the curing speed of 1C-PUR.

Today, special approvals are required for load-bearing elements such as glued-laminated timber or cross-laminated timber. Bonding for load-bearing elements is done with adhesives with increased moisture resistance approved by the Deutsches Institut für Bautechnik (DIN) in Berlin. For this purpose, the Materialprüfungsanstalt – "Otto-Graf-Institute" at the University of Stuttgart publishes a list of adhesives that contains approved adhesives for selected types of wood (a list of certified adhesives used in timber construction. The companies must be approved for production, and quality control must be carried out.

They must possess a gluing permit to produce glued-laminated timber (proof of suitability for producing glued timber components according to DIN 1052-12 (DIN 2008), certificate A or B) and have been familiarised with the construction method through training or instruction.

Adhesives from European accredited tests are now also used for special adhesives, *e.g.* special types of one-component polyurethane (1C-PUR).

For timber construction, Service Classes are defined according to the EN 1995-1-1 standard (CEN 2004). Service Class 1 means that the bond-line is tested at 20 °C and a relative humidity (RH) lower than 65%, corresponding to an equilibrium moisture content (EMC) of about 12%. Service Class 2: 20 °C, RH only a few weeks higher than 85%, EMC lower than 20%, and in Service Class 3: a higher EMC than in Service Class 2 must be reached.

Other requirements apply to non-load-bearing elements (parquet, laminated wood), *e.g.* in the EN 204 and EN 205 standards (CEN 2016b,c). For PVAc adhesives, depending on the moisture resistance (D Classes): D1 (dry area), D2 (short-term exposure to water), D4 frequent short-term exposure to water, D (can be used indoors and outdoors, for frequent and long-term exposure to moisture) apply similarly. For thermosetting adhesives and duroplastic adhesives, EN 12765 (CEN 2016d) groups C1-C4 are applied. An overview can be found in Zeppenfeld and Grunwald (2005), Niemz and Sonderegger (2017), and Frihart and Konnerth (2023). Figure 12.1 shows important factors influencing bonding (Hänsel *et al.* 2022).

Figure 12.2 shows possible bonding variants for engineered wood products. Longitudinal (finger jointing) and surface bonding (cross-laminated timber, glued-laminated timber) dominate, and edge gluing is often used for cross-laminated timber. The boards are glued width-wise to create surfaces, which are then glued crosswise in a second step to form multi-layer panels. The longitudinal gluing of the lamellas is also essential for the use of fasteners such as screws during assembly. In some cases, there is no longitudinal bonding in the middle layer, and spacing between the lamellas is also standard to reduce stresses caused by changes in humidity.

It should be noted that it is challenging to detect bonding errors afterwards. A high degree of care is therefore required during installation. The following essential standards must be taken into consideration:

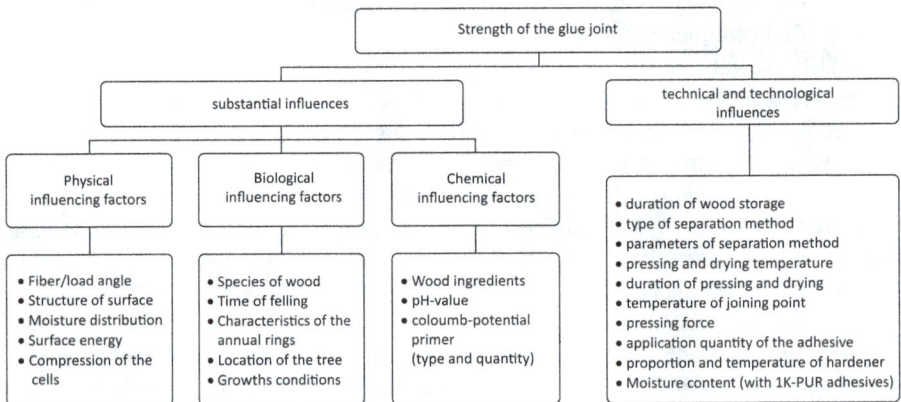

FIGURE 12.1 Overview of key factors influencing the strength of glue-bonded joints.

(Courtesy of A. Hänsel.)

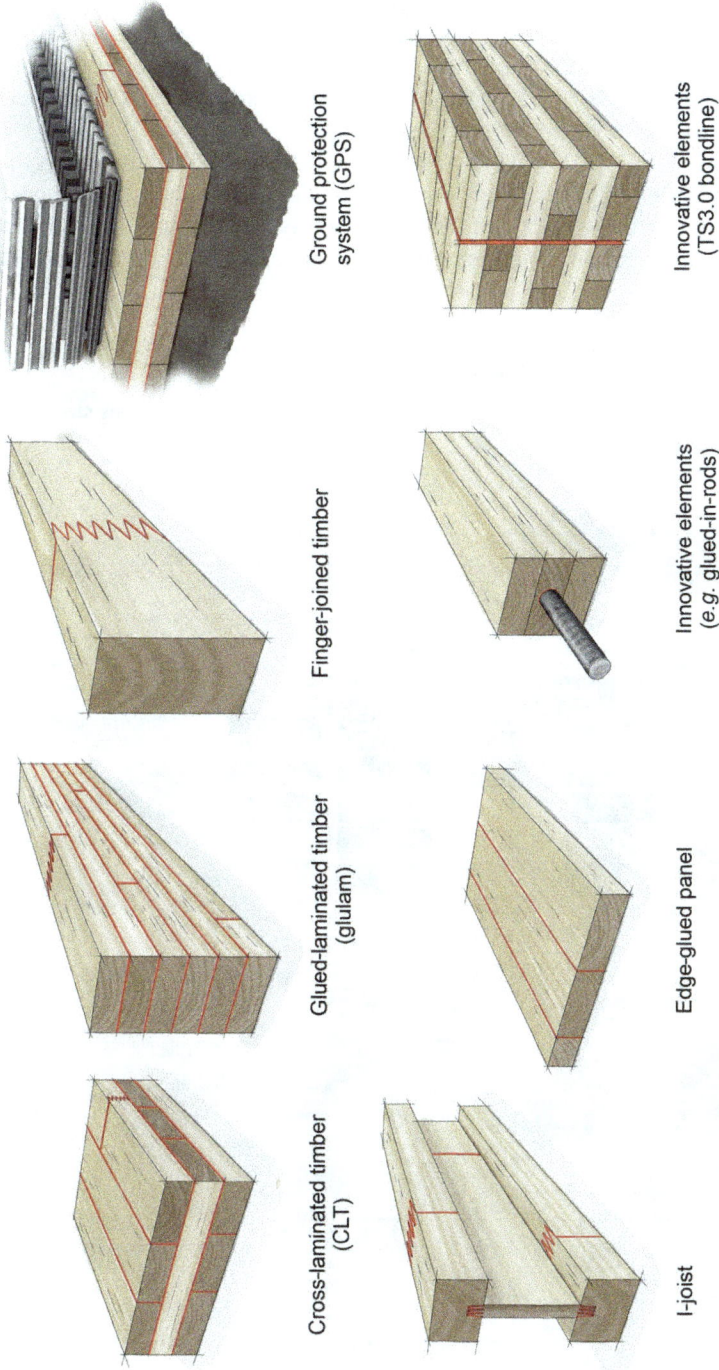

FIGURE 12.2 Examples of glued wood products (engineered wood products), adhesive joint shown in red colour in each case. (Courtesy of P. Niemz.)

- EN 301 (CEN 2023a): Adhesives, phenol plastics, and amino plastics for load-bearing timber structures – Classification and performance requirements.
- EN 15425 (CEN 2017a): Adhesives – One-component polyurethane (PUR)-based adhesives for load-bearing timber structures – Classification and performance requirements.
- DIN 68141 (DIN 2022): Wood adhesives – Determination of the drying time and evaluation of wetting and how easily and uniformly the adhesive can be applied or spread over a surface (spreadability). The adhesive's viscosity, consistency, and the type of surface it's being applied to can influence the spreadability.

Executing companies must prove their suitability for bonding load-bearing timber components according to the DIN 1052-10 standard (DIN 2012).

Figure 12.3 overviews the central European and North American standards for bonding in timber construction.

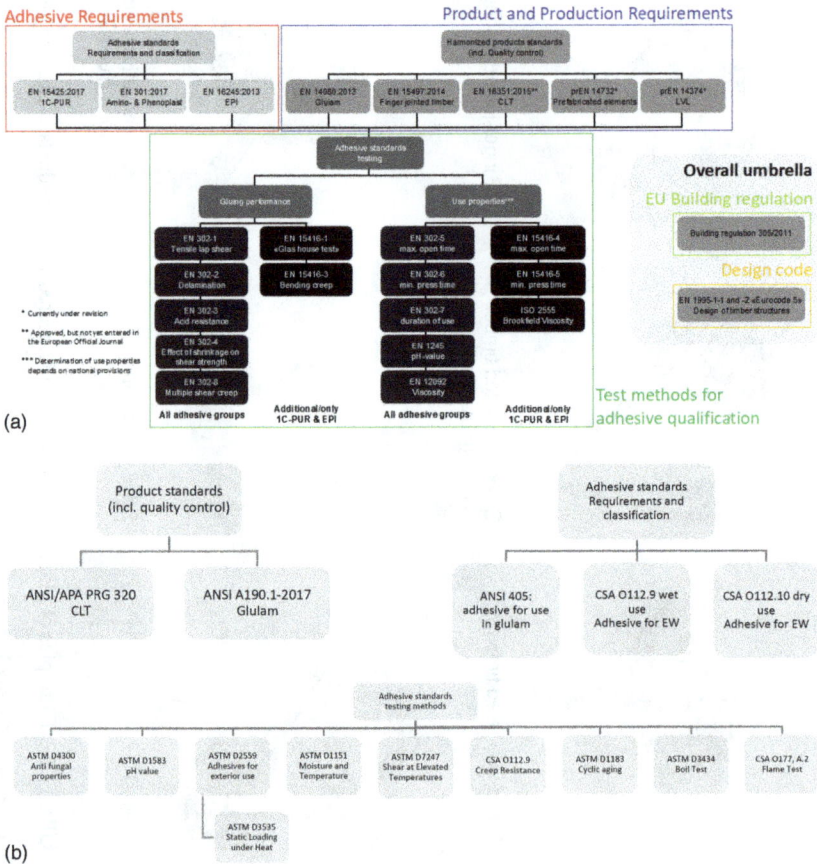

FIGURE 12.3 Standards for adhesives for engineered wood products: (a) European and (b) North American.

(Henkel Company, courtesy of P. Niemz.)

Adhesives used for load-bearing purposes in the construction industry must have building authority approval. The manufacture of bonded components requires trained personnel and quality monitoring. The Service Class regulates the moisture load on the components.

In addition to the purely mechanical load, moisture resistance (Service Classes 2 and 3) is fundamental in the construction industry and the indoor climate. For Service Classes 2 and 3, moisture-resistant adhesives are required for load-bearing components. Their suitability is determined by mechanical loading, *e.g.* by a tensile shear test according to the EN 302-1 standard (CEN 2023c) and delamination testing according to the EN 302-2 standard (CEN 2017b). This test is carried out in conjunction with moisture stress, *e.g.* boiling (tensile shear test EN 302-1) or water storage and rapid drying of bonded glued-laminated timber or cross-laminated timber components after water storage at various temperatures in the delamination test according to the EN 302-2 standard.

High temperature, solar radiation *e.g.* in conservatories or areas of the home with little shade (behind balcony doors, windows raking down to the floor), air movement, and also solar radiation (temperature) can lead to additional moisture transport in the building component (Keylwerth 1966, 1969). Very low relative humidity (20-30%), as occurs in heated rooms in winter, can also lead to cracking and delamination, especially if the wood is installed in very damp conditions. The wood moisture content must be adjusted during production to the moisture content in subsequent use (Niemz and Sonderegger 2021).

Table 12.1 shows a percentage breakdown of the adhesives used for load-bearing timber construction. MUF/MF and EPI dominate for glued-laminated timber and one-component polyurethane (1C-PUR) for cross-laminated timber.

However, the proportion of 1C-PUR also increases in glued-laminated timber. This is often due to the ease of processing, the colourless glue joints, and the fact that it is formaldehyde-free. Some types of wood are difficult to bond due to special extractives (*e.g.* larch, robinia, selected extract-rich tropical hardwoods). Therefore, priming is sometimes necessary, *e.g.* for 1C-PUR in combination with most hardwoods and larch (Kläusler 2014, Böger *et al.* 2022, 2024, Hänsel *et al.* 2022).

12.2.1 Emissions from Glued-laminated Timber

Glued-laminated timber is manufactured using melamine resin or mixtures of MF and UF (MUF), moisture-curing one-component polyurethane (1C-PUR), emulsion

TABLE 12.1
Used Adhesives in Timber Construction for Structural Application (Henkel & Cie. AG, Switzerland)

% in Different Adhesive Groups	1C-PUR	MUF/MF	PRF/PF	EPI
Cross-laminated timber (CLT)	69	18	—	13
Glued-laminated timber (glulam)	8	48	9	35
I-joist	1	6	78	15

Note: 1C-PUR – one-component polyurethane.

polymer isocyanate (EPI), or phenolic resorcinol resin adhesives. Glue joints made with melamine resin, PUR, and EPI adhesives are colourless/light. Phenolic resorcinol resin adhesives are dark (brown). The adhesive content for glued-laminated timber with a lamella thickness of 40 mm is about 0.3% to 0.5% (mass percentage, based on wood), glued-laminated timber bonded with 1C-melamine (MF), modified melamine (MUF) resins, and phenol resorcinol resins contains formaldehyde. Since the adhesive quantity is meagre (related to the wood) and particularly low formaldehyde adhesives are used, the room air concentrations to be expected are well below the limits for wood-based panels.

12.3   ADHESIVES FOR VENEER, PARTICLE-, AND FIBRE-BASED MATERIALS

Depending on the application area, various UF/MUF or PF resins are used for veneer-based wood materials (LVL, plywood). For particle materials, UF, MUF, PMDI, and PF are used to a lesser extent. UF and MUF are used in dry indoor climates and rooms with increased moisture exposure (cellars, some floors, bathrooms). Formaldehyde emissions have been drastically reduced in recent decades; formaldehyde emissions have now been practically eliminated (Figure 12.3). The use of adhesives for glued-laminated timber and cross-laminated timber is low compared to particle-based materials.

Wood-based materials are classified according to the EN 13986 standard (CEN 2015) into

- non-load-bearing (normal load-bearing) and
- load-bearing constructions (high load-bearing)

and according to the moisture load (dry area: (Service Class 1), damp area (Service Class 2), and Service Class 3 (outdoor area). Chipboard and fibreboard are unsuitable for Service Class 3. The wet area requires moisture-resistant adhesives such as PMDI, PF, or MUF/MF resins. Service Class 3 (outdoor area) is not approved for organically bonded particle materials. With PMDI, the adhesive content about the weight of dried wood is lower than with the other systems (approximately 2-6% in relation to the wood mass). For UF/MF resins, 6-10% (often somewhat higher for very low-density boards), 10-12%, and for PF resins, 6-8%. 90% of European adhesives are generally approved for particleboard, MDF/HDF, OSB, and plywood.

Polyurethane adhesives are formaldehyde-free. However, since formaldehyde is a natural wood substance, glued-laminated timber joined with polyurethane adhesive contains small amounts of formaldehyde (Roffael 2017). In the case of amino-plastic-based resins, formaldehyde release has been drastically reduced in recent decades (Figure 12.4). Regarding volume, amino plastics are the most widely used adhesives in the wood industry. Adhesive consumption in Europe amounts to around 6 million tons per year (Dunky 2022).

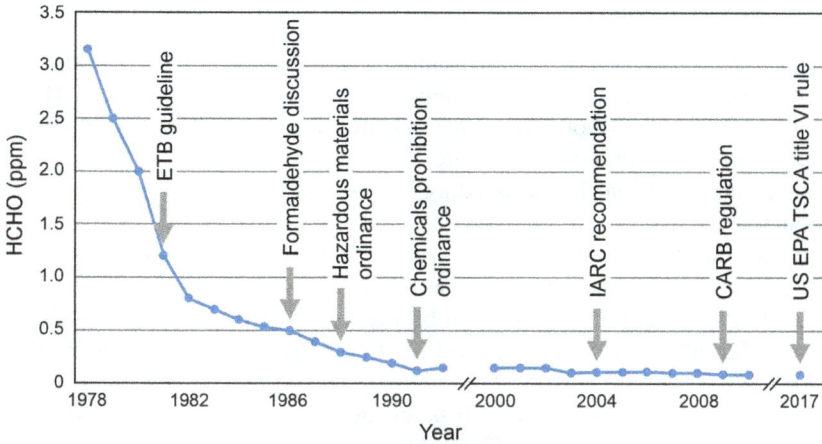

FIGURE 12.4 Development of formaldehyde emissions from wood-based materials since 1978.

(Tobisch *et al*. 2023, courtesy of Springer Nature.)

12.4 APPLICATIONS OF ADHESIVES

Various adhesive systems are used for parquet flooring, windows, and doors, with the elasticity of the adhesives varying depending on the application. Several systems are employed, including silanes, low-swelling PVAc systems, and more flexible PUR adhesives. The most common types include:

- rigid adhesives (shear strength equal to or greater than 3 MPa),
- rigid elastic adhesives (shear strength of 2 MPa), with a slide of 0.5 mm at a 1 mm joint gap, and
- elastic adhesives (shear strength of 1–2 MPa), with sliding of 1 mm at a 1 mm joint gap.

PVAc adhesives are commonly used for non-load-bearing components, such as partition walls, longitudinally bonded rod-shaped products, and panels. Isocyanate is added to PVAc adhesives to enhance moisture resistance (D Classes D3 and D4). This is especially important for applications such as windows and conservatories. In situations where there is increased moisture exposure, such as during ship transport, D4 adhesives are often preferred.

This also applies to bonding windows and doors, lamella and finger-jointed connections in non-load-bearing applications, partition wall and element production, general wood joints, and laminate bonding in kitchens, laboratories, and interior fittings. PVAc adhesives can be used outdoors, provided the joints are protected from direct weather exposure. Additionally, PVAc adhesives are suitable for press systems with radiofrequency curing.

Other typical applications for adhesives include the following.

12.4.1 BLOCK GLUING

When larger cross-sections of glued-laminated timber beams are required, as is often the case for high-rise buildings, several beams are bonded together. This process is known as block bonding. It necessitates specialised technology for adhesive application and pressing, and careful control of the adhesive quantity (joint thickness) and selection. Block bonding can be used to produce various building components.

Suppose multiple glued-laminated timber components are joined to create a larger, cavity-free element. In that case, this is referred to as a composite component (with a rectangular cross-section by the EN 14080 standard (CEN 2013) or with a different cross-section according to the DIN 1052-10 standard (DIN 2012)).

Rib or box cross-sections can also be created with thick adhesive joints (block joints), as specified in the DIN 1052-10 standard (DIN 2012).

12.4.2 SCREW GLUING AND OTHER APPLICATIONS

In some cases, screws are used to generate the pressing pressure required for bonding. Experimental and computational studies on this are available in Aicher *et al.* (2021).

If the bonding process is not carried out correctly, such as improper surface treatment, incorrect adhesive selection, incorrect joint thickness, or insufficient pressing pressure, the joints may later open due to changes in humidity, even in dry climates. Adhesives are also employed in the construction industry for specialised applications, such as

- PUR foam for gluing and sealing windows and doors,
- Epoxy resin or 2C-PUR for glued rods,
- adhesives for the elastic bonding of parquet to the substrate (*e.g.* concrete), such as silanes, low-swelling PVAc adhesives (objective: to create a relatively elastic bond with the substrate, reduce swelling and shrinkage, and provide certain impact sound insulation), and
- PUR adhesives for gluing windows to window frames.

12.5 CONNECTIONS IN TIMBER CONSTRUCTION

Corner joints, such as carpenter's joints and various traditional timber-to-timber connections, are still commonly used (Figures 12.5a,b). As timber constructions have increased in size, more rigid connectors have become necessary, leading to the introduction of steel. Steel provides the required ductility. Steel sheets, which are sometimes partly screwed and partly glued, are used (Figure 12.5c). For the connection of plate-shaped and beam-shaped components, steel fasteners are employed, including special screws, which are occasionally used to reinforce the wood against transverse pressure loads (Frihart *et al.* 2023).

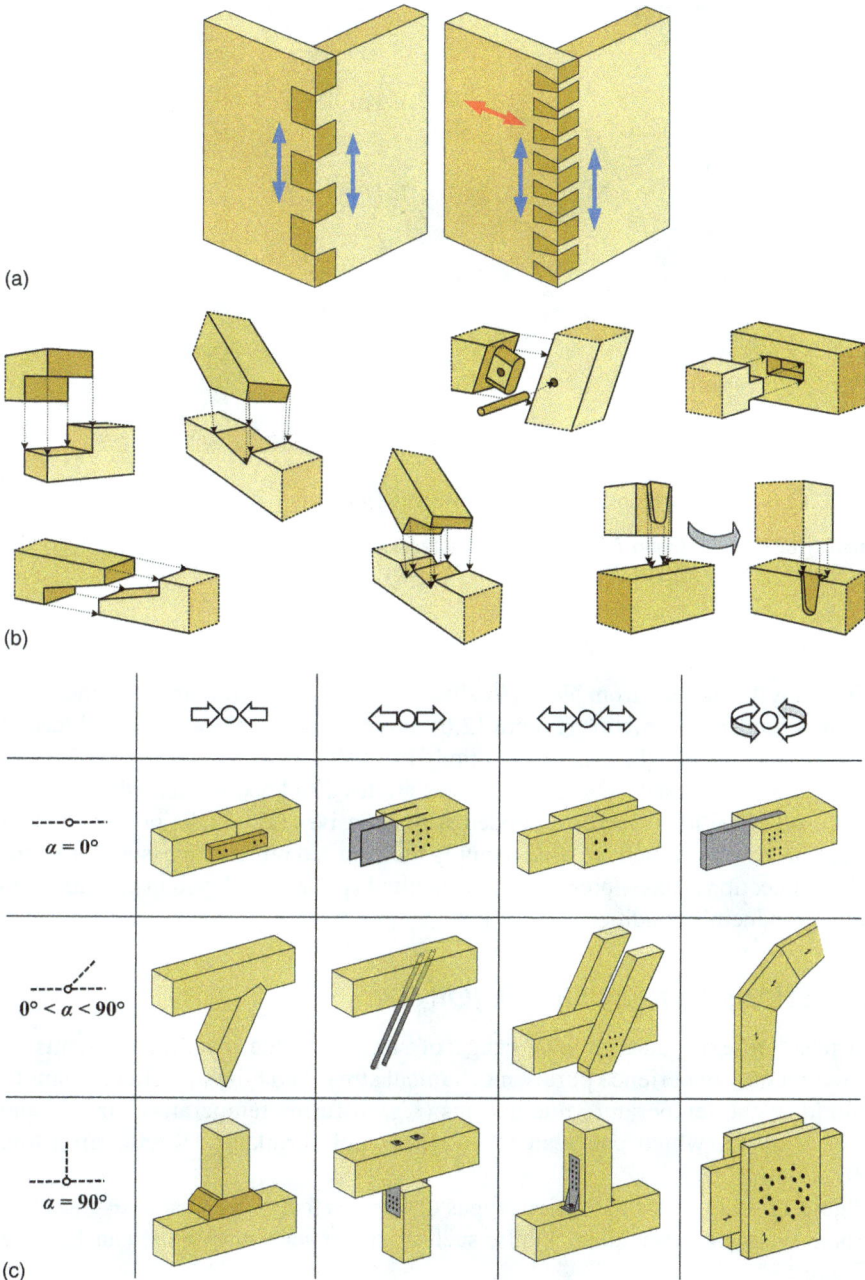

FIGURE 12.5 Examples of timber joints (Frihart et al., 2023): (a) traditional carpentry joints used in furniture making, (b) typical traditional timber-engineering connections that transfer forces through contact between members, and (c) principles of connecting timber and steel for various loading conditions.

(Courtesy of P. Niemz.)

FIGURE 12.6   GSA Technologie produced by neue Holzbau.

(Courtesy of neue Holzbau AG, Lungern, Switzerland.)

The GSA technology from Neue Holzbau AG is a high-performance fastener used in modern timber construction (Figure 12.6). It is a timber-steel joint where threaded rods and steel connect glued-laminated timber or LVL.

Both epoxy resin and 2C-PUR adhesives are used to bond the threaded rods. In some cases, panel joints are also bonded with adhesives (*e.g.* PUR) in combination with special screws, which are used to apply the required pressing pressure for bonding the connections (screw pressing). This method is often employed in operations to ensure reproducible conditions.

12.6   STRESSES ON ADHESIVE JOINTS

Adhesives are exposed to a wide range of stresses in the installed elements. A connection may experience purely mechanical stress and climate-related changes in humidity and temperature fluctuations (*e.g.* differing temperatures inside and outside a door), which can lead to swelling and shrinkage (Keylwerth 1966, 1969).

Figure 12.7 illustrates the various types of stresses that occur in adhesive joints.

Some types of longitudinal joints, such as finger and butt joints, can be seen in Figure 12.8.

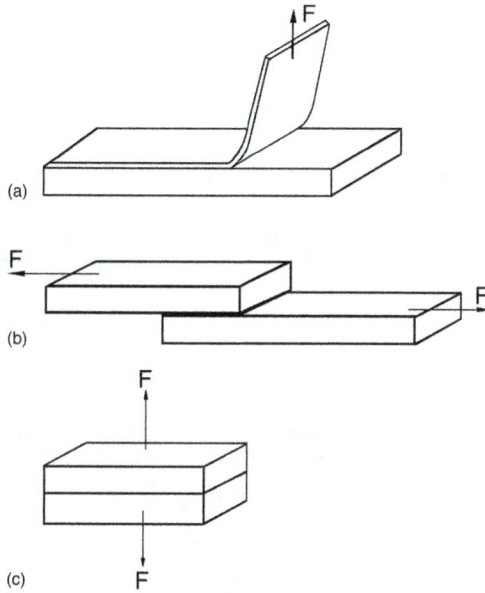

FIGURE 12.7 Different types of loads in adhesive bonding: (a) peeling, (b) shearing, and (c) tensile loading.

(Courtesy of P. Niemz and D. Sandberg.)

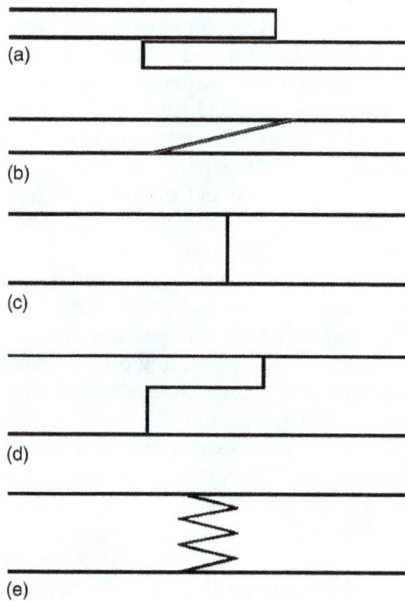

FIGURE 12.8 Common timber joints: a) overlap, b) shaft, c) butt-joint, d) offset overlap, e) finger-jointing.

(Courtesy of P. Niemz and D. Sandberg.)

REFERENCES

Aicher S., Zisa N. & Simon K. (2021). Screw-gluing of ribbed timber elements – effects of screw spacing and plate stiffness on bond line cramping pressure. *Otto Graf Journal*, 20, 9–38.

Böger T., Richter K. & Sanchez-Ferrer A. (2022). Hydroxymethylated resorcinol (HMR) primer to improve the performance. *International Journal of Adhesion and Adhesives*, 113, Article ID: 103070.

Böger T., Engelhardt M., Tangwa Suh F., Richter K. & Sanchez-Ferrer A. (2024). Wood water interactions of primers to enhance wood-polyurethane bonding performance. *Wood Science and Technology*, 58(5/6), 135–160.

CEN (2004). EN 1995-1-1: 2004+A 1: Eurocode 5: Design of timber structures - Part 1-1: General Common rules and rules for buildings. *The European Committee for Standardization (CEN)*, Brussels, Belgium.

CEN (2013). EN 14080: Timber structures - Glued laminated timber and glued solid timber - Requirements. *The European Committee for Standardization (CEN)*, Brussels, Belgium.

CEN (2015). EN13986: Wood-based panels for use in construction - Characteristics, evaluation of conformity and marking. *The European Committee for Standardization (CEN)*, Brussels, Belgium.

CEN (2016a). EN 338: Structural timber - Strength classes. *The European Committee for Standardization (CEN)*, Brussels, Belgium.

CEN (2016b). EN 204: Classification of thermoplastic wood adhesives for non-structural applications. *The European Committee for Standardization (CEN)*, Brussels, Belgium.

CEN (2016c). EN 205: Adhesives - Wood adhesives for non-structural applications - Determination of tensile shear strength of lap joints. *The European Committee for Standardization (CEN)*, Brussels, Belgium.

CEN (2016d). EN 12765: Classification of thermosetting wood adhesives for non-structural applications. *The European Committee for Standardization (CEN)*, Brussels.

CEN (2017a). EN 15425: Adhesives - One component polyurethane (PUR) for load-bearing timber structures - Classification and performance requirements. *The European Committee for Standardization (CEN)*, Brussels, Belgium.

CEN (2017b). EN 302-2: Adhesives for load-bearing timber structures - Test methods - Part 2: Determination of resistance to delamination. *The European Committee for Standardization (CEN)*, Brussels, Belgium.

CEN (2023a). EN 301: Adhesives, phenolic and aminoplastic, for load-bearing timber structures - Classification and performance requirements. *The European Committee for Standardization (CEN)*, Brussels, Belgium.

CEN (2023b). EN 16254: Adhesives - Emulsion polymer isocyanate (EPI) for load-bearing timber structures - Classification and performance requirements. *The European Committee for Standardization (CEN)*, Brussels, Belgium.

CEN (2023c). EN 302-1: Adhesives for load-bearing timber structures - Test methods - Part 1: Determination of longitudinal tensile shear strength. *The European Committee for Standardization (CEN)*, Brussels, Belgium.

DIN (2008). DIN 1052-12: Design of timber structures - General rules and rules for buildings. *Deutsches Institut für Normung (DIN)*, Berlin, Germany.

DIN (2012). DIN 1052-10: Design of timber structures - Part 10: Additional provisions. *Deutsches Institut für Normung (DIN)*, Berlin, Germany.

DIN (2022). DIN 68141: Wood adhesives - Determination of the open drying time and evaluation of wetting and brushability. *Deutsches Institut für Normung (DIN)*, Berlin, Germany.

Dunky M. (2022). Chemistry in wood-based panels: actual situation, achievements, and challenges. *12th European Wood-Based Panels Symposium*, 12–14 October, Hamburg, Germany.

Dunky M. & Mittal K. (Eds.) (2023). *Biobased Adhesives. Sources, Characteristics and Applications.* Wiley, Hoboken (NJ), USA, (768 pp.).

Frihart C.R., Konnerth J., Frangi A., Gottlöber C., Jockwer R. & Pichelin F. (2023). Joining and Reassembling of Wood. In: Niemz P., Teischinger A. & Sandberg D. (Eds.), *Springer Handbook of Wood Science and Technology.* Springer Nature, Cham, Switzerland, (XXV+2069 pp.).

Frühwald A. (1972). *Ein Beitrag zur Kenntnis des diffusionstechnischen Verhaltes von Furnierplatten und kunstharzbeschichtetem Holz.* Doctoral Thesis, Universität Hamburg, Hamburg, Germany.

Habenicht G. (2017). *Kleben – erfolgreich und fehlerfrei.* (7th Ed.) Vieweg, Wiesbaden, Germany, (288 pp.).

Hänsel A., Sandak J., Sandak A., Niemz P. & Mai J. (2022). Selected previous findings on the factors influencing the gluing quality of solid wood products in timber construction and possible developments: A review. *Wood Material Science & Engineering*, 17(3), 221–232.

Keylwerth R. (1966). Temperatur- und Feuchtigkeitsgefälle in Holzbauteilen. *Holz als Roh- und Werkstoff*, 24(10), 452–454.

Keylwerth R. (1969). Praktische Untersuchungen zum Holzfeuchtigkeits-Gleichgewicht. *Holz als Roh- und Werkstoff*, 27(8), 285–290.

Kharazipour A., Hüttermann A. & Lüdemann H. (1997). Enzymatic activation of wood fibres as a means for the production of wood composites. *Journal of Adhesion Science and Technology*, 3(11), 419–427.

Kläusler O.F. (2014). *Improvement of one-component polyurethane bonded wooden joints under wet conditions.* Doctoral Thesis. ETH Zürich, Zurich, Switzerland.

Niemz P. & Sonderegger W. (2017) Fichtenholz - Physikalisch-mechanische Eigenschaften. *Holz-Zentralblatt*, 741–742.

Niemz P. & Sonderegger W. (2021). Holzphysik. *Physik des Holzes und der Holzwerkstoffe* (2nd Ed.) Hanser Verlag, München, Germany, (580 pp.).

Niemz P., Teischinger A. & Sandberg D. (Eds.) (2023). *Springer Handbook of Wood Science and Technology.* Springer Nature, Cham, Switzerland, (XXV+2069 pp.).

Roffael E. (2017). *Formaldehyd in der Natur, im Holz und in Holzwerkstoffen.* (2nd Ed.), DRW Verlag, Stuttgart, Germany, (288 pp.).

Sonderegger W.U. (2011). *Experimental and theoretical investigations on the heat and water transport in wood and wood-based materials.* Doctoral Thesis, ETH Zürich, Zurich, Switzerland.

Tobisch S., Dunky M., Hänsel A., Krug D., & Wenderdel C. (2023). Survey of wood-based materials. In: Niemz P., Teischinger A. & Sandberg D. (Eds.), Springer Handbook of Wood Science and Technology. Springer Nature, Cham, Switzerland, (XXV+2069 pp.).

Wagenführ A. (1988). *Praxisrelevante Untersuchungen zur Nutzung biotechnologischer Wirkprinzipien bei der Holzwerkstoffherstellung.* Doctoral Thesis, TU Dresden, Dresden, Germany.

Zeppenfeld G. & Grunwald D. (2005). *Klebstoffe für die Holz- und Möbelindustrie.* (2nd Ed.), DRW-Verlag Weinbrenner, Leinfelden-Echterdingen, Germany, (16+352 pp.).

13 Guidelines for the Use of Timber in Construction

M. Klippel and Y. Plüss

13.1 TIMBER SPECIES FOR LOAD-BEARING PRODUCTS

The design of timber structures requires in-depth knowledge and a fundamental understanding of timber and wood-based materials as modern, sustainable building materials, which includes components, load-bearing systems made of timber, and load-bearing connections in timber structures or between timber and other building materials. The structural design and dimensioning of timber constructions cannot be seen in isolation from the design rules in standards. However, applying design standards will not lead to successful and economical constructions.

A good understanding of wood anatomy is essential for understanding the mechanical and physical properties of timber and, thus, for designing or optimising the use of structural timber.

The following sections summarise essential parameters for the structural design of timber members and discuss their influence on the design.

13.1.1 RELEVANT DESIGN PARAMETERS

13.1.1.1 Moisture Content

Almost all properties of wood are influenced by the amount of water in the wood, *i.e.* the moisture content. This is due to its structural composition, which is more pronounced below the fibre-saturation point than above. With increasing amount of water in wood:

- the strength of the wood decreases, *cf.* Figure 9.5,
- the creep deformation increases under long-term loading,
- the thermal conductivity increases, and
- the susceptibility of the wood to fungal attack increases significantly at high moisture content.

An equilibrium moisture content develops in the wood after sufficiently long periods at certain ambient conditions, *i.e.* relative humidity (RH), temperature (T), and air pressure (p). It is specific to the kind of wood species. It is an essential parameter for timber construction engineers, as the moisture content of the timber elements to be installed depends on the ambient conditions expected after installation. The aim is to minimise moisture-induced distortion in the timber (minor swelling and shrinkage).

DOI: 10.1201/9781003411994-13

Wood moisture also has a decisive influence on assessing susceptibility to durability (*cf.* Chapter 10).

The equilibrium moisture content of timber with large cross-sections takes a long time to reach, and the capillary and diffusion processes only ensure fast moisture changes in the surface regions.

Only timber with a maximum moisture content of 20% shall be installed to avoid cracks, unacceptable dimensional changes, and distortion and to ensure the durability of timber constructions. It is recommended that timber members are installed close to their expected equilibrium moisture content in use (Figure 13.1).

In the European structural timber design standard FprEN 1995-1-1 (CEN 2024a), a distinction is made between four Service Classes, of which three consider the influences on wood moisture resulting from the installation of typical timber structures. Service Class 1 is characterised by a climate with a temperature of approximately 20 °C and a relative humidity not exceeding 65%, except under a few weeks a year. Under these conditions, most softwoods have an average equilibrium moisture content of up to approximately 12% (Table 13.1). In this installation situation, the wood moisture content is usually below 10% and can be as low as 6%. Components in closed and heated rooms are categorised in Service Class 1. Service Class 2 includes all components in open, roofed structures that are not directly exposed to the weather. These components achieve an average equilibrium moisture content of approximately 20% (20 °C/85% RH), but this is dependent on the regional climate. Service Class 3 is assigned to all components that are freely exposed to the weather, and an average equilibrium moisture content of more than 20% may occur due to the installation situation. This includes, for example, bridges, masts, and similar structures.

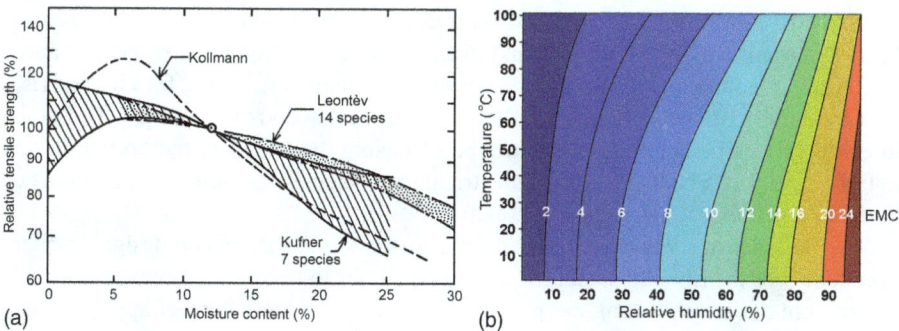

FIGURE 13.1 The effect of change in moisture and relative humidity: (a) influence of the wood moisture content on tensile strength parallel to the grain (100% is the reference value for timber at 12% moisture content) (Gerhards 2007), and (b) the equilibrium moisture content (EMC) of wood depends on relative humidity and air temperature.

(Data from Glass and Zelinka 2010, courtesy of U.S. Department of Agriculture, USDA, Forest Products Laboratory.)

TABLE 13.1

Service Classes for Timber Elements Based on Eurocode 5 (CEN 2024a)

Service Class	Upper limit[a]		Yearly average	
	RH[b]	WMC[c]	RH[b]	WMC[c]
1	65%	12%	50%	10%
2	85%	20%	75%	16%
3	95%	24%	85%	20%
4	[d]	Saturated	[d]	Saturated

Notes:

[a] The upper limit of relative humidity and/or the upper limit of the average moisture content should not be exceeded for more than a period of a few consecutive weeks per year.

[b] Relative humidity of surrounding air at a temperature of 20°C

[c] The equilibrium moisture content in commonly used softwood of a thikness of approximately 50 mm.

[d] The moisture content of members in SC 4 is affected by the surrounding element (eg. soil or water).

When categorising components into Service Classes, the existing installation situation must be guaranteed over the entire service life of the structure. If this is not the case, a higher Service Class should be used to determine the design strength and stiffness.

13.1.1.2 Density

The bulk density of timber is an essential parameter in characterising wood. This is due to the correlation between density and mechanical properties (Niemz and Sonderegger 2003). Most mechanical properties show a positive correlation with the bulk density. With increasing density, the strength and stiffness properties of timber members increase.

In timber technology and engineering, the main parameters used for characterisation are the dry density ρ_0 and the bulk density at 12% wood moisture ρ_{12}. The moisture content can be determined (*cf.* Chapter 5) using the oven-dry method described in EN 13183-1 (CEN 2002a) or the electrical resistance measuring method described in the EN 13183-2 standard (CEN 2002b).

The bulk-density values (ρ_{12}) according to the structural timber design standard Eurocode 5 or the EN 338 standard (CEN 2016) refer to mass and volume at the 12% equilibrium moisture content, *i.e.* ρ_{12}, which occurs at a temperature of approximately 20 °C and a relative humidity of 65%. The bulk density values according to Eurocode 5 or EN 338 (CEN 2016), therefore, refer either to the mean bulk density $\rho_{12,mean}$ or to the characteristic bulk density $\rho_{12,k}$ defined as the 5% quantile of the population.

13.1.1.3 Temperature

The influence of thermal expansion can usually be neglected in timber construction. However, constraining forces can occur in composite structures with steel or concrete members.

Thermal expansion is much less critical than the swelling and shrinkage of timber. However, a temperature change may cause a change in the moisture content, causing considerable swelling and shrinkage deformations. The swelling and shrinkage dimensions perpendicular to the grain are approximately ten times larger than the thermal changes in length.

Timber's strength decreases with increasing temperature. Section 13.2 provides further information on the behaviour of wood and wood-based materials in the event of fire.

13.1.1.4 Rheological Properties

Material's flow and deformation properties are called rheological properties (*cf.* Chapter 8). Wood exhibits a combined elastic and viscous time-dependent behaviour, referred to as a visco-elastic material. The visco-elastic properties are usually categorised as creep or relaxation.

The most critical aspect for timber engineers is timber's creep behaviour. Creep is the increase in deformations over time under constant loading. The decrease in stresses with constant deformations over time is known as relaxation.

The parameters influencing the creep behaviour in timber are:

- wood moisture and changes in the moisture content,
- load-duration,
- temperature, and
- type and level of applied stresses.

The first two factors most influence creep. The moisture content is included in the design via the Service Class to be selected, and the load duration is included via the Load Duration Class. The creep behaviour is also very different for different materials. For example, wood-based panels generally creep more than sawn timber, and the size of the timber component is essential; the smaller the timber components, the more pronounced the creep.

13.1.1.5 Durability

When referring to the durability of timber structures, the structural engineer's main concern is durability in terms of water and moisture-related effects. Although this is not exhaustive, most possible strengths and stiffness related to durability problems are directly or indirectly associated with the moisture content. Many factors must be considered regarding the durability of wood. If possible, the wood should be installed with the equilibrium moisture content that will occur during the use phase of the building, so that only the seasonal fluctuations in moisture affect the construction. If this is not observed, considerable cracks can occur in the timber surface after installation. The risk of cracking is most significant in timber members exposed to direct weathering or where fluctuations in humidity are high. Cracks allow water and fungal spores to penetrate and insects to lay eggs inside the timber. The structural engineer needs to consider that changes in moisture content are caused by:

- direct ingress of water, and
- changes in the humidity and temperature of the ambient air.

Water in its liquid state penetrates the wood primarily in the direction of the fibres. For this reason, end faces must be arranged or covered in a way that no water can penetrate through capillary action. One structural wood protection method is to cover exposed components, especially the end-grain (cross-sections). The DIN 68800-2 standard (DIN 2022) explains preventive structural measures in building construction.

In addition, preventive fungal and insect protection must be correctly addressed depending on the type of wood and geography, for example, the DIN 68800-3 standard (DIN 2020). Corrosion protection for fasteners is also an essential element in the context of the durability of timber structures. Eurocode 5 contains examples of minimum corrosion protection requirements for different Service Classes. Increased corrosion protection measures are required, for example, in warehouses for chemical products, salt, and artificial fertilisers, or in phosphoric acid factories, where it is essential to use bolts, rod dowels, and steel plates made of special stainless steels.

While relevant standards primarily address the durability issue during the building's use phase, measures for moisture protection during the construction phase remain insufficiently regulated. Guidance can be found in CODIFAB (2020) and Informationsdienst Holz (2024).

13.1.1.6 Stiffness

The modulus of elasticity (MOE) measures a material's stiffness. It determines the ratio of stress to elastic strain, quantifying how a material deforms under a given stress. The modulus of elasticity of timber is influenced by:

- the density, whereby the modulus of elasticity increases with increasing density,
- the moisture content, whereby the modulus of elasticity decreases with increasing moisture content, and
- the angle between the force and fibre direction.

The stiffness of timber products is usually given parallel and perpendicular to the grain direction.

Under long-term loads, the deformation increases over time (creep), usually, this can be taken into account by decreasing the design value of the modulus of elasticity or by introducing a multiplication factor for the instantaneous deformations. This behaviour must be considered when designing load-bearing timber members.

The stiffness of a timber member (order of magnitude of 10,000 N/mm²) is significant for (1) the design of serviceability (SLS) and (2) the design addressing the stability of a timber member, for example, a timber column. The design of serviceability is often the decisive design of timber members that define the cross-section, as the stiffness of timber members is comparably low. The verification of the SLS ensures that a structure or component performs its intended function under normal usage conditions throughout its lifespan regarding its usability, comfort, or appearance. Serviceability considerations typically include the design for maximum deflections, vibrations and deformations perpendicular to the grain. In summary, designing for serviceability is about ensuring a structure remains fit for use, not just safe from collapse or failure.

13.1.1.7   Strength

The following factors essentially determine the strength of timber:

- modulus of elasticity,
- bulk density,
- knottiness, and
- the angle between force and grain direction.

As most of these factors depend on the growing conditions under which a tree grows (location, climate), wood's strength is subject to wide variation. To use timber in structural applications, wood is graded into strength classes, which provide standardized mechanical properties the engineer can rely on.

The timber used for load-bearing or bracing purposes is strength graded following the EN 14081-1 standard (CEN 2019a), in which general requirements for visually and mechanically graded structural timber for load-bearing purposes are specified. The characteristic strength, stiffness, and density values for the design by the EN 1995-1-1 standard (CEN 2024a) are given in EN 338 (CEN 2016). A distinction is made between C classes for softwood based on upright bending tests, T classes for softwood based on tensile tests, and D classes for hardwood based on upright bending tests.

Wood is inhomogeneous due to the natural growth of trees. Knots, resin pockets, and other growth-related features significantly influence strength and, therefore, cause a considerable spread of strength properties within a timber member. If large pieces of wood are sawn into smaller pieces and reassembled by gluing, the characteristics are distributed within the material, and the scattering of material properties is reduced. The greater load-bearing capacity of glued-laminated timber (glulam) compared to sawn timber does not result from a higher average load-bearing capacity but from a reduction in the scatter of strength properties, which results in higher characteristic strength. A similar effect can be achieved by peeling logs and then gluing the individual veneers together. This process can enhance both the strength and, to a lesser extent, the stiffness of the timber member. For beech LVL (in German: Baubuche), a comparatively high characteristic bending strength of 75 N/mm^2 can be achieved, while a Norway spruce glued-laminated timber, as a typically used linear timber member, achieves a characteristic bending strength of about 24 N/mm^2.

For the design of timber members, bending, tensile, and compression strength are usually relevant strength properties. Tensile and compression strength depend on the direction of loading, for example, parallel or perpendicular to the grain direction. Additionally, timber products' shear strength and rolling shear strength are usually given. Any strength properties are essential for the ultimate limit state (ULS) design. The ultimate limit state is the condition beyond which a structure or its components no longer fulfils the required safety or performance criteria. This is the point at which the structure would fail due to excessive load, leading to collapse or significant structural damage. Designing for the ultimate limit state ensures the structure can support the maximum expected loads with an appropriate safety margin.

13.1.2 LOAD-BEARING TIMBER PRODUCTS

Typical load-bearing elements of timber can be categorised as follows:

- sawn timber,
- components of sawn timber, *i.e.* in length (finger-jointed) and/or transversely adhesively bonded sawn timber with well-defined dimensions and properties,
- glued-laminated timber (glulam), and
- cross-laminated timber (CLT).

Sawn timber is obtained from round timber by sawing or profiling. European softwoods such as common larch, Douglas fir, fir, Norway spruce, and Scots pine are mainly used, with a much smaller proportion of hardwoods such as common beech and oak. Based on numerous research projects in recent years, it can be expected that other European hardwoods, such as common ash, will also be more frequently used to design timber buildings.

So-called engineered wood products (EWPs) or wood-based materials are other frequently used products. Wood-based materials are produced by pressing together variously sized wood components, such as sawn timber, rods, veneers, veneer strips, chips, wood particles, or fibres, using adhesives or mineral binders (*cf.* Chapter 2). Wood fibreboards produced through the wet process can be made without adhesive. This process also utilises wood residues and uncontaminated recycled wood. The manufacturing process enhances and homogenises the raw material. As a result, wood defects and features that reduce strength, such as knots, cracks, and spiral grain, are common in wood and are either insignificant or eliminated in wood-based materials. The high degree of homogeneity also results in slight variations in the properties of the panels, leading to more favourable 5th percentiles, which are critical for determining characteristic strength and stiffness values.

Strategically arranging the individual wood components can optimise the load-bearing capacity in a specific direction. Panel-shaped wood-based materials typically exhibit less swelling and shrinking compared to sawn timber. Another advantage of panel-shaped wood-based materials is their large surface area, while rod-shaped wood-based materials can be manufactured in long lengths and large cross-sections. These materials are produced or offered in standard dimensions, which is highly beneficial for planning, inventory management, and reducing assembly times.

Most wood-based materials used in construction (*cf.* Chapters 2 and 3) are regulated by the European harmonised standard EN 13986 (CEN 2015a):

- solid-wood panels, according to EN 13353 (CEN 2022),
- plywood according to EN 636-1-3 (CEN 2021a),
- laminated veneer lumber (LVL) according to EN 14374 (CEN 2004a),
- oriented strand board (OSB) according to EN 300 (CEN 2021b),
- particleboard according to EN 312 (CEN 2020a),

- cement-bonded particleboard according to EN 634-2 (CEN 2007),
- hardboard according (HDF) to EN 622-2 (CEN 2004b),
- medium-hardboard according to EN 622-3 (CEN 2004c),
- porous fibreboards (LDF) according to EN 622-4 (CEN 2019b),
- medium-density fibreboard (MDF) according to EN 622-5 (CEN 2019c).

13.1.3 RELEVANT PRODUCT STANDARDS

Various standards regulate the building materials used in timber construction. In such standards, general performance requirements are presented depending on the application of the product. These performance requirements relate to using the products in dry, damp, and outdoor areas for load-bearing or non-load-bearing purposes. In addition, requirements are given for the physical properties, the durability of the products, and their environmental compatibility (*e.g.*, formaldehyde release). Furthermore, the principles of internal and external monitoring of manufacturing companies and the labelling of construction products are regulated. The following relevant standards can be summarised in this context:

- EN 13986 (CEN 2015a) for wood-based materials,
- EN 14080 (CEN 2013) for glued-laminated timber (glulam) and laminated beams,
- EN 14081 (CEN 2019a) for structural timber with rectangular cross-section, and
- EN 16351 (CEN 2015b) for cross-laminated timber (CLT).

The primary aim of the product standards is to define the products so that they can be classified and traded. Property values, as required for structural design, are not always included in these product standards. These are sometimes specified in additionally developed standards. The design of load-bearing timber structures can be done following EN 1995-1-1 (CEN 2024a).

13.1.4 CONNECTIONS

The construction of spatial structures from primarily linear timber components, shaped naturally or from planar-engineered wood products, necessitates reliable joining techniques or bending to a certain degree. Historically, wood was used in half-timbered buildings or log construction (*cf.* Chapter 1). Recently, small- to medium-sized buildings have predominantly used lightweight designs using frame construction. However, with the increased use of engineering structures, skeleton and mass-timber construction methods are playing an increasingly significant role. Various fastening and joining techniques are available, each tailored to meet specific design criteria and the characteristics of the wood products employed.

Each joining method possesses unique attributes regarding load-bearing capacity, stiffness, on-site feasibility, and long-term behaviour. Therefore, engineers must select the most appropriate technique based on the structural requirements of the project. The choice of connection method directly impacts the manufacturing process and the depth of prefabrication. Especially in multi-storey buildings, where horizontal loads play an essential role, and therefore, the bracing of the building needs in-depth analysis, the accurate consideration of rotational and translational stiffness of the connections is crucial and must be considered when designing the timber structure.

In multi-storey buildings or constructions with large spans, the load distribution can be heavily influenced by the stiffness of the connections. When relying mainly on finite-element analysis (FEA) to design statically indeterminate structures, the connection stiffness knowledge is crucial to get realistic results of the internal force distribution. Furthermore, in the design of seismic actions, the ductility of a structure is a key aspect in determining earthquake loads. Due to the brittle failure modes generally determinant for timber structures, ductility must be achieved within the connections.

Simultaneously, ensuring adequate weather protection is critical for the durability of connections in timber structures. Each type of connection requires specific protective measures to maintain its performance throughout the structure's lifespan. Mechanical connections, such as screws, nails, and bolts, must be protected against corrosion, particularly in environments exposed to moisture. To maintain structural integrity, adhesive bonds must be suited for moisture and temperature fluctuations. Traditional wood joinery, such as mortise and tenon joints, must be carefully protected from direct water exposure or oriented to allow for efficient water run-off. Overall, the protection of connections is essential for the long-term durability of timber structures, as unprotected joints are vulnerable to degradation from mechano-sorptive stress or decay due to excessive moisture content.

13.1.5 ADHESIVES

In modern structural timber construction, only curing synthetic resin adhesives are used. These adhesives can be categorised into three main types used for surface bonding and finger-jointing:

- phenoplastic and amino-plastic adhesives for example, melamine-formaldehyde (MF), melamine-urea-formaldehyde (MUF), phenol-resorcinol-formaldehyde (PRF) by EN 301 (CEN 2023a),
- moisture-curing one-component adhesives based on polyurethane (PUR) by EN 15425 (CEN 2023b), and
- emulsion polymer isocyanate adhesives (EPI) according to EN 16254 (CEN 2023c).

Ensuring that the adhesive used for gluing is suitable for use in the intended Service Class is essential. Regarding their suitability for use under different climatic conditions, adhesives are classified into two types:

- type I suitable for all Service Classes according to EN 1995-1-1 (CEN 2024a), and
- type II suitable for use only in Service Class 1.

Additionally, the suitability of adhesives depends on the maximum test temperature and the maximum adhesive joint thickness in use. The required pressure for the gluing depends on several factors, such as the finger-joint profile, wood species, moisture content, and the cross-section area of the timber. Manufacturers provide guidelines for these parameters, which are made available by the producers of the finger-jointing equipment.

For laminated veneer lumber (LVL), a phenolic resin adhesive is usually used for bonding. Modified melamine resin adhesives are also employed when bonding face veneers. The bonding process is generally carried out under high pressure and heat.

The load-bearing capacity of adhesive joints is influenced by:

- moisture content,
- grain orientation and load direction,
- adhesive application, adhesive age, and pressing duration,
- climatic conditions during the curing process,
- stresses resulting from moisture and temperature fluctuations, and
- additional stresses (shear) and the size of the adhesive surface (non-uniform stress distribution).

Therefore, the design of adhesive joints can only be limitedly specified. In the structural design according to the EN 1995-1-1 standard (CEN 2024a), the same characteristic values as for non-finger-jointed wood, as defined in EN 14081 (CEN 2019a), can be assumed if it is verified through quality control that the bending strength of the finger-jointed wood is equal to or greater than that of the non-finger-jointed wood. Regarding their load-bearing behaviour, bonded joints are considered rigid joints, *i.e.* no deformations occur when loaded. For face-glued engineered wood products such as cross-laminated timber (CLT) and glued-laminated timber (glulam), the adhesive might influence their charring performance when exposed to fire, see Section 13.2.

13.1.6 Timber-hybrid Buildings

Large and complex buildings are increasingly being constructed using timber as a primary structural material or through hybrid timber constructions. In recent years, hybrid timber buildings have gained significant popularity. Combining wood with steel or concrete leads to more efficient construction methods. The advantages include, among others, (1) increased strength and stiffness: The combination of timber with steel or concrete enhances the overall structural strength and stiffness, allowing for larger spans and taller buildings; (2) optimised material use: Combining materials allows for the best properties of each to be utilised, leading to more efficient use of resources and potentially lower material costs. Notably, timber-concrete composite slab systems have integrated into the state of the art of timber constructions (Dias 2018).

The significant advancements in the last two decades can be attributed to the rapid development of fasteners, timber construction materials, and production technologies, which have propelled timber construction to the forefront of some regions of the construction industry. These advancements are highly dynamic and are expected to continue evolving in the coming years, driven in part by the shift from the predominant use of spruce to the growing adoption of hardwoods in construction.

Spruce, in particular, is facing significant challenges due to the changing growing conditions brought on by climate change. However, with an abundant supply of round timber, mainly beech and ash, the development of hardwood and hybrid timber materials has gained momentum. These hardwoods, which offer greater strength and stiffness, are a valuable addition to the range of wood and wood-based materials. Examples include glulam made from oak, ash, and beech, hybrid glulam combining spruce and ash, and laminated veneer lumber made from beech.

The research and use of eucalyptus for engineered wood products have also increased in recent years, as this species is, for example, locally available in Africa and Australia. However, challenges in the production (*e.g.*, cutting, drying, and gluing) of engineered wood products using eucalyptus must be addressed and well-researched before such products can be economically designed and used.

In addition, timber is being used together with other natural-based materials to design timber hybrid structures. Recently, the combination of timber and clay has received much attention, and innovative floor and wall structures combining both materials have been developed.

13.2 FIRE PERFORMANCE AND DESIGN

The design objectives in the event of a fire are guided by regulatory requirements and the building's fire safety strategy. Fire design is considered an accidental load case, addressing situations where a structure is exposed to extreme or unforeseen conditions, as described in the EN 1991-1-2 standard (CEN 2024b). The structure must be designed to maintain its integrity for a specified period of fire, *i.e.*, to ensure it does not collapse or lose its load-bearing capacity. This is a crucial aspect of safety design and is treated as a separate load case, distinct from normal or service loads.

Designing timber members for standard fire situations, for example, according to the EN 1363-1 standard (CEN 2020b) requires an assessment of the reduction in cross-section due to charring and the impact of heat on the strength and stiffness of the remaining cross-section. Charring can be influenced by protective claddings (*e.g.*, gypsum, cement, or timber-based) and cavity insulation. For engineered timber members, the integrity of the bond-line during a fire can also affect the charring behaviour and load-bearing capacity. Any charring of structural or non-structural timber members contributes to the fuel load within the fire compartment.

The following sections introduce engineers and architects to three key topics for designing a fire-safe timber building: (1) fire resistance, (2) reaction to fire, and (3) fire dynamics.

13.2.1 FIRE RESISTANCE

The fire resistance of building elements, including timber members, is defined and assessed according to various standards depending on the region and country. The fire-resistance regulations are typically defined based on the height of the building, its use, and the number of occupants, among other factors. In Europe, the EN 13501-2 standard (CEN 2023d) specifies the classification of construction products and building elements based on their fire resistance. It outlines the criteria for determining and classifying fire resistance performance.

It is well established that timber and wood-based products char at a relatively consistent rate when exposed to a standard fire as defined by the EN 1363-1 standard (CEN 2020b). Many national and international codes provide charring rates for various wood species and wood-based products, and these charring rates are well-documented. The charring rates for standard fire exposure according to EN 1363-1 standard (CEN 2020b) are specified in Eurocode 5 according to the EN 1995-1-2:2004 (CEN 2004d) and prEN 1995-1-2 (CEN 2023e) standards. The density of the timber member primarily influences the charring rate, though the moisture content can also affect charring performance, often with this factor being neglected in design. The adhesive influence on the charring performance can be determined for glued wood products based on the newly developed bond-line integrity in fire (BLIF) method presented in Annex B of prEN1995-1-2:2024 (CEN 2024a). The technique is compared in a fire-resistance test, applying EN 1363-1 (CEN 2020b) standard fire exposure to the charring performance of the glued timber member with a solid-wood panel (Figure 13.2).

FIGURE 13.2 Testing of the bond-line integrity in fire for a cross-laminated timber (CLT) panel according to Annex B of prEN1995-1-2:2024.

(Courtesy of M. Klippel, Y. Plüss.)

Structures' fire resistance can be evaluated through fire testing or calculations. Calculation methods are generally expected to provide conservative results compared to fire testing. Design parameters for timber and protective materials are required to use the calculation methods effectively.

The fire resistance R of load-bearing timber members usually includes the determination of the residual load-bearing cross-section, which is slightly different in various regions. Following Eurocode 5, in the first step, the residual cross-section is determined by reducing the original cross-section with the thickness of the char layer d_{char} (Figure 13.3). In the second step, the reduction in strength and stiffness due to elevated temperatures between 20 °C and 300 °C behind the char layer is accounted for by further reducing the cross-sectional area by the so-called zero-strength layer (ZSL, d_0). The remaining effective cross-section is assumed to retain its original strength properties, which means that the load-bearing capacity of this effective cross-section is then determined with the original timber properties. It shall be mentioned

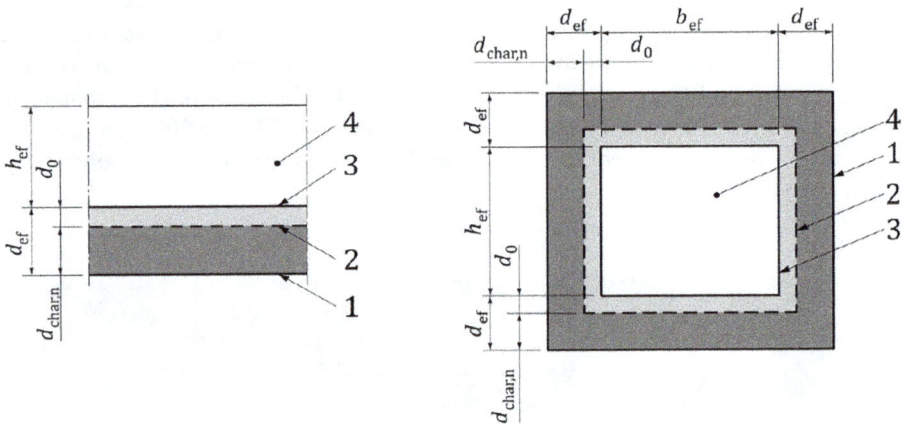

One-dimensional charring   Two-dimensional charring

Key

1	fire exposed side(s) or fire exposed perimeter
2	border-line of the residual cross-section
3	border-line of the effective cross-section
4	effective cross-section
d_{ef}	effective charring depth
$d_{char,n}$	notional charring depth
d_0	zero-strength layer depth
b_{ef}	width of the effective cross-section
h_{ef}	depth of the effective cross-section

FIGURE 13.3 Determination of effective cross-section according to Eurocode 5 prEN 1995 1 2:2024.

(Based on CEN 2024a, courtesy of M. Klippel, Y. Plüss.)

that a timber member that is protected by a fire protection lining might be subjected to heat, which reduces the load-bearing capacity. Thus, also for fire-protected timber members, the application of an appropriate zero-strength layer might be required.

The boundary conditions of a structural system can change during a fire exposure. If a structural member is boarded at ambient temperature but the boarding fails during the fire, the member must be considered unprotected in the fire safety design. Elements stabilising a building, such as wood-based panels or gypsum plasterboards in wall or floor diaphragms, often lose their racking resistance during a fire unless adequately protected. This effect on the overall structural system must be considered. Accounting for potential premature failure may benefit redundant structural systems, provided an alternative load path is available.

Load-bearing timber members such as cross-laminated timber (CLT) or timber frame assemblies wall and floor structures might also form a compartmentation line, meaning those plane members have a fire resistance EI. The integrity E is the ability of a building element to maintain its structural integrity and prevent the passage of flames and hot gases during a fire. Insulation I is the ability of a building element to limit the temperature rise on the unexposed side, thus preventing heat transfer that could endanger the safety of occupants or adjacent structures or lead to the ignition of combustible elements at the fire unexposed side of the building element providing compartmentation EI. The EI rating indicates how long a structural element can maintain both its integrity and insulation performance during a fire. For example, an EI 60 rating means the element can withstand a (standard) fire for 60 minutes while maintaining its integrity and insulation properties. Compartmentation in buildings (*e.g.*, dividing the building into different fire compartments) is one of the most effective passive fire protection measures for life safety and property protection.

Load-bearing timber connections and compartmentation joints must achieve the same fire resistance as the connecting members. The fire design of load-bearing connections is often decisive in determining the timber members' cross-section. When metallic components of a connection are exposed to fire, the charring rates at the connection are more significant compared to other parts of the timber member. The design of glued connectors shall consider the potential loss of strength and stiffness of the resin/adhesive at comparable low temperatures (*e.g.*, 50 °C to 80 °C). Connection solutions that specify fire resistance must be validated through verification by an accepted standard, first principles analysis, and fire test data or advanced simulations, for example, finite-element (FE) simulations (Figure 13.4).

The combination of cross-laminated timber (CLT) supported on a steel frame structure has become increasingly popular in recent years. The design of such structures not only includes the verification of the fire resistance but also the design of the steelwork for an allowed maximum temperature. This is required as the typically used design charring temperature of timber is 300 °C, which is below the typically used critical fire design temperature for steel members of 550 °C. Therefore, defining the allowed (critical) steel temperature is an essential element that design teams need to determine. Even with protection from intumescent coating or fire-rated panels, the steel section might still transfer heat to the supported CLT, leading to uncontrolled charring and thus to premature failure of the load-bearing capacity R or the compartmentation EI of the structure.

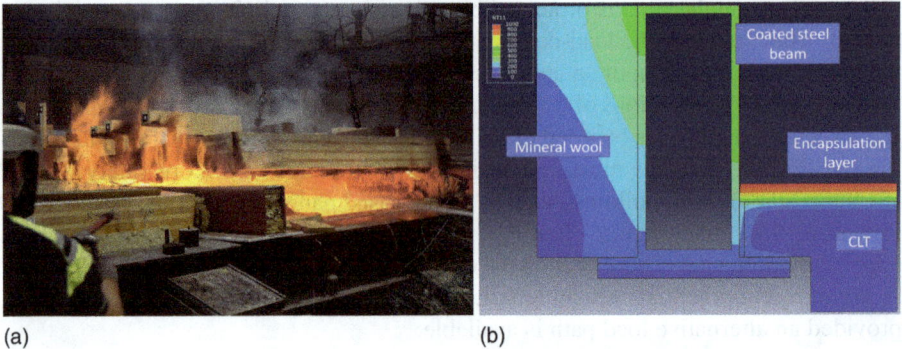

(a)　　　　　　　　　　　　　　　　　　　(b)

FIGURE 13.4　Fire-resistance validation: (a) Loaded fire test of a timber-hybrid structure with a steel beam supporting a cross-laminated timber (CLT) floor panel. (b) Advanced design with thermal finite-element simulations of a timber-hybrid structure with a steel beam supporting a CLT floor panel.

(IGNIS – Fire Design Consulting, courtesy of M. Klippel, Y. Plüss.)

13.2.2　REACTION TO FIRE

The reaction to Fire Class refers to a classification system used to assess how materials react when exposed to fire. This classification is crucial for determining the fire performance of building materials and ensuring they meet safety standards. The system evaluates factors such as:

- flammability, *i.e.* how easily the material ignites and burns,
- heat release, *i.e.* the amount of heat released by the material when it burns,
- smoke production, *i.e.* the amount and density of smoke produced during combustion, and
- toxicity, *i.e.* the potential release of harmful gases or substances.

Materials are categorised into different classes based on their performance in standardised fire tests. These classes help to ensure that construction materials contribute to overall fire safety and comply with regulations. For example, in Europe, the reaction to Fire Classes ranges from A1 (non-combustible) to F (no performance determined), as defined by the EN 13501-1 standard (2023f).

It is essential to recognise that their composition influences the reaction to fire characteristics of wood products, as the three main components (cellulose, hemicelluloses, and lignin) each have distinct thermal degradation properties. Thermogravimetric analysis (TGA) demonstrates that these components decompose and release flammable volatiles at different temperature ranges: cellulose decomposes between 240 °C and 350 °C, hemicelluloses between 200 °C and 260 °C, and lignin between 280 °C and 500 °C (Roberts 1970). As a result, the thermal degradation of wood shifts to higher temperatures with increasing lignin content, which explains why the ignition temperature is notably higher for softwoods than for hardwoods. Additionally, only about 50% of the lignin (typically 18–35% of the wood by mass) decomposes into volatiles, so a higher lignin content generally leads to increased char formation. Besides these main components, wood contains extractives (organic compounds that comprise 4% to 10% of the wood) and small amounts of inorganic minerals (less than 1%). Extractives can negatively impact the wood's flammability. Petterson (1984) provides comprehensive chemical composition data for wood species from the USA and other regions.

Wood products can be treated in various ways to enhance their durability, with traditional toxic treatments being gradually replaced by more environmentally friendly alternatives. These methods include different types of wood modifications such as acetylation (treatment with acetic acid), furfurylation (treatment with furfuryl alcohol), and thermal modification (Gérardin 2016, Sandberg *et al.* 2021). These treatments alter the chemistry of wood and reduce the amount of water the cell walls can absorb. As a result, the fire performance of the material may be affected, a finding that several studies have supported (Morozovs and Bukšāns 2009, Dong *et al.* 2015).

Internationally standardised methods for assessing the reaction to fire performance of wall linings and ceiling materials include the Room/Corner test described in the ISO 9705-1 standard (ISO 2016) and the cone calorimeter test according to the ISO 5660-1 standard (ISO 2015). The cone calorimeter test is the most widely used method for evaluating building product reaction to fire performance. This advanced small-scale testing device uses the oxygen consumption technique to measure materials and products' heat release rate (HRR) under various thermal exposure conditions. Additional data provided by cone calorimeter tests includes time to ignition, mass loss rate, smoke-production rate, and effective heat of combustion. During a test, a 100×100 mm specimen is placed on a load cell and subjected to a pre-set radiant heat flux from an electric heater. The heater, shaped like a truncated cone, can deliver heat fluxes ranging from 0 to 100 kW/m^2 to the specimen (Figure 13.5).

Engineered wood products (EWPs) for structural use, such as cross-laminated timber (CLT), glued-laminated timber (glulam), or laminated veneer lumber (LVL), may feature visible wood surfaces and require documented evidence of their declared reaction to fire performance based on testing.

Most wood products with a density higher than 300 kg/m^3 and a thickness of more than 9 mm to 12 mm, depending on how they are mounted, meet Class D requirements (Östman *et al.* 2010). Classes A1 and A2 are designated for non-combustible products, while classes E and F, which only require testing according to the small flame test described in the EN ISO 11925-2 standard (CEN 2020c), are rarely used in buildings.

Fire-retardant treatments, such as chemical modification, can significantly enhance the reaction to fire performance of wood products, allowing them to achieve the highest fire classifications for combustible materials, such as Group 1 in Australia and New Zealand, Euroclass B in Europe, or Class A in the USA for wall linings and ceiling materials. This makes it possible to use exposed wood more widely for interior wall and ceiling linings and exterior cladding, such as on façades. The primary purpose of flame retardants for wood is to delay ignition and reduce the heat released during combustion. The durability of the fire-retardant treatment is a crucial factor to consider, and the European standard EN 16755 (CEN 2017) has been established to determine the durability of the reaction-to-fire performance (DRF) classes.

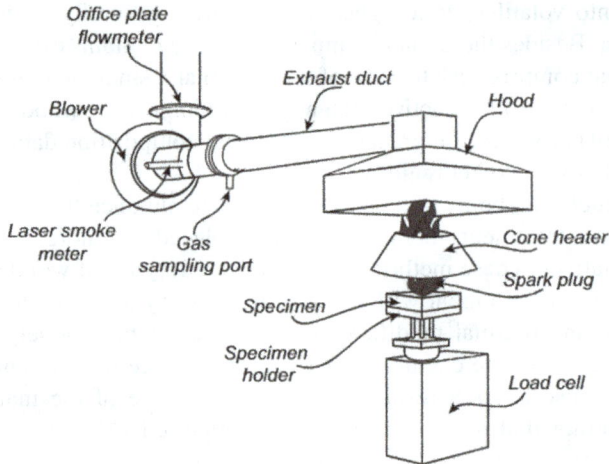

FIGURE 13.5 Principle of cone-calorimeter test setup according to the ISO 5660-1 standard (ISO 2015).

(Courtesy of M. Klippel, Y. Plüss.)

13.2.3 FIRE DYNAMICS

Designers should be mindful that while standard fire-resistance tests have limitations for all buildings, timber buildings have additional considerations, especially when structural timber members are exposed or inadequately protected (*e.g.*, by encapsulating materials). In such cases, the total fire load will include the combustible building contents and fittings (movable fuel load) and any contribution from exposed or insufficiently protected timber surfaces in the compartment. This fact might need to be addressed in the fire strategy concept for tall and complex buildings with a higher consequence class, or disregarded for buildings with a lower consequence class, see EN 1991-1-7 (CEN 2006).

National regulations usually define if and how exposed structural timber members shall be considered in the design. Prescriptive regulations generally define a maximum share of exposed structural timber in the room (*e.g.*, Germany or the USA), or only linear timber members are permitted (*e.g.*, Switzerland). Suppose the additional fuel load from structural timber is considered in engineering assessments. In that case, a performance-based design route is typically applied to design a building structure that provides a reasonable likelihood of surviving the entire duration of the fire (including the decay phase). Thus, the structural timber members do not collapse. For buildings where the consequences of failure are more severe, such as those involving prolonged evacuation of the users or requiring significant fire service resources deep within or at height in the building, regulations may require designing the structure with a reasonable likelihood of surviving the entire duration of a fire, also for the case that sprinklers (if present) are not operational and there is a delayed firefighting intervention.

It is generally recommended that non-life safety goals, such as insurance require-ments, be reviewed and incorporated on a project-by-project basis when developing the fire safety strategy. It is advisable to contact insurance and warranty providers early in the design process of a mass-timber structure, as they may impose require-ments beyond national fire safety regulations and requirements.

The presence of significantly exposed timber within a compartment may lead to higher HRRs, faster flashover times, larger fires, and prolonged burning after flash-over compared to a compartment without timber contribution. Moreover, flames emerging from unprotected openings, such as windows and doors, may be more sig-nificant and persist longer, increasing the risk of vertical fire spread to upper floors and horizontal spread to neighbouring buildings.

Figure 13.6 presents the general trendlines of HRR over fire duration time in a compartment with exposed structural timber. For complex and high-rise timber buildings, some building regulations require verifying that the HRR decays when the room content is consumed in case sprinklers do not work (behaviour 1). To achieve this behaviour, the compartment configuration needs to be designed accordingly to limit the share of exposed timber, have sufficient no-rated openings (windows), and use the correct timber products, among others. If wrong timber products are used, a re-growth in the decay phase might be possible due to a

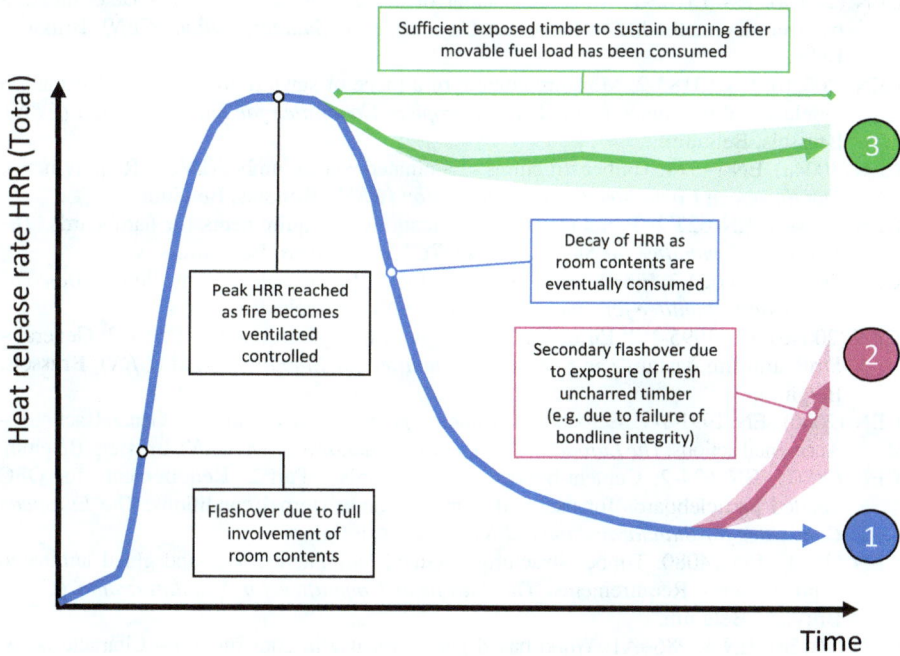

FIGURE 13.6 General trendlines of heat release rate (HRR) vs. time plots from experimen-tal data of ventilation-controlled fires within exposed mass-timber compartments.

(Adapted after Arup 2024, courtesy of Arup Group, London, UK.)

failure of the bond-line integrity, which means that charring lamellas might fall off, or the encapsulation layer falls off and fresh unburnt timber is exposed again (behaviour 2). Suppose too much timber is exposed in the compartment, and the ventilation conditions are unfavourable. In that case, the fire might continue with sustained smouldering in structural elements, leading possibly to a structural failure (behaviour 3).

Designers must thoroughly understand fire dynamics in compartments constructed from various materials, particularly timber, to address their unique challenges effectively. This requires considering the specific characteristics of the building and its occupants, including compartment size, geometry, height, ventilation, use, the amount and location of exposed timber surfaces, and the required or expected performance and fire safety strategy adopted in its design. The choice of engineered wood products and the adhesive's performance in a fire should also be addressed in such a design.

Addressing fire dynamics and performance-based design also involves designing connection and joint details that reduce the likelihood of smouldering and glowing combustion. A competent fire engineer with a clear record of experience is required to address the fire dynamics in a building with exposed timber members correctly.

REFERENCES

Arup (2024). *Fire-safe Design of Mass Timber Buildings*. Arup Group, London, UK, (157 pp.)

CEN (2002a). EN 13183-1: Moisture content of a piece of sawn timber – Determination by oven dry method. *The European Committee for Standardization (CEN)*, Brussels, Belgium.

CEN (2002b). EN 13183-2: Moisture content of a piece of sawn timber - Part 2: Estimation by electrical resistance method. *The European Committee for Standardization (CEN)*, Brussels, Belgium.

CEN (2004a). EN 14374: Timber structures - Laminated veneer lumber (LVL) - Requirements. *The European Committee for Standardization (CEN)*, Brussels, Belgium.

CEN (2004b). EN 622-2: Fibreboards - Specifications - Requirements for hardboards. *The European Committee for Standardization (CEN)*, Brussels, Belgium.

CEN (2004c). EN 622-3: Fibreboards - Specifications - Requirements for medium boards. *The European Committee for Standardization (CEN)*, Brussels, Belgium.

CEN (2004d). EN 1995-1-2: Eurocode 5 – Design of timber structures – Part 1-2: General – Structural fire design. *The European Committee for Standardization (CEN)*, Brussels, Belgium.

CEN (2006). EN 1991-1-7: Eurocode 1: Actions on structures – Part 1-7: General actions – Accidental actions. *The European Committee for Standardization (CEN)*, Brussels, Belgium.

CEN (2007). EN 634-2: Cement-bonded particleboards - Part 2: Requirements for OPC bonded particleboards for use in dry, humid, and external conditions. *The European Committee for Standardization (CEN)*, Brussels, Belgium.

CEN (2013). EN 14080: Timber structures – Glued laminated timber and glued laminated solid timber – Requirements. *The European Committee for Standardization (CEN)*, Brussels, Belgium.

CEN (2015a). EN 13986+A1: Wood-based panels for use in construction – Characteristics, evaluation of conformity and marking. *The European Committee for Standardization (CEN)*, Brussels, Belgium.

CEN (2015b). EN 16351: Timber structures – Cross-laminated timber – Requirements. *The European Committee for Standardization (CEN)*, Brussels, Belgium.

CEN (2016). EN 338: Structural timber - Strength classes. *The European Committee for Standardization (CEN)*, Brussels, Belgium.

CEN (2017). EN 16755: Durability of reaction to fire performance. Classes of fire-retardant treated wood products in interior and exterior end use applications. *The European Committee for Standardization (CEN)*, Brussels, Belgium.

CEN (2019a). EN 14081-1+A1: Timber structures. Strength graded structural timber with rectangular cross section. *Part 1: General requirements. The European Committee for Standardization (CEN)*, Brussels, Belgium.

CEN (2019b). EN 622-4: Fibreboards - Specifications - Requirements for softboards. *The European Committee for Standardization (CEN)*, Brussels, Belgium.

CEN (2019c). EN 622-5: Fibreboards - Specifications - Requirements for dry-process boards (MDF). *The European Committee for Standardization (CEN)*, Brussels, Belgium.

CEN (2020a). EN 312: Particleboards - Specifications. *The European Committee for Standardization (CEN)*, Brussels, Belgium.

CEN (2020b). EN 1363-1: Fire resistance tests – Part 1: General requirements. *The European Committee for Standardization (CEN)*, Brussels, Belgium.

CEN (2020c). EN ISO 11925-2: Reaction to fire tests – Ignitability of products subjected to direct impingement of flame – Part 2: Single-flame source test. *International Organization for Standardization, Geneva. The European Committee for Standardization (CEN)*, Brussels, Belgium.

CEN (2021a). EN 636: Plywood - Specifications. *The European Committee for Standardization (CEN)*, Brussels, Belgium.

CEN (2021b). EN 300: Oriented strand boards (OSB) - Definitions, classification, and requirements. *The European Committee for Standardization (CEN)*, Brussels, Belgium.

CEN (2022). EN 13353: Solid wood panels (SWP) – Requirements. *The European Committee for Standardization (CEN)*, Brussels, Belgium.

CEN (2023a). EN 301: Adhesives, phenolic and aminoplastic, for load-bearing timber structures – Classification and performance requirements. *The European Committee for Standardization (CEN)*, Brussels, Belgium.

CEN (2023b). EN 15425: Adhesives – One component polyurethane (PUR) for load-bearing timber structures – Classification and performance requirements. *The European Committee for Standardization (CEN)*, Brussels, Belgium.

CEN (2023c). EN 16254: Adhesives – Emulsion polymer isocyanate (EPI) for load-bearing timber structures – Classification and performance requirements. *The European Committee for Standardization (CEN)*, Brussels, Belgium.

CEN (2023d). EN 13501-2: Fire classification of construction products and building elements – Part 2: Classification using data from fire resistance and/or smoke control tests, excluding ventilation services. *The European Committee for Standardization (CEN)*, Brussels, Belgium.

CEN (2023e). prEN 1995-1-2, Eurocode 5: Design of timber structures – Part 1-2: General – Structural fire design. *The European Committee for Standardization (CEN)*, Brussels, Belgium.

CEN (2023f). EN 13501-1: Fire classification of construction products and building elements. Part 1: Classification using data from reaction to fire tests. *The European Committee for Standardization (CEN)*, Brussels, Belgium.

CEN (2024a). FprEN 1995-1-1: Eurocode 5 – Design of timber structures – Part 1-1: General rules and rules for buildings. *The European Committee for Standardization (CEN)*, Brussels, Belgium.

CEN (2024b). EN 1991-1-2: Eurocode 1: Actions on structures – Part 1-2: General actions – Actions on structures exposed to fire. *The European Committee for Standardization (CEN)*, Brussels, Belgium.

CODIFAB (2020). Construction bois et gestion de l'humidité en phase chantier. [Wood construction and humidity management during the construction phase.] *Comité Professionnel de Développement des Industries Françaises de l'Ameublement et du Bois (CODIFAB)*, Paris, France.

Dias A., Schänzlin J. & Dietsch P. (Eds.) (2018). Design of timber-concrete composite structures. *A state-of-the-art report by COST Action FP1402 /WG 4. European Cooperation in Science and Technology (COST)*, Shaker Verlag Aachen, Germany.

DIN (2020). DIN 68800-3: Wood preservation - Part 3: Preventive protection of wood with wood preservatives. *Deutsches Institut für Normung (DIN)*, Berlin, Germany.

DIN (2022). DIN 68800-2: Wood preservation – Part 2: Preventive constructional measures in buildings. *Deutsches Institut für Normung (DIN)*, Berlin, Germany.

Dong Y., Yan Y., Zhang S., Li J. & Wang J. (2015). Flammability and physical–mechanical properties assessment of wood treated with furfuryl alcohol and nano-SiO2. *Journal of Wood and Wood Products*, 73(4), 457–464.

Gérardin P. (2016) New alternatives for wood preservation based on thermal and chemical modification of wood – a review. *Annals of Forest Science*, 73(3), 559–570.

Gerhards C. (2007). Effect of moisture content and temperature on the mechanical properties of wood: An analysis of immediate effects. *Wood and Fiber Science*, 14(1), 4–36.

Glass S.V. & Zelinka S.L. (2010). Moisture Relations and Physical Properties of Wood. In: Ross B. (Ed.), *Wood Handbook: Wood as an Engineering Material. (Centennial Ed.), General technical report FPL GTR-190*. U.S. Dept. of Agriculture, Forest Service, Forest Products Laboratory, Madison (WI), USA, (pp. 4.1–4.19).

Informationsdienst Holz (2024). *Feuchtemanagement, Witterungsschutz in der Bauphase*. Holzbau Handbuch, Holzbau Deutschland - Institut e.V., Berlin, Germany, (32 pp.).

ISO (2015). ISO 5660-1: Reaction-to-fire tests - Heat release, smoke production and mass loss rate - Part 1: Heat release rate (cone calorimeter method) and smoke production rate (dynamic measurement). *International Organization for Standardization (ISO)*, Geneva, Switzerland.

ISO (2016). ISO 9705-1: Reaction to fire tests — Room corner test for wall and ceiling lining products — Part 1: Test method for a small room configuration. *International Organization for Standardization (ISO)*, Geneva, Switzerland.

Morozovs A. & Bukšāns E. (2009). Fire performance characteristics of acetylated ash (*Fraxinus excelsior*) wood. *Wood Material Science & Engineering*, 4(1/2), 76 –79.

Niemz P. & Sonderegger W.U. (2003). Untersuchungen zur Korrelation ausgewählter Holzeigenschaften untereinander und mit der Rohdichte unter Verwendung von 103 Holzarten. *Schweizerische Zeitschrift Für Forstwesen*, 154(12), 489–493.

Petterson R. (1984). The Chemical Composition of Wood. In: Rowell M. (Ed.), *The Chemistry of Solid Wood, Advances in Chemistry Series 207*, American Chemical Society, Washington (DC), USA, (pp. 57–84).

Roberts A. (1970). A review of kinetic data for the pyrolysis of wood and related substances. *Combustion and Flame*, 14(2), 261–272.

Sandberg D., Kutnar A., Karlsson O. & Jones D. (2021). *Wood Modification Technologies. Principles, Sustainability, and the Need for Innovation*. CRC Press, Boca Raton (FL), USA, (432 pp.).

Östman B., Mikkola E., Stein R., Frangi A., König J., Dhima D., Hakkarainen T. & Bregulla, J. (2010). *Fire Safety in Timber Buildings: Technical Guideline for Europe. SP Report 2010:19. SP Trätek*, Stockholm, Sweden, (210 pp.).

Appendix
Additional Supporting Data

A1 BOOKS AND STANDARDS FOR TIMBER USE

Below are some books in Timber Engineering that are suggested for further reading. Niemz *et al.* (2023) provide a more comprehensive compilation of literature in chemical engineering, energy from biomass, pulp and paper, timber engineering, wood science, and wood-based materials.

BASICS OF TIMBER ENGINEERING

Anderson D., Wu Q. & Han, G. (2013). *Introduction to Wood and Natural Fiber Composites.* Wiley Series in Renewable Resources, John Wiley & Sons Ltd.

Bodig J. & Jayne B. (1993). *Mechanics of Wood and Wood Composites.* Krieger Publishing Company.

Gong M. (Ed.) (2021). *Wood and Engineered Wood Products: Stress and Deformations.* IntechOpen.

Ross R. (2021). *Wood Handbook - Wood as an Engineering Material.* Forest Products Laboratory, Madison, USA.

Dinwoodie J.M. (2000). *Timber: Its Nature and Behaviour.* Taylor & Francis.

Fengel D. & Wegener G. (1984). *Wood: Chemistry*, Ultrastructure, Reactions. W. de Gruyter.

Kollmann F. & Côte W. (1984). *Principles of Wood Science and Technology: I.* Solid wood. Springer-Verlag.

Niemz P. & Sonderegger W. (2021) *Holzphysik: Eigenschaften und Kennwerte*, Hanser Verlag.

Niemz P., Teischinger A. & Sandberg D. (Eds.) (2023). *Springer Handbook of Wood Science and Technology.* Springer Nature.

Rowell R.M. (Ed.). (2012). *Handbook of Wood Chemistry and Wood Composites.* CRC Press.

Shmulsky R. & Jones P.D. (2011). *Forest Products and Wood Science – An Introduction.* Wiley-Blackwell.

Thelandersson S. & Larsen H. (Eds.) (2003). *Timber Engineering.* Wiley.

STRUCTURAL TIMBER DESIGN

Blaß H.-J. & Sandhaas C. (2017). *Timber Engineering – Principles for Design.* KIT Scientific Publishing.

Borgström E. (Ed.) (2016). *Design of Timber Structures. 1.* Structural Aspects of Timber Construction. Swedish Wood.

Borgström E. (Ed.) (2016). *Design of Timber Structures.* 2. Rules and Formulas According to Eurocode 5. Swedish Wood.

Borgström E. (Ed.) (2016). *Design of Timber Structures. 3. Examples.* Swedish Wood.

Breyer D., Fridley K., Cobeen K. & Pollock D. (2006). *Design of Wood Structures - ASD/ LRFD.* McGraw-Hill.

Buchanan A. & Östmann B. (Eds) (2022). *Fire Safe Use of Wood in Buildings.* Global Design Guide. CRC Press, Taylor & Francis Group.

Kermani A. (1999). *Structural Timber Design.* Blackwell Science.

Kolb J., Kolb H. & Müller A. (2024). *Holzbau mit System.* Birkhäuser Verlag.

Larsen H. & Enjily V. (2009). *Practical Design of Timber Structures to Eurocode 5.* ICE Publishing.

Madsen B. (1992). *Structural Behaviour of Timber.* Timber Engineering, Vancouver.

Madsen B. (2000). *Behaviour of Timber Connections.* Timber Engineering, Vancouver.

McKenzie W. & Zhang B. (2007). *Design of Structural Timber to Eurocode 5.* Palgrave.

Ozelton E.C. & Baird J.A. (2002). *Timber Designers' Manual.* Blackwell Science.

Porteous J. & Kermani A. (2013). *Structural Timber Design to Eurocode 5.* Wiley-Blackwell.

Porteous J. & Ross P. (2013). *Designers' Guide to Eurocode 5: Design of timber buildings: EN 1995-1-1.* Eurocode Designers' Guide, ICE Publishing.

Stalnaker J.J. & Harris E.C. (1997). *Structural Design in Wood.* Kluwer Academic Publishers.

Winter S. & Peter M. (Eds.) (2021). *Holzbautaschenbuch.* Ernst und Sohn (in German).

STANDARDS

Chapter 38.7 of the *Springer Handbook of Wood Science and Technology* (Niemz *et al.* 2023) compiles key wood standards from worldwide regions. The digital version provides additional information, including standards for wood properties, materials, additives, and quality testing. Standards concerning the use of wood in construction, such as wood protection, fire protection, structural design, and more, are also included.

Standardisation bodies interact in various ways. Many national and transnational standardisation organisations are members of ISO and contribute to developing ISO standards. Unlike in Europe, where member countries of CEN are required to incorporate CEN standards into national standards, it should be noted that not all ISO member states necessarily adopt ISO standards.

Users of standards can quickly determine whether a transnational standardisation body has adopted a standard. Standards are encoded using the following system:

- the code of the national standardisation body and
- the code of the transnational standardisation body from which the standard was adopted, followed by the standard's number, year, and month of publication.

A detailed list is omitted here; however, international and national standards apply.

There are standards for:

- Physical and mechanical properties: *e.g.* grading by load-bearing capacity, testing of small defect-free specimens and components for strength, moisture, swelling, *etc.*,
- Wood protection and fire protection,
- Bonding for structurally loaded elements,
- Surface treatment, and
- Timber construction and constructive designs (particularly Eurocode 5).

A selection of standards in these areas is available, *e.g.*, Niemz and Sonderegger (2021).

A2   PROPERTIES OF SOME EUROPEAN WOOD SPECIES AND WOOD-BASED MATERIALS

A short compilation of the properties of important European wood species will show the variability between the properties. This applies to mechanical-physical properties and durability, which is important for the construction industry (especially in the case of increased moisture stress).

The *Springer Handbook of Wood Science and Technology* (Niemz *et al.* 2023) describes the material properties of wood species worldwide.

A3   USE CLASSES AND SERVICE CLASSES FOR TIMBER

In timber classification, Service Classes and Use Classes refer to different aspects of how timber is used and how it should be treated to ensure it performs well under certain conditions. Service Classes focus on the environment where timber is used (moisture and climate conditions). In contrast, Use Classes focus on the timber's intended application and its exposure to moisture or decay. Service Classes help inform the timber's durability based on environmental factors, whereas Use Classes guide how the timber is treated for its specific use case.

TABLE A1

Properties of Clear Specimens of Solid Wood at 12% Moisture Content (Wagenführ and Wagenführ 2022)

Property	Unit	Common Ash	Common Beech	English Oak	Norway Spruce	Scots Pine
Density	kg/m³	450–510–610	490–680–880	390–650–930	300–430–640	300–400–860
Bending strength //	N/mm²	44–94–172	74–123–210	60–94–100	42–66–116	35–87–206
Tensile strength //	N/mm²	55–94–140	57–135–180	50–90–180	40–90–245	35–104–196
Compression strength //	N/mm	31–55–77	41–62–99	41–55–59	30–43–67	30–47–80
Maximum shrinkage:	%					
radial		4.3	5.8	4.0	3.6	4.0
tangential		9.3	11.8	7.8	7.8	7.7
longitudinal		0.4	0.3	0.4	0.3	0.4
Calorific value	MJ/kg	14.2	15.0	14.5	16.3	15.9
Thermal conductivity ⊥	W/mK	0.121	0.160	0.163	0.121	0.140

// – parallel and ⊥ – perpendicular to the fibre direction of the specimen. Values: min.–mean–max.

TABLE A2
Strengths of Wood in the Principal Directions at 12% Moisture Content (Measurements ETH Zurich)

| Species | Mode of Loading | Strength (N/mm²): | | | Relation |
		Longitudinal (L)	Radial (R)	Tangential (T)	T : R : L
Cherry	Tension	109	17.3	10.8	1 : 1.6 : 10.1
	compression	53.5	14.4	9.5	1 : 1.5 : 5.6
Common ash	Tension	130.0	12.5	10.1	1 : 1.2 : 12.9
	compression	43.4	10.5	10.0	1 : 0.95 : 4.3
Common beech	Tension	96.7	14.7	8.9	1 : 1.7 : 10.9
	compression	45.0	11.0	6.0	1 : 1.8 : 7.5
English oak	Tension	73.0	6.0	7.8	1 : 0.8 : 9.4
	compression	47.9	10.6	9.0	1 : 1.2 : 5.3
European walnut	Tension	89.1	10.8	8.9	1 : 1.2 : 10.0
	compression	60.4	13.4	11.9	1 : 1.13 : 5.1
Norway maple	Tension	112.0	16.2	8.9	1 : 1.8 : 12.6
	compression	61.5	15.4	10.3	1 : 1.5 : 6.0
Norway spruce	Tension	87.2	4.0	3.1	1 : 1.3 : 28.4
	compression	40.2	4.1	4.2	1 : 0.98 : 9.6

TABLE A3
Strength of Wood-Based Panel Materials (ETH Zürich)

| Material | | Strength in the Plane of the Panel: | | Internal Bond | Relation |
		In Production Flow Direction (x) (N/mm²)	Perpendicular to the Production Flow (y) (N/mm²)	Thickness Direction (z) (N/mm²)	z : y : x
Particleboard	tension	6.3	5.7	0.45	1: 12.7 : 14
(660 kg/m³)	compression	10.7	10.6		
MDF	tension	20.6	20.3	0.60	1: 33.8 : 4.3
(742 kg/m³)	compression	20.3	20.4		

x, y – in plane direction of the panel, z – in the thickness direction of the panel.

TABLE A4
Strength Properties of Bamboo, Solid-wood, and Wood-based Panel Materials

Material	Bending Strength (N/mm²)	Tensile Strength // Grain/Plane (N/mm²)	Tensile Strength ⊥ Grain/Plane (N/mm²)	Compression Strength // Grain/Plane (N/mm²)	Compression Strength ⊥ Grain/Plane (N/mm²)
Bamboo	84–270			20–40	
Black poplar	60	67	2.3	34	
Common ash	120	165	7.0	52	11.0
English oak	94	90	4.0	60	11.0
Norway spruce	78	90	2.7	43	5.8
Scots pine	87	105	3.0	55	7.7
High-density fibreboard	45–50	20–24	0.8–1.0	23–26	
Particleboard	15–25	8–10	0.35–0.40	8–16	
Plywood	30–60	30–60		20–40	

// – parallel and ⊥ – perpendicular to the fibre direction of the specimen.

TABLE A5
Properties of Multi-layer Solid-wood Panels

Property	Unit	Panel Thickness (mm)			
		12–20	20–30	30–42	>42
Density	kg/m³	410	410	410	410
Bending strength:					
// with the fibre direction of the top layer	N/mm²	35	30	16	12
⊥ to the fibre direction of the top layer	N/mm²	5	5	9	9
Modulus of elasticity in bending:					
// with the fibre direction of the top layer	N/mm²	8500	7000	6500	6000
⊥ to the fibre direction of the top layer	N/mm²	470	470	1300	1300

// – parallel and ⊥ – perpendicular to the fibre (length) direction of the top layer of the panel.

TABLE A6

Properties of Wood-based Panel Materials (Permissible Stresses) when Bent Perpendicular to the Board Plane (Informationsdienst Holz, Germany)

Material	Bending Strength Load ⊥ Panel Plane (N/mm²)	MOE (N/mm²)
Solid wood panel three-layer, 20 mm)		
// with the fibre direction of the top layer	3–17	6000–9000
⊥ to the fibre direction of the top layer	4–11	1000–5000
Particleboard (EN 312-7, CEN 1997)	2.0–4.5	1200–3200
MDF HSL (12–19 mm)	4.4	3200
OSB (16–22 mm)		
// with the direction of surface-particle orientation	8.0	6500
⊥ to the direction of surface-particle orientation	3.6	2800
Plywood (common beech with more than five layers)		
// with the fibre direction of the top layer	18–29	5900–9600
⊥ to the fibre direction of the top layer	5–17	650–4000
LVL Kerto S (Norway spruce)		
// with the fibre direction of the top layer	20	13,000
⊥ to the fibre direction of the top layer		0
LVL Kerto Q (Norway spruce)		
// with the fibre direction of the top layer	11	10,000
⊥ to the fibre direction of the top layer	4	2000
LSL	10.4	9500
Parallam	21	14,500

MOE – modulus of elasticity, // – parallel and ⊥ – perpendicular to the fibre (length) direction of the top layer of the panel.

TABLE A7

Properties of Beech Laminated Veneer Lumber (LVL from Pollmeier) according to Deutsches Institut für Bautechnik (DIBT) Approval Z-9.1-837:2013

Strength Class	GL 70 (N/mm²)
Characteristic flexural strength when bent flat-edged	70
Characteristic flexural strength when bent on edge	70
Characteristic tensile strength in fibre direction	55
Characteristic tensile strength perpendicular to the fibre direction	1.2
Characteristic compressive strength parallel to the fibre direction	49.5
Characteristic compressive strength perpendicular to the fibre direction	8.3

(Continued)

TABLE A7 (Continued)

Strength Class	GL 70
	(N/mm²)
Characteristic shear strength	4.0
Stiffness parameters	
Average E-modulus in fibre direction	16,700
5% quantile of the modulus of elasticity in the fibre direction	15,300
Average modulus of elasticity perpendicular to the fibre direction	470
5% quantile of the modulus of elasticity perpendicular to the fibre direction	400
Average shear modulus	850
5% quantile of the shear modulus	760
The characteristic value of the bulk density (kg/m³)	680

TABLE A8

Characteristic Properties of Beech Laminated Veneer Lumber (LVL from Pollmeier) according to Deutsches Institut für Bautechnik (DIBT) approval Z-9.1-837:2013

	Beech LVL	
	Only Longitudinal Veneers	Cross-wise Veneers Occur
Properties	(N/mm²)	(N/mm²)
Nominal thickness (mm)	$20 \le t \le 120$	$20 \le t \le 100$
Density ρ_k (kg/m³)	680	
Strength (flat-wise loading):		
bending $f_{m,0,k}$	66	40
compression $f_{c,90,k}$	10	10
rolling shear $f_{v,k}$	3.3	
Strength (edge-wise loading):		
bending[a] $f_{m,0,k}$	70	60
tension parallel to fibre $f_{t,0,k}$	70	40
tension perpendicular to fibre $f_{t,90,k}$	1.5	17.0
compression parallel to fibre $f_{c,0,k}$	41.6	24.2
compression perpendicular to fibre $f_{c,90,k}$	14	14
Shear $f_{v,k}$	9	
Stiffness parameters:		
MOE $E_{0,mean}$	16,800	11,800
MOE l $E_{0,05}$	14.900	10,700
MOE $E_{90,mean}$	470	3700
Shear modulus upright loaded G_{mean}	760	890
Shear module flat loaded G_{mean}	850	430

[a] Values apply to h≤300mm, for 300<h≤100 the characteristic strength value is to be multiplied by the coefficient $k_h=(300/h) \times 0.12$, where h is the dimension of the cross-section in mm that is decisive for the total bending stress.

TABLE A9

Characteristic Values of Kerto Laminated Veneer Lumber (Norway spruce) in N/mm² and Characteristic Bulk Density in kg/m³ According to Deutsches Institut für Bautechnik (DIBT) DIBT Z-9.1-100.2011 Approval (Finno Forest Oy)

Properties	Kerto S (N/mm²)	Kerto Q (N/mm²)	
Nominal thickness (mm)	20≤ t ≤75	21≤ t ≤24	27≤ t ≤69
Strength (flat-wise loading):			
bending parallel[a] $f_{m,0,k}$	50	32	36
bending perpendicular[a] $f_{m,0,k}$	-	9[b]	9
compression $f_{c,90,k}$	2	2	
rolling shear $f_{v,k}$	2,3	1,5	
Strength (edge-wise loading):			
bending[a] $f_{m,0,k}$	48	32	36
tension parallel to fire $f_{t,0,k}$	38	20	27
tension perpendicular to fibre $f_{t,90,k}$	0,8	6	
compression perpendicular to fibre $f_{c,0,k}$	38	20	27
compression perpendicular to fibre $f_{c,90,k}$	6	9	
Shear $f_{v,k}$	4,4	4,8	
Stiffness parameters:			
MOE $E_{0,mean}$	13.800	10.000	10.500
MOE $E_{0,05}$	11.600	8.500	
MOE $E_{90,mean}$	300	1.000	2.500
Shear modulus G_{mean}	500	500	

[a] Values apply to h≤300 mm; for 300<h≤100, the characteristic strength value is to be multiplied by the coefficient $k_h=(300/h) \times 0.12$, where h is the dimension of the cross-section in mm that is decisive for the total bending stress

[b] For t=21 mm the structure I-III-I may be accepted for $f_{m,90,k}$ = 16 N/mm² or 2500 N/mm²

TABLE A10
Diffusion Resistance for Solid Wood and Wood-based Panel Materials According to DIN EN ISO 10456:2010-05 (DIN 2010)

Material	Density (kg/m³)	dry cup (-)	wet cup (-)
		Diffusion Resistance Factor, µ:	
Solid wood	450–500	50	20
	700	200	50
Plywood, LVL	300	150	50
	500	200	70
	700	220	90
	1000	250	110
Particleboard	300	50	10
	600	50	15
	900	50	20
OSB	650	50	30
Cement-bonded particleboard	1200	50	30
Insulation fibreboard	40–250	5	3
Fibreboard (MDF, HDF)	400	10	5
	600	20	12
	800	30	20

TABLE A11
Swelling and Shrinkage of Wood-based Panel Materials (von Halász and Scheer 1986)

Materials	In-plane	Perpendicular to the Plane
	Swelling/Shrinking (% per 1% Unit Change in MC)	
Plywood	0.020	0.300
Particleboard bonded with PF	0.025	0.450
Particleboard bonded with UF	0.035	0.700
Fibreboard (MDF, HDF)	0.040	0.800

MC – moisture content, PF – phenol-formaldehyde and UF – urea-formaldehyde adhesives.

TABLE A12
Equilibrium Moisture Content (EMC) for Solid Wood and Wood-based Panel Materials at Different Relative Humidity (RH) at 20 °C (von Halász and Scheer 1995)

Material	EMC at 20 °C and RH		
	30%	65%	85%
Wood		12	
Plywood	4–6	8–12	12–18
MDF, HDF	3–5	6–8	10–14
Particleboard (PF adhesive)	4–8	9–12	13–18
Gypsum-bonded panels		2–3	
Cement-bonded panels		2–4	

PF – phenol formaldehyde.

TABLE A13
Recommended Equilibrium Moisture Content (EMC) for Solid Wood and Wood-based Materials in Different Applications (Scheer *et al.* 2004)

Use	EMC (%)
Closed buildings with heating	4–12
Closed buildings without heating	9–15
Components in open, roofed constructions	12–20
Exposed components	> 20
Interior design	6–10
Windows and exterior doors	10–15
Floors, parquet	(5) 7–11 (13)
Sitting-room furniture	8–10

SERVICE CLASSES

Service Classes relate to the environmental conditions that timber will be exposed to during its service life. These include factors like moisture levels, temperature, and exposure to biological threats (e.g., fungi and insects).

The Service Class helps determine the type of wood protection required and the durability of the timber needed for a particular situation.

Structures shall be assigned to one of the Service Classes given below. In construction design, the service class system is mainly used to assign strength values and calculate deformations under defined environmental conditions.

Service Class 1 is characterised by the moisture content of the materials corresponding to a temperature of 20 °C with the relative humidity of the surrounding air exceeding 65% only for a few weeks per year. In Service Class 1, the average moisture content in most softwoods does not exceed 12%.

Service Class 2 is characterised by the moisture content of the materials corresponding to a temperature of 20 °C with the relative humidity of the surrounding air exceeding 85% only for a few weeks per year.

Service Class 3 is characterised by climatic conditions leading to higher moisture content than Service Class 2.

USE CLASSES

Use Classes are related to the timber's intended use or application and the level of protection required, considering how exposed the timber will be to moisture or biological threats.

Use Classes describe the conditions the timber will face practically, determining what kind of treatment or preservative is necessary to ensure its longevity.

The European standard EN 335 (CEN 2013) defines five Use Classes representing different service situations to which wood and wood-based products can be exposed. This standard also indicates the biological agents relevant to each situation. A Use Class is not a "performance class" and does not give guidance for how long wood and wood-based products will last in service. The differences between the Use Classes are based on differences in environmental exposures that can make the wood or wood-based products susceptible to biological deterioration. The risk of decay increases, mainly if the exposure is prolonged. A more detailed description of the Use Classes and typical examples in use are shown in Table A14.

TABLE A14
Preventive Protection of Structural Timber and Joinery Timber

Use Class	Service Situations	Typical Use	Leach Risk
1	The wood used in building internals in permanently dry environments (relative humidity below 70%)	Internal fittings (furniture, wood panelling, parquet flooring) and internal joinery timber where the moisture content of the wood is permanently below 20%	No
2	Wood is not in ground contact and is not generally exposed to weathering or leaching. Occasional risk of wetting.	Structural timber, roof structures, … where the moisture content of the wood may occasionally exceed 20% Glued-laminated timber components where the moisture content of the wood occasionally exceeds 20%	low
3	Wood is not in ground contact but is exposed to weathering or condensation.	Solid wood or glued-laminated timber components exposed to weathering or condensation. Exterior joinery timber.	significant
4	Wood in permanent ground contact Wood in permanent contact with fresh water	Piles, poles, stakes, solid wood, or glued-laminated timber components in ground contact. Wood immersed in freshwater Cooling towers	very high
5	Wood immersed in salt water	Port structures, wharves, breakwaters	very high

A4 LIST OF WOOD SPECIES MENTIONED IN THE BOOK

Ordered by the Common Name

Common Name	Scientific Name	Alternative Common Names
abachi	*Triplochiton scleroxylon* K.Schum.	African whitewood, abachi, obeche (in Nigeria), wawa (in Ghana), ayous (in Cameroon) and sambawawa (in Ivory Coast)
African ebony	*Diospyros* spp. *e.g. Diospyros crassiflora* Hiern *or Diospyros mespiliformis* Hochst. ex A. DC.	*D. crassiflora*: Gaboon ebony, African ebony, Cameroon ebony, Nigeria ebony, West African ebony, and Benin ebony *D. mespiliformis*: jackalberry, African ebony

(Continued)

Common Name	Scientific Name	Alternative Common Names
African teak	A common name for several plants and may refer to: *Baikiaea plurijuga* Harms, native to the northern Kalahari *Milicia excelsa* (Welw.) C.C. Berg, native to Africa from the Ivory Coast to Ethiopia and south to Angola and Mozambique *Pericopsis elata* (Harms) van Meeuwen, native to western Africa from the Ivory Coast to the Democratic Republic of the Congo (DRC) *Pterocarpus angolensis* DC., native to southern Africa from Tanzania and the DRC south to South Africa	*Baikiaea plurijuga*: African teak, Mukusi, Rhodesian teak, Zambian teak, or Zambesi redwood *Milicia excelsa*: iroko, oji, African teak, iroko, intule, kambala, moreira, mvule, odum, and tule *Pericopsis elata*: Afrormosia, afromosia, kokrodua, and assamela *Pterocarpus angolensis*: wild teak
alder	*Alnus* spp.	
American mahogany	*Swietenia* spp.	Mahogany is wood from any of three tree species: Honduran or big-leaf mahogany (*Swietenia macrophylla* King), West Indian or Cuban mahogany (*Swietenia mahagoni* (L.) Jacq.), and *Swietenia humilis* Zuccarini.
Asian teak	see teak	
Atlas cedar	*Cedrus atlantica* (Endl.) Manetti ex Carrière	
Austrian black pine	*Pinus nigra* J.F. Arnold	Austrian pine
azobé	*Lophira alata* Banks ex Gaertn.	bubinga, ekki, red ironwood
balsa	*Ochroma pyramidale* (Cav. ex Lam.) Urb.	
Bangkirai	*Shorea* spp.	yellow balau, bangkirai, Selangan batu (Brunei); Phchok (Cambodia); Sal (India); Anggelam, Balau, Bangkirai and Dammar laut (Indonesia); Mai chik khok (Laos); Thitya (Myanmar); Gisok, Malayakal and Yakal (Philippines); and Aek, Ak, Balao, Takhian-samphon, Teng and Rang (Thailand).
basswood	*Tilia* spp.	*T. americana* L.: American basswood *T. caroliniana* Mill.: Carolina basswood *T. heterophylla* Mill.: white basswood
bilinga	*Nauclea diderrichii* (De Wild.) Merr.	Aloma (Germany), Opepe (England)
bintangor	*Calophyllum* spp.	The most notable species is *Calophyllum inophyllum* L. commonly called tamanu, oil-nut, mastwood, beach calophyllum, or beautyleaf

(Continued)

Common Name	Scientific Name	Alternative Common Names
birch spp.	*Betula* spp.	
black locus	*Robinia pseudoacacia* L.	robinia,
black locus	*Robinia pseudoacacia* L.	robinia, locusts
black poplar	*Populus nigra var. pyramidalis* L. (*Populus nigra var. Betulifolia*)	cottonwood poplar
bongossi	*Lophira alata* Banks ex Gaertn.	red ironwood
boxwood	*Buxus sempervirens* L.	common box, European box
burflower-tree	*Neolamarckia cadamba* (Roxb.) Bosser	laran (UK), Leichhardt pine (UK), kadamba, kadam, cadamba
cherry	*Prunus* spp.	
chestnut	*Castanea* spp.	
common ash	*Fraxinus excelsior* L.	ash, European ash
common beech	*Fagus sylvatica* L.	European beech
common oak	*Quercus robur* L.	English oak, pedunculate oak
copper beech	*Fagus sylvatica* f. purpurea	purple beech
Corsican pine	*Pinus nigra var. Maritima*	European black pine
cumarú	*Dipteryx odorata* (Aubl.) Willd.	cumaru, kumaru, Brazilian teak
cypress	*Cupressus* spp.	
Dahurian larch	*Larix gmelinii* (Rupr.) Rupr.	Gmelin larch
dark-red meranti	*Shorea argentifolia* Symington	
dark-red meranti	*Shorea* spp.	lauan, luan, lawaan, meranti, seraya, balau, bangkirai, Philippine mahogany (Subgen. Rubroshorea: a particular subgenus within Shorea, which has certain shared characteristics distinguishing it from other subgenera, typically reddish-coloured wood and distinctive features in terms of leaf and flower morphology.)
Douglas fir	*Pseudotsuga menziesii* (Mirbel) Franco	bigcone spruce, Douglas spruce, Oregon pine, Columbian pine
doussie	*Afzelia* spp.	Afzelia, doussie, xylay, chanfuta, pod mahogany, *Afzelia africana*, *Afzelia bipindensis*, *Afzelia bella*, *Afzelia pachyloba*
eastern white pine	*Pinus strobus* L.	
English oak	*Quercus robur* L.	common oak
eucalyptus	*Eucalyptus* spp.	
European beech	*Fagus sylvatica* L.	common beech
European chestnut	*Castanea sativa* Mill.	sweet chestnut, Spanish chestnut, Portuguese chestnut
European larch	*Larix decidua* Mill.	
European walnut	*Juglans regia* L.	common walnut, English walnut, Persian walnut, Royal walnut

(*Continued*)

Common Name	Scientific Name	Alternative Common Names
European yew	*Taxus baccata* L.	common yew, English yew (North America)
fir	*Abies alba* Mill.	European silver fir, silver fir
giant sequoia	*Sequoiadendron giganteum* (Lindl.) J.Buchh.	giant sequoia, giant redwood, Sierra redwood, Wellingtonia
goat willow	*Salix caprea* L.	pussy willow, great sallow
grand fir	*Abies grandis* (Douglas) Lindley	grand fir, giant fir, lowland white fir, great silver fir, western white fir, Vancouver fir, or Oregon fir
Ipé	*Handroanthus* spp.	Brazilian Walnut, Lapacho
iroko	*Milicia excelsa* Sim *and Milicia regia* Sim	
Japanese cedar	*Cryptomeria japonica* (L.f.) D.Don	Japanese redwood
juniper	*Juniperus* spp.	
kasai	*Pterocarpus* spp. *(specifically Pterocarpus soyauxii* Taub.*)*	padauk (or padouk), mukwa, narra; (African padauk, African coralwood)
Lignum vitæ	*Guaiacum*, specifically *G. sanctum* L., but also *G. officinale* L. and *G. coulteri* L.	pockenholz, lignum guaiacum, pockenholz, holywood lignum-vitæ, roughbark lignum-vitæ, *Lignum vitæ*
lime	*Tilia* spp.	lime tree
maçaranduba	*Manilkara* spp.	bulletwood, balatá, ausubo, massaranduba, quinilla
mahogany	*Swietenia* spp.	Mahogany is wood from any of three tree species: Honduran or big-leaf mahogany (*Swietenia macrophylla* King), West Indian or Cuban mahogany (*Swietenia mahagoni* (L.) Jacq.), and *Swietenia humilis* Zuccarini.
makoré	*Tieghemella heckelii* (A.Chev.) Pierre ex Dubard	baku, cherry mahogany
maobi	*Pterocarpus marsupium* Kurz	Burma padauk
maple	*Acer* spp.	
maritime pine	*Pinus pinaster* Aiton	cluster pine, pinaster pine
massaranduba	*Manilkara bidentata* (A.DC.) A.Chev.	bulletwood, balatá, ausubo, massaranduba, quinilla, cow-tree
meranti	*Shorea* spp.	see dark-red meranti
merbau	*Intsia* spp. *(I. bijuga* (Colebr.) Kuntze*, I. palembanica* Miq.*)*	*I. bijuga*: Borneo teak, ipil, Johnstone River teak, and kwila *I. palembanica*: Borneo teak, Malacca teak, merbau, Moluccan ironwood
Montezuma bald cypress	*Taxodium mucronatum* Ten.	Montezuma cypress, ahuehuete
Norwegian maple	*Acer platanoides* L.	
Norwegian spruce	*Picea abies* (L.) Karst.	European spruce

(Continued)

Common Name	Scientific Name	Alternative Common Names
Patagonian cypress	*Fitzroya cupressoides* I.M.Johnst.	alerce, lahuén
pitch pine	*Pinus palustris* Mill.	
pockenholts	*Guaiacum* (specific *G. sanctum* L., but also *G. officinale* L. and *G. coulteri* A.Gray)	pockenholz, lignum guaiacum, *Lignum vitæ*, holywood lignum-vitæ, roughbark lignum-vitæ, boxwood
poplar	*Populus* spp.	
radiata pine	*Pinus radiata* D.Don	
rauli	*Nothofagus alpina* Popp. & Endl.	raulí beech
red ironwood	*Lophira alata* Banks ex Gaertn.	bongossi, red ironwood tree
red oak	*Quercus rubra* L.	northern red oak
robinia	*Robinia pseudoacacia* L.	black locus, locusts
roble	*Nothofagus obliqua* (Mirb.) Oerst.	Patagonian oak, pellín, roble pellín, and hualle, roble beech
rosewood	*Dalbergia latifolia* Roxb.	beete, Bombay blackwood, East Indian rosewood, Indian palissandre, Java palissandre, reddish-brown rosewood, rosea rosewood, satisal, sitsal.
rubber wood	*Hevea brasiliensis* Müll. Arg.	Pará rubber tree, rubber tree, rubber plant, seringueira, sharinga tree
Scots elm	*Ulmus glabra* Huds.	wych elm
Scots pine	*Pinus sylvestris* L.	Baltic pine, European red pine
shining gum	*Eucalyptus nitens* (H.Deane & Maiden) Maiden	silvertop
Siberian larch	*Larix sibirica* Ledeb.	Russian larch
silver birch	*Betula pendula* Roth.	East Asian white birch, European white birch, warty birch
silver fir	*Abies alba* Mill.	European silver fir
sipo	*Entandrophragma utile* (Dawe & Sprague) Sprague	sipo mahogany, utile
sipo mahogany	*Entandrophragma utile* (Dawe & Sprague) Sprague	ipo, utile
Sitka spruce	*Picea sitchensis* (Bong.) Carr.	
Southern yellow pine	*Pinus* spp.	*A number of conifer species which tend to grow in similar plant communities in the USA and yield similar strong wood.*
sugar maple	*Acer saccharum* Marshall	
sweet chestnut	*Castanea sativa* Mill.	chestnut, Spanish chestnut
sycamore	*Acer pseudoplatanus* L.	sycamore maple (US)
sycamore maple	*Acer pseudoplatanus* L.	sycamore (UK)
teak	*Tectona grandis* L. f.	Asian teak
walnut	*Juglans* spp.	see English walnut
wenge	*Millettia laurentii* De Wild.	awong, bokonge, dikela, faux ebony, mibotu
western hemlock	*Tsuga heterophylla* (Raf.) Sarg.	western hemlock-spruce
western red cedar	*Thuja plicata* Donn ex D.Don	giant cedar, giant arborvitae, just cedar, pacific red cedar, shinglewood, western arborvitae, western red cedar (UK)

(Continued)

Common Name	Scientific Name	Alternative Common Names
white meranti	*Anthoshorea symingtonii* (G.H.S.Wood) P.S.Ashton & J.Heck.	
wild cherry	*Prunus* spp.	*Prunus avium* L.: wild cherry (British Isles) *Prunus serotina* Ehrh.: wild cherry (North America) *Prunus cerasus* L.: sour cherry, tart cherry
yellow birch	*Betula alleghaniensis* Britt.	golden birch, swamp birch
yellow cedar	*Cupressus nootkatensis* D.Don	Alaska cedar, Alaska cypress, Alaska yellow cedar, Nootka cedar, Nootka cypress, yellow cedar, yellow cypress
yellow meranti	*Richetia faguetiana* (F.Heim.) P.S.Ashton & J.Heck.	
yew	*Taxus* spp.	

Ordered by the Scientific (Binomial) Name

Scientific Name	Common Names
Abies alba Mill.	European silver fir, fir, silver fir
Abies grandis (Douglas) Lindley	grand fir, giant fir, lowland white fir, great silver fir, western white fir, Vancouver fir, or Oregon fir
Acer platanoides L.	Norwegian maple
Acer pseudoplatanus L.	sycamore (UK), sycamore maple (US)
Acer saccharum Marshall	sugar maple
Acer spp.	maple
Afzelia spp.	Afzelia, doussie, xylay, chanfuta, pod mahogany, *Afzelia africana*, *Afzelia bipindensis*, *Afzelia bella*, *Afzelia pachyloba*
Alnus spp.	alder
Anthoshorea symingtonii (G.H.S.Wood) P.S.Ashton & J.Heck.	white meranti
Baikiaea plurijuga Harms	*Baikiaea plurijuga*: African teak (native to the northern Kalahari), Mukusi, Rhodesian teak, Zambian teak, or Zambesi redwood
Betula alleghaniensis Britt.	golden birch, swamp birch, yellow birch
Betula pendula Roth.	East Asian white birch, European white birch, silver birch, warty birch
Betula spp.	birch spp.
Buxus sempervirens L.	boxwood, common box, European box
Calophyllum spp.	The most notable species is *Calophyllum inophyllum* L. commonly called: beach calophyllum or beautyleaf, bintangor, mastwood, oil-nut, tamanu,
Castanea sativa Mill.	chestnut, European chestnut, Portuguese chestnut, Spanish chestnut, sweet chestnut
Cedrus atlantica (Endl.) Manetti ex Carrière	Atlas cedar

(Continued)

Scientific Name	Common Names
Cryptomeria japonica (L.f.) D.Don	Japanese cedar, Japanese redwood
Cupressus nootkatensis D.Don	Alaska cedar, Alaska cypress, Alaska yellow cedar, Nootka cedar, Nootka cypress, yellow cedar, yellow cypress
Dalbergia latifolia Roxb.	beete, Bombay blackwood, East Indian rosewood, Indian palissandre, Java palissandre, reddish-brown rosewood, rosea rosewood, rosewood, satisal, sitsal.
Diospyros spp. e.g. *Diospyros crassiflora* Hiern *or Diospyros mespiliformis* Hochst. ex A. DC.	*D. crassiflora*: Gaboon ebony, African ebony, Cameroon ebony, Nigeria ebony, West African ebony, and Benin ebony *D. mespiliformis*: jackalberry, African ebony
Dipteryx odorata (Aubl.) Willd.	cumarú, kumaru, Brazilian teak
Entandrophragma utile (Dawe & Sprague) Sprague	ipo, sipo mahogany, utile
Eucalyptus nitens (H.Deane & Maiden) Maiden	shining gum, silvertop
Eucalyptus spp.	eucalyptus
Fagus sylvatica f. purpurea	copper beech, purple beech
Fagus sylvatica L.	common beech, European beech
Fitzroya cupressoides I.M.Johnst.	alerce, lahuén, Patagonian cypress
Fraxinus excelsior L.	common ash, ash, European ash
Guaiacum (specific *G. sanctum* L., but also *G. officinale* L. and *G. coulteri* A.Gray)	*Lignum vitæ*, lignum guaiacum, holywood lignum-vitæ, pockenholts, pockenholz, roughbark lignum-vitæ.
Handroanthus spp.	Brazilian Walnut, Ipé, Lapacho
Hevea brasiliensis Müll. Arg.	Pará rubber tree, rubber plant, rubber tree, rubber wood, seringueira, sharinga tree
Intsia bijuga (Colebr.) Kuntze	Borneo teak, ipil, Johnstone River teak, kwila, merbau
Intsia palembanica Miq.	Borneo teak, Malacca teak, merbau, Moluccan ironwood
Juglans regia L.	common walnut, English walnut, European walnut, Persian walnut, Royal walnut, walnut
Juniperus spp.	juniper
Larix decidua Mill.	European larch
Larix gmelinii (Rupr.) Rupr.	Dahurian larch, Gmelin larch
Larix sibirica Ledeb.	Siberian larch, Russian larch
Lophira alata Banks ex Gaertn.	azobé, bongossi, bubinga, ekki, red ironwood
Manilkara bidentata (A.DC.) A.Chev.	bulletwood, balatá, ausubo, Maçaranduba, quinilla, (ambiguously) "cow-tree".
Milicia excelsa (Welw.) C.C. Berg	*Milicia excelsa*: iroko, ọji, African teak, iroko, intule, kambala, moreira, mvule, odum and tule *Milicia regia*: iroko
Millettia laurentii De Wild.	awong, bokonge, dikela, faux ebony, mibotu, wenge
Nauclea diderrichii (De Wild.) Merr.	Aloma (Germany), bilinga, Opepe (UK)
Neolamarckia cadamba (Roxb.) Bosser	burflower-tree, laran (UK), Leichhardt pine (UK), kadamba, kadam, cadamba
Nothofagus alpina Popp. & Endl.	rauli, raulí beech
Nothofagus obliqua (Mirb.) Oerst.	Patagonian oak, pellín, roble pellín, and hualle, roble beech
Ochroma pyramidale (Cav. ex Lam.) Urb.	balsa
Pericopsis elata (Harms) van Meeuwen	African teak, Afrormosia, afromosia, kokrodua, assamela

(Continued)

Scientific Name	Common Names
Picea abies (L.) Karst.	European spruce, Norwegian spruce
Picea sitchensis (Bong.) Carr.	Sitka spruce
Pinus nigra J.F. Arnold	Austrian pine, Austrian black pine
Pinus nigra var. Maritima	Corsican pine, European black pine
Pinus palustris Mill.	pitch pine
Pinus pinaster Aiton	cluster pine, maritime pine, pinaster pine
Pinus radiata D.Don	radiata pine
Pinus strobus L.	eastern white pine
Pinus sylvestris L.	Baltic pine, European red pine, Scots pine
Populus nigra var. pyramidalis L. (*Populus nigra var. Betulifolia*)	black poplar, cottonwood poplar
Prunus avium L.	wild cherry (British Isles)
Prunus serotina Ehrh	wild cherry (North America)
Prunus cerasus L.	sour cherry, tart cherry
Pseudotsuga menziesii (Mirbel) Franco	bigcone spruce, Douglas fir, Douglas spruce, Oregon pine, Columbian pine
Pterocarpus angolensis DC.	*Pterocarpus angolensis*: wild teak African teak, native to southern Africa from Tanzania and the DRC south to South Africa
Pterocarpus marsupium Kurz	maobi, Burma padauk
Pterocarpus spp. (specifically Pterocarpus soyauxii Taub.)	padauk (or padouk), kasai, mukwa, narra; (African padauk, African coralwood)
Quercus robur L.	common oak, English oak, pedunculate oak
Quercus rubra L.	northern red oak, red oak
Richetia faguetiana (F.Heim.) P.S.Ashton & J.Heck.	yellow meranti
Robinia pseudoacacia L.	black locus, locusts, robinia
Salix caprea L.	great sallow, goat willow, pussy willow
Sequoiadendron giganteum (Lindl.) J.Buchh.	giant sequoia, giant redwood, Sierra redwood, Wellingtonia
Shorea spp.	Bangkirai, yellow balau, bangkira, Selangan batu (Brunei); Phchok (Cambodia); Sal (India); Anggelam, Balau, Bangkirai and Dammar laut (Indonesia); Mai chik khok (Laos); Thitya (Myanmar); Gisok, Malayakal and Yakal (Philippines); and Aek, Ak, Balao, Takhian-samphon, Teng and Rang (Thailand). other names are lauan, luan, lawaan, meranti, seraya, balau, bangkirai, Philippine mahogany (Subgen. Rubroshorea: a particular subgenus within Shorea, which has certain shared characteristics distinguishing it from other subgenera, typically reddish-coloured wood and distinctive features in terms of leaf and flower morphology.)
Shorea centifolia Symington	dark-red meranti or meranti
Swietenia spp.	American mahogany, mahogany: Mahogany is wood from any of three tree species: Honduran or big-leaf mahogany (*Swietenia macrophylla* King), West Indian or Cuban mahogany (*Swietenia mahagoni* (L.) Jacq.), and *Swietenia humilis* Zuccarini.

(Continued)

Scientific Name	Common Names
Taxodium mucronatum Ten.	ahuehuete, Montezuma cypress, Montezuma bald cypress
Taxus baccata L.	common yew, English yew (North America), European yew, yew
Tectona grandis L. f.	Asian teak, teak
Thuja plicata Donn ex D.Don	giant cedar, giant arborvitae, just cedar, pacific red cedar, shinglewood, western arborvitae, western red cedar (UK)
Tieghemella heckelii (A.Chev.) Pierre ex Dubard	baku, cherry mahogany, Makoré
Tilia americana L.	American basswood, basswood
Tilia caroliniana Mill.	Carolina basswood, basswood
Tilia heterophylla Mill.	white basswood, basswood
Tilia spp.	lime, lime tree
Triplochiton scleroxylon K.Schum.	African whitewood, abachi, obeche (in Nigeria), wawa (in Ghana), ayous (in Cameroon), and sambawawa (in Ivory Coast)
Tsuga heterophylla (Raf.) Sarg.	western hemlock, western hemlock-spruce
Ulmus glabra Huds.	Scots elm, wych elm

REFERENCES

Anon. (1975). *Werkstoffe aus Holz und andere Werkstoffe der Holzindustrie*. Fachbuchverlag, Leipzig, Germany, (928 pp.).

CEN (1997). *EN 312-7: Particleboards - Specifications - Part 7: Requirements for Heavy Duty Load-bearing Boards for use in Humid Conditions*. The European Committee for Standardization (CEN), Brussels, Belgium.

CEN (2013). *EN 335: Durability of Wood and Wood-based Products - Use Classes: Definitions, Application to Solid Wood and Wood-based Products*. The European Committee for Standardization (CEN), Brussels, Belgium.

DIN (2010). *DIN EN ISO 10456:2010-05: Building Materials and Products - Thermal and Humidity Properties - Tabulated Design Values and Methods for Determining Nominal and Rated Thermal Insulation Values*. Deutschen Institut für Normung (DIN), Berlin, Germany.

Dunky M. & Niemz P. (2002). *Holzwerkstoffe und Leime: Technologie und Einflussfaktoren*. Springer, Berlin, Germany.

Gong M. (Ed.) (2021). *Engineered Wood Products for Construction*. IntechOpen. London, UK, (358 pp.).

Greubel O. & Drewes H. (1987). Ermittlung der Sorptionsisothermen von Holzwerkstoffen bei verschiedenen Temperaturen mit einem neuen Messverfahren. *Holz als Roh- und Werkstoff*, 45(7), 289–295.

Niemz P., Teischinger A. & Sandberg D. (Eds.) (2023). *Springer Handbook of Wood Science and Technology*. Springer Nature, Cham, Switzerland, (XXV+2069 pp.).

Niemz P. & Sonderegger W. (2021). *Holzphysik. Physik des Holzes und der Holzwerkstoffe*. (2nd Ed.) Carl Hanser Verlag, Leipzig, Germany, (580 pp.).

Scheer C., Peter M. & Stöhr S. (2004). *Holzbau-Taschenbuch. Bemessungsbeispiele nach DIN 1052, Ausgabe 2004*. (10th Ed.), Ernst und Sohn, Berlin, Germany, (396 pp.).

von Halász R. & Scheer C. (1986). *Holzbau-Taschenbuch. Band 1: Grundlagen, Entwurf und Konstruktionen*. (2nd Ed.), Ernst & Sohn, Berlin, Germany, (712 pp.).

von Halász R. & Scheer C. (1995). *Holzbau-Taschenbuch: Grundlagen, Entwurf, Bemessung und Konstruktionen*. (9th Ed.), Ernst & Sohn, Berlin, Germany, (1087 pp.).

Wagenführ R. & Wagenführ A. (2022). *Holzatlas*. (7th Ed.), Carl Hanser Verlag, München, Germany, (928 pp.).

Index

Pages in *italics* refer to figures and pages in **bold** refer to tables.

For Product Safety Concerns and Information please contact our EU
representative GPSR@taylorandfrancis.com
Taylor & Francis Verlag GmbH, Kaufingerstraße 24, 80331 München, Germany

www.ingramcontent.com/pod-product-compliance
Lightning Source LLC
Chambersburg PA
CBHW060743220326
41598CB00022B/2308

*9 7 8 1 0 3 2 5 3 3 6 0 5 *